COMBINATORICS '84

NORTH-HOLLAND MATHEMATICS STUDIES 123
Annals of Discrete Mathematics (30)

NORTH-HOLLAND – AMSTERDAM • NEW YORK • OXFORD • TOKYO

COMBINATORICS '84

Proceedings of the International Conference on Finite Geometries and Combinatorial Structures Bari, Italy, 24-29 September, 1984

edited by

A. BARLOTTI

Università di Firenze, Firenze, Italy

M. BILIOTTI

Università di Lecce, Lecce, Italy

A. COSSU

Università di Bari, Bari, Italy

G. KORCHMAROS

Università della Basilicata, Potenza, Italy

G. TALLINI

Università 'La Sapienza', Rome, Italy

1986

NORTH-HOLLAND – AMSTERDAM • NEW YORK • OXFORD • TOKYO

ISBN: 0 444 87962 5

Publishers:
ELSEVIER SCIENCE PUBLISHERS B.V.
P.O. Box 1991
1000 BZ Amsterdam
The Netherlands

Sole distributors for the U.S.A. and Canada:
ELSEVIER SCIENCE PUBLISHING COMPANY, INC.
52 Vanderbilt Avenue
New York, N.Y. 10017
U.S.A.

Library of Congress Cataloging-in-Publication Data

International Conference on Finite Geometries and Combinatorial Structures (1984 : Bari, Italy)
Combinatorics '84 : proceedings of the International Conference on Finite Geometries and Combinatorial Structures, Bari, Italy, 24-29 September 1984.

(Annals of discrete mathematics ; 30)
(North-Holland mathematics studies ; 123)
Includes bibliographies.
1. Combinatorial geometry--Congresses. I. Barlotti, A. (Adriano), 1923-   . II. Title. III. Title: Combinatorics eighty-four. IV. Series. V. Series: North-Holland mathematics studies ; 123.
QA167.I58 1984   511'.6   85-31121
ISBN 0-444-87962-5

PRINTED IN THE NETHERLANDS

PREFACE

Every year, since 1980, an International Combinatoric Conference has been held in Italy: Trento, October '80; Rome, June '81; La Mendola, July '82; Rome, at the Istituto Nazionale di Alta Matematica, May '83.

The International Conference *Combinatorics '84*, held in Giovanazzo (Bari) in September '84 is part of the well established tradition of annual conferences of Combinatorics in Italy. Like the previous ones, this Conference was really successful owing to the number of participants and the level of results.

The present volume contains a large part of these scientific contributions. We are indebted to the University of Bari and to the *Consiglio Nazionale delle Ricerche* for financial support. We are profoundly grateful to the referees for their assistance.

A. BARLOTTI
M. BILIOTTI
A. COSSU
G. KORCHMAROS
G. TALLINI

Intervento di Apertura del Prof. G. Tallini al Convegno "COMBINATORICS 84"

Vorrei dare, a nome del Comitato scientifico e mio, il benvenuto ai molti partecipanti che dall'Italia e dall'estero sono qui convenuti per prendere parte a questo convegno. Esso si ricollega e fa seguito ai congressi internazionali di combinatoria tenuti a Roma nel giugno del 1981, a La Mendola nel luglio del 1982, a Roma presso l'Istituto Nazionale di Alta Matematica nel maggio del 1983. Questi incontri, ormai annuali in Italia e che spero possano continuare, s'inquadrano nell'ampio sviluppo che la combinatoria va acquistando a livello internazionale.

Come è noto il mondo moderno si va indirizzando ed evolvendo sempre di più verso la programmazione e l'informatica, al punto che un paese oggi è tanto più progredito, importante e all'avanguardia quanto più è avanzato nella scienza dei computers. Il ramo della Matematica che è più vicino a questi indirizzi e che ne è la base teorica è proprio la combinatoria. Essa al gusto astratto del ricercatore, del matematico, associa appunto le applicazioni più concrete. Ciò spiega il prepotente affermarsi di questa scienza nel mondo e ne prova il fervore di studi e di ricerca che si effettuano in quest'ambito, le pubblicazione dei molti periodici specializzati, i numerosi convegni internazionali al riguardo.

Vorrei ringraziare gli Enti che hanno permesso la realizzazione di questo convegno, tutti i partecipanti, in particolare gli ospiti stranieri che numerosi hanno accolto il nostro invito e tra i quali sono presenti insigni scienziati.

Concludo con l'augurio che questo convegno segni una tappa da ricordare nello sviluppo della nostra scienza.

CONTENTS

Annals of Discrete Mathematics 30 (1986) 1–8

TRANSLATION PLANES WITH AN AUTOMORPHISM GROUP ISOMORPHIC TO SL(2,5)

Vito Abatangelo Bambina Larato
Università di Bari
Italia

In this paper translation planes of odd order q, $5|q^2-1$, are constructed. Their main interest consists in the fact that their translation complement contains a group isomorphic to $SL(2,5)$. At first these planes were obtained in other ways by O. Prohaska in the case $5|q+1$ ([10],1977) and by G. Pellegrino and G. Korchmàros in the case $5|q-1$ ([9],1982), but in both papers the Authors did not establish the previous group property. Moreover we show that Pellegrino and Korchmàros plane is not a near-field plane of order 11^2.

1. – AN AUTOMORPHISM GROUP OF THE AFFINE DESARGUESIAN PLANE OF ORDER q^2, $5|q^2-1$, ISOMORPHIC TO SL(2,5)

Set $K = GF(q^2)$, q odd. We may assume that the elements of K can be written in the form $\xi + t\eta$ with $\xi, \eta \in F = GF(q)$ and $t^2 = s$, where s is a non-square element of F.

Let π be the affine Desarguesian plane coordinatized by K : points are pairs (x,y) of elements of K and lines are sets of points satisfying equations of the form $y = mx+b$ or $x = c$ with m,b,c elements of K. The affine subplane π_0 coordinatized by F is an affine Baer subplane; the image of π_0 under a composition of a linear transformation with a translation of π is also taken to be an affine Baer subplane. The lines at infinity of Baer subplanes are called Baer sublines at infinity. By standard arguments (similar to those of [7], p. 80-91) one can show the following facts: Baer sublines at infinity are sets of elements of $K \cup \{\infty\}$ of the form:

$$\{ a\mu+b \mid a,b \in K ,\ \mu \text{ runs over } F \cup \{\infty\} \} \tag{1.1}$$

or

$$\{ m \mid (m-v)^{q+1} = r^{q+1} ,\ v \in K ,\ r \in K-\{0\} \} \tag{1.2}$$

Let v and r be any two elements of K. For any $d \in K$ and $g \in K$, such that $g^{q+1} = 1$, the set of all points (x,y) for which

$$y = xv + rx^q g + d \tag{1.3}$$

Research partially supported by M.P.I. (Research project "Strutture Geometriche Combinatorie e loro Applicazioni").

is the point-set of an affine Baer subplane. Its Baer subline at infinity has equation (1.2) as the line $y = xm+d$ intersects (1.3) either in q points or in the only point $(0,d)$ according as (1.2) holds or does not hold. The affine Baer subplanes with equations of the form (1.3) constitute a net on the affine points of π.

The line at infinity becomes the Miquelian inversive plane $M(q)$ if we take as circles the sets (1.1) and (1.2) of elements of $K \cup \{\infty\}$.

In order to consider an automorphism group of π isomorphic to $SL(2,5)$ we distinguish two cases, according as $5|q+1$ or $5|q-1$.

CASE $5|q+1$

By assumption $5|q+1$ there exists an element a of K such that $a^5 = 1$ (cf. [4]). We put $b = (a-a^q)^{-1}$ and c such that $c^{q+1} + b^{q+1} = 1$. After this let us consider the following affine mappings of π

$$\alpha : x' = ax \quad , \quad y' = a^q y$$
$$\beta : x' = bx + cy \quad , \quad y' = -c^q x + b^q y$$

and the identity $\varepsilon : x' = x \quad , \quad y' = y$.
Note that $<\alpha, \beta> \simeq SL(2,5)$, (cf. [4], p. 199).

Now our purpose is to study the action of $<\alpha, \beta>$ on the line at infinity, i.e. the action of $\Gamma = <\alpha,\beta>/<-\varepsilon> \simeq PSL(2,5)$ on $M(q)$.

PROPOSITION 1. - In $M(q)$ the group Γ maps the circle $C_0 : m^{q+1} = -1$ onto itself and leaves the set $\mathscr{C} = \{C_1, C_2, \dots, C_6\}$ invariant, where $C_1 : m^{q+1} = 1$ and

$$C_{i+2} : (m-2a^{i(1-q)}b^q c(2b^2+1)^{-1})^{q+1} = (2b^2+1)^{-2} \quad , (i = 0, \dots, 4).$$

PROOF - Some long and easy calculations prove that α and β send C_0 onto itself and act on the set $\mathscr{C}$ as follows

$$\alpha : (C_1)(C_2C_3C_4C_5C_6) \quad , \quad \beta : (C_1C_2)(C_3C_6)(C_4)(C_5) \quad .$$

PROPOSITION 2. - The group Γ acts on $\mathscr{C}$ as $PSL(2,5)$ acts in its usual two-transitive representation on six objects.

PROOF - Assuming B. Huppert's terminology, [4],
(1.4) let $\Delta = GF(5) \cup \{\infty\}$ be the set consisting of six objects;
and
(1.5) let $PSL(2,5)$ be the permutation group on Δ of the form
$$x' = (ax+b)(cx+d)^{-1} \quad , \quad ad-bc \in \{1,4\} \quad , \quad a,b,c,d \in GF(5) \quad .$$

We identify C_1 with the symbol ∞, C_2 with 0, C_3 with 3, C_4 with 1, C_5 with 4 and C_6 with 2; then $\alpha' : x' = x+3$ and $\beta' : x' = x^{-1}$ act on Δ as α and β, respectively, act on $\mathscr{C}$. As $<\alpha', \beta'> \simeq PSL(2,5)$, we get our assert.

PROPOSITION 3. - Any two circles of $\mathscr{C}$ have two or zero common points according as $q \equiv 3 \pmod 4$ or $q \equiv 1 \pmod 4$.

PROOF - By Prop. 2 it is sufficient to prove that the circles C_1 and C_2 satisfy our assert.

STEP 1 - If C_1 and C_2 have any common point, its coordinates would satisfy both their equations which form a system which is equivalent to the following

$$(1.6) \qquad \begin{cases} m^{q+1} = 1 \\ c^q m - c m^q = 0 \end{cases} .$$

As $\{ m = ch \mid h \in GF(q) \}$ is the set of solutions of the second equation of (1.6), the whole system (1.6) has any solution if and only if the equation $c^{q+1} h^2 = 1$ is solvable, i.e. if and only if c^{q+1} is a square in F.

STEP 2 - c^{q+1} is a square of F or not according as $q \equiv 3 \pmod 4$ or $q \equiv 1 \pmod 4$. In order to prove this, let x_1 and x_2 be elements of F such that $a = x_1 + tx_2$; note that

$$(1.7) \qquad a^{q+1} = x_1^2 - sx_2^2 \in F$$

and $(a^{q+1})^5 = 1$, so

$$(1.8) \qquad a^{q+1} = 1$$

(because in F there is no element of order 5) and therefore $5x_1^4 + 10x_1^2x_2^2 s + x_2^4 s^2 = 0$. By means of (1.7) and (1.8), we get $16x_1^4 - 12x_1^2 + 1 = 0$, i.e. $\pm\sqrt{5} = 8x_1^2 - 3$: this proves that 5 is a square in F. Now $c^{q+1} = [(a-a^q)^2 + 1](a-a^q)^{-2} = [4(1+sx_2^2) - 3](4x_2^2 s)^{-1} = (4x_1^2-3)(4x_2^2 s)^{-1} = -1\, s^{-1}(1 \pm\sqrt{5})^2 (4x_2)^{-2}$

and therefore c^{q+1} is a square in F if and only if -1 is not a square in F, i.e. if and only if $q \equiv 3 \pmod 4$.

PROPOSITION 4. - C_0 is disjoint from any circle of $\mathscr{C}$.

PROOF - It is easy to check that C_0 and C_1 are disjoint. Our assertion follows by means of Prop. 2.

CASE $5 | q-1$

By assumption $5|q-1$ there exists an element a of F_1 such that $a^5 = 1$. Let b, c, d be elements of K such that $b = (a-a^{-1})^{-1}$, $cd = -1-b^2$. Now let us consider the following affine mappings of π

$$\gamma \; : x' = a^{-1}x \;, \quad y' = (2+t)b^{-1}x + ay$$

$$\delta \; : x' = [-b+(2+t)d]x + dy \;, \quad y' = [2(2+t)b+c-(2+t)^2 d]x + [b-(2+t)d]y$$

and note that $< \gamma, \delta > \simeq SL(2,5)$ (cf. [4], p. 198).

We shall now look at the action of $<\gamma,\delta>$ on the line at infinity, i.e. the action of the group $\Lambda = <\gamma,\delta>/<-\varepsilon>$ on $M(q)$.

PROPOSITION 5. - In $M(q)$ the group Λ maps the circle $D_0 : 2tm^{q+1}-m+m^q=0$ onto itself and leaves the set $\{D_1, \dots, D_6\}$ invariant where $D_1 : 4m^{q+1}+m+m^q=0$ and

$$D_{i+2} : [(4-t^2) - 4a^{2i}(1+2b^2)(2bd)^{-1}-a^{-i}cd^{-1}]\, m^{q+1} + [(2+t)-a^{2i}(1+2b^2)(2bd)^{-1}]\,m + [(2-t) - a^{2i}(1+2b^2)(2bd)^{-1}]\, m^q + 1 = 0 \quad , \quad (i = 0,1, \dots ,4).$$

PROOF - Long and easy calculations prove that γ and δ map D_0 onto itself and act on the set $\{D_1, D_2, \dots, D_6\}$ as follows

$$\gamma : (D_1)(D_2 D_3 D_4 D_5 D_6) \quad \text{and} \quad \delta : (D_1 D_2)(D_3 D_6)(D_4)(D_5) .$$

PROPOSITION 6. - The group Λ acts on $\{D_1, D_2, \dots, D_6\}$ in the same way as $PSL(2,5)$ acts in its usual two-transitive representation on six objects.

PROOF - The proof is similar to that one of Prop. 2, provided we identify D_1 with the symbol ∞, D_2 with 0, D_3 with 1, D_4 with 2, D_5 with 3 and D_6 with 4 and moreover γ and δ with γ' and δ', respectively, where

$$\gamma' : x' = x + 1 \quad \text{and} \quad \delta' : x' = 4x .$$

Assume now $\mathcal{D} = \{D_0, D_1, \dots, D_6\}$; by direct calculations and by Prop.6 we get the following

PROPOSITION 7. - (i) Any two circles of $\mathcal{D}$ have two common points; (ii) no three circles of $\mathcal{D}$ have a common point.

2. - DERIVED AND ß-DERIVED PLANES

In this section we discuss briefly the processes of derivation and multiple derivation due to T.G. Ostrom and of ß-derivation and multiple ß-derivation due to A.A. Bruen. The reader can find some details in Hughes-Piper [3], Ostrom [7,8] and Bruen [1].

Let us assume $q > 3$; every circle C of $M(q)$ is a derivation set, i.e. every pair of affine points lies in a unique Baer affine subplane, whose subline at infinity is C . From now on we will consider circles C with equation (1.2). An affine plane can be obtained by replacing lines whose equation is $y = mx+b$, for each m satisfying (1.2), with the $q+1$ Baer affine subplanes, called also components, with equation $y = xv + rx^q g$ where g runs over $A = \{ g \in K \mid g^{q+1} = 1\}$, together with their translates. This construction is said derivation by C and the obtained plane is a translation plane of order q^2.
Derivation can be repeated many times, when there is a set of circles $C_1, C_2, C_3, \dots, C_k$ each disjoint from the other; in this case the process is said multiple derivation with respect to the circles $C_1, C_2, \dots, C_k$.

About ß-derivation, suppose that $M(q)$ admits a chain of circles, i.e. a family $\mathcal{C}$ of circles satisfying the following properties:

(i) any two circles of $\mathcal{C}$ have two common points;
(ii) no three circles of $\mathcal{C}$ have a common point;
(iii) $\mathcal{C}$ consists of $(q+3)/2$ circles.

Let I denote the point set covered by the circles of $\mathcal{C}$. It follows that I contains exactly $(q+1)(q+3)/4$ points and each point of I belongs exactly to two circles of $\mathcal{C}$. Hence any two points on a line with its ideal point on I belong exactly to two affine Baer subplanes whose sublines at infinity represent circles of $\mathcal{C}$.

The affine Baer subplanes with Baer sublines at infinity representing circles of $\mathcal{C}$ do not form a net on the affine points of π. However, for small q $(5 \leq q \leq 13)$, several Authors have constructed a net N on the affine points of π by taking a suitable half of such affine Baer subplanes; namely, for each circle C of $\mathcal{C}$, $(q+1)q^2/2$ affine Baer subplanes with the same Baer subline representing C. If such a net N exists, we can obtain a translation plane on the affine points of π by replacing the affine lines of π whose ideal points lie on I with N. The resulting translation plane is said to be ß-derived from π.

Of course if there is a set of disjoint chains of circles on $M(q)$ and the corresponding nets exist, then the above method can be repeated: the resulting translation plane is said to be multiple ß-derived from π.

3. - A CLASS OF TRANSLATION PLANES OF ORDER q^2, $5|q+1$, CONTAINING IN THEIR TRANSLATION COMPLEMENT AN AUTOMORPHISM GROUP ISOMORPHIC TO SL(2,5)

In this section we assume that $q \equiv 1 \pmod 4$. By Propositions 3 and 4 we get three planes π_j $(j = 1,2,3)$: π_1 by derivation from π with respect to the circle C_0 (so π_1 is the well-known Hall plane of order 81, cf. [2], p. 225), π_2 by derivation with respect to the circles $C_1, C_2, \dots, C_6$, π_3 by derivation with respect to the circles $C_0, C_1, \dots, C_6$; each of them contains in its translation complement the group $<\alpha,\beta> \simeq SL(2,5)$.

Now we prove the following propositions

PROPOSITION 8. - The group $<\alpha,\beta>$ leaves invariant each of the $q+1$ components corresponding with the derivation set C_0.

PROOF - The $q+1$ components are the Baer subplanes B_g with equations B_g : $y = rx^q g$ where $r^{q+1} = -1$ and g runs over A. A straightforward calculation shows that α as well as β leaves each B_g invariant.

PROPOSITION 9. - The group $<\alpha,\beta>$ splits the set of the $6(q+1)$ components corresponding with the multiple derivation set $C_1 \cup C_2 \cup \dots \cup C_6$ into $(q+1)/2$ orbits each of length 12.

PROOF - Let H_i $(i = 1,2, \dots ,6)$ be the set consisting of the $q+1$ components which correspond with the derivation set C_i. Then $<\alpha,\beta>$ acts on the set $\{H_1, \dots ,H_6\}$ in the same way as on the set $\{C_1, \dots ,C_6\}$. By Prop. 2, $<\alpha,\beta>$ acts transitively on $\{H_1, \dots ,H_6\}$. The stabilizer of H_1 in $<\alpha,\beta>$ is $<\alpha,\lambda>$ with $\lambda = \beta\alpha^3\beta\alpha^2\beta$, i.e.

$$\lambda : x' = c(2b^2-ac^{q+1}-a^q b^2)y \quad , \quad y' = -c^q(2b^2-a^q c^{q+1}-ab^2)x \quad .$$

The $q+1$ components belonging to H_1 are the Baer subplanes E_g with equations $y = x^q g$ where g runs over A. A straightforward calculation shows that α leaves each E_g invariant and λ maps E_g onto E_{-g}.

PROPOSITION 10.- The line-orbit of $<\alpha,\beta>$ containing the line joining the origin to Y_∞ has length 12.

PROOF - The vertical line through 0 is left invariant by α and $-\varepsilon$ but not by λ. Since each subgroup of $<\alpha,\beta>$ containing properly $<\alpha,-\varepsilon>$ contains also λ, it follows that the stabilizer of the vertical line through 0 has order 5. This proves our assert.

We point out that for $q = 9$, which is the first non trivial case, Propositions 8,9 and 10 yield:

PROPOSITION 11.- (i) The group $\Gamma = <\alpha,\beta>/<-\varepsilon>$ splits the line at infinity of π_2 into one orbit of length ten and six orbits of length twelve;
(ii) the group $\Gamma = <\alpha,\beta>/<-\varepsilon>$ splits the line at infinity of π_3 into six orbits of length twelve and ten orbits of length one.

Now we state the following

PROPOSITION 12.- Let σ be a plane obtained by derivation from π_j ($j = 2,3$). If $SL(2,5)$ is an automorphism group of σ, then σ coincides with π_2 or π_3.

PROOF - As it is well known, the number of disjoint circles of $M(q)$ is $q-1$ and when it occurs they form a linear flock by a theorem due to W.J. Orr (cf. [6]). In our situation $C_1, C_2, \dots, C_6$ belong to no linear flock. So if C is any circle which determines a derivation of π_j, C must coincide with some circle C_i ($i = 0,1, \dots, 6$) or C must not intersect each of them. If $C \in \{C_0, C_1, \dots, C_6\}$, then necessarily $C = C_0$; on the other hand C cannot stay on H_3 because, by Prop. 8, no orbit of $SL(2,5)$ is long 10 or less than 10 in H_3.

4. - A TRANSLATION PLANE OF ORDER q^2, $5|q-1$, CONTAINING IN ITS TRANSLATION COMPLEMENT AN AUTOMORPHISM GROUP ISOMORPHIC TO SL(2,5)

In the previous section 1 we determine the set of circles $\mathcal{D}$ which is a family satisfying properties (i) and (ii) of chains of circles. Moreover, when $q = 11$, $\mathcal{D}$ satisfies property (iii) and, therefore, is a chain of circles.

By means of the automorphism

$$\omega \; : \; x' = (2+t)x \quad , \quad y' = (8+7t)x + (2+t)y$$

of $M(11)$, we can check that $\mathcal{D}$ is equivalent to the following chain:

$$C'_\infty : m - m^{11} = 0 \quad , \quad C'_0 : m + m^{11} = 0 \quad ,$$

$$C'_i : (m - 2^{-2i})^{11} = (-2)^{1+i} \quad , \quad i = 1,2, \dots ,5 \; ,$$

which was studied by G. Pellegrino and G. Korchmàros. So $\mathcal{D}$ determines a translation plane (cf. [9]).

Pellegrino and Korchmàros used a geometrical construction and so they cannot notice that the translation plane associated to the chain $\mathcal{D}$ admits $SL(2,5)$ as automorphism group.

Finally we want to remark that Pellegrino and Korchmàros plane surely is not a near-field plane of order 11^2 (cf. [5], p. 88), though it satisfies the same group property. The near-field planes have only two orbits on the line at infinity: the first has length 2 and the other consists of all the remaining points. In the present case the orbit length are 2 and 120, while the Pellegrino and Korchmàros plane has an orbit of length $42 = |H_2|$ on its line at infinity.

REFERENCES

[1] A.A. Bruen, Inversive geometry and some new translation planes I, Geom. Dedic., 7 (1977), 81-98.

[2] P. Dembowski, Finite geometries (Springer-Verlag, Berlin-Heidelberg-New York, 1968).

[3] D.R. Hughes-F. Piper, Projective planes (Springer-Verlag, Berlin-Heidelberg-New York, 1973).

[4] B. Huppert, Endliche gruppen I (Springer-Verlag, Berlin-Heidelberg-New York, 1967).

[5] H. Lüneburg, Translation planes (Springer-Verlag, Berlin-Heidelberg-New York, 1980).

[6] W.J. Orr, A characterization of subregular spreads in finite 3-space, Geom. Dedic., 5 (1976), 43-50.

[7] T.G. Ostrom, Finite translation planes (Springer-Verlag, Berlin-Heidelberg-New York, 1970).

[8] T.G. Ostrom, Lectures on finite translation planes, Conf. Sem. Mat. Univ. Bari, n. 191, 1983.

[9] G. Pellegrino-G. Korchmàros, Translation planes of order 11^2, Annals of Discrete Math., 14 (1982), 249-264.

[10] O. Prohaska, Konfigurationen einander meidender kreise in Miquelschen Möbiusebenen ungerader ordnung, Arch. Math. (Basel), 28 (1977), n. 5, 550-556.

V. Abatangelo-B. Larato
Dipartimento di Matematica
Via Giustino Fortunato
Università degli Studi
70125 - B A R I

Annals of Discrete Mathematics 30 (1986) 9–14

SYMPLECTIC GEOMETRY, QUASIGROUPS, AND STEINER SYSTEMS

Lucien Bénéteau
UER-M.I.G.
Université Paul Sabatier
31062 TOULOUSE - CEDEX
FRANCE

Zassenhaus's process of construction of Hall Triple Systems can be generalized. It turns out that there is a canonical correspondence between equivalence classes of non zero alternate trilinear forms of V(n,3) and isomorphism classes of rank (n+1) HTSs whose order is $3^{(n+1)}$. Thus the problem of classifying these designs and the related Steiner quasigroups may be presented as a special case of a more general classification problem of exterior algebra. As an illustration of these ideas we shall deal completely with the case $n\leqslant 6$. For n=6 one obtains exactly 5 isomorphism classes of HTSs.

1-INTRODUCTION -

Section 2 gives a brief introduction to the Hall Triple Systems (HTSs) and to the related groups and quasigroups. There are two statements giving precisions about the correspondence between the HTSs on one side, and the cubic hypersurface quasigroups and the Fischer groups on the other side. We refer the reader to the literature for the connections with other parts of algebra and design theory ([7,10,1]).

Further on a process of explicit construction of HTSs is recalled (section 3). This process is not canonical. But it allows to get all the non affine HTSs whose 3-order s equals the rank ρ. As usual the rank is to be understood as the minimum possible cardinal number of a generator subset. The equality $s=\rho$ corresponds to an extremal situation, the non affine HTSs obeying $s\geqslant\rho$, while the affine ones obey $s=\rho-1$ (see [1]). It is the classification of non affine HTSs of given rank whose order is minimal that led us to a problem of symplectic geometry.

Given some vector space V, there is a natural action of GL(V) on the set of symplectic trilinear forms of V. We shall be counting orbits in some special cases. For further investigations the most important result is some process of translation in case the field is GF(3) : there is then a one-to-one correspondence between the orbits of the non-zero forms and the isomorphism classes of some HTSs. This will be used here to obtain an exhaustive list of the HTSs of order $\leqslant 2187$ whose ranks are $\neq 6$. We shall also classify the non affine HTSs admitting a codimension 1 affine subsystem.

2-HALL TRIPLE SYSTEMS, MANIN QUASIGROUPS AND FISCHER GROUPS -

A <u>Steiner Triple System</u> is a 2-(v,3,1) design, namely it is a pair (E,L) where E is a set of "points" and L a collection of 3-subsets of E, called "lines", such that ony two distinct points lie in exactly one line $\ell \in L$. The corresponding Steiner quasigroup consists of the same set E under the binary law : $E^2 \to E$; $x,y \longmapsto xoy$ defined by xox=x and, whenever $x\neq y$, xoy=z, the third point of the line through x and y. The Steiner quasigroups can be algebraically characterized by the fact that the law is idempotent and symmetric. Recall that a law is said to be symmetric when any equality of the form $xoy=z$ is invariant under any permutation of x,y,z ; this is equivalent to the conjunction of the commutativity and the identity $xo(xoy)=y$. For a fixed set E, to endow E with a family of lines L such that (E,L) be a Steiner Triple System is equivalent to provide E with a structure of Steiner quasigroup. So in what follows we shall identify (E,L) with (E,o).

A Hall Triple System (HTS) is a Steiner Triple System in which any subsystem that is generated by three non collinear points is an affine plane $\simeq AG(2,3)$. This additional assumption is equivalent to the fact that the corresponding Steiner quasigroup is distributive ($a\,o\,(xoy)=(aox)o(aoy)$ identically ; see Marshall Hall Jr. [6]). Therefore the HTSs are identified with the distributive Steiner quasigroups.

Let K be a commutative field. Consider an absolutely irreducible cubic hypersurface V of the projective space $\mathbf{P}_n(K)$. Let E be the set of its non-singular K-points. Three points x,y,z of V will be said to be collinear (notation : L(x,y,z)) if there exists a line ℓ containing x,y,z such that either $\ell \subset V$ or $x+y+z=\ell.V$ (intersection cycle).

The best known case is when dim V=1, and n=2 : V is then a plane curve, it does not contain any line and overall for any x,y in E, there is exactly one point z in E such that L(x,y,z). The corresponding law $x,y \mapsto xoy=z$ is obviously symmetric. The set of the idempotent points of (E,o) is the set of flexes ; it is isomorphic to AG(t,3) with $t\leqslant 2$ (endowed with the mid-point law). Lastly for any fixed u in E, $x,y \mapsto x*y=uo(xoy)$ makes E into an abelian group.

Let us now consider the case dim V>1. Assume that K is infinite. We have the following fact that we mention here without all the required definitions (for a more complete account see Manin [9] pp. 46-57, especially theorems 13.1 and 13.2):

Theorem of Manin : If V admits a point of "general type", then in a suitable factor set $\bar{E}$ of E, the three-place relation of collinearity gives rise to a symmetric law obeying $(aox)o(aoy)=a^2o(xoy)$ and $x^2ox^2=x^2$ identically.

As a relatively easy consequence we have :

Corollary : The square mapping $x \mapsto x^2=p(x)$ is an endomorphism. The set of the idempotent elements of $(\bar{E},o)$ is I=Im p ; it is a distributive Steiner quasigroup. All the fibres $A=p^{-1}(e)$ of p are isomorphic elementary abelian 2-groups, and

$$(\bar{E},o) \simeq I \times A \text{ (direct product)}.$$

Let us say that a Fischer group is a group of the form G=<S> where S is a conjugacy class of involutions of G such that $O(xy)\leqslant 3$ for any two elements x and y from S (in other terms the dihedral group generated by any two elements of S has order $\leqslant 6$). In case we have O(xy)=3 for any x,y S, $x\neq y$, G is, say a special Fischer group. In any special Fischer group G there is just one class of involutions S_G (namely, the set of all the involutions from G), and S_G may be provided with a structure of HTS by setting $xoy=x^y=yxy(=xyx)$. We call (S_G,o) the HTS corresponding to G. This group-theoretic construction of HTSs is canonical. More precisely :

Theorem : Given any HTS E, the (non-empty) family $\mathfrak{F}$ of special Fischer groups whose corresponding HTS is E admits :
(i) a universal object U ; any G in $\mathfrak{F}$ is of the form G=U/C where $C\triangleleft Z(U)$.
(ii) a smallest object I=U/Z(U), which is also the unique centerless element of $\mathfrak{F}$.

3-A PROCESS OF EXPLICIT CONSTRUCTION : Let E be a vector space over GF(3) with dim E=n+1. Pick up some basis $e_1, e_2, \ldots e_n, e_{n+1}$. Besides choose a non-zero sequence of $\binom{n}{3}$ elements from GF(3), say :

$$\sigma=(\lambda_{ijk})_{1\leqslant i<j<k\leqslant n}$$

Define a binary law in E by setting that, if $x=\sum_{i=1}^{i=n+1} x^i e_i$ and $y=\sum_{j=1}^{j=n+1} y^j e_j$ where $x^i, y^j \in GF(3)$, then :

$$x \circ y = -x-y+\Big(\sum_{1\leqslant i<j<k\leqslant n} \lambda_{ijk}(x^i-y^i)(x^j y^k - x^k y^j)\Big)\, e_{n+1}$$

Proposition : (E, o_σ) is a HTS of rank n+1. Any rank n+1 HTS of order 3^{n+1} arises in this manner.

For n=3, σ is just a non-zero scalar : ±1 ; up to isomorphism, one gets only one HTS which is the order 81 distributive Steiner quasigroup discovered by Zassenhaus (see Bol [5]).

The equalities : $t(e_i, e_j, e_k) = \lambda_{ijk}$, $1\leqslant i<j<k\leqslant n$ détermine a unique alternate trilinear form $t : V^3 \to GF(3)$ with $V=\langle e_1, e_2, \ldots e_n\rangle$. Since several different sequences $\sigma'=(\lambda'_{ijk})$ yield isomorphic HTSs, there are several such forms related to the same HTS (E, o_σ). In fact all these trilinear forms are "equivalent" in a sense that we are going to precise.

4-THE PROBLEM OF CLASSIFICATION OF ALTERNATE TRILINEAR FORMS : Consider a n-dimensional vector space V=V(n,K) over a commutative field K. Recall that a trilinear form $t:V^3 \to K$, $(x,y,z) \longmapsto t(x,y,z)$ is alternate (or : symplectic, skew-symmetric) if $t(\pi(x),\pi(ty),\pi(z))=\mathrm{sign}(\pi)\,.\,t(x,y,z)$ for every permutation π of x,y,z. Any two trilinear forms t and u are equivalent if there exists ℓ in $GL_K(V)$ such that $u(x,y,z)=t(\ell(x),\ell(y),\ell(z))$ identically. Designate as A(n,K) the family of the non-vanishing alternate trilinear forms of V and $\tilde{A}(n,K)$ the corresponding set of equivalence classes. We shall be concerned with the classification of the trilinear forms up to equivalence in two special cases : first when there is a totally isotropic vector space, second when n⩽6. In the first situation, n can be any integer ⩾3 but we restrict attention to the forms that vanish identically on M^3 where M is a (n-1)-dimensional subspace. When n⩽6 we obtain an almost complete classification, and in fact a complete one in case K=GF(3). Now any result concerning the special case K=GF(3) can be interpreted in terms of HTSs in view of the following result :

Theorem 4.1 - (translation theorem) : There is a one-to-one correspondence between $\tilde{A}$(n, GF(3)) and the isomorphism classes of HTSs of rank n+1 whose order is 3^{n+1}.

This translates the problem of the classification of non affine HTSs of maximal rank into an exterior algebraic classification problem. We shall study two special cases where the number of classes can be specified. As examples let us give a couple of statements consisting first of classification theorems of symplectic geometry and second of an applications to the case K=GF(3) through the translation theorem.

Theorem 4.2 - For any commutative field K the elements of A(n,K), n⩾3, admitting a totally isotropic codimension 1 subspace form $[^{n-1}/2]$ complete equivalence classes. If n is odd there is only one such class for which V is non singular ; if n is even there is none.

Corollary - The rank n+1 HTSs of order 3^{n+1}, n⩾3, admitting an affine subsystem of order 3^n form exactly $[^{n-1}/2]$ complete isomorphism classes. If n is odd there is exactly one such system without non trivial affine direct factor ; if n is even there is none.

For instance if n=9, there are 4 rank 10 HTSs of order 3^{10} admitting an affine order 3^9 subspace ; they can be described in the process of section 3 by setting $\lambda_{123}=1$, $\lambda_{145}=\alpha$, $\lambda_{167}=\beta$, $\lambda_{189}=\gamma$ and, for $\{i,j,k\} \notin \{\{1,2,3\}, \{1,4,5\}, \{1,6,7\}, \{1,8,9\}\}$ $\lambda_{ijk}=0$, where α,β,γ are 0 or 1. The non singular case arises when one takes $\alpha=\beta=\gamma=1$.

Theorem 4.3 - For any commutative field K, $|\mathcal{A}(3,K)|=1=|\mathcal{A}(4,K)|$ and $|\mathcal{A}(5,K)|=2$. Besides $|\mathcal{A}(6, GF(3))|=5$.

Corollary - Up to isomorphism there are exactly (i) one rank 4 (resp. 5) HTS of order 3^4 (resp. 3^5), (ii) two rank 6 HTSs of order 3^6, and (iii) five rank 7 HTSs of order 3^7.

The link between theorem 4.3 and the corollary follows from the translation theorem. It gives a quick proof of (i) and (ii) that had been previously established [6,8,1,2,3,4]. The determination of $\mathcal{A}(6, GF(3))$ was obtained in collaboration with J. Lacaze.

The following table gives the number of isomorphism classes of HTSs corresponding to a given 3-order $s \leq 13$ and a given rank $\rho \leq 8$.

ρ \ s	3	4	5	6	7	8	9	10	11	12
4	1	1^{J_1}								∅
5		1	1	1	1	4	?	?	?	1
6			1	2^{J_2}	?	?	?		?	?
7				1	5	?	?	?		
8		∅			1	J_3 1	?			

The last corollary settles the case $(s,\rho)=(7,7)$. It turns out that there are as many HTSs corresponding to this case as the total number of non affine HTSs with $s \leq 6$. In the cells corresponding to $s=\rho>7$ the complete classification seems difficult. But the HTSs with an affine maximal sub-system are now determined, and for $s=2n+2=\rho$ there is just one such system J_n indecomposable.

BIBLIOGRAPHY

[1] L. Bénéteau ; Topics about Moufang loops and Hall triple systems, Simon Stevin 54 (1980) 107-124.

[2] L. Bénéteau ; Une classe particulière de matroïdes parfaits, in "Combinatorics 79", Annals of Discrete Math. 8 (1980) 229-232.

[3] L. Bénéteau ; Les Systèmes Triples de Hall de dimension 4, European J. Combinatorics (1981) 2, 205-212.

[4] L. Bénéteau ; Hall Triple systems and related topics. Proc. Interna. Conf. on Combinatorial Geometries and their applications, 1981 ; Annals of Discrete Math. 18 (1983) 55-60.

[5] G. Bol ; Gewebe und Gruppen, Math. Ann. 114 (1937) 414-431 ; Zbl. 16, 226.

[6] Marshall HALL Jr. ; Automorphisms of Steiner Triple systems, IBM J. Res. Develop. (1960), 406-472. MR 23A#1282 ; Zbl. 100, p. 18.

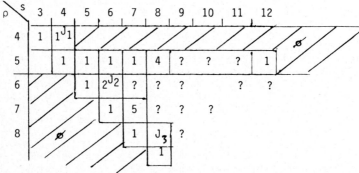

[7 Marshall Hall Jr. ; Group theory and block designs, in Proc. Intern. Conf. on the Theory of Groups, Camberra 1965, Gordon and Breach, New-York. MR 36#2514 ; Zbl. 323.2011.

[8 T. Kepka ; Distributive Steiner quasigroups of order 3^5, Comment. Math. Univ. Carolinae 19,2 (1978) ; MR 58 # 6032.

[9 Yu. I. Manin ; Cubics forms, North-Holland Publishing Company, Amsterdam-London ; Amercian Elsevier Publishing Company, Inc. New-York (1974).

[10 H.P. Young ; Affine triple systems and matroïd designs, Math. Z. 132 (1973) 343-366. MR. 50 # 142.

Annals of Discrete Mathematics 30 (1986) 15–30

ON A TEST OF DOMINANCE, A STRATEGIC DECOMPOSITION AND STRUCTURES T(t,q,r,n)

Walter Benz

Mathematisches Seminar der Universität Hamburg

1. A task in psychology or economics is to compare different groupings $G_1, \ldots, G_r$ of objects (people, goods) concerning a property E (or properties $E_1, \ldots, E_n$) and to order them concerning E starting with the dominant grouping. This can be done by considering interactions (i.e. r-sets intersecting every grouping in one element) and by judging all the possible interactions: If $I = \{g_1, \ldots, g_r\}$, $g_i \in G_i$, is an interaction one looks for an ordering $g_{i_1} \geq \ldots \geq g_{i_r}$, where " $\geq$ " stands for "is E-better than" or "is E-equal to". Hence every interaction leads to a position number for a certain grouping and one has to add up these numbers in order to get the E-ordering for all the groupings.

Usually there are too many interactions, and the problem is to find a balanced subclass of interactions on which the judgement should be based.

When I became confronted with this problem in a special case, my proposal was to use generalized Laguerre geometries in case of one single property E (and generalized Minkowski geometries for several properties $E_1, \ldots, E_n$): The parallel classes of points of a plane Laguerre geometry may for instance represent the groupings, and the selected interactions could be given by the blocks of the geometry. The fact that through three pairwise non parallel points there is exactly one block could serve as property of balance concerning the chosen subclass of interactions. As a matter of fact those generalized Laguerre geometries are studied in the literature under different names like orthogonal arrays (Bush [2]), optimal geometries (Halder, Heise [4]) and they play an important rôle in connection with optimal codes in coding theory (Halder, Heise [4]).

In section 2 we like to solve a practical problem which comes up in carrying out a test under consideration: To find a disjoint decomposition of the set of blocks of an optimal geometry such that all the components of the decomposition are partitions of the set of points. By using such a decomposition in a practical case one can divide the whole test in a number of subtests such that all objects are involved in a subtest.

In section 3 we like to deal with a simultaneous test of a set of objects concerning properties $E_1, \ldots, E_n$. The combinatorial structures T(t,q,r,n) which we offer in this connection are generalizations of certain chain geometries ([1]). The class of optimal geometries can be identified with the class of T(t,q,r,1). The Minkowski-m-structures are the structures T(t,q,q,2) with t = m+2. - In Theorem 3 we show that a necessary condition for the existence of T(t,q,r,n) is that r^{n-1} is a divisor of q. In Theorem 4 we determine the number of blocks of a T(t,q,r,n) and also the number of the so called global interactions. In Theorem 5,6 we characterize the $T(t,\lambda r^{n-1}, r,n)$ (cases $\lambda=1$, $\lambda>1$) by applying permutation sets and special classes of functions.

2. Let q,r,t be integers such that $q > 1$ and $2 \leq t \leq r$. Consider the matrix

$$M = \begin{pmatrix} (1,1) & \dots & (r,1) \\ & \cdot & \\ & \cdot & \\ & \cdot & \\ (1,q) & \dots & (r,q) \end{pmatrix}$$

where the ordered pairs (i,j) are called points (or objects), and where we put (i,j) = (i',j') iff i=i' and j=j'. The columns of M are also called groupings. An interaction of M is an r-set containing one element of every column. There are q^r interactions of M. By I(M) we denote the set of all interactions of M. Consider now a subset B(t) of I(M) and call the interactions of B(t) blocks. We are then interested in the following property of balance

(*) To every t-set S having a non-empty intersection with t distinct groupings of M there is exactly one block containing S.

By T(t,q,r) (or T(t,q,r,1) with respect to section 3) we denote a structure (M,B(t)) satisfying (*). Many examples of structures T(t,q,r) for certain t,q,r and also non existence statements for certain t,q,r are known (s. for instance Halder, Heise [4], Heise [5], Heise, Karzel [6]). Two structures T(t,q,r), T(t',q',r') are called isomorphic iff there is a bijection (called isomorphism) of the set of points of T(t,q,r) onto the set of points of T(t',q',r') such that the blocks of the first structure are mapped onto blocks of the second structure. Since two distinct points of T(t,q,r) are in the same grouping iff there is no block joining them (note $t \geq 2$) isomorphisms map columns onto columns. Obviously, isomorphic T(t,q,r), T(t',q',r') coincide in the parameters, i.e. t=t', q=q', r=r'.

In [1] we have studied chain geometries. The finite chain geometries of Laguerre type ([1], p. 144) $\Sigma(K,L)$ are structures T(t,q,r). Here K is a Galois field $GF(\gamma)$ and $L > K$ is a finite local ring with $L/N \cong K$, where N denotes the maximal ideal of L. The parameters are given by $t = 3, q = \# N$, $r = \gamma+1$. The class of chain geometries of Laguerre type $\Sigma(K,L)$, $L = K[\varepsilon]/_{<\varepsilon^2>}$, coincides with the class of miquelian Laguerre planes. Two Laguerre geometries $\Sigma(K,L)$, $\Sigma'(K',L')$ with char $K \neq 2 \neq$ char K' are isomorphic iff there is an isomorphism $\sigma: L \to L'$ such that $\sigma|K$ is an isomorphism of K onto K' ([1], p. 176,

Satz 3.1). Consider a Galois field $GF(\gamma)$ with $2 \nmid \gamma$ and put $K=GF(\gamma)=K'$. Let $n \geq 3$ be an integer and V be the vector space of dimension $n-1$ over K. Define local rings $L := K[\varepsilon]/\langle\varepsilon^n\rangle$ and $L' := \{(k,v) \mid k \in K, v \in V\}$ with

$$(k_1, v_1) + (k_2, v_2) := (k_1 + k_2, v_1 + v_2)$$

$$(k_1, v_1) \cdot (k_2, v_2) := (k_1 k_2, k_1 v_2 + k_2 v_1).$$

Hence $L \not\cong L'$ because of $N^{n-1} \neq 0$, $(N')^2 = 0$, where respectively N, N' are the maximal ideals of L, L'. We thus get two non isomorphic structures $T(3,\gamma^{n-1}, \gamma+1)$, $T'(3,\gamma^{n-1}, \gamma+1)$.

The strategic decomposition we have announced in section 1 concerns the following class of structures $T(t,q,r)$ which is a subclass of those structures defined in Halder, Heise [4] on pages 268, 269 by using linear forms. Let K be a Galois field $GF(\gamma)$ and let V be a vector space over K with $1 < \dim V < \infty$. For an integer t such that $3 \leq t < \gamma + 1$ now define $T(t, \# V, \gamma+1)$ as follows: The set of points is given by $K' \times V$ with $K' := K \cup \{\infty\}$ and the blocks are given by

$$b(v_1,\ldots,v_t) := \{(\alpha, \sum_{\nu=1}^{t} v_\nu \alpha^{t-\nu}) \mid \alpha \in K\} \cup \{(\infty, v_1)\}.$$

By using Vandermonde's determinant it is easy to check that the property of balance (*) is satisfied for t.

Theorem 1: Let Δ be the Galois field $GF(\gamma^{t-1})$ and let $f \in K[x]$ be the minimal polynomial of a primitive element δ of Δ over K. Assume $v_1,\ldots,v_{t-1} \in V$. Then the set $B(v_1,\ldots,v_{t-1})$ of blocks

$$\{(\alpha, vf(\alpha) + \sum_{\nu=1}^{t-1} v_\nu \alpha^{t-1-\nu}) \mid \alpha \in K\} \cup \{(\infty, v)\}, \quad v \in V,$$

is a partition of the set of points and

$$\bigcup_{(v_1,\ldots,v_{t-1}) \in V\times\ldots\times V} B(v_1,\ldots,v_{t-1})$$

is a disjoint decomposition of $B(t)$. (Notice that the degree of f is $t-1$).

Proof. Let v, w be two distinct elements of V. We like to show that the two blocks

$$\{(\alpha, vf(\alpha) + \sum_{\nu=1}^{t-1} v_\nu \alpha^{t-1-\nu}) \mid \alpha \in K\} \cup \{(\infty, v)\},$$

$$\{(\beta, wf(\beta) + \sum_{\nu=1}^{t-1} v_\nu \beta^{t-1-\nu}) \mid \beta \in K\} \cup \{(\infty, w)\},$$

of $B(v_1,\ldots,v_{t-1})$ have no point in common. Assume to the contrary that (ξ, x), $\xi \in K'$, $x \in V$, is a point in both blocks. This implies $\xi \neq \infty$ because of $v \neq w$. Hence

$$vf(\xi) + \sum_{\nu=1}^{t-1} v_\nu \xi^{t-1-\nu} = wf(\xi) + \sum_{\nu=1}^{t-1} v_\nu \xi^{t-1-\nu},$$

i.e. $f(\xi) = 0$ which is not true since f is irreducible over K.- The set $B(v_1,\dots,v_{t-1})$ contains as many blocks as there are elements in V. The number of points on a block is $\# K' = \gamma+1$. Hence $B(v_1,\dots,v_{t-1})$ contains $(\gamma+1)\cdot \# V$ many points and is thus a partition of the set of points.

Now we like to show

$$B(v_1,\dots,v_{t-1}) \cap B(w_1,\dots,w_{t-1}) = \emptyset$$

in case that the two ordered $(t-1)$-plets $(v_1,\dots,v_{t-1})$, $(w_1,\dots,w_{t-1})$ are distinct. Assume to the contrary that the blocks

$$\{(\alpha,\ vf(\alpha) + \sum_{\nu=1}^{t-1} v_\nu \alpha^{t-1-\nu}) \mid \alpha \in K\} \cup \{(\infty, v)\}$$

$$\{(\beta,\ wf(\beta) + \sum_{\nu=1}^{t-1} w_\nu \beta^{t-1-\nu}) \mid \beta \in K\} \cup \{(\infty, w)\}$$

are equal. This implies $v=w$ and hence

$$\sum_{\nu=1}^{t-1} (w_\nu - v_\nu)\alpha^{t-1-\nu} = 0$$

for all $\alpha \in K$. Because of $t < \gamma + 1$ there exist pairwise distinct elements $\alpha_1,\dots,\alpha_{t-1}$ in K. We hence have in matrix notation

$$(w_1-v_1 \ \dots \ w_{t-1}-v_{t-1}) \begin{pmatrix} \alpha_1^{t-2} & \dots & \alpha_{t-1}^{t-2} \\ \vdots & & \vdots \\ \alpha_1 & \dots & \alpha_{t-1} \\ 1 & \dots & 1 \end{pmatrix} = 0 .$$

The Vandermonde matrix P here is regular because of $\#\{\alpha_1,\dots,\alpha_{t-1}\} = t-1$ and by multiplying the matrix equation with P^{-1} from the right we get $(w_1-v_1 \dots w_{t-1}-v_{t-1}) = 0$ which is not true.- There are $(\# V)^{t-1}$ many sets $B(v_1,\dots,v_{t-1})$. Every $B(v_1,\dots,v_{t-1})$ contains $\# V$ many blocks. Since there are q^t many blocks in a structure $T(t,q,r)$ the set

$$\bigcup_{(v_1,\dots,v_{t-1}) \in V \times \dots \times V} B(v_1,\dots,v_{t-1})$$

contains all the blocks.

Example. Consider $K = GF(3)$, $V = K$ and $t = 3$. A required decomposition here is given by

aApP	aBrP	aCqP
bBqQ	bCpQ	bArQ
cCrR	cAqR	cBpR
aArR	aBqR	aCpR
bBpP	bCrP	bAqP
cCqQ	cApQ	cBrQ
aAqQ	aBpQ	aCrQ
bBrR	bCqR	bApR
cCpP	cArP	cBqP

3. Let q,r,t,n be integers greater than 1 and such that $2 \leq t \leq r$. Let P be a set of cardinality qr. The elements of P are called points (or objects). Consider moreover n matrices

$$M_i = \begin{pmatrix} (1,1,i) & \dots & (r,1,i) \\ \cdot & & \\ \cdot & & \\ \cdot & & \\ (1,q,i) & \dots & (r,q,i) \end{pmatrix}, \quad i=1,\dots,n,$$

such that

$$P = \{(1,1,i),\dots,(r,1,i),\dots,(1,q,i),\dots,(r,q,i)\}$$

for all $i=1,\dots,n$. A global interaction of $M_1,\dots,M_n$ is a r-set which is an interaction for all the matrices $M_1,\dots M_n$. Two points are called competitors if they are not in a common column for all the matrices $M_1,\dots,M_n$. By $G(M_1,\dots,M_n)$ we denote the set of all global interactions of $M_1,\dots,M_n$. We are now interested in a set $B(t) \subset G(M_1, \dots,M_n)$ such that the following two conditions are satisfied (the elements of B(t) are called blocks)

(i) Through t distinct points which are parwise competitors there is exactly one block .

(ii) For every integer j with $1 \leq j \leq n$ the following holds true: If D_1 is the point intersection of j distinct columns (of $M_1,\dots,M_n$) such that no two of them belong to the same M_ν and if D_2 is another such intersection of j columns then $\# D_1 = \# D_2$.

We denote a structure $(M_1,\dots,M_n; B(t))$ by $T(t,q,r,n)$. Conditions (i), (ii) serve as properties of balance.

Example. Assume that four firms are offering each a comparable collection of four wines and that four other firms are offering each a comparable collection of four bottles manufactured to be filled up with wine. The question is to test simultaneously the quality of the wine collections and that of the bottle collections. We like to do this with a T(3,4,4,2):

$$M_1 = \begin{pmatrix} aApP \\ bBqQ \\ cCrR \\ dDsS \end{pmatrix}, \qquad M_2 = \begin{pmatrix} aApd \\ BqbQ \\ rcCs \\ SRPD \end{pmatrix}.$$

The columns of M_1 represent the four wine collections and the columns of M_2 the bottle collections. Now the bottle collection A,q,c,R for instance is filled up in A,q,c,R with wine of the 2.,3.,1.,4. wine producer respectively. Now check both qualities along the following set of blocks:

bBqQ	rAPd	bqDS	bBRs
ApaQ	BqPd	cpSD	CaRs
rcCQ	brRD	cBPs	CqSd
brAQ	paRD	aAPs	ApSd
qCaQ	bAsS	pBdR	aqDP
BcpQ	cCsS	rCdR	rcDP.

(This example is the miquelian Minkowski plane of order 3.)

Theorem 2: Let K be the Galois field GF(γ) and let $n > 1$ be an integer. Denote by L_n the ring $K \times \ldots \times K$ with n factors. The chain geometry $\Sigma(K, L_n)$ is then a structure $T(3, (\gamma+1)^{n-1}, \gamma+1, n)$.

Proof. a) Consider the following maximal ideals

$$J_i := \{(k_1, \ldots, k_n) \in L_n \mid k_i = 0\}$$

of L_n for $i=1,\ldots,n$. A point of $\Sigma(K,L_n)$ is given by

$$R(p_1, p_2) := \{(rp_1, rp_2) \mid r \in R\},$$

where R denotes the group of units of L_n and where p_1,p_2 are elements of L_n such that the ideal generated by p_1,p_2 is the whole ring L_n. For two points $P = R(p_1,p_2)$ $Q = R(q_1,q_2)$ we define

$$P \,||_i\, Q \Leftrightarrow \begin{vmatrix} p_1 & p_2 \\ q_1 & q_2 \end{vmatrix} \in J_i, \quad i \in \{1,\ldots,n\}.$$

b) The points R(a,b) with $a = (a_1,\ldots,a_n)$, $b = (b_1,\ldots,b_n)$ can be identified with the ordered n-plets $(K(a_1,b_1),\ldots, K(a_n,b_n))$ of points $K(a_i,b_i)$ of the projective line π over K. Moreover: $R(a,b) \,||_i\, R(c,d)$ iff $K(a_i,b_i) = K(c_i,d_i)$. Hence $||_i$ is an equivalence relation on the set of points and there are $(\gamma+1)^{n-1}$ many ordered n-plets $(P_1,\ldots,P_n)$, $P_1,\ldots,P_n \in \pi$, such that P_i = const. Thus the number of points in an equivalence class concerning $||_i$ is $(\gamma+1)^{n-1}$. The number of equivalence classes concerning $||_i$ is $\gamma+1$ since there are $\gamma+1$ points $K(a_i,b_i)$ in π.

c) We now define the matrix M_i ($i \in \{1,\ldots,n\}$) with $(\gamma+1)^{n-1}$ rows and $\gamma+1$ columns. The matrix M_i can be chosen arbitrarily up to the fact that the columns are supposed to be the $||_i$-equivalence classes.

d) Two points $R(a_1, a_2)$, $R(b_1,b_2)$ are obviously competitors iff $\begin{vmatrix} a_1 & a_2 \\ b_1 & b_2 \end{vmatrix} \in R$. - In chain geometry [1] it is proved that through three points A,B,C such that

$$\begin{vmatrix} a_1 & a_2 \\ b_1 & b_2 \end{vmatrix}, \begin{vmatrix} b_1 & b_2 \\ c_1 & c_2 \end{vmatrix}, \begin{vmatrix} c_1 & c_2 \\ a_1 & a_2 \end{vmatrix} \in R$$

there is exactly one chain and that for two distinct points P,Q the determinant $\begin{vmatrix} p_1 & p_2 \\ q_1 & q_2 \end{vmatrix}$ must be an element of R. Chains are hence global interactions (note that any chain contains $\gamma+1$ points) of $M_1,\ldots,M_n$. Define the set of chains to be the set B(3). Then (i) holds true for t=3. In order to verify (ii) let j be an integer with $1 \leq j \leq n$ and let $i_1,\ldots,i_j \in \{1,\ldots,n\}$ be j distinct integers. Consider equivalence classes $E(i_1),\ldots,E(i_j)$ of the relations $\| i_1,\ldots, \| i_j$ respectively. Then

$$\# \; [E(i_1) \cap \ldots \cap E(i_j)] = (\gamma+1)^{n-j}$$

for j=1,...,n. For if we fix the components $i_1,\ldots,i_j$ in $(P_1,\ldots,P_n)$, $P_i \in \pi$, then the number of the remaining n-plets is $(\gamma+1)^{n-j}$.

Remark: A bijection of the set of points of a structure T(t,q,r,n) is called an automorphism iff images and inverse images of blocks are blocks. As a special case of a theorem of Schaeffer [10] the automorphism group of $\Sigma(K,L_n)$ is known for L_n semilocal, $\#K > 3$ and char $K \neq 2$ (in case that K is finite, obviously, L_n must be semilocal): This is the group $p\Gamma L_K(2,L_n)$. -
The $\Sigma(K,L_2)$ are the miquelian Minkowski planes, K an arbitrary field.

Theorem 3: Consider a structure T(t,q,r,n). Then r^{n-1} must be a divisor of q. Moreover: If D is the point intersection of $j (1 \leq j \leq n)$ distinct columns (of $M_1,\ldots,M_n$) such that no two of them belong to the same M_ν then

$$\alpha_j := \# D = \frac{q}{r^{j-1}}$$

Proof: The formula is true for j=1. Assume now $2 \leq j \leq n$. Let E_ν be a column of M_ν for $\nu=1,\ldots,j-1$ and let $C_1,\ldots,C_r$ be the columns of M_j. Observe that the $C_1,\ldots,C_r$ are pairwise disjoint and that their union is the whole set of points. Hence

$$E_1 \cap \ldots \cap E_{j-1} = \bigcup_{\nu=1}^{r} [C_\nu \cap (E_1 \cap \ldots \cap E_{j-1})]$$

and thus $\alpha_{j-1} = r \; \alpha_j$. This proves the theorem.

At the beginning of section 3 we require n>1 for a structure T(t,q,r,n). But obviously the structures T(t,q,r) of section 2 can be considered as structures T(t,q,r,1), since (ii) plays no rôle in case n=1.

The structures T(t,q,r,2) have been studied extensively in the literature in case q=r. See for instance the results in Ceccherini [3],

Heise, Karzel [6], Heise, Quattrocchi [7], Quattrocchi [8], where Minkowski-m-structures are considered.
Concerning the real case $\Sigma(\mathbb{R}, \mathbb{R} \times \mathbb{R} \times \mathbb{R})$ compare Samaga [9].

Theorem 4: Let $(M_1,\dots,M_n;\ B(t))$ be a structure $T(t,q,r,n)$. Then

$$\# B(t) = \left(\frac{q}{r^{n-1}}\right)^t \cdot \left(t!\binom{r}{t}\right)^{n-1}$$

and the cardinality of the set of all global interactions of $M_1,\dots,M_n$ is given by

$$\left(\frac{q}{r^{n-1}}\right)^r \cdot (r!)^{n-1}$$

Proof. If b is a point, denote by $[b]_i$ the column of M_i through b. Consider t distinct columns $C_1,\dots,C_t$ of M_1. We now like to define sets $D_1,\dots,D_t$. Those sets (but not their cardinalities) will depend on certain points $a_1, a_2, \dots$. Put $D_1 := C_1$. In case $D_\nu (1 \leq \nu \leq t-1)$ is defined, take $a_\nu \in D_\nu$ and put

$$D_{\nu+1} := C_{\nu+1} \setminus \bigcup_{\mu=2}^{n} ([a_1]_\mu \cup \dots \cup [a_\nu]_\mu).$$

We then have

$$\# B(t) = \prod_{\nu=1}^{t} (\# D_\nu)$$

since the number of blocks must be equal to the number of $(b_1,\dots,b_t)$, $b_i \in C_i$, such that the b's are pairwise competitors.
Put

$$B_{\nu\mu} := C_\nu \cap ([a_1]_\mu \cup \dots \cup [a_{\nu-1}]_\mu)$$

for $\nu \in \{2,\dots,t\}$ and $\mu \in \{2,\dots,n\}$. For $\mu_1 \neq \mu_2$ we get

$$\text{(o)}\quad B_{\nu\mu_1} \cap B_{\nu\mu_2} = (C_\nu \cap [a_1]_{\mu_1} \cap [a_1]_{\mu_2}) \cup \dots \cup (C_\nu \cap [a_{\nu-1}]_{\mu_1} \cap [a_{\nu-1}]_{\mu_2})$$

and hence by Theorem 3

$$\# (B_{\nu\mu_1} \cap B_{\nu\mu_2}) = (\nu-1)^2 \alpha_3 = (\nu-1)^2 \frac{q}{r^2}$$

since the sumands of the right hand side of (o) are pairwise disjoint. Similary we have

$$\# (B_{\nu\mu_1} \cap \dots \cap B_{\nu\mu_s}) = (\nu-1)^s \cdot \frac{q}{r^s}$$

for pairwise distinct $\mu_1,\dots,\mu_s$ in $\{2,\dots,n\}$. Because of

$$D_\nu = C_\nu \setminus \bigcup_{\mu=2}^{n} B_{\nu\mu}$$

and

$$\# \left(\bigcup_{\mu=2}^{n} B_{\nu\mu}\right) = \sum_{1<i\le n} \# B_{\nu i} - \sum_{1<i<j\le n} \#(B_{\nu i} \cap B_{\nu j}) + \sum_{1<i<j<k\le n} \#(B_{\nu i} \cap B_{\nu j} \cap B_{\nu k})$$

$$\mp \ldots + (-1)^n \# (B_{\nu 2} \cap \ldots \cap B_{\nu n})$$

$$= q\left[\binom{n-1}{1}\frac{\nu-1}{r} - \binom{n-1}{2}\left(\frac{\nu-1}{r}\right)^2 \mp \ldots + (-1)^n \binom{n-1}{n-1}\left(\frac{\nu-1}{r}\right)^{n-1}\right]$$

we get

$$\# D_\nu = q\cdot\left[1 - \binom{n-1}{1}\frac{\nu-1}{r} \pm \ldots + (-1)^{n-1}\binom{n-1}{n-1}\left(\frac{\nu-1}{r}\right)^{n-1}\right]$$

$$= q\cdot\left(1 - \frac{\nu-1}{r}\right)^{n-1} \quad \text{for} \quad \nu = 2,\ldots,t.$$

(This especially implies that $D_\nu \neq \emptyset$.) Hence

$$\# B(t) = q\cdot q\left(1-\frac{1}{r}\right)^{n-1} \ldots q\left(1-\frac{t-1}{r}\right)^{n-1}$$

$$= \left(\frac{q}{r^{n-1}}\right)^t \left(t!\binom{r}{t}\right)^{n-1}.$$

In order to get the cardinality of the set of all global interactions we determine

$$\# B(r) = \left(\frac{q}{r^{n-1}}\right)^r (r!)^{n-1}.$$

Remarks: 1) In case $\Sigma(K,L_n)$ with $K = GF(\gamma)$ and $n > 1$ we get for the number of blocks by Theorem 4

$$\# B(3) = \left[6\cdot\binom{\gamma+1}{3}\right]^{n-1} = (\gamma^3 - \gamma)^{n-1}.$$

The number of global interactions in this case is given by

$$[(\gamma+1)!]^{n-1}.$$

2) If a_1, a_2 are competitors of a structure $T(t,q,r,n)$ then according to the construction in the proof of Theorem 4 there is a block (notice $t \geq 2$) containing a_1, a_2. On the other hand: Two distinct points of a block must be competitors. We thus have: Two distinct points are competitors iff there is a block joining them.

In Theorem 2 we have presented structures $T(3, (\gamma+1)^{n-1}, \gamma+1, n)$, where $n > 1$ is an integer and where γ is a prime power. We now like to define further structures $T(t,q,r,n)$.

A. Consider n matrices $M_1,\ldots,M_n$ all with $q > 1$ rows and $r > 1$ columns and all filled up with the elements of P, $\# P = qr$, such that property

(ii) is satisfied. Taking then the set $B(r)$ of all global interactions we get a $T(r,q,r,n)$.

Let $H = \{h_1,\dots,h_r\}$ be a set containing $r \geq 2$ elements. Define P to be the cartesian product

$$P = H \times \dots \times H$$

with n factors for a given integer $n > 1$. In order to define matrices M_i, $i \in \{1,\dots,n\}$, put

$$C_{i\nu} := \{(x_1,\dots,x_n) \in P \mid x_i = h_\nu\}$$

for the columns $C_{i1},\dots,C_{ir}$ of M_i. Then property (ii) is satisfied for the matrices $M_1,\dots,M_n$.- Let t be an integer with $2 \leq t \leq r$ and let $\Gamma_2,\dots,\Gamma_n$ be sets of permutations of H such that every Γ_i is sharply t-transitive on H. We then like to define a set $B(t)$ of global interactions for $M_1,\dots,M_n$: For given permutations $\pi_\mu \in \Gamma_\mu$, $\mu = 2,\dots,n$, call

$$b(\pi_2,\dots,\pi_n) := \{(h, \pi_2(h),\dots, \pi_n(h)) \mid h \in H\}$$

a block, i.e. an element of $B(t)$. Since $h_1 \neq h_2$ implies $\pi_\nu(h_1) = \pi_\nu(h_2)$ ($\nu = 2,\dots,n$) we are sure that two distinct elements of $b(\pi_2,\dots,\pi_n)$ are competitors. Consider now t distinct points $(h_\nu, x_{2\nu},\dots,x_{n\nu})$, $\nu = 1,\dots,t$, such that no two of them are competitors. There is exactly one permutation $\pi_\mu \in \Gamma_\mu$ with $\pi_\mu(h_\nu) = x_{\mu\nu}$ for $\nu = 1,\dots,t$. There is thus exactly one block, $b(\pi_2,\dots,\pi_n)$, joining the t points. Hence we get a $T(t,r^{n-1}, r,n)$. The Mathieu groups M_{11}, M_{12} for instance lead to structures $T(4,11^{n-1}, 11,n)$ (put $\Gamma_2 = \dots \Gamma_n = M_{11}$) and $T(5,12^{n-1}, 12,n)$ for every $n > 1$.- In case that H is the projective line over $K = GF(\gamma)$ and $\Gamma_2 = \dots = \Gamma_n$ is its projective group we get $\Sigma(K,L_n)$ up to an isomorphism.

Theorem 5: Every structures $T(t,r^{n-1},r,n)$, $n > 1$, can be described as was done before by an r-set H and permutation sets $\Gamma_2,\dots,\Gamma_n$ on H which are sharply t-transitive on H.

Proof. Let $H := \{h_1,\dots,h_r\}$ be a block of $T(t,r^{n-1},r,n)$ and let x be an arbitrary point. Put $x_i := h_\nu$ iff x, h_ν are in the same column of M_i. We like to show that x is completely determined by $(x_1,\dots,x_n)$ and that every $(y_1,\dots,y_n)$, $y_i \in H$, occurs as a point: This is a consequence of the fact that n columns, no two of them in the same matrix, intersect in $\alpha_n = \dfrac{r^{n-1}}{r^{n-1}} = 1$ points. Consider now a block

$$(x_{\rho 1},\dots,x_{\rho n}), \quad \rho = 1,\dots,r, \quad \text{of } T(t,r^{n-1},r,n).$$

Then for a given $j \in \{2,\dots,n\}$

$$\pi_j(x_{\rho 1}) = x_{\rho j} \qquad (\rho = 1,\dots,r)$$

is a permutation of H. Define Γ_j to be the set of all π_j stemming from blocks, j=2,...,n. It is Γ_{j_o} for every $j_o \in \{2,...,n\}$ sharply t-transitive on H: Consider t distinct elements $y_{11}, y_{21},...,y_{t1}$ of H and moreover t distinct elements $y_{1j_o},...,y_{tj_o}$ of H. Let

$$(x_{\rho 1},...,x_{\rho n}), \quad \rho=1,...,t,$$

be t distinct points such that no two of them are competitors and such that

$$x_{\rho 1} = y_{\rho 1}, \quad x_{\rho j_o} = y_{\rho j_o} \quad \text{for} \quad \rho=1,...,t.$$

Since there is exactly one joining block there must be exactly one $\pi_{j_o} \in \Gamma_{j_o}$ such that

$$\pi_{j_o}(y_{\rho 1}) = y_{\rho j_o} \quad \text{for} \quad \rho=1,...,t.$$

B. Let us have matrices $M_1,...,M_n$ as was described at the beginning of A. Suppose n > 1. Then also $M_1,...,M_{n-1}$ satisfy (ii) and a structure T(r,q,r,n) thus leads to a structure T(r,q,r,n-1).

C. Derivation process. Consider a structure T(t,q,r,n) such that $t \geq 3$ and $r \geq 3$. Let $M_1,...,M_n$ be the underlying matrices, P be the point set and B(t) be the set of blocks. Take a point $p \in P$. In every matrix M_ν cancel the column $[p]_\nu$ through p getting the matrix M'_ν. Denote

$$[p]_1 \quad ... \quad [p]_n$$

by Δ and cancel in every remaining column of $M_1,...,M_n$ the points of Δ. Define

$$B_p(t) := \{ \beta \in B(t) \mid p \in \beta\}$$

Then $M'_1,...,M'_n$ together with

$$B'(t-1) := \{ \beta \setminus \{p\} \mid \beta \in B_p(t)\}$$

is a structure T(t-1,q',r-1,n) with $q' = \frac{q}{r^{n-1}} \cdot (r-1)^{n-1}$.

Remark. Starting with the structure $T(3,(\gamma+1)^{n-1}, \gamma+1,n)$ n > 1, of Theorem 2, we get a structure $T(2,\gamma^{n-1}, \gamma, n)$ by the derivation process. This structure can be described as follows: Take the $K^n, K = GF(\gamma)$, and define the columns of M_i to be the hyperplanes $x_i = \text{const}$ (i=1,...,n). The blocks are given by the lines

$$(x_1,...,x_n) = (p_1,...,p_n) + \lambda(v_1,...,v_n), \quad \lambda \in K,$$

such that $\prod_{\nu=1}^{n} v_\nu \neq 0$.

D. For integers t,λ,r such that $2 \leq t \leq r$ and $\lambda \geq 2$ we denote by

$\Phi(t,\lambda,r)$ a set of functions

$$\varphi: \{1,...,r\} \to \{1,...,\lambda\}$$

satisfying the following condition:

(*) For given distinct $i_1,...,i_t \in \{1,...,r\}$ and for given $j_1,...,j_t \in \{1,...,\lambda\}$ there is exactly one $\varphi \in \Phi$ such that $\varphi(i_\nu) = j_\nu$ for $\nu=1,...,t$.

Examples:

1) $\Phi(2,2,3)$:

$\varphi_1(1) = 1,\quad \varphi_1(2) = 1,\quad \varphi_1(3) = 1$

$\varphi_2(1) = 1,\quad \varphi_2(2) = 2,\quad \varphi_2(3) = 2$

$\varphi_3(1) = 2,\quad \varphi_3(2) = 1,\quad \varphi_3(3) = 2$

$\varphi_4(1) = 2,\quad \varphi_4(2) = 2,\quad \varphi_4(3) = 1$

2) $\Phi(3,2,4)$:

1	2	3	4
1	1	1	1
1	1	2	2
1	2	1	2
1	2	2	1
2	1	1	2
2	1	2	1
2	2	1	1
2	2	2	2

Remark. Consider a class of functions $\Phi(t,\lambda,r)$ satisfying (*). Then obviously

$$M = \begin{pmatrix} (1,1) & . \; . \; . & (r,1) \\ & \vdots & \\ (1,\lambda) & . \; . \; . & (r,\lambda) \end{pmatrix}$$

with the blocks

$$b(\varphi) := \{(i,\varphi(i)) \mid i \in \{1,...,r\}\} \quad , \quad \varphi \in \Phi(t,\lambda,r),$$

is a $T(t,\lambda,r,1)$.- On the other hand consider a $T(t,\lambda,r,1)$. Define to every block

$$b = \{(1,j_1),...,(r,j_r)\}$$

the function $\varphi_b : \{1,\ldots,r\} \to \{1,\ldots,\lambda\}$ by $\varphi_b(i) = j_i$ $(i=1,\ldots,r)$. Then it is clear that the class of functions φ_b satisfies (*).

Let $H = \{h_1,\ldots,h_r\}$ be a set containing $r \geq 2$ elements. Consider integers t,λ,n greater than 1 such that $2 \leq t \leq r$. Let $\Gamma_2,\ldots,\Gamma_n$ be permutation sets which are sharply t-transitive on H and let $\Phi(t,\lambda,r)$ be a class of functions as defined above. We then like to define a structure $T(t,\lambda r^{n-1},r,n)$. The points are the ordered (n+1)-plets $(x_1,\ldots,x_n,\nu)$ such that $x_1,\ldots,x_n \in H$ and $\nu \in \Lambda := \{1,\ldots,\lambda\}$. In order to define matrices M_i, $i \in \{1,2,\ldots,n\}$, put

$$C_{i\nu} := \{(x_1,\ldots,x_n,\mu) \in P \mid x_i = h_\nu\}$$

for the columns $C_{i1},\ldots,C_{ir}$ of M_i. Then property (ii) is satisfied for the matrices $M_1,\ldots,M_n$. The set B(t) of blocks is defined as follows: For given permutations $\pi_\rho \in \Gamma_\rho$, $\rho=2,\ldots,n$, and for a given $\varphi \in \Phi$ call

$$b(\pi_2,\ldots,\pi_n,\varphi) := \{(h_i,\pi_2(h_i),\ldots,\pi_n(h_i),\varphi(i)) \mid i \in \{1,\ldots,r\}\}$$

a block, i.e. an element of B(t). This construction leads to a $T(t,\lambda^{n-1},r,n)$.

Theorem 6. Every structure $T(t,\lambda r^{n-1},r,n)$ $n > 1$, $\lambda > 1$, can be described as was done before by an r-set H, by permutation sets $\Gamma_2,\ldots,\Gamma_n$ on H which are sharply t-transitive on H and by a function set $\Phi(t,\lambda,r)$ satisfying (*).

Proof. For given points a,b write $a \sim b$ iff they are in the same columns of $M_1,\ldots,M_n$. This is an equivalence relation and since $\#(C_1 \cap \ldots \cap C_n) = \lambda$, C_i a column of M_i $(i=1,\ldots,n)$ the equivalence classes contain exactly λ points. Let $H := \{h_1,\ldots,h_r\}$ be a block of $T(t,\lambda r^{n-1},r,n)$ and let x be an arbitrary point. Put $x_i := h_\nu$ iff x,h_ν are in the same column of M_i. The n-plet $(x_1,\ldots,x_n)$ does not determine the point x. But there are λ points equivalent to x. We call them $(x_1,\ldots,x_n,\nu)$, $\nu=1,\ldots,\lambda$. We construct the permutation sets $\Gamma_2,\ldots,\Gamma_n$ the same way it was done in the proof of Theorem 5. To every block

$$(x_{\rho 1},\ldots,x_{\rho n},\nu_\rho), \qquad \rho=1,\ldots,r,$$

of $T(t,\lambda r^{n-1},r,n)$ we like to associate a function $\varphi : \{1,\ldots,r\} \to \{1,\ldots,\lambda\}$: Put $\varphi(i) = \nu_\rho$ in case $x_{\rho 1} = h_i$. Call $\Phi(t,\lambda,r)$ the set of all such functions stemming from blocks. Given now distinct $i_1,\ldots,i_t \in \{1,\ldots,r\}$ and elements $j_1,\ldots,j_t \in \{1,\ldots\lambda\}$. Let then

$$(h_{i_1},\ldots,j_1),\ldots(h_{i_t},\ldots,j_t)$$

be t distinct points such that no two of them are competitors. Since there is exactly one joining block there must be hence exactly one function φ in $\Phi(t,\lambda,r)$ such that $\varphi(i_1) = j_1,\ldots,\varphi(i_t) = j_t$.

Remarks. 1) With the examples $\Phi(2,2,3)$, $\Phi(3,2,4)$ one can construct structures $T(2,2\cdot 3^{n-1},3,n)$ (put $\Gamma_2 = \ldots = \Gamma_n =: S_3$), $T(3,2\cdot 4^{n-1},4,n)$ (put $\Gamma_2 = \ldots = \Gamma_n =: S_4$) in case $n > 1$.

2) Because of the connection of function classes $\Phi(t,\lambda,r)$ and geometries $T(t,\lambda,r,1)$ and because of the non-existence of $T(3,\lambda,\lambda+2)$, λ odd, (s. Heise [5]), there do not exist $T(3,\lambda\cdot(\lambda+2)^{n-1},\lambda+2,n)$ for λ odd and $n \in N$ according to Theorem 6.

3) By applying the derivation process it is easy to verify that there do not exist $T(t,\lambda,r)$ in case $\lambda^2 < (\lambda-1)(r-t+2)$. (For instance there does not exist a $T(3,10,13)$). This implies by Theorem 6 that there do not exist $T(t,\lambda r^{n-1},r,n)$ in case $n \in N$ and $\lambda^2 < (\lambda-1)(r-t+2)$.

4) As far as the number of blocks of a $T(t,\lambda r^{n-1},r,n)$ is concerned, Theorem 5,6 lead to a new proof of Theorem 4: A sharply t-transitive permutation set on a r-set contains $r\cdot(r-1)\ldots(r-t+r) = t!\,\binom{r}{t}$ many elements. Hence

$$\# B(t) = [\, t!\, \binom{r}{t}]^{n-1} \cdot \lambda^t$$

because of $\# \Phi(t,\lambda,r) = \lambda^t$.- This remark does not concern the number of global interactions of arbitrary $M_1,\ldots,M_n$ satisfying (ii) which is determined by the proof of Theorem 4.

References

[1] W. Benz, Vorlesungen über Geometrie der Algebren. Springer-Verlag, Berlin-New York 1973.

[2] K.A. Bush, Orthogonal arrays of index unity. Ann. Math. Stat. 23 (1952), 426-434.

[3] P.V. Ceccherini, Alcune osservazioni sulla teoria delle reti. Rend. Acc. Naz. Lincei, 40 (1966), 218-221.

[4] H.R. Halder, W. Heise, Kombinatorik. Hanser Verlag, München - Wien 1976.

[5] W. Heise, Es gibt keinen optimalen (n+2,3)-Code einer ungeraden Ordnung n. Math. Z. 164 (1978), 67-68.

[6] W. Heise, H. Karzel, Laguerre und Minkowski-m-Strukturen. Rend. Ist. Mat. Univ. Trieste IV (1972).

[7] W. Heise, P. Quattrocchi, Survey on Sharply k-Transitive Sets of Permutations and Minkowski-m-Structures. Atti Sem. Mat. Fis. Univ. Modena 27 (1978), 51-57.

[8] P. Quattrocchi, On a theorem of Pedrini concerning the non-existence of certain finite Minkowski-m-structures. Journ. Geom. 13 (1979), 108-112.

[9] H. -J. Samaga, Dreidimensionale reelle Kettengeometrien. Journ. Geom. 8 (1976), 61-73.

[10] H. Schaeffer, Das von Staudtsche Theorem in der Geometrie der Algebren. J. reine angew. Math. 267 (1974), 133-142.

Annals of Discrete Mathematics 30 (1986) 31–38

ON n-FOLD BLOCKING SETS

Albrecht Beutelspacher and Franco Eugeni

Fachbereich Mathematik der Universitat Mainz
Federal Republic of Germany
Istituto Matematica Applicata Facolta' Ingegneria
L'Aquila , Italia

An n-fold blocking set is a set of n-disjoint blocking sets. We shall prove upper and lower bounds for the number of components in an n-fold blocking set in projective and affine spaces.

INTRODUCTION

A blocking set of an incidence structure $\mathcal{I} = (P, \mathcal{L}, I)$ is a set B of points such that any element of $\mathcal{L}$ (any "line" or "block") contains a point of B and a point off B. An n-fold blocking set of $\mathcal{I}$ is a set $B = \{B_1, B_2, \dots, B_n\}$ of n mutually disjoint blocking sets of P. Any set B_i is said to be a component of B. While blocking sets have been studied for a long time (cf.for istance [1],[6],[12],[15],[17],[18]), there are not many papers dealing with n-fold blocking sets.

Generalizing a theorem of Harary [9] (which was already known to Von Newmann and Morgenstern [19]), Kabell [14] recently, proved the following assertion. (We shall use our above terminology).

RESULT. If a projective plane of order q has an n-fold blocking set, then $n \leq q-1$. Any n-fold blocking set of an affine plane of order q satisfies $n \leq q-2$.

In Section 2 we shall prove a theorem which unifies, generalizes and improves this result. Later on, we shall consider blocking sets in projective and affine spaces. Let Σ be a projective or affine space. A set B of Σ is said a t-blocking set if any t-dimensional subspace of Σ contains at least a point of B and a point of Σ-B. A set $B = \{B_1, \dots, B_n\}$ of n mutually disjoint t-blocking sets is said an n-fold t-blocking set of Σ.

In Section 3 we shall deal with the maximal number n of components of an n-fold t-blocking set in PG(r,q) or AG(r,q). We shall prove upper and lower bounds for this maximal number.

In Section 4 we shall construct examples of n-fold blocking sets. In particular, we shall prove the following fact:Given a positive integer n, there is an integer q_0 such that any projective or affine plane of order $q \geq q_0$ has an n-fold blocking set.

We want to remark that we use the word "n-fold blocking set" in a lightly different meaning as Hill and Mason [10] .

2. BLOCKING SETS IN STEINER SYSTEMS.

We begin with the following

2.1 THEOREM. Let S be an $S(2,k,v)$ Steiner system. If S admits an n-fold blocking set, then $n \leq k-2$.

PROOF. Denote by $\{B_1, \ldots, B_n\}$ an n-fold blocking set of S. Consider a point x outside a classe B_i. Since any line through x is incident with at least one point of B_i, it results $|B_i| \geq r$, where $r=(v-1)/(k-1)$ is the number of lines through x. Suppose we have $|B_i| = r$. Then any line through a point x outside B_i meets B_i in just one point. In other words, any line joining two points of B_i is totally contained in B_i. Since B_i has at least two points, $\{B_1, \ldots, B_n\}$ cannot be an n-fold blocking set. So, $|B_i| \geq r+1$ for any $i \in \{1,2, \ldots, n\}$. (This is also a consequence of Theorem 1 in [7]). Therefore,

$$v = \sum_{i=1}^{n} |B_i| \geqq (r+1)n.$$

On the other hand, we have $v-1 = r(k-1)$. Toghether we get

$$n(r+1) \leqq v=r(k-1)+1$$

Hence $n \leq k-1-(k-2)/(r+1) < k-1$ and so $n \leq k-2$.

EXAMPLES. There exist n-fold blocking sets in some $S(2,k,v)$ with $n=k-2$: The projective plane of order 3 and the affine plane of order 4 have blocking sets; these blocking sets form, together with their respective complements, 2-fold blocking sets. Also, the projective plane of order 4 admits a partition in 3 Baer subplanes. This is a 3-fold blocking set.

We remark also that no $S(2,4,v)$ Steiner system has a 3-fold blocking set. (Assume to the contrary that an $S(2,4,v)$ has a 3-fold blocking set $\{B_1, B_2, B_3\}$. Since any B_i is a blocking set and since by [18, (2.13)] any blocking set in $S(2,4,v)$ has at least $(v-\sqrt{v})/2$ points, we have $3(v-\sqrt{v})/2 \leq v$, i.e. $v \leq 9$, a contradiction.) A similar argumentation holds in $S(3,4,v)$ since (cf. [18]) if a blocking set there exists then it contains exactly $v/2$ points.

Now, we consider projective planes. The following result is a substantial improvement of Kabell's result.

2.2 THEOREM. Denote by P a projective plane of order q, and let $B = \{B_1, \ldots, B_n\}$ be an n-fold blocking set of P. Then $n \leqq q-\sqrt{q}+1$ with equality if and only if any B_i is a Baer subplane.

PROOF. Any class $B_i \in B$ is a blocking set in the usual sense. Hence, the theorem of Bruen [6] implies $|B_i| \geq q+\sqrt{q}+1$. Hence

$$n(q+\sqrt{q}+1) \leqq q^2+q+1 = (q+\sqrt{q}+1)(q-\sqrt{q}+1)$$

implying the inequality of assertion. If $n=q-\sqrt{q}+1$, then any B_i has precisely $q+\sqrt{q}+1$ points. Again using Bruen's result, B_i is a Baer subplane. The other direction is trivial.

We remark that any cyclic (so, in particular, any desarguesian) projective plane of square order has a partition in Baer subplanes (see Hirschfeld [11] 4.3.6). So, the bound in Theorem 2.2 is sharp.

3. THE PROJECTIVE AND AFFINE CASE

Let Σ be an r-dimensional projective space of order q. In this Section, we are interested in the question, how many components an n-fold t-blocking set of Σ can have. Clearly, the existence of an n-fold t-blocking set implies the existence of an m-fold blocking set for any $m \leq n$. Therefore, we may define $n_p = n_p(t,r,\Sigma)$ as the greatest integer with the property that Σ has an n-fold t-blocking set for any $n \leq n_p(t,r,\Sigma)$. If Σ is desarguesian of order q, then we write also $n_p(t,r,q)$ instead of $n_p(t,r,\Sigma)$.
Similarly, the functions $n_a(t,r,A)$ and $n_a(t,r,q)$ for an affine space A of dimension r and order q are defined.

In 2.2. Theorem we have already shown that $n_p(1,2,q) \leq q-\sqrt{q}+1$. Now we shall deal with the higher dimensional case. First, we shall state some easy-to-prove upper bounds for n_p. By $\vartheta_r(q)=q^r+\ldots+1$ we denote the number of points in PG(r,q).

3.1 THEOREM.(a) If $r \geqq 2t$, then $n_p(t,r,q) \leqq \vartheta_r(q)/(\vartheta_{r-t}(q)+\sqrt{q}\ \vartheta_{r-t-1}(q))$.
(b) If $r<2t$, then $n_p(t,r,q) \leqq \vartheta_r(q)/\vartheta_{r-t}(q)$ with equality if and only if r-t+1 divides t.

PROOF. Let $B=\{B_1,\ldots,B_n\}$ be an n-fold t-blocking set in PG(r,q).
(a) By ([2],2.11), any t-blocking set B_i in PG(r,q) satisfies

$$|B_i| \geqq \vartheta_{r-t}(q)+\sqrt{q}\ \vartheta_{r-t-1}(q)$$

Therefore,

$$\vartheta_r(q) \geq \sum_{i=1}^{n} |B_i| \geqq n\left[\vartheta_{r-t}(q)+\sqrt{q}\ \vartheta_{r-t-1}(q)\right].$$

(b) Since $r<2t$, by [4] , any t-blocking set B_i in PG(r,q) has at least $\vartheta_{r-t}(q)$ points, equality holds if and only if B_i is the point set of an (r-t)-dimensional subspace. Therefore:

$$\vartheta_r(q)=\sum_{i=1}^{n} |B_i| \geq n\,\vartheta_{r-t}(q).$$

If equality holds, then $\vartheta_{r-t}(q)$ divides $\vartheta_r(q)$, which implies that $r-t-1 \mid r+1$, and so $r-t+1 \mid t$.

REMARK. Suppose $r<2t$ and $r-t+1 \mid t$. Then in PG(r,q) a total (r-t)-spread (see [8]) is an n-fold t-blocking set with $n=\vartheta_r(q)/\vartheta_{r-t}(q)$.

Similarly, the first assertion of the following theorem follows.

3.2 THEOREM. (a) If M(t,r,q) denotes the maximal cardinality of a partial t-spread in PG(r,q), then $n_p(t,r,q) \geqq M(r-t,r,q)$.
(b) Put $r=a(r-t+1)+b$, where a and b are integers with $a>0$ and $0 \leqq b \leqq r-t-1$. Then

$$n_p(t,r,q) \geqq \sum_{i=1}^{a-1} q^{i(r-t+1)+b+1}+1.$$

PROOF. By ([3], Theor. 4.2), there exists a partial (r-t)-spread with the cardinality in question.

Now, we consider the case r=2t.

3.3 Theorem. Denote by P=PG(2t,q) the projective space of dimension $2t \geqq 4$ and order q. Suppose that q is a square. Define the positive integer s by s= [t+2/2] . Then

$$n_p(t,2t,q) \geqq \vartheta_{2(t-s+1)}(q) \left[\vartheta_{2(t-s+1)}(\sqrt{q})\right]^{-1} > q^{t-s+1} - \sqrt{q}\, q^{t-s}.$$

PROOF. Let B be a subspace of dimension 2(t-s+1) of P. (Note that in view of $t \geqq 2$, the definition of s implies $t-s \geqq 0$.) Denote by R a complement of B in P. Then R is a subspace of dimension 2t-2(t-s+1)-1 = 2s-3. Let $B_1,\ldots,B_n$ be a partition of B in Baer subspaces of dimension 2(t-s+1). (It is well known that such a partition exists; for a proof see for example [11] 4.3.6 Theor.) Then

$$n = \vartheta_{2(t-s+1)}(q) \left[\vartheta_{2(t-s+1)}(\sqrt{q})\right]^{-1} < q^{t-s+1}.$$

Since an (s-2)-spread of R has exactly q^{s-1} elements, by our hypothesis $s \geqq (t+2)/2$, there exist subspaces $R_1,\ldots,R_n$ of R of dimension s-2 which are mutually skew.
Consider the "Baer cones"

$$\mathscr{C}_i = \mathscr{C}(R_i,B_i) = \bigcup_{x \in B_i} \langle x, R_i\rangle, \qquad (i=1,\ldots,n).$$

By [13] it follows in particular that these Baer cones are t-blocking sets. Since B and R are skew, the sets $\mathscr{C}_i$ are mutually disjoint. Hence $n_p(t,2t,q) \geqq n$.

EXAMPLE. Theorems 3.1 and 3.3 imply for instance

$$q-\sqrt{q}+1 \leqq n_p(2,4,q) \leqq q^2 - q\sqrt{q}+q-\sqrt{q}+1,$$

if the prime power q is a square.

Now we consider the affine case.

3.4 THEOREM. $n_a(t,r,q) \leqq q^r \left[(t+1)q^{r-t} - t\right]^{-1}$.

PROOF. Any component B_i of an n-fold t-blocking set of $A = AG(r,q)$ (with $n \geqq 2$) is a t-blocking set of A. So, by [2], Cor 2.23, we have

$$|B_i| \geqq (t+1)q^{r-t} - t.$$

Hence, the assertion follows.

As a consequence we have

3.5 COROLLARY. $n_a(1,r,q) \leq q/2$.

It is well known [15] that there exists a function $b_p = b_p(t,q)$ (and a function $b_a = b_a(t,q)$) such that there exists a t-blocking set in PG(r,q) (or AG(r,q)) if and only if $r \leq b_p$ (or $r \leq b_a$, respectively). These functions have been called the Mazzocca-Tallini functions. By [18] we have $b_a(t,q) \leq b_p(t,q)$. If a projective or affine space contains an n-fold t-blocking set (with $n \geq 2$), then it has also a t-blocking set. Consequently, there exist functions $b_p(n,t,q)$ and $b_a(n,t,q)$ such that PG(r,q) (or AG(r,q)) contains an n-fold t-blocking set if and only if $r \leq b_p(n,t,q)$ (or $r \leq b_a(n,t,q)$) $(n \geq 2)$.
Clearly, $b_p(n',t,q) \leq b_p(n,t,q)$ and $b_a(n',t,q) \leq b_a(n,t,q)$ for $n' \geq n$.
Now we prove a generalization of the mentioned theorem of Tallini.

3.6 THEOREM. $b_a(n,t,q) \leqq b_p(n,t,q)$.

PROOF. Assume to the contrary that $b_a \geq b_p + 1$. Fix in $P = PG(b_p+1,q)$ a hyperplane H, and denote by B an n-fold t-blocking set of H. By our assumption, there exists an n-fold t-blocking set B' in the affine space P-H. Consequently $B \cup B'$ would be an n-fold t-blocking set of P =PG(b_p+1,q), a contradiction.

We finish this Section by determining a particular class of values $b_p(n,t,q)$, when q is a square.

3.7 PROPOSITION. If the prime-power q is a square, then

$$b_p(q-\sqrt{q}+1,1,q)=2.$$

PROOF. It is known that in the desarguesian projective plane PG(2,q) there exists an $(q-\sqrt{q}+1)$-fold blocking set. Assume that PG(3,q) contains a $(q-\sqrt{q}+1)$-fold 1-blocking set. Then, by 3.1(a),

$$q-\sqrt{q}+1=n\leq(q^3+q^2+q+1)\cdot(q^2+q\sqrt{q}+q+\sqrt{q}+1)^{-1}<q-\sqrt{q}+1,$$

a contradiction.

4. EXAMPLES

In this Section, we shall construct some examples of n-fold t-blocking sets in projective and affine spaces. First, we shall deal with projective spaces of small order.

4.1. THEOREM. Let r be a positive integer with $r\geq 3$.

(a) $n_p(1,r,3) = 0$
(b) $n_p(1,r,4) \leq 2$
(c) $n_p(1,r,5) \leq 3$
(d) $n_p(1,r,7) \leq 4$
(e) $n_p(1,r,8) \leq 5$

PROOF. (a) By [17], there does not exist a 1-blocking set in PG(r,3) for $r\geq 3$; (b) and (c) follows immediately from 3.1 (a).
Now we prove (d). A blocking set of PG(2,7) has at least 12 points (cf. [11] 13.4.8 Th.). Suppose that S is an irreducible blocking set in PG(r,7).Hence, $\forall x\in S$ there exists at least a line L with $|L\cap S|=1$ Each of the $(7^{r-1}-1)/6$ planes through L contains at least 11 points of $S-\{x\}$. So, $|S|\geq 1+11(7^{r-1}-1)/6$. Then

$$n_p(1,r,7)\leqq(7^{r+1}-1)/(11\cdot 7^{r-1}-5)<5.$$

Now we prove (e). A blocking set of PG(2,8) has at least 13 points (cf. [11] 13.4.9 Th.). By a similar argumentation as above, it follows $|S|\geq 1+12(8^{r-1}-1)/7$. So

$$n_p(1,r,8)\leq(8^{r+1}-1)/(12\ 8^{r-1}-5)<6.$$

4.2 THEOREM. Any projective plane π_q of order $q\geq 7$ has a 3-fold blocking set.

PROOF. Consider three lines a,b,c through a point C_1 .Fix the points A_1, $A_2\in a-\{C_1\}$ and $B_1,B_2\in b-\{C_1\}$. Define

$$S:=A_1B_1\cap c\quad,\quad C_2:=A_1B_2\cap c\quad,\quad Q_1:=A_1B_2\cap A_2B_1,$$

$$Q_2:=A_1B_1\cap A_2B_2\quad,\quad T_1:=C_1Q_2\cap C_2B_1.$$

Finally, fix a point T_2 on $B_1C_2-\{B_1,C_2\}$ which is not on a.
Then

$$\underline{B}_1 := a\cup b\cup\{Q_1,Q_2\}-\{A_1,A_2,B_1,B_2\}$$

$$\underline{B}_2 := c\cup A_1B_1\cup\{B_2,A_1,T_1,T_2\}-\{C_1,C_2,B_1,Q_2\}$$

are disjoint blocking sets.
Since $A_2, B_1 \notin \underline{B}_1 \cup \underline{B}_2$ and $q \geq 7$, also $\underline{B}_1 \cup \underline{B}_2$ forms a blocking set. Hence $\underline{B}_1, \underline{B}_2$ and the complement of $\underline{B}_1 \cup \underline{B}_2$ form a 3-fold blocking set of π_q.

Now we shall deal with the following question. Given a positive integer n, is there an integer q_0 such that any projective plane of order $q \geq q_0$ has an n-fold blocking set? The following theorem answers this question in the affirmative.

4.3 THEOREM. If π_q denotes a projective plane of order q, then π_q has an n-fold blocking set with $n \leq \sqrt[3]{(q+24)/4}$.
In other words, $n_p(1,2,q) \geq \sqrt[3]{(q+24)/4}$.

PROOF. Let n be a positive integer with $n \leq \sqrt[3]{(q+24)/4}$. Fix a line l in π_q, and let $P_1, P_2, \ldots, P_n$ be n points on l.
Step 1. For any $i (1 \leq i \leq n)$ there are two $a_i, b_i \neq l$ through P_i such that through no point of π_q pass three of the lines $a_1, a_2, \ldots, a_n$, $b_1, b_2, \ldots, b_n$.
In order to choose a_i and b_i, at most $\sum_{j=1}^{i-1} 4(j-1) = 2(i-1)(i-2)$ of the q lines different from l through P_i are forbidden. By the hypothesis $2(n-1)(n-2) \leq q-2$, hence for every $i \leq n$ one can find two lines a_i, b_i satisfying the assumption.

Now, we define for any i with $1 \leq i \leq n$

$$\underline{B}_i' = \Big\{ a_i \cup b_i - a_i \cap a_{i+1}, \ldots, a_i \cap a_n, b_i \cap b_{i+1}, \ldots$$
$$\ldots, b_i \cap b_n, a_i \cap b_{i-1}, \ldots, a_i \cap b_1, b_i \cap a_{i-1}, \ldots, b_i \cap a_1 \Big\}.$$

Step 2. Any line a_i or b_i contains a point of each of the sets $\underline{B}_1', \underline{B}_2', \ldots, \underline{B}_n'$.
Namely: Let j be an integer with $1 \leq j \leq n$. In order to show that a_i and b_i have a point of $\underline{B}_j'$, we may assume that $j \neq i$.
If $j > i$, then a_i contains the point $a_i \cap a_j \in \underline{B}_j'$, and b_i is incident with $b_i \cap b_j \in \underline{B}_j'$. On the other hand, if $j < i$, then $b_j \cap a_i$ lies in $\underline{B}_j' \cap a_i$, and b_i passes through $a_j \cap b_i \in \underline{B}_j'$.

Now, we embedd $\underline{B}_i'$ in a blocking set $\underline{B}_i$. In view of Step 2, we must adjoin points in order to block exactly $4(n-1)^2$ lines connecting a point of b_i with a point of b_j. Any such line shall contain just one point of $\underline{B}_i$. Moreover, we want to do this in such a way that the resulting blocking sets $\underline{B}_1, \underline{B}_2, \ldots, \underline{B}_n$ are mutually disjoint. Clearly, $\underline{B}_1'$ can be enlarged to a blocking set $\underline{B}_i$ in the way described above. Now, let i be an integer with $1 < i \leq n$, and suppose that the blocking sets $\underline{B}_1, \underline{B}_2, \ldots, \underline{B}_n$ have been constructed. On any line x which is not blocked by $\underline{B}_i'$, there are at most

$$2i + (i-1)4(n-1)^2 \leq 2n + (n-1)\,4(n-1)^2 = 4n^3 - 12n^2 + 4n - 4$$

points which are contained in $\underline{B}_1, \underline{B}_2, \ldots, \underline{B}_{i-1}$. By our hypothesis this number is at most

$$q + 24 - 12n^2 + 14n - 4 = q - (12n^2 - 14n - 20) \leq q.$$

Therefore, it is possible to adjoin a point X to $\underline{B}_i'$ in order to block the line x.

The remainder of this Section is devoted to the affine plane-case.

4.4 THEOREM. $n_a(1,2,\alpha_q) \leq [n_p(1,2,\pi_q)-1]/2$, where α_q is an affine plane of order q with projective closure π_q.

PROOF. Let $B = \{\underline{B}_1, \underline{B}_2, \dots \underline{B}_n\}$ be an n-fold blocking set in the projective plane π_q. Define

$$B' = \{\underline{B}_1 \cup \underline{B}_2, \underline{B}_3 \cup \underline{B}_4, \dots, \underline{B}_{n-1} \cup \underline{B}_n\}$$

if n is even, and

$$B' = \{\underline{B}_1 \cup \underline{B}_2 \cup \underline{B}_3, \underline{B}_4 \cup \underline{B}_5, \dots, \underline{B}_{n-1} \cup \underline{B}_n\}$$

if n is odd. Then B' is a set of $m \geq (n-1)/2$ sets $\hat{\underline{B}}_1, \hat{\underline{B}}_2, \dots, \hat{\underline{B}}_n$ with the property that any set $\hat{\underline{B}}_j$ has at least 2 points on any line of π_q. Consequently, B' induces an n-fold blocking set in α_q.

As corollaries we have the following two theorems.

4.5 THEOREM. Let α_q be the desarguesian affine plane of order q. If q is a square, then $n_a(1,2,\alpha_q) \geq (q-\sqrt{q})/2$.

PROOF. By the remark after 2.2 Theorem, $n_p(1,2,\pi_q) = (q-\sqrt{q}+1)/2$.

4.6 THEOREM. Any affine plane of order q has an n-fold blocking set with $n \geq \sqrt[3]{(q/24+1)} - 1/2$.

The proof follows by 4.4 and 4.3 .

REFERENCES

[1] Berardi, L.,and Eugeni, F., On blocking sets in affine planes, J.Geom. 22 (1984) 167-177.

[2] Berardi,L.,Beutelspacher, A.,and Eugeni,F., On (s,t;h)-blocking sets in finite projective and affine spaces, Atti Sem. Mat. Fis.Modena 31 (1983) 130-157.

[3] Beutelspacher, A., Partial spreads in finite projective spaces and partial designs, Math. Z. 145 (1975) 211-230.

[4] Beutelspacher, A.,Blocking sets and partial spreads in finite projective spaces, Geom. Dedicata 9 (1980) 425-449.

[5] Bose, R.C.,and Burton, R.C., A characterization of flat spaces in a finite geometry and the uniqueness of the Hamming and the Mac Donald codes, J. Combin. Theory 1 (1966) 96-104.

[6] Bruen, A., Blocking sets in finite projective planes, SIAM J. Appl. Math. 21 (1971) 380-392.

[7] Bruen, A.,and Thas, J.A., Blocking sets, Geom. Dedicata 6 (1977) 192-203.

[8] Dembowsky, P., Finite geometries (Springer, Berlin, 1968).

[9] Harary, F., Generalized Ramsey theory XII: Achievement and avoidance games on finite geometries and configurations. (Unpublished manuscript).

[10] Hill, R.,and Mason, J.R.M., On (k;n)-arcs and the falsity of the Lunelli-Sce conjecture, in: Finite Geometries and Design, L.M.S. Lectures notes 49 (Cambridge Univ.Press, 1981) 52-61.

[11] Hirschfeld, J.W.P., Projective geometries over finite fields (Clarendon Press, Oxford 1979).

[12] Hoffmann, A.J.,and Richardson, M., Block design games, Canad. J. Math. 13 (1961) 110-128.

[13] Huber, M., A caracterization of Baer cones in finite projective spaces, University of Mainz, Preprint 1984.

[14] Kabell, J., A note on colorings of finite plane, Discrete Math. 44 (1983) 319-320.

[15] Mazzocca, F.,and Tallini, G., On the non existence of blocking sets in PG(n,q) and AG(n,q) for all large enough n, Simon Stevin (to appear).
[16] Richardson, M., On finite projective games, Proc. Amer. Math. Soc. 7 (1956) 456-465.
[17] Tallini, G., k-insiemi e blocking sets in PG(r,q) e in AG(r,q), Quaderno n.1 Ist.Mat.Applicata Univ. L'Aquila (1982).
[18] Tallini, G., Blocking sets nei sistemi di Steiner e d-blocking sets in PG(r,q) e AG(r,q), Quaderno n.3 Ist.Mat.Applicata Univ. L'Aquila (1983).
[19] Von Newmann, J.,and Morgestern, O., Theory of games and economic behaviour (University Press, Princeton 1972).

Annals of Discrete Mathematics 30 (1986) 39–56

EMBEDDING FINITE LINEAR SPACES IN PROJECTIVE PLANES

Albrecht Beutelspacher and Klaus Metsch

Fachbereich Mathematik der Universität
Saarstr. 21
D-6500 Mainz
Federal Republic of Germany

It is shown that a finite linear space in which all points have degree n+1 can be embedded in a projective plane of order n, provided that the line sizes are big enough.

INTRODUCTION

A *linear space* is an incidence structure $\mathbf{S} = (p, L, I)$ of points, lines and incidences such that any two distinct points are on a unique line and any line contains at least two points. A great variety of linear spaces can be obtained as follows. Let $\mathbf{P}$ be a projective plane and denote by x a set of points of $\mathbf{P}$ with the property that outside x there are some non-collinear points. Then the incidence structure $\mathbf{S} = \mathbf{P}-x$ whose *points* are the points of $\mathbf{P}$ outside x and whose *lines* are the lines of $\mathbf{P}$ which have at least two points outside x is a linear space. Any linear space isomorphic to such an $\mathbf{P}-x$ is called *embeddable* in the projective plane $\mathbf{P}$.

An old result of M. HALL [4] asserts that any linear space can be embedded in a projective plane, which is usually infinite. There is a conjecture dating back to the time of M. HALL's paper which says that any *finite* linear space can be embedded in a *finite* projective plane. This seems to be an extremely difficult problem.

We want to deal, however, with an even more difficult question. For a finite linear space $\mathbf{S}$ denote by $n+1$ the maximal number of lines which pass through a common point; the so-defined integer n is called the *order* of $\mathbf{S}$. Clearly, the order of any projective plane in which $\mathbf{S}$ is embedded is at least n. The question we are interested in is the following: When can a finite linear space of order n be embedded in a projective plane of the same order n ? One of our answers can be formulated as follows (cf. corollaries 3.3 and 3.4 below).

Let $\mathbf{S}$ be a finite linear space in which any point has degree $n+1$. Denote by $n+1-a$ the minimal number of points on a line. If

$$n > \frac{1}{8}(a^2-1)(5a^2-3a+20),$$

then $\mathbf{S}$ is embeddable in a projective plane of order n.

In other words: For a given a, there is at most a finite number of such linear spaces, which cannot be embedded in a projective plane of order n.

As corollaries we obtain the theorem of THAS and DE CLERCK [6] on the embedding of the pseudo-complement of a maximal arc and the theorem of MULLIN and VANSTONE [5] on the embedding of the pseudo-complement of a pencil of lines.

This paper contains the main results of the second author's Diplomarbeit.

1. THE MAIN IDEA

Throughout this section, $\mathbf{S} = (\mathcal{P},\mathcal{L},I)$ denotes a finite incidence structure consisting of points, lines and incidences such that any two distinct points are incident with at most one common line. For a point p (or a line L), its *degree* is the number r_p (or k_L) of lines through p (or points on L, respectively.)

Two lines are called *parallel*, if they have no point in common. For two distinct lines L_1 and L_2, $h(L_1,L_2)$ denotes the number of lines parallel to both, L_1 and L_2. Throughout, we denote the number of points of $\mathbf{S}$ by v, the number of lines of $\mathbf{S}$ by b. Also, $\mathbf{S}$ has maximal point degree n+1, and z is the integer defined by $z = b - (n^2+n+1)$.

The theorems which we are going to prove in this section will play a crucial role in the embedding of linear spaces in projective planes. Our method generalizes the method introduced by BRUCK [3] and BOSE [2].

THEOREM 1.1. Fix a line H of $\mathbf{S}$, and let a, c, d, e and x be integers with the following properties:
(1) The degree of H is $n+1-d < n+1$.
(2) The number of lines parallel to H is $nd+x > 0$.
(3) For every parallel L of H we have $n+a \le h(L,H) \le n+c$.
(4) For any two intersecting lines L_1, L_2 parallel to H we have

$$h(L_1,L_2) \le e.$$

(5) $$2n > (d+1)(de-d-2a-2) + 2x.$$

(6) $$n > (2d-1)(c+1) + e - 1 - 2x.$$

(7) At least one of the following assertions is true:
(i) On every parallel of H there is a point of degree n+1, or
(ii) $n(d-s) > s(c+1) - x$ for every integer s with $0 \le s \le d-1$.

Then we have:

(a) There is an integer m with the following property: If $\mathcal{M}_1,\ldots,\mathcal{M}_t$ are all maximal sets of mutually parallel lines with

$$H \in \mathcal{M}_i \quad \text{and} \quad |\mathcal{M}_i| \ge m,$$

then t = d and every line parallel to H is contained in exactly one $\mathcal{M}_i$. (If $d \ge 2$, we can take $m = n - (d-1)(c+1) + x + 1$.)

(b) If $\mathbf{S}$ is a linear space such that any point outside H has degree n+1, then the sets $\mathcal{M}_i$ are parallel classes (i.e., every $\mathcal{M}_i$ induces a partition of the point set of $\mathbf{S}$).

The proof of this theorem will be prepared by several Lemmas. From now on, we suppose that $\mathbf{S}$ fulfilles the hypotheses of Theorem 1.1.

By a *claw* we mean a set $\mathcal{S}$ of lines with the following properties:
(a) $H \notin \mathcal{S}$, and H is parallel to every line of $\mathcal{S}$,
(b) any two lines of $\mathcal{S}$ intersect.

A claw $\mathcal{S}$ is called *normal*, if $|\mathcal{S}| = d$. The *order* of a claw is the number of its elements.

A *clique* is a set of mutually parallel lines of $\mathbf{S}$. The clique $\mathcal{M}$ is called *maximal*, if
(a) $H \in \mathcal{M}$,
(b) $\mathcal{M}$ is a maximal set of mutually parallel lines, and
(c) $|\mathcal{M}| > n - (d-1)(c+1) + x$.

Now we can state our first Lemma.

LEMMA 1. Let $\mathcal{S}$ be a claw of order s. Denote by $\mathcal{T}$ the set of lines parallel to H which do not lie in $\mathcal{S}$. For $0 \leq y \leq s$ let $f(y)$ be the number of lines in $\mathcal{T}$ which have exactly y parallels in $\mathcal{S}$. Then $s \leq d$. Moreover,

$$f(0) - \sum_{y=1}^{s} f(y)(y-1) \geq n(d-s) + x - s(1+c), \text{ and}$$

$$\sum_{y=0}^{s} f(y) = nd + x - s.$$

PROOF. From our assumption (2) we get

$$\text{(i)} \qquad \sum_{y=0}^{s} f(y) = |\mathcal{T}| = nd + x - s.$$

By double counting we have

$$\sum_{y=0}^{s} f(y)y = \sum_{L\in\mathcal{S}} h(L,H).$$

Using hypothesis (3) we obtain therefore

$$\text{(ii)} \qquad \sum_{y=0}^{s} f(y)y \leq s(n+c) \quad \text{and} \quad \text{(ii)'} \quad \sum_{y=0}^{s} f(y)y \geq s(n+a).$$

Now let L_1 and L_2 be two distinct lines of $\mathcal{S}$. Since H is parallel to L_1 and L_2, there are at most $h(L_1,L_2)-1$ lines in $\mathcal{T}$ which are parallel to L_1 and L_2. Hence

$$\sum_{y=0}^{s} f(y)y(y-1) \leq \sum_{\substack{K,L\in\mathcal{S}\\ K\neq L}} (h(K,L)-1),$$

and from assumption (4) we get

$$\sum_{y=1}^{s} f(y)y(y-1) \leq s(s-1)(e-1).$$

Together with (i) and (ii)' we conclude

$$\text{(iii)} \qquad 0 \leq f(0) + \frac{1}{2}\cdot\sum_{y=1}^{s} f(y)(y-1)(y-2)$$
$$\leq n(d-s) + x - s(1+a) + \frac{1}{2}s(s-1)(e-1).$$

If we assume $s = d+1$, then by (iii) we would obtain the following contradiction to (5):

$$n = n(s-d) \leq x - s(1+a) + \frac{1}{2}s(s-1)(e-1)$$
$$= \frac{1}{2}(d+1)(de-d-2a-2) + x.$$

Hence there exists no claw of order $d+1$, therefore any claw has at most d elements.□

LEMMA 2. For every line L parallel to H, there exists a normal claw containing L.

PROOF. Consider first the case that there is a point p of degree $n+1$ on L. Let $\mathcal{S}$ be the set of lines through p parallel to H. Since H has degree $n+1-d$, we have $|\mathcal{S}| = d$. Obviously, $\mathcal{S}$ is a claw containing L. Thus, we may suppose that hypothesis (7)(ii) is true. Let $\mathcal{S}$ be a claw of maximal order with $L \in \mathcal{S}$. With the notation of Lemma 1 we get

$$f(0) \geq n(d-s) + x - s(1+c).$$

Assume now $s < d$. Then (7)(ii) implies $f(0) > 0$. Therefore, there is a line L' in $\mathcal{T}$ which intersects every line of $\mathcal{S}$. But then $\mathcal{S}' = \mathcal{S}\ \ \{L'\}$ is a claw with $L \in \mathcal{S}'$ and $|\mathcal{S}'| > |\mathcal{S}|$. This contradiction proves Lemma 2.□

LEMMA 3. Let L be a line parallel to H, and denote by $\mathcal{S}$ a normal claw containing L. Moreover, let $\mathcal{M}$ be the set of all lines $L' \notin \mathcal{S}-\{L\}$ which are parallel to H and intersect every line of $\mathcal{S}-\{L\}$. Then $\mathcal{M} \cup \{H\}$ is contained in a maximal clique through L.

PROOF. Clearly, $\mathcal{S}' = \mathcal{S}-\{L\}$ is a claw of order $d-1$. If $L_1, L_2 \in \mathcal{M}$, then $|\mathcal{S}' \cup \{L_1,L_2\}| = d+1$, and therefore $\mathcal{S}' \cup \{L_1,L_2\}$ is not a claw. This shows that L_1 and L_2 are parallel and that $\mathcal{M} \cup \{H\}$ is a clique. Lemma 1 applied to $\mathcal{S}'$ gives

$$|\mathcal{M}| = f(0) \geq n - (d-1)(c+1) + x.$$

Therefore, $\mathcal{M} \cup \{H\}$ is contained in a maximal clique. Since $L \in \mathcal{M} \cup \{H\}$, Lemma 3 is proved.□

LEMMA 4. If $\mathcal{M}_1$ and $\mathcal{M}_2$ are different maximal cliques, then $\mathcal{M}_1 \cap \mathcal{M}_2 = \{H\}$.

PROOF. Because $\mathcal{M}_1$ and $\mathcal{M}_2$ are maximal cliques, there exist intersecting lines $L_1 \in \mathcal{M}_1$ and $L_2 \in \mathcal{M}_2$. From our hypothesis (4) we get

$$|\mathcal{M}_1 \cap \mathcal{M}_2| \leq h(L_1,L_2) \leq e.$$

Assume now $|\mathcal{M}_1 \cap \mathcal{M}_2| \geq 2$. Then there is a line $L \neq H$ in $\mathcal{M}_1 \cap \mathcal{M}_2$. From hypothesis (3) we obtain therefore

$$|\mathcal{M}_1 \cup \mathcal{M}_2| \leq h(L,H) + |\{L,H\}| \leq n+c+2, \text{ and}$$

$$|\mathcal{M}_1| + |\mathcal{M}_2| = |\mathcal{M}_1 \cup \mathcal{M}_2| + |\mathcal{M}_1 \cap \mathcal{M}_2| \leqslant n+c+2 + e.$$

On the other hand, we have

$$|\mathcal{M}_1| + |\mathcal{M}_2| \geq 2(n - (d-1)(c+1) + x + 1);$$

together we get a contradiction to our hypothesis (6).□

Now we are ready for the proof of theorem 1.1.

(a) If $d = 1$, then the statement is obvious. Therefore we may suppose $d \geq 2$. Since there are lines parallel to H, there exists a normal claw $\mathcal{S}$ (cf. Lemma 2). We shall use the notation of Lemma 1 for $\mathcal{S}$. Because there exists no claw of order $d+1$, we have $f(0) = 0$. From Lemma 1 we get

$$\sum_{y=1}^{d} f(y) = nd + x - d, \text{ and } -\sum_{y=2}^{d} f(y)(y-1) \geq x - d(1+c).$$

So,

$$f(1) = nd + x - d - \sum_{y=2}^{d} f(y) \geq nd + x - d - \sum_{y=2}^{d} f(y)(y-1)$$
$$\geq nd + x - d + x - d(1+c) = nd - d(c+2) + 2x.$$

Put $\mathcal{S} = \{L_1,\dots,L_d\}$, and define $\mathcal{M}_i'$ as the set of all lines $L \in$ $\mathcal{S}-\{L_i\}$ which are parallel to H and intersect every line of $\mathcal{S}-\{L_i\}$. Obviously, the sets $\mathcal{M}_i'$ are distinct; therefore,

$$|\mathcal{M}_1-\{H\}| + \dots + |\mathcal{M}_d-\{H\}| \geq |\mathcal{M}_1'| + \dots + |\mathcal{M}_d'|$$
$$= |\mathcal{M}_1-\{L_1\}| + \dots + |\mathcal{M}_d-\{L_d\}| = f(1) + d.$$

Assume that there is another maximal clique $\mathcal{M}_{d+1}$. Then

$$f(1) + d + n - (d-1)(c+1) + x$$
$$\leq f(1) + d + |\mathcal{M}_{d+1}-\{H\}| \leq |\mathcal{M}_1-\{H\}| + \dots + |\mathcal{M}_{d+1}-\{H\}|.$$

On the other hand, Lemma 4 yields

$$|\mathcal{M}_1-\{H\}| + \dots + |\mathcal{M}_{d+1}-\{H\}| = |(\mathcal{M}_1 \quad \dots \quad \mathcal{M}_{d+1})-\{H\}|$$
$$\leq \text{number of parallels of } H \leq nd + x.$$

Together we get

$$nd + x \geq f(1) + d + n - (d-1)(c+1) + x.$$

Since $f(1) \geq nd-d(c+2)+2x$, we conclude $n \leq (2d-1)(c+1)-2x$. Hence condition (6) implies $e \leq 0$. But in view of $d \geq 2$ and $h(L_1,L_2) \geq |\{H\}| = 1$, this contradicts condition (4).

Therefore, there are exactly d maximal cliques. Now Lemma 3 and 4 show that every parallel of H is contained in exactly one of the sets $\mathcal{M}_i$. This proves (a); (b) is obvious.□

In the following corollary, we handle an important particular case.

COROLLARY 1.2. Let **S** be a finite linear space of order n, and let H be a line with $k_H \leq n$ such that every point outside H has degree $n+1$. Let the integers d, x, z be defined in the following way: The number of lines of **S** is $b = n^2+n+1+z$, $k_H = n+1-d$, and H has exactly $nd+x+z$ parallels in **S**.
Suppose that there exists positive integers $\underline{d}$ and $\bar{d}$ with the following properties:

1) $n+1-\bar{d} \leq k_L \leq n+1-\underline{d}$ for every parallel L of H.

2) $2n > (d+1)(d\bar{d}^2 + d - 2d\underline{d} + 2\underline{d} - 2) - 2dx + d(d-1)z$.

3) $n > (2d-1)(d-1)(\bar{d}-1) + \bar{d}^2 - 1 + (2d-3)x + 2(d-1)z$.

Then assertion (a) of Theorem 1.1 is true. Furthermore, the sets $\mathcal{M}_i$ are parallel classes of **S**.

PROOF. Define $a = d\underline{d}-d-\underline{d}+x+z$, $c = d\bar{d}-d-\bar{d}+x+z$ and $e = \bar{d}^2$, $x' = x+z$. Using 2) and 3) we see that (5) and (6) of Theorem 1.1 are satisfied (note that x' replaces x and x is replaced by $x+z$). Obviously, the conditions (1), (2), (7) of 1.1 are fulfilled.
Now, let L be a parallel of H with $k_L = n+1-d'$. Since L intersects $k_L(d-1)$ parallels of H and H has $nd+x+z$ parallel lines, we get

$$h(L,H) = nd + x + z - k_L(d-1) - 1 = n + (dd'-d-d') + x + z.$$

So, condition (3) of Theorem 1.1 is true (note that $\underline{d} \le d' \le \bar{d}$).

Let L_1 and L_2 be two different intersecting lines parallel to H, and put $k_{L_i} = n+1-d_i$ $(i = 1,2)$. Then

$$h(L_1,L_2) = d_1d_2 + z \le e,$$

which shows that also condition (4) of Theorem 1.1 is fulfilled. Therefore, the corollary follows from Theorem 1.1.□

REMARK. If $p_1,\dots,p_{n+1-d}$ are the points on H and if we denote the degree of p_i by $n+1-d_i$, then we have $x = d_1+\dots+d_{n+1-d}$ in the corollary. In particular, $x = 0$, if every point of **S** has degree n+1.

2. CONSTRUCTION OF THE PROJECTIVE PLANE

In this section, $\mathbf{S} = (\mathcal{P},\mathcal{L},I)$ denotes a finite linear space of order n with $b = n^2+n+1+z$ lines. First, we show the following theorem.

THEOREM 2.1. Suppose that **S** satisfies the following conditions:
(a) $b \ge n^2$.
(b) For every line L of **S** there is an integer $t(L)$ with the following property: If $k_L = n+1-d$, then there are exactly d maximal sets $\mathcal{M}$ of mutually parallel lines with $L \in \mathcal{M}$ and $|\mathcal{M}| \ge t(L)$. Furthermore, every line parallel to L appears in exactly one of these d sets $\mathcal{M}$.

Then **S** is embeddable in a projective plane of order n.

For the proof of this theorem we shall use the following notation. A *clique* is a maximal set $\mathcal{M}$ of mutually parallel lines with $|\mathcal{M}| \ge t(L)$ for at least one line $L \in \mathcal{M}$. A clique $\mathcal{M}$ is called *normal*, if $|\mathcal{M}| = n$. By $\tilde{\mathcal{P}}$ we denote the set of all cliques of **S**.

For $p \in \mathcal{P}$ and $\mathcal{M} \in \tilde{\mathcal{P}}$ we define

$$p \sim \mathcal{M}, \text{ if } p \mathrel{I} L \text{ for at least one line } L \text{ of } \mathcal{M}.$$

For $\mathcal{M} \in \tilde{\mathcal{P}}$ we put

$$\mathcal{G}_1(\mathcal{M}) = \{\mathcal{M}' \mid \mathcal{M}' \in \tilde{\mathcal{P}},\ \mathcal{M} \cap \mathcal{M}' = \emptyset\},$$

$$\bar{\mathcal{G}}_1(\mathcal{M}) = \mathcal{G}_1(\mathcal{M}) \cup \{\mathcal{M}\},\quad \mathcal{G}_2(\mathcal{M}) = \{p \mid p \in \mathcal{P},\ p \nsim \mathcal{M}\}.$$

For every normal clique $\mathcal{M}$ we define

$$\mathcal{G}(\mathcal{M}) = \bar{\mathcal{G}}_1(\mathcal{M}) \cup \mathcal{G}_2(\mathcal{M}) \quad\text{and}\quad \tilde{\mathcal{L}} = \{\mathcal{G}(\mathcal{M}) \mid \mathcal{M} \in \tilde{\mathcal{P}},\ |\mathcal{M}| = n\}.$$

Now we can define the incidence structure $\mathbf{S}' = (\mathcal{P} \cup \tilde{\mathcal{P}}, \mathcal{L} \cup \tilde{\mathcal{L}}, I')$ in the following way:

$$\begin{aligned}
&p \mathrel{I'} L \Leftrightarrow p \mathrel{I} G && \text{for all } p \in \mathcal{P},\ L \in \mathcal{L};\\
&p \mathrel{I'} \mathcal{G}(\mathcal{M}) \Leftrightarrow p \in \mathcal{G}(\mathcal{M}) && \text{for all } p \in \mathcal{P},\ \mathcal{G}(\mathcal{M}) \in \tilde{\mathcal{L}};\\
&\mathcal{M} \mathrel{I'} L \Leftrightarrow L \in \mathcal{M} && \text{for all } \mathcal{M} \in \tilde{\mathcal{P}},\ L \in \mathcal{L};\\
&\mathcal{M} \mathrel{I'} \mathcal{G}(\mathcal{M}') \Leftrightarrow \mathcal{M} \in \mathcal{G}(\mathcal{M}') && \text{for all } \mathcal{M} \in \tilde{\mathcal{P}},\ \mathcal{G}(\mathcal{M}') \in \tilde{\mathcal{L}}.
\end{aligned}$$

As in section 1, we shall prepare the proof by several lemmas. From now on we suppose that **S** satisfies the hypotheses of Theorem 2.1.

LEMMA 1. (a) A line of degree $n+1-d$ is contained in exactly d cliques.
(b) If L_1 and L_2 are parallel lines, then there is a unique clique $\mathcal{M}$ with $L_1,L_2 \in \mathcal{M}$.

PROOF. Let L be a line of degree $n+1-d$. Then there are exactly d cliques $\mathcal{M}_i$ with $L \in \mathcal{M}_i$ and $|\mathcal{M}_i| \geq t(L)$. Furthermore, every line parallel to L appears in exactly one of theses cliques.
Assume that there is another clique $\mathcal{M}$ with $L \in \mathcal{M}$. Then $|\mathcal{M}| \leq t(L)$. By definition, $\mathcal{M}$ contains a line L' with $|\mathcal{M}| \geq t(L')$. In $L' \neq L$, and L' is parallel to L. Let j be the index with $L' \in \mathcal{M}_j$. Then we have $L, L' \in \mathcal{M}_j, \mathcal{M}$, and $|\mathcal{M}_j|, |\mathcal{M}| \geq t(L')$. Now condition (b) of Theorem 2.1 gives $\mathcal{M} = \mathcal{M}_j$ contradicting $|\mathcal{M}| < t(L) \leq |\mathcal{M}_j|$.□

LEMMA 2. (a) Let L be a line, and denote by p a point off L. If $k_L = n+1-d$ and $r_p = n+1-y$, then there are exactly $d-y$ cliques $\mathcal{M}$ with $L \in \mathcal{M}$ and $p \sim \mathcal{M}$.
(b) If p is a point of degree $n+1$, then $p \sim \mathcal{M}$ for every clique $\mathcal{M}$.
(c) We have $|\tilde{p}| + v = n^2+n+1$.

PROOF. (a) There are exactly $r_p - k_L = d-y$ parallels of L through p. Therefore, the assertion follows from Lemma 1 (b).
(b) Let L be a line with $p \not I L$. From (a) and Lemma 1(a) we infer $p \sim \mathcal{M}$ for every clique $\mathcal{M}$ with $L \in \mathcal{M}$.
(c) Let p be a point of degree $n+1$, and let $L_1, \ldots, L_{n+1}$ be the lines through p. If the degree of L_i is $n+1-d_i$, then we have

$$v-1 = \sum_{i=1}^{n+1} (k_{L_i} - 1) = (n+1)n - \sum_{i=1}^{n+1} d_i .$$

In view of (b) and Lemma 1 we conclude

$$|\tilde{p}| = |\{\mathcal{M} \mid \mathcal{M} \in \tilde{p},\ L_i \in \mathcal{M} \text{ for some } i\}|$$

$$= \sum_{i=1}^{n+1} |\{\mathcal{M} \mid \mathcal{M} \in \tilde{p},\ L_i \in \mathcal{M}\}| = \sum_{i=1}^{n+1} d_i .$$

Together, our assertion follows.□

LEMMA 3. Let $\mathcal{M}$ be a normal clique, and denote by L a line with $L \notin \mathcal{M}$. Put $k_L = n+1-d$, and let t denote the number of points p on L with $p \nsim \mathcal{M}$. Then there are exactly $1-t$ cliques which are disjoint to $\mathcal{M}$ and contain L. In particular, we have $0 \leq t \leq 1$.

PROOF. L has $k_L - t$ points p with $p \sim \mathcal{M}$. Therefore, there are exactly

$$m := |\mathcal{M}| - (k_L - t) = d + t - 1$$

lines $L_1, \ldots, L_m$ in $\mathcal{M}$ which are parallel to L. Let $\mathcal{M}_i$ be the clique which contains L and L_i. Since $L \notin \mathcal{M}$, we have $\mathcal{M} \neq \mathcal{M}_i$ for all i. Therefore, by Lemma 1(b), $\mathcal{M}_i \neq \mathcal{M}_j$ for $i \neq j$. Obviously,

$$|\{\mathcal{M}' \mid \mathcal{M}' \in \tilde{p},\ \mathcal{M} \cap \mathcal{M}' \neq \emptyset,\ L \in \mathcal{M}'\}| = m.$$

Now, Lemma 1(a) shows

$$|\{\mathcal{M}' \mid \mathcal{M}' \in \tilde{p},\ \mathcal{M} \cap \mathcal{M}' = \emptyset,\ L \in \mathcal{M}'\}| = d - m = 1 - t.□$$

LEMMA 4. Denote by $\mathcal{M}$ a normal clique. Then
(a) $|\mathcal{G}_2(\mathcal{M})| \leq 1$. In other words: There is at most one point p with $p \nsim \mathcal{M}$.
(b) $|\mathcal{G}(\mathcal{M})| = n+1$.

PROOF. (a) Assume that there are two distinct points p and q with $p, q \nsim \mathcal{M}$. Let L be the line which passes through p and q, and define t as in the preceding lemma. Then $t \geq 2$, contradicting our Lemma 3.

(b) If $\mathcal{M} = \{L_1,\dots,L_n\}$ and $k_{L_i} = n+1-d_i$, then

$$|\mathcal{G}_2(\mathcal{M})| = |\{p \in \mathcal{P} \mid p \sim \mathcal{M}\}| = v - \sum_{i=1}^{n} (n+1-d_i).$$

From Lemma 1 we get

$$|\mathcal{G}_1(\mathcal{M})| = |\tilde{\mathcal{P}}| - |\{\mathcal{M}\}| - \sum_{i=1}^{n} (d_i-1),$$

and the assertion follows in view of $|\tilde{\mathcal{P}}| + v = n^2+n+1$.□

LEMMA 5. Let $\mathcal{M}$ be a normal clique.
(a) If L is a line, then L and $\mathcal{G}(\mathcal{M})$ intersect in **S'** in a unique point; i.e. one of the following cases occurs:
(i) There is a unique clique in $\bar{\mathcal{G}}_1(\mathcal{M})$ containing L. If $\mathcal{G}_2(\mathcal{M}) \neq \emptyset$, then L is not incident with the point of $\mathcal{G}_2(\mathcal{M})$.
(ii) No clique of $\bar{\mathcal{G}}_1(\mathcal{M})$ contains L, $|\mathcal{G}_2(\mathcal{M})| = 1$ and L is incident with the point of $\mathcal{G}_2(\mathcal{M})$.
(b) Any two cliques of $\bar{\mathcal{G}}_1(\mathcal{M})$ are disjoint.

PROOF. (a) We may suppose $L \in \mathcal{M}$. Using the notation of Lemma 3 we get $t \in \{0,1\}$ by Lemma 3. Moreover, Lemma 3 implies that (i) (or (ii)) occurs if and only if $t = 0$ (or $t = 1$, respectively).
(b) is a consequence of (a).□

LEMMA 6. Let $\mathcal{M}_1$ and $\mathcal{M}_2$ be two distinct normal cliques. Then

(a) $\quad \mathcal{G}(\mathcal{M}_1) = \mathcal{G}(\mathcal{M}_2) \Leftrightarrow \mathcal{M}_1 \cap \mathcal{M}_2 = \emptyset.$

(b) $\quad |\mathcal{G}(\mathcal{M}_1) \quad \mathcal{G}(\mathcal{M}_2)| = 1 \Leftrightarrow \mathcal{M}_1 \cap \mathcal{M}_2 \neq \emptyset.$

PROOF. (a) One direction is obvious. Suppose therefore $\mathcal{M}_1 \cap \mathcal{M}_2 = \emptyset$. Then $M_2 \in \mathcal{G}_1(\mathcal{M}_1)$; hence, by Lemma 5(b), $\bar{\mathcal{G}}_1(\mathcal{M}_1) \subseteq \bar{\mathcal{G}}_1(\mathcal{M}_2)$. Similarly, we have $\bar{\mathcal{G}}_1(\mathcal{M}_2) \subseteq \bar{\mathcal{G}}_1(\mathcal{M}_1)$, hence equality.
In view of Lemma 4 we may assume without loss of generality that $\mathcal{G}_2(\mathcal{M}_i) = \{p_i\}$ with points p_1 and p_2. Lemma 5(a) says that a line L is incident with p_i if and only if no clique of $\bar{\mathcal{G}}_1(\mathcal{M}_i)$ contains L. Since $\bar{\mathcal{G}}_1(\mathcal{M}_1) = \bar{\mathcal{G}}_1(\mathcal{M}_2)$, this shows that a line is incident with p_1 if and only if it is incident with p_2. Since $r_{p_i} \geq 2$, we have $\mathcal{G}_2(\mathcal{M}_1) = \mathcal{G}_2(\mathcal{M}_2)$.
(b) In view of (a), one direction is obvious, Let us suppose $\mathcal{M}_1 \cap \mathcal{M}_2 \neq \emptyset$. Then $\mathcal{M}_1$ and $\mathcal{M}_2$ intersect in a line L.Define

$$\mathcal{T} = \{\mathcal{M} \mid \mathcal{M} \in \tilde{\mathcal{P}},\ \mathcal{M} \cap \mathcal{M}_1 = \emptyset \neq \mathcal{M} \cap \mathcal{M}_2\}.$$

Since $\mathcal{M}_1, \mathcal{M}_2 \notin \mathcal{G}(\mathcal{M}_1) \quad \mathcal{G}(\mathcal{M}_2)$, we have

$$\begin{aligned}
&|\mathcal{G}(\mathcal{M}_1) \cap \mathcal{G}(\mathcal{M}_2)| \\
&= \{\mathcal{M} \in \tilde{\mathcal{P}} \mid \mathcal{M} \cap \mathcal{M}_1 = \emptyset = \mathcal{M} \cap \mathcal{M}_2\}| + |\mathcal{G}_2(\mathcal{M}_1) \cap \mathcal{G}_2(\mathcal{M}_2)| \\
&= \{\mathcal{M} \in \tilde{\mathcal{P}} \mid \mathcal{M} \cap \mathcal{M}_1 = \emptyset\}| - |\mathcal{T}| + |\mathcal{G}_2(\mathcal{M}_1) \cap \mathcal{G}_2(\mathcal{M}_2)| \\
&= |\mathcal{G}_1(\mathcal{M}_1)| - |\mathcal{T}| + |\mathcal{G}_2(\mathcal{M}_1) \cap \mathcal{G}_2(\mathcal{M}_2)|.
\end{aligned}$$

Now we distinguish three cases.

Case 1. $\mathcal{G}_2(\mathcal{M}_1) = \emptyset$.

Then by Lemma 5(a) every line is contained in a unique clique of $\bar{\mathcal{G}}_1(\mathcal{M}_1)$. Because no clique of $\bar{\mathcal{G}}_1(\mathcal{M}_1)$ contains two lines of $\mathcal{M}_2$, we have $|\mathcal{T}| = |\mathcal{M}_2 - \{L\}| = n-1$, and so

$$|\mathcal{G}(\mathcal{M}_1) \cap \mathcal{G}(\mathcal{M}_2)| = |\mathcal{G}_1(\mathcal{M}_1)| - |\mathcal{T}| = n - (n-1) = 1,$$

since $|\mathcal{G}_1(\mathcal{M}_1)| + 1 = |\bar{\mathcal{G}}_1(\mathcal{M}_1)| = |\mathcal{G}(\mathcal{M}_1)| = n+1$.

Case 2. $|\mathcal{G}_2(\mathcal{M}_1)| = 1$ and $\mathcal{G}_2(\mathcal{M}_1) = \mathcal{G}_2(\mathcal{M}_2)$.
Since $\mathcal{G}_2(\mathcal{M}_1) = \mathcal{G}_2(\mathcal{M}_2)$, no line of $\mathcal{M}_2$ is incident with the point of $\mathcal{G}_2(\mathcal{M}_1)$. As in case 1, this implies $|\mathcal{T}| = |\mathcal{M}_2 - \{L\}| = n-1$. Since $|\mathcal{G}_2(\mathcal{M}_1)| = 1$ we now have $|\mathcal{G}_1(\mathcal{M}_1)| = n-1$, and this implies

$$|\mathcal{G}(\mathcal{M}_1) \cap \mathcal{G}(\mathcal{M}_2)| = |\mathcal{G}_1(\mathcal{M}_1)| - |\mathcal{T}| + |\mathcal{G}_2(\mathcal{M}_1) \cap \mathcal{G}_2(\mathcal{M}_2)| = 1.$$

Case 3. $|\mathcal{G}_2(\mathcal{M}_1)| = 1$ and $\mathcal{G}_2(\mathcal{M}_1) \neq \mathcal{G}_2(\mathcal{M}_2)$.
Since $\mathcal{G}_2(\mathcal{M}_1) \neq \mathcal{G}_2(\mathcal{M}_2)$, there is a unique line L' in $\mathcal{M}_2$ which is incident with the point of $\mathcal{G}_2(\mathcal{M}_1)$. In view of $L \in \mathcal{M}_1$ we have $L \neq L'$, and so $|\mathcal{T}| = |\mathcal{M}_2 - \{L, L'\}| = n-2$. This shows again

$$|\mathcal{G}(\mathcal{M}_1) \cap \mathcal{G}(\mathcal{M}_2)| = |\mathcal{G}_1(\mathcal{M}_1)| - |\mathcal{T}| = 1.$$

Since $|\mathcal{G}_2(\mathcal{M}_i)| \leq 1$, we have handled all cases. Thus. Lemma 6 is proved.□

LEMMA 7. Any two distinct lines of $\mathbf{S'}$ intersect in a unique point of $\mathbf{S'}$.

PROOF. If one of the two lines is an element of $\tilde{\mathcal{L}}$, we already proved the assertion in Lemma 6(a) and Lemma 7. If both lines are elements of $\mathcal{L}$, the assertion follows from Lemma 1(b).□

Now we are ready for the proof of theorem 2.1.

Let $\mathbf{S}^* = (\mathcal{L} \cup \tilde{\mathcal{L}}, p \cup \tilde{p}, I^*)$ be the dual incidence structure of $\mathbf{S'}$. By Lemma 7, $\mathbf{S}^*$ is a linear space with n^2+n+1 lines (Lemma 2(c)) and at least n^2 points (hypothesis (a)).Furthermore, in view of Lemmas 1(a) and 4(b), any point of $\mathbf{S}^*$ has degree $n+1$. Now, by the theorem of VANSTONE [7], $\mathbf{S}^*$ is embeddable in a projective plane of order n. But then also $\mathbf{S'}$ is embeddable in a projective plane $\mathbf{P}$ of order n.

This completes the proof of Theorem 2.1.□

We remark that $\mathbf{S'} = \mathbf{P}$ if $b > n^2$. (Assume to the contrary that $\mathbf{S'} \neq \mathbf{P}$. Since $|p| + |\tilde{p}| = n^2+n+1$, there is a line L of $\mathbf{P}$ which is not a line of $\mathbf{S'}$; so, $L \notin \mathcal{L} \cup \tilde{\mathcal{L}}$. Because $\mathbf{S}$ is a linear space, at most one of the points $p_1, \ldots, p_{n+1}$ incident with L is a point of p. Every line of $\mathbf{S}$ is in $\mathbf{P}$ incident with exactly one of the points p_i. Since $b > n^2$, this shows that there are at least two points among the p_i's which are incident with n lines of $\mathbf{S}$. But one of these two points, say p_1, is an element of $\tilde{p}$. Hence p_1 is a normal clique of $\mathbf{S}$, and $\mathcal{G}(p_1) = \{p_1, \ldots, p_{n+1}\} = L$, contradicting $L \notin \tilde{\mathcal{L}}$.)

The following theorem is probably the main result of this paper.

THEOREM 2.2. Suppose that the hypotheses of part (a) of Theorem 1.1 or its corollary are satisfied for every line of $\mathbf{S}$ which has not degree $n+1$. If $b \geq n^2$, then $\mathbf{S}$ is embeddable in a projective plane of order n.

PROOF. We show that for every line L of $\mathbf{S}$ there exists an integer $t(L)$ such that the hypothesis (b) of Theorem 2.1 is fulfilled. If $k_L = n+1$, we put $t(L) = 2$. If $k_L \neq n+1$, then Theorem 1.1 (or its corollary) show that such an $t(L)$ exists. Therefore Theorem 2.2 follows from Theorem 2.1.□

3. LINEAR SPACES WITH CONSTANT POINT DEGREE

Let A be a finite set of nonnegative integers. We say that the linear space $\mathbf{S}$ is A-*semiaffine*, if $r_p-k_L \in A$ for every non-incident point-line pair (p,L) of $\mathbf{S}$. The linear space $\mathbf{S}$ is called A-*affine*, if it is A-semiaffine, but not A'-semiaffine for every proper subset A' of A.

Throughout this section, $\mathbf{S}$ will denote an A-affine linear space in which every point has degree $n+1$. Because the $\{0\}$-affine linear spaces are the projective planes, we will assume $A \neq \{0\}$ throughout.

Denote by $\bar{a}$ the maximal and by $\underline{a}$ the minimal element in $A-\{0\}$. The integer z is defined by $b = n^2+n+1+z$. The following two facts shall be used frequently.

For any line L whose degree is not $n+1$ we have $n+1-\bar{a} \leq k_L \leq n+1-\underline{a}$. If L is a line which has at least one parallel line, then $k_L < n+1$.

LEMMA 3.1. (a) We have $z \geq -\underline{a}\bar{a}$.
(b) If $n > \underline{a}\bar{a}(\bar{a}-\underline{a})$, then $z \leq (\bar{a}-\underline{a}-1)\underline{a}$.

PROOF. (a) Since $\underline{a} \in A$ and since every point has degree $n+1$, there is a line L of degree $n+1-\underline{a}$. Let L' be a line intersecting L at a point q. Then every point of L' other than q is on precisely $\underline{a}$ lines parallel to L. Thus, L has at least

$$(k_{L'}-1)\underline{a} \geq (n-\bar{a})\underline{a}$$

parallels. On the other hand, L intersects exactly $k_L \cdot n = (n+1-\underline{a})n$ lines. Hence,

$$n^2+n+1+z = b \geq (n-\bar{a})\underline{a} + (n+1-\underline{a})n + 1 = n^2+n+1 - \bar{a}\underline{a},$$

i.e. $z \geq -\bar{a}\underline{a}$.

(b) If $\mathbf{S}$ has a line of degree $n+1$, then $b = n^2+n+1$, and so $z=0$. Therefore, we may assume $n+1-\bar{a} \leq k_X \leq n+1-\underline{a}$ for every line X. Let L be a line of degree $n+1-\underline{a}$, and denote by $\mathcal{M}$ the set of all lines parallel to L. Then

$$|\mathcal{M}|(n+1-\bar{a}) \leq \sum_{X\in\mathcal{M}} k_X = \sum_{p\notin L} (r_p-k_L) = (v-k_L)\underline{a}.$$

Because every line has at most $n+1-\underline{a}$ points, we have

$$v \leq k_L + n(n-\underline{a}).$$

Together it follows

$$|\mathcal{M}| \leq \frac{(v-k_L)\underline{a}}{n+1-\bar{a}} \leq \frac{n(n-\underline{a})\underline{a}}{n+1-\bar{a}} = n\underline{a} + (\bar{a}-\underline{a}-1)\underline{a} + \frac{(\bar{a}-1)\underline{a}(\bar{a}-\underline{a}-1)}{n+1-\bar{a}}.$$

Our hypothesis yields

$$n+1-\bar{a} > \underline{a}\bar{a}(\bar{a}-\underline{a}) + 1 - \bar{a} \geq (\bar{a}-1)\underline{a}(\bar{a}-\underline{a}-1),$$

therefore $|\mathcal{M}| \leq n\underline{a} + (\bar{a}-\underline{a}-1)\underline{a}$. It follows

$$b = 1 + k_L \cdot n + |\mathcal{M}| \leq n^2+n+1 + (\bar{a}-\underline{a}-1)\underline{a},$$

i.e. $z \leq (\bar{a}-\underline{a}-1)\underline{a}$. □

THEOREM 3.2. Suppose $b \leq n^2+n+1$ and assume that $\mathbf{S}$ satisfies the following conditions:

(1) $n > \frac{1}{2}(\bar{a}^2-1)(\bar{a}^2+\bar{a}-2\underline{a}+2) + \frac{1}{2}\underline{a}(\underline{a}-1)z$,

(2) $n > 2(\bar{a}-1)(\bar{a}^2-\bar{a}+1) + 2(\underline{a}-1)z$,

(3) $b \geq n^2$ or $n \geq \underline{a}\bar{a} - 1$.

Then **S** is embeddable in a projective plane of order n.

(b) If $b > n^2+n+1$, then one of the following inequalities holds:

$$n \leq \frac{1}{2}(\bar{a}^2-1)(\bar{a}^2+\bar{a}-2\underline{a}+2) + \frac{1}{2}\bar{a}(\bar{a}-1)z, \text{ or}$$

$$n \leq 2(\bar{a}-1)(\bar{a}^2-\bar{a}+1) + 2(\bar{a}-1)z.$$

PROOF. (a) If $n \geq \underline{a}\bar{a}-1$, then $b = n^2+n+1 + z \geq n^2$ by Lemma 3.1. Hence we have $b \geq n^2$ in any case. In view of Theorem 2.2 it suffices to show that for every line L with $k_L < n+1$ the hypotheses of Corollary 1.2 are fulfilled.

Consider therefore a line L_- of degree $n+1-a \leq n$. Put $d = a$, $\bar{d} = \bar{a}$, $\underline{d} = \underline{a}$ and $x = 0$. Then $\bar{d} \geq d \geq \underline{d}$, and our hypothesis (1) shows

$$n > \frac{1}{2}(\bar{d}^2-1)(\bar{d}^2+\bar{d}-2\underline{d}+2) + \frac{1}{2}\underline{d}(\underline{d}-1)z$$

$$= \frac{1}{2}(\bar{d}+1)(\bar{d}^3-2\bar{d}\underline{d}+\bar{d}+2\underline{d}-2) + \frac{1}{2}\underline{d}(\underline{d}-1)z$$

$$\geq \frac{1}{2}(d+1)(d\bar{d}^2-2d\underline{d}+d+2\underline{d}-2) + \frac{1}{2}d(d-1)z,$$

since $z \leq 0$. Hypothesis (2) shows

$$n > 2(\bar{d}-1)(\bar{d}^2-\bar{d}+1) + 2(\underline{d}-1)z$$

$$= (2\bar{d}-1)(\bar{d}-1)^2 + \bar{d}^2 - 1 + 2(\underline{d}-1)z$$

$$\geq (2d-1)(d-1)(\bar{d}-1) + \bar{d}^2 - 1 + 2(d-1)z.$$

Therefore, the hypotheses of part (a) of Corollary 1.2 are fulfilled. Hence the assertion follows in view of Theorem 2.2.

(b) Assume that our statement is false. Then, as in part (a), we would be able to embed **S** in a projective plane of order n, contradicting $b > n^2+n+1$.□

COROLLARY 3.3. If $b \leq n^2+n+1$ and $n > \frac{1}{2}(\bar{a}^2-1)(\bar{a}^2+\bar{a}-2\underline{a}+2)$, then **S** is embeddable in a projective plane of order n.□

COROLLARY 3.4. If $b > n^2+n+1$, then $n \leq \frac{1}{2}(\bar{a}^2-1)(\bar{a}^2+\bar{a}-2\underline{a}+2) + \frac{1}{2}\bar{a}(\bar{a}-1)(\bar{a}-\underline{a}-1)\underline{a}$.

PROOF. Since $b > n^2+n+1$, there is no line of degree $n+1$. Assume first $\underline{a} = \bar{a}$. Then every line has degree $n+1-\underline{a}$, we have

$$v = 1 + (n+1)(n-\underline{a}) \quad \text{and} \quad v(n+1) = b(n+1-\underline{a}).$$

We obtain $b \leq n^2+n+1$, a contradiction.
Hence we may suppose $1 \leq \underline{a} < \bar{a}$. Assume that our statement is false. Then

$$n > \frac{1}{2}(\bar{a}^2-1)(\bar{a}^2+\bar{a}-2\underline{a}+2) \geq \frac{1}{2}(\bar{a}^2-1)(\bar{a}^2-\bar{a}) \geq \bar{a}(\bar{a}-1)^2$$

$$\geq \underline{a}\bar{a}(\bar{a}-\underline{a}),$$

and from Lemma 3.1 we get $z \leq (\bar{a}-\underline{a}-1)\underline{a}$. In view of $z > 0$ we have $\bar{a} \geq \underline{a}+2$. Now we get

$$n > \frac{1}{2}(\bar{a}^2-1)(\bar{a}^2+\bar{a}-2\underline{a}+2) + \frac{1}{2}\bar{a}(\bar{a}-1)(\bar{a}-\underline{a}-1)\underline{a}$$

$$\geq 2(\bar{a}-1)(\bar{a}^2-\bar{a}+1) + 2(\bar{a}-1)z.$$

This is a contradiction to Theorem 3.2(b).□

In the remainder of this last section we shall study the case $|A| = 2$. Let a and c be non-negative integers with $a < c$, and denote by $\mathbf{S}$ an $\{a,c\}$-affine linear space in which every point has degree $n+1$. Then every line has either $n+1-a$ or $n+1-c$ points. We call a line of degree $n+1-a$ *long*; the other lines are said to be *short*. The number of long lines (or short lines) of $\mathbf{S}$ is denoted by b_a (or b_c, respectively). Let t be the (constant) number of long lines through a point. Then we have

$$v-1 = t(n-a) + (n+1-t)(n-c) = t(c-a) + (n+1)(n-c).$$

The proof of the following assertion is straightforward and will be omitted here.

LEMMA 3.5. (a) We have

$$b_a = (n-c+a)t + \frac{((c-a)(t-a)+1-a)t}{n+1-a}$$

and

$$b_c = n^2+n+1 - c - (n-c+a)t - \frac{(t(c-a)+1-c)(t-c)}{n+1-c}.$$

(b) If $b = n^2+n+1$, then

$$(c-a)^2t^2 - (c-a)[(n+1)(c-a)+n]t + cn(n+1-a) = 0,$$

or

$$t = t_o \pm \frac{\sqrt{D}}{2(c-a)},$$

where

$$t_o = \frac{n+1}{2} + \frac{n}{2(c-a)}$$

and

$$D = [(c-a+1)^2-4c]n^2 + 2(c^2+a^2-c-a)n + (c-a)^2. \square$$

COROLLARY 3.6. Let c be a positive integer, and denote by $\mathbf{S}$ a finite $\{0,c\}$-affine linear space in which every point has degree $n+1$. If

$$n > \frac{1}{2}(c^2-1)(c^2-c+2),$$

then $\mathbf{S}$ is the complement of a maximal c-arc in a projective plane of order n. In particular, c divides n.

PROOF. Since there is a line of degree $n+1$, we have $b = n^2+n+1$. By corollary 3.3, $\mathbf{S}$ is embeddable in a projective plane $\mathbf{P}$ of order n. Hence there is a set $\mathcal{C}$ of points of $\mathbf{P}$ such that $\mathbf{S} = \mathbf{P}-\mathcal{C}$. It follows that $\mathcal{C}$ is a set of class $\{0,c\}$ of $\mathbf{P}$, so $\mathcal{C}$ is a maximal c-arc of $\mathbf{P}$. $\square$

REMARK. Corollary 3.6 is a slight generalisation of a theorem of THAS and DE CLERCK [6].

COROLLARY 3.7. Let $c \neq 1$ be a positive integer, and denote by $\mathbf{S}$ a finite $\{1,c\}$-affine linear space in which every point has degree $n+1$. If

$$n > \frac{1}{2}c(c-1)(c^2+3c-1),$$

then $\mathbf{S}$ is embeddable in a projective plane $\mathbf{P}$ of order n. Moreover, one of the following cases occurs:

(a) $\mathbf{S}$ is the complement of c concurrent lines of $\mathbf{P}$.

(b) There is a maximal $(c-1)$-arc $\mathcal{C}$ and a line L in $\mathbf{P}$ such that L does not contain a point of $\mathcal{C}$. $\mathbf{S}$ is obtained by removing the line L and the points of $\mathcal{C}$ from $\mathbf{P}$.

(c) $b = n^2+n+1$, $c-1$ divides n, and $c(c-4)n^2 + 2c(c-1)n + (c-1)^2$

is a perfect square.

PROOF. First we show that **S** is embeddable in a projective plane of order n. From our hypothesis we get

(*) $$n > \frac{1}{2}(c^2-1)(c^2+c) + \frac{1}{2}c(c-1)(c-2).$$

Therefore, in view of Corollary 3.3, we may assume $b > n^2+n+1$, i.e. $z > 0$. Hence, in view of Theorem 3.2, it suffices to show that

(1) $$n > \frac{1}{2}(c^2-1)(c^2+c) + \frac{1}{2}c(c-1)z$$

and

(2) $$n > 2(c-1)(c^2-c+1) + 2(c-1)z.$$

Let L be a long line. Then through every point outside L there is precisely one line which is parallel to L. Hence L together with its parallel lines forms a parallel class Π of **S**. Since $b = k_L \cdot n + |\Pi|$, we have $|\Pi| = n+1+z$. Since $t \leq n$ we have

$$(n+1+z)(n+1-c= = |\Pi|(n+1-c) \leq v = 1 + t(c-1) + (n+1)(n-c)$$
$$\leq n^2+1-c.$$

Now we claim $z \leq c-2$. (Otherwise, we would have

$$(n+c)(n+1-c) \leq n^2+1-c,$$

so

$$n \leq c^2-2c+1 = (c-1)^2,$$

contradicting (*).) Now (1) and (2) follow immediately from (*).

Hence **S** is embeddable in a projective plane **P** of order n. In particular, $b \leq n^2+n+1$. Let L and Π be as above. We distinguish two cases.

Case 1. All long lines are contained in Π.
Then $t \leq 1$ and so $t = 1$. From Lemma 3.5 we get

$$b = b_a + b_c = n^2+n+1 - c.$$

Hence there are exactly c lines $L_1,\ldots,L_c$ in **P** which are not lines of **S**. Since $b = n^2 + |\Pi|$, we have $|\Pi| = n+1-c$. Now it is easy to see that **S** is the complement of the c concurrent lines $L_1,\ldots,L_c$.

Case 2. There is a long line L' outside Π.
Since every point of L' is on a unique line of Π, we have $|\Pi| \geq n$, and so $b \geq n^2+n$. This means $b \in \{n^2+n, n^2+n+1\}$.

Consider first the possibility $b = n^2+n$. Then there is exactly one line X in **P** which is not a line of **S**. Because all the points of **S** have degree $n+1$, none of the points of X is a point of **S**. Adding X to **S** we get a $\{1,c-1\}$-affine linear space **S'**, in which every point has degree $n+1$. Corollary 3.6 shows that **S'** is the complement of a maximal $(c-1)$-arc.

Suppose finally $b = n^2+n+1$. Then $|\Pi| = n+1$. Let s be the number of lomg lines in Π. Then

$$v = sn + (|\Pi|-s)(n+1-c) = s(c-1) + (n+1)(n+1-c).$$

On the other hand, we have $v = 1 + t(c-1) + (n+1)(n-c)$. Together we get $n = (t-s)(c-1)$. Therefore, $c-1$ divides n. From Lemma 3.5(b) we obtain furthermore that $c(c-4)n^2+2c(c-1)n+(c-1)^2$ is a perfect square. Thus, Corollary 3.7 is proved completely.□

REMARKS. 1. Corollary 3.7 has already been proved by BEUTELSPACHER and KERSTEN [1] under the additional hypothesis $b \leq n^2+n+1$.

2. Case (a) of Corollary 3.7 can be obtained from the theorem of MULLIN and VANSTONE [5].

3. If **P** is a projective plane of order $(c-1)^2$, and c is a Baer-subplane of **P**, then **P**$-c$ meets the conditions of 3.7(c).

COROLLARY 3.8. Let a and c be positive integers with $2 \leq a < c$, and let **S** be a finite $\{a,c\}$-affine linear space in which every point has degree $n+1$. Suppose that **S** satisfies the following conditions:

$$(1) \qquad n > \frac{1}{2}(c^2-1)(c^2+c-2a+2) + \frac{c^3-c^2+4}{8c}(c-a+1)^2 - \frac{c^3-c^2}{2}$$

and

$$(2) \qquad n > \frac{1}{2}(c^2-1)(c^2+c-2a+2).$$

Then **S** is embeddable in a projective plane **P** of order n and one of the following cases occurs:

(a) There is a positive integer x with $a = x^2+1$ and $c = x^2+x+1$; **S** is the complement of a subplane of order x in a projective plane of order n.

(b) $n^2+n+1-a \leq b \leq n^2+n$, $c-a$ divides n and $c-1$. In particular, $c \leq 2a-1$.

(c) $b = n^2+n+1$, $c-a$ divides n, and $[(c-a+1)^2-4c]n^2+2(c^2+a^2-c-a)n+(c-a)^2$ is a perfect square.

PROOF. From Lemma 3.5 we get $b = n^2+n+1+z = n^2+n+1-c+f(t)$, where

$$(3) \qquad f(t) = \frac{[(t-a)(c-a)+1-a]t}{n+1-a} - \frac{[t(c-a)+1-c](t-c)}{n+1-c}.$$

Obviously, $f(t)$ is a polynomial of second degree with negative coefficient in t^2, which has its maximum in

$$\bar{t} = \frac{(n+1)(c-a)+n}{2(c-a)}.$$

From (2) we get

$$(4) \qquad f(0) = -\frac{c(c-1)}{n+1-c} > -1, \text{ and}$$

$$f\left(\frac{n}{c-a}\right) = f(n+1) = c-a-\frac{a(a-1)}{n+1-a} > c-a-1.$$

First we show that **S** is embeddable in a projective plane of order n. If $z \leq 0$, this follows from (2) and Corollary 3.3. Therefore, we may assume $z > 0$. Then

$$\begin{aligned}(n+1-a)(n+1-c)f(\bar{t}) &\leq (n+1)\bar{t}(c-a)(c-a+1) - \bar{t}^2(c-a)^2 \\ &= \frac{1}{4}(n+1)^2(c-a)^2 - \frac{n^2}{4} + \frac{1}{2}(n+1)\left[(n+1)(c-a)+\frac{n+1}{2}\right] + \frac{1}{2}(n+1)\frac{n-1}{2} \\ &\leq \frac{1}{4}(n+1)^2(c-a)^2 + \frac{1}{2}(n+1)\left[(n+1)(c-a)+\frac{n+1}{2}\right] \\ &= \frac{1}{4}(n+1)^2(c-a+1)^2.\end{aligned}$$

From (2) we get

$$n > \frac{1}{2}(c^2-1)(c^2-c+4) = \frac{1}{2}(c^4-c^3+3c^2+c-4),$$

and so

$$\frac{(n+1)^2}{(n+1-a)(n+1-c)} < \frac{n+1}{n+1-a-c} < \frac{n+1}{n+1-2c} \leq$$

$$\leq \frac{\frac{1}{2}(c^4-c^3+3c^2+c-4)+1}{\frac{1}{2}(c^4-c^3+3c^2+c-4)+1-2c} \leq \frac{c^3-c^2+4}{c^3-c^2} .$$

Thus,

$$f(\bar{t}) \leq \frac{(n+1)^2(c-a+1)^2}{4(n+1-a)(n+1-c)} < \frac{c^3-c^2+4}{4(c^3-c^2)}(c-a+1)^2,$$

and

$$(5) \qquad z = f(t) - c \leq f(\bar{t}) - c < \frac{c^3-c^2+4}{4(c^3-c^2)}(c-a+1)^2 - c.$$

In view of (1), this implies

$$n > \frac{1}{2}(c^2-1)(c^2+c-2a+2) + \frac{1}{2}c(c-1)z.$$

On the other hand, Theorem 3.2(b) yields

$$n \leq 2(c-1)(c^2-c+1) + 2(c-1)z.$$

Together, we have $a = 2$, $c = 3$ and $z \geq 20$, which contradicts (5). Therefore, $b \leq n^2+n+1$, and **S** is enbeddable in a projective plane **P** of order n.

Denote by x the set of points of **P** which are not points of **S.** Consider a point p of x. Since the lines of **S** through p constitute a parallel class of **S**, every line L of degree $n+1-d$ is contained in exactly d parallel classes $\Pi_1(L),\ldots,\Pi_d(L)$. Furthermore, every line parallel to L lies in exactly one parallel class $\Pi_i(L)$. It follows in particular

$$b = 1 + (n+1-d)n + |\Pi_1(L)| + \ldots + |\Pi_d(L)|.$$

Now we distinguish two cases.

Case 1. There is a parallel class Π of **S** having exaczly $n+1$ lines.
If s denotes the number of long lines in Π, then

$$|\Pi|(n+1-c) + s(c-a) = v = 1 + (n+1-t)(n-c) + t(n-a).$$

Hence $s(c-a) = t(c-a) - n$, so $c-a$ divides n. Furthermore,

$$t = s + \frac{n}{c-a} \geq \frac{n}{c-a} ;$$

hence (4) implies $b \geq n^2+n+1-a$. If $b = n^2+n+1$, then Lemma 3.5(b) shows that we are in case (c) of Corollary 3.8. Therefore we may assume that

$$n^2+n+1-a \leq b \leq n^2+n.$$

Let X be a line of **P** which is not a line of **S.** Since every point of **S** has degree $n+1$, each point p_i of X lies in x ($i \in \{1,\ldots,n+1\}$). Let h_i be the number of lines of **S** through p_i. Then $h_i \leq n$ and $h_1 + \ldots + h_{n+1} = b$. Since $b \geq n^2+n+1-a$, there is a $j \in \{1,\ldots,n+1\}$ with $h_j = n$. Therefore the lines of **S** through p_j form a parallel class Π' with exactly n elements. If s' denotes the number of long lines in Π', then

$$n(n+1-c) + s'(c-a) = v = 1 + (n+1-t)(n-c) + t(n-a).$$

Hence $s'(c-a) = t(c-a) - (c-1)$. Consequently, $c-a$ is a divisor of $c-1$, and now we are in case (b) of Corollary 3.8.

Case 2. Every parallel class of **S** has at most n elements.
Consider a short line L of **S.** Then

$$n^2+n+1-c \leq b = 1 + k_L n + \sum_{i=1}^{c}(|\Pi_i(L)|-1) \leq n^2+n+1-c.$$

Hence $b = n^2+n+1-c$ and

(6) $|\pi_i(L)| = n \quad (i \in \{1,\ldots,c\})$ for every short line L.

In particular we have $f(t) = 0$. By (4) we obtain $t \leq \frac{n}{c-a}$.
Consider now o long line L of **S**. We get

$$n^2+n+1-c = b = 1 + k_L n + \sum_{i=1}^{a} (|\pi_i(L)|-1) \leq n^2+n+1-a.$$

Hence there exists a parallel class π which contains L and has fewer than n elements. Since π is a parallel class corresponding to a point of x, it follows in view of (6) that every line of π is long. Hence, $n+1-a$ divides $v\ (= 1+(n+1)(n-c)+t(c-a))$; so,

(7) $n+1-a \mid (t-a)(c-a) + 1 - a$.

Since $t \leq \frac{n}{c-a}$, we have $(t-a)(c-a)+1-a < n+1-a$. On the other hand, we get from (2) that

$$(t-a)(c-a)+1-a \geq -a(c-a)+1-a > -c(c-1)+1-c = -c^2+1 > -(n+1-a),$$

so, $|(t-a)(c-a)+1-a| < n+1-a$. Now, (7) implies $(t-a)(c-a)+1-a = 0$, in particular

$$v = (n+1-a)(n-c+a), \text{ and } t = \frac{a(c-a)-1+a}{c-a}.$$

Together with

$$0 = f(t) = \frac{t^2(c-a)^2 - t(c-a)[(c-a)(n+1)+n] + c(c-1)(n+1-a)}{(n+1-a)(n+1-c)}$$

we get

$$n[a(c-a)^2+2(a-1)(c-a)-(c-1)^2] = (a-1)[a(c-a)^2+2(a-1)(c-a)+(c-1)^2].$$

Since $n \neq a-1$, we obtain

$$a(c-a)^2 + 2(a-1)(c-a) - (c-1)^2 = 0,$$

and so $c = a + \sqrt{a-1}$. Therefore, there is a positive integer x satisfying $a = x^2+1$, $c = x^2+x+1$, $v = (n-x^2)(n-x)$, $b = n^2+n+1 - (x^2+x+1)$, $b_a = (x^2+x+1)(n-x)$, and $b_c = n^2+n+1 - (x^2+x+1) - b_a$. Moreover, if π is a parallel class corresponding to a point of **P** outside **S**, then one of the following possibilities occurs:

(I) $|\pi| = n$, and π contains x^2 long and $n-x^2$ short lines;

(II) $|\pi| = \frac{v}{n+1-a}$, and π consists of lomg lines only.

Using these properties it is now easy to see that we are in case (a) of our corollary.□

REMARKS. 1. Suppose $a = 2$ and $c = 3$. If we are in case (a) of the above corollary and if $n > 42$, then **S** is the complement of a triangle in a projective plane of order n. (This result has been proved in case $n > 7$ by DE WITTE [8].)
2. The existence of strectures in case (b) and (c) satisfying (1) and (2) is not known to the authors.

COROLLARY 3.9. If **S** is a finite $\{2,4\}$-semiaffine linear space in which every point has degree $n+1$, then $n \in \{5,7,13\}$.

PROOF. Since a short line has $n-3$ points, we have $n \geq 5$. By 3.5 we get

(8) $b_2 = (n-2)t + \frac{(2t-5)t}{n-1}$, $b_4 = n^2+n-3 - \frac{(2t-3)(t-4)}{n-3}$,

and

$$b = n^2+n-3 + f(t) \text{ with } f(t) = \frac{(2t-5)t}{n-1} - \frac{(2t-3)(t-4)}{n-3}.$$

Obviously, f is a polynomial of second degree with negative leading coefficient, which takes its maximum at $\bar{t} = (3n+2)/4$. Since $n \geq 5$, we have

$$f(t) \leq f(\bar{t}) = \frac{9n^2-36n+52}{4(n-1)(n-3)} < 4;$$

therefore $f(t) \leq 3$.

From $n-1 \mid (2t-5)t$ and $n \geq 5$ we get $t \geq 4$; consequently,

$$f(t) \geq f(4) = \frac{12}{n-1} > 0 \quad \text{or} \quad f(t) \geq f(n+1) = 2 - \frac{1}{n-1} > 1.$$

Hence $f(t) = s$ with $s \in \{1,2,3\}$. By (1), we obtain

$$t = \frac{3n+2\pm\sqrt{D}}{4} \quad \text{with} \quad D = (9-4s)n^2 + (16s-36)n + 52-12s.$$

Now we distinguish three cases.

Case 1. $s = 3$.
Then $D = -3n^2 + 12n + 16 \geq 0$, and therefore $n = 5$.

Case 2. $s = 2$.
Then $D = n^2 - 4n + 28 = (n-2)^2 + 24$. Since D is a perfect square, we get $n = 7$.

Case 3. $s = 1$.
Then $D = 5(n^2-4n+8)$ and $b = n^2+n-2$. Assume $n > 135$. Then, by Corollary 3.3, **S** is embeddable in a projective plane of order n. Because $b < n^2+n+1$, there is a parallel class Π of **S** with n elements. If s denotes the number of long lines in Π, then

$$v = |\Pi|(n-3) + 2s = n^2 - 3n + 2s.$$

On the other hand, we have

$$v = 1 + (n+1)(n-4) + 2t = n^2 - 3n - 3 + 2t.$$

Together we get $2(t-s) = 3$, a contradiction.

Consequently $n \leq 135$. Since D is a perfect square, it follows $n \in \{6,13,31,78\}$. In view of $t = (3n+2\pm\sqrt{D})/4$ and $t \leq n$ we get $n \neq 6,31,78$. So, $n = 13$.□

REMARK. The authors do not know, whether the structures considered in 3.9 exist in the cases $n = 7$ or $n = 13$. For $n = 5$ we give the following example.
Let **A** be an affine plane of order 4, and let **S'** be the linear space which is obtained by removing one of the points of **A**. Then there are five lines of degree 3 in **S'**. Replacing each of these lines by three lines of degree 2, we get a $\{2,4\}$-affine linear space **S** of order 5 with 15 lines of degree 4 and 15 lines of degree 2.

REFERENCES

[1] Beutelspacher, A. and Kersten, A., Finite semiaffine linear spaces, Arch. Math. 44 (1985), 557-568.

[2] Bose, R.C.: Strongly regular graphs, partial geometries and partially balanced designs, Pacific J. Math. 13 (1963), 389-419.

[3] Bruck, R.H.: Finite nets II. Uniqueness and imbedding, Pacific J. Math. 13 (1963), 421-457.

[4] Hall, M.: Projective planes, Trans. Amer. Math. Soc. 54 (1943), 229-277.

[5] Mullin, R.C. and Vanstone, S.A.: A generalization of a theorem of Totten, J. Austral. Math. Soc. A 22 (1976), 494-500.

[6] Thas, J.A. and De Clerck, F.: Some applications of the fundamental characterization theorem of R.C. Bose on partial geometries, Lincei - Rend. Sc. fis. mat. e nat. 59 (1975), 86-90.

[7] Vanstone, S.A., The extendability of (r,1)-designs, in: Proc. third Manitoba conference on numerical math. 1973, 409-418.

[8] De Witte, P., On the complement of a triangle in a projective plane, to appear.

Annals of Discrete Mathematics 30 (1986) 57–68

VERONESE QUADRUPLES

Alessandro Bichara

Dipartimento di Matematica
Istituto "G. Castelnuovo"
Università di Roma "La Sapienza"
I-00185 - Rome, Italy

ABSTRACT. The classical Veronese variety representing the conics in a projective plane is generalized starting from Buekenhout ovals. This leads to the definition of a Veronese quadruple which is completely characterized as a proper irreducible partial linear space containing two disjoint families of suspaces satisfying suitable axioms.

1. INTRODUCTION

The pair $\underline{S} = (\mathcal{P}, \mathcal{L})$ is said to be a proper irreducible partial linear space (PLS) if $\mathcal{P}$ is a non-empty set, whose elements are called points, $\mathcal{L}$ is a proper family of subsets of $\mathcal{P}$, lines, and the following hold [3]:

(i) Through any point of $\underline{S}$ there is at least one line.

(ii) Any two lines have at most one point in common.

(iii) Any line of $\underline{S}$ is on at least three points.

(iv) There exist two distinct points such that no line contains both of them.

Through this paper ,$\underline{S} = (\mathcal{P}, \mathcal{L})$ denotes a partial linear space.

Two distinct points p and q in S are said to be collinear, if they lie on a common line; in this case we write $p \sim q$.

A subset H of $\mathcal{P}$ is said to be a proper subspace of $\underline{S}$, if H consist of collinear points, at least three of which are not on the same line.

Now we construct an irreducible proper PLS $\underline{S}$ containing proper subspaces. Let $P = (\mathcal{I}, \mathcal{B})$ be an irreducible projective plane of order greater than three and denote by $\mathcal{P}$ the set of all unordered pairs $[l,s]$ of lines in P. For a line l in P, we define

$$\pi_l = \{[l,s] : s \in \mathcal{B}\}.$$

Such a π_l is naturally endowed with the structure of a projective plane (three pairs $[l,s_1]$, $[l,s_2]$, $[l,s_3]$ are said to be collinear if the three lines s_1, s_2, s_3 are concurrent in P). Such a plane is isomorphic to the dual plane of P.

Define

$$\mathcal{P}_1 = \{\pi_l : l \in \mathcal{B}\}.$$

Next, for $p \in \mathcal{I}$ define

$$\alpha_p = \{[l,s],\ p \in l,\ p \in s,\ l,\ s \in \mathcal{B}\}.$$

If a Buekenhout oval [2] B(p) is defined on the pencil F(p) of lines in P through p, then the structure of linear space (B(p)) can be given to α_p as follows: the three pairs $[l_1,s_1]$, $[l_2,s_2]$, $[l_3,s_3]$ are collinear if either an involution of B(p) interchanges l_i ans s_i $(i = 1,2,3)$, or $l_1 = l_2 = l_3$. Therefore, the linear spaces $\alpha(B(p))$ is isomorphic to the dual space of the one containing B(p).

In what follows, it will be assumed that for any point p in $\mathcal{I}$ a Buekenhout oval B(p) is given. THen we define $\mathcal{P}_2 = \{\alpha(B(p)) : p \in \mathcal{I}\}$.

Denote by $\mathcal{L}$ the family of lines belonging to some elment in $\mathcal{P}_1 \cup \mathcal{P}_2$. Hence the pair $\underline{S} = (\mathcal{P},\mathcal{L})$ is a proper irreducible PLS containing the collections $\mathcal{P}_1$ and $\mathcal{P}_2$ of proper subspaces.

The quadruple $(\mathcal{P},\mathcal{L},\mathcal{P}_1,\mathcal{P}_1)$ will be said to be the Veronese space of P associated with the family $\{B(p) : p \in \mathcal{I}\}$ of Buekenhout ovals.

In order to characterize the Veronese space of a plane associated with a collection of Buekenhout ovals, we define a Veronese quadruple as a quadruple $(\mathcal{P},\mathcal{L},\mathcal{P}_1,\mathcal{P}_2)$ satisfying the following conditions.

(i) The pair $\underline{S} = (\mathcal{P},\mathcal{L})$ is a proper irreducible PLS.

(ii) $\mathcal{P}_1$ and $\mathcal{P}_2$ are two disjoint families of proper subspaces such through any line of $\underline{S}$ there is at least one subspace of $\mathcal{P}_1 \cup \mathcal{P}_2$ and

(1.1) For any $\pi \in \mathcal{P}_1$, and any $\alpha \in \mathcal{P}_2$ either $\alpha \cap \pi = \emptyset$ or any line in α meets π.

(1.2) Any two distinct subspaces in $\mathcal{P}_i$ $(i = 1,2)$ have precisely one point in common.

(1.3) Through any point there are at least three elements of $\mathcal{P}_1 \cup \mathcal{P}_2$ and at most two elements of $\mathcal{P}_1$.

(1.4) If a point p is on two elements of $\mathcal{P}_2$, then p is on exactly one element of $\mathcal{P}_1$.

(1.5) Any three elements in $\mathcal{P}_2$ meeting the same element in $\mathcal{P}_1$ have a common point.

(1.6) Any three elements of $\mathcal{P}_1$ meeting the same element in $\mathcal{P}_2$ meet every element in $\mathcal{P}_1$ in collinear points.

(1.7) If three elements in $\mathcal{P}_1$ meet a subspace in $\mathcal{P}_1$ in collinear points then there exists an element in $\mathcal{P}_2$ having a non-trivial intersection with each of them.

An isomorphism between two Veronese quadruples $(\mathcal{P},\mathcal{L},\mathcal{P}_1,\mathcal{P}_2)$ and $(\mathcal{P}',\mathcal{L}',\mathcal{P}'_1,\mathcal{P}'_2)$ is a bijection $f : \mathcal{P} \to \mathcal{P}'$ such that

(1.8) Both f and f^{-1} map lines onto lines.

(1.9) Both f and f^{-1} preserve the two collections of proper subspaces.

It is easy to check that the Veronese space $(\mathcal{P},\mathcal{L},\mathcal{P}_1,\mathcal{P}_2)$ of a projective plane $P = (\mathcal{I},\mathcal{B})$ associated with a family $\{B(p) : p \in \mathcal{S}\}$ of Buekenhout ovals is a Veronese quadruple. Furthermore, if P can be coordinatised by a (commutative) field K and each B(p) is associated with a conic in P, then $(\mathcal{P},\mathcal{L},\mathcal{P}_1,\mathcal{P}_2)$ is isomorphic to that part of the cubic surface M_4^3 in PG(5,K), representing the conics in P which split into two lines in P [1].

In this paper the following results will be proved.

I. - If $Q = (\mathcal{P},\mathcal{L},\mathcal{P}_1,\mathcal{P}_2)$ is a Veronese quadruple, then there exists a projective plane of order greater then three such that for each point p in P a Buekenhout oval B(p) is defined on the pencil of lines through p. Furthermore, Q is isomorphic to the Veronese space of P associated with the family $\{B(p) : p \in \mathcal{P}\}$

When $\mathcal{P}$ is finite, axiom (1.7) will be shown to be a consequence of the remaining ones.

II. - Let $Q = (\mathcal{P},\mathcal{L},\mathcal{P}_1,\mathcal{P}_2)$ be a quadruple in which $(\mathcal{P},\mathcal{L})$ is an irreducible PLS and $\mathcal{P}_1$ and $\mathcal{P}_2$ are two families of proper subspaces such that through any line of $\underline{S}$ there is at least one subspace of $\mathcal{P}_1 \cup \mathcal{P}_2$. Suppose moreover that Q fulfils axioms (1.1),...,(1.6). If $\underline{S}$ is finite, then also (1.7) holds.

2. SOME PROPERTIES OF VERONESE QUADRUPLES

Let $Q = (\mathcal{P}, \mathcal{L}, \mathcal{P}_1, \mathcal{P}_2)$ be a Veronese quadruple.

III. Denote by V the set of all points in $\underline{S}$ through which exactly one element in $\mathcal{P}_1$ passes and by A the set of all points in $\underline{S}$ through which precisely two elements pass of $\mathcal{P}_1$. Then both A and V are non-empty and $\mathcal{P} = A \cup V$.

Proof. By (1.3), through any point p in $\underline{S}$ at most two elements pass of $\mathcal{P}_1$. If $p \notin A$, i.e. through p at most one element passes of $\mathcal{P}_1$, then two elements exist in $\mathcal{P}_2$ containing it; therefore, through p precisely one subspace passes of $\mathcal{P}_1$ (see (1.4)) and $p \in V$. Consequently, $\mathcal{P} = A \cup V$.

Next, $A \neq \emptyset$ will be proved. Take $q_1 \in \mathcal{P}$ and π_1 a subspace in $\mathcal{P}_1$ through q_1 (by the previous argument such a subspace exists in $\mathcal{P}_1$). Since $(\mathcal{P},\mathcal{L})$ is a properer PLS and π_1 is a subspace a point q_2 exists in $\mathcal{P} \setminus \pi_1$. Let π_2 be an element of $\mathcal{P}_1$ through q_2. Obviously, $\pi_1 \neq \pi_2$ and by (1.2) the point $\pi_1 \cap \pi_2$ belongs to A; thus, $A \neq \emptyset$.

Finally, $V \neq \emptyset$ will be proved. Through both q_1 and q_2 at least one element $\alpha_i \in \mathcal{P}_2$, $i = 1,2$, passes (see (1.3)). If $\alpha_1 \neq \alpha_2$, then $\alpha_1 \cap \alpha_2$ is a point in V. Assume $\alpha_1 = \alpha_2$ and let α be an element in $\mathcal{P}_2$ through a point q off α_1; thus $\alpha \cap \alpha_1 \in V$ and the statement is proved.

The next proposition is a strightforward consequence of axioms (1.3) and (1.4).

IV. Through any point in V at least two distinct subspaces of $\mathcal{P}_2$ pass and through any point in A precisely two subspaces of $\mathcal{P}_1$ and one of $\mathcal{P}_2$ pass.

V. If $\pi \in \mathcal{P}_1$; then $|\pi \cap V| \leq 1$.

Proof. Assume π contains two distinct points in V, say p_1 and p_2. By prop. IV, through p_1 two distinct subspaces α_1 and α_1' of $\mathcal{P}_2$ pass which share just the point p_1 (see (1.2)); hence, they have a non-empty intersection with π. Similarly, two subspaces α_2 and α_2' exists in $\mathcal{P}_2$ meeting at p_2 and not skew with π. The subspaces α_i, α'_i, $i = 1,2$, of $\mathcal{P}_2$ share a point by (1.5). Therefore, $p_1 = \alpha_1 \cap \alpha_1'$ belongs to $\alpha_2 \cap \alpha_2'$ so that $\alpha_2 \cap \alpha_2' \supset \{p_1, p_2\}$; a contradiction, since $p_1 \neq p_2$. The statement follows.

VI. If $\pi \in \mathcal{P}_1$, $\alpha \in \mathcal{P}_2$, then either $\pi \cap \alpha = \emptyset$ or $|\pi \cap \alpha| \geq 3$.

Proof. Assume $\alpha \cap \pi \neq \emptyset$; thus, a point p exists in $\alpha \cap \pi$. Let l be a line in α through p and q a point in α not on l (since α is a proper subspace it is not coincident with l; hence, q exists) and q' a point on l. By (1,1) the line through q and q' meets π at a point $p' \in S$ and - obviously - $p' \neq p$. The subspaces π and α of $(\mathcal{P},\mathcal{L})$ meet at a subspace which contains the line through its distinct points p and p'. Since $(\mathcal{P},\mathcal{L})$ is irreducible, $|l| \geq 3$.

VII. If $\pi \in \mathcal{P}_1$, $\alpha \in \mathcal{P}_2$, then either $\pi \cap \alpha = \emptyset$ or $\pi \cap \alpha \in \mathcal{L}$.

Proof. By prop.s V and VI, if $\pi \cap \alpha \neq \emptyset$ then at least two distinct points p_1 and p_2 of A lie on $\pi \cap \alpha$. The line l in $\mathcal{L}$ through them is contained in $\pi \cap \alpha$, as two subspaces meet at a subspace. To prove that $\pi \cap \alpha = l$, take a point p_3 in $\pi \cap \alpha \cap A$. By prop. IV, through p_i, $i = 1,2,3$, precisely one subspace π_i of $\mathcal{P}_1$ passes other than π. Moreover, $\pi_1 \cap \pi = p_1$. The three subspaces π_i, $i = 1,2,3$ all meet α (at least at p_i, resp.) and by axiom (1.6) meet π in collinear points; hence, p_1, p_2, p_3 are collinear and $\pi \cap \alpha \cap A \subset l$. If $\pi \cap \alpha = \pi \cap \alpha \cap A$, then the statement is proved. On the other hand, if there exists a point q of V in $\pi \cap \alpha$, then, by prop. V, $\pi \cap \alpha \cap V = q$, whence $\pi \cap \alpha = l \cup \{q\}$. The subspace $\pi \cap \alpha$ of $(\mathcal{P},\mathcal{L})$ is an irreducible linear space, therefore, $q \in l$ otherwise any line in $\pi \cap \alpha$ joining q with a point p on l would consists of just p and q, a contradiction. The statement follows.

3. THE PLANE $(\mathcal{P}_2,\mathcal{B})$

Let π be an element in $\mathcal{P}_1$. By prop. VII, every subspace α of $\mathcal{P}_2$ meeting π at a point meets it at a line in $\mathcal{L}$. Thus, the following subset of $\mathcal{P}_2$ is defined

$$B(\pi) = \{\alpha \in \mathcal{P}_2 : \alpha \cap \pi \in \mathcal{L}\}.$$

VIII. If $\pi \in \mathcal{P}_1$, then $B(\pi)$ is distinct from $\mathcal{P}_2$ and contains at least two elements in $\mathcal{P}_2$; hence, $B(\pi)$ is a proper subset of $\mathcal{P}_2$.

Proof. By prop. IV, through any point p_1 in π at least one element α_1 of $\mathcal{P}_2$ passes which meets π at a line l in $\mathcal{L}$(see prop. VII); thus, α_1 $B(\pi)$. Through a point p_2 in $\pi \setminus l$ there is a subspace α_2 other than α_1 of $\mathcal{P}_2$ ($p_1 \in \alpha_1$ and $p_2 \notin \alpha_1$). Of course, $\alpha_2 \in B(\pi)$ and since $\alpha_1 \in B(\pi)$, $|B(\pi)| \geq 2$.

To prove that $B(\pi) \neq \mathcal{P}_2$, assume, on the contrary, $B(\pi) = \mathcal{P}_2$. Thus, any subspace of $\mathcal{P}_2$ meets π in a line in $\mathcal{L}$, so that, by axiom (1.5); a point q exists in V through which all the elements in $\mathcal{P}_2$ pass. Therefore, since $|B(\pi)| \geq 2$, by axiom (1.4), through q precisely one element $\pi' \in \mathcal{P}_1$ passes. π' contains the point q which is on all elements in $\mathcal{P}_2$; hence, is met by every subspace of $\mathcal{P}_2$ at a line in $\mathcal{L}$ through q (see prop. VII). Next, let q_1 and q_2 be two points in π' other than q and non-collinear with it. By prop.V, q_1 and q_2 belong to A and through then the subspaces π_1 and π_2, resp. of $\mathcal{P}_1$ pass; moreover, $\pi_1, \pi_2 \neq \pi'$ and $\pi_1 \neq \pi_2$. Through the point $q' = \pi_1 \cap \pi_2$ precisely one element $\alpha \in \mathcal{P}_2$ passes. The subspace α meets π_1 at a line l_1 in $\mathcal{L}$ and is not skew with π' (since no element in $\mathcal{P}_2$ is disjoint from π'). Thus, by axiom (1.1), the line l_1 on α meets π' at a point. Since l_1 is contained in π_1 it passes through the point $q_1 = \pi' \cap \pi_1$. Consequently, q_1 is on l_1 and since $l_1 \subset \alpha$, q_1 lies on α. By a similar argument, $q_2 \in \alpha$. The subspace $\alpha \in \mathcal{P}_2$ passes through q, the point on all the elements in $\mathcal{P}_2$, thus it contains q, q_1, and q_2. These three points are also on π'; hence, $\{q, q_1, q_2\} \subset \alpha \cap \pi'$, a contradiction since these points are non-collinear (see prop. VII). The statement follows.

IX. If $\pi_1, \pi_2 \in \mathcal{P}_1$, $\pi_1 \neq \pi_2$, then $|B(\pi_1) \cap B(\pi_2)| = 1$.

Proof. Through the point $p = \pi_1 \cap \pi_2$ a unique element $\alpha \in \mathcal{P}_2$ passes (see prop. IV). Obviously, $\alpha \in B(\pi_1) \cap B(\pi_2)$. Take $\alpha' \in B(\pi_1) \cap B(\pi_2)$; α' meets π_1 at a line $l_1 \in \mathcal{L}$ and π_2 at a line $l_2 \in \mathcal{L}$. l_1 is contained in α' and by axiom (1.1) meets π_2 at a point $p' \in \pi_2$. Since $p' \in l_1$ and $l_1 \subset \pi_1$, $p' \in \pi_1$ whence $p' \in \pi_1 \cap \pi_2$. Since π_1 and π_2 meet just at p, $p' = p$. Therefore, α' passes through p and is coincident with α the unique element through p in $\mathcal{P}_2$. Thus, $\alpha = B(\pi_1) \cap B(\pi_2)$ and the statement is proved.

By prop. IX, the family $\mathcal{B} = \{B(\pi) : \pi \in \mathcal{P}_1\}$ of proper subsets of $\mathcal{P}_2$ is defined.

X. The pair $(\mathcal{P}_2, \mathcal{B})$ is a projective plane.

Proof. By prop.s VIII and IX, $\mathcal{B}$ is a proper collection of proper subsets of $\mathcal{P}_2$ and any two distinct elements in $\mathcal{B}$ share precisely one element in $\mathcal{P}_2$.

Next, we prove that two distinct elements α_1 and α_2 in $\mathcal{P}_2$ are contained in a unique element of $\mathcal{B}$. Set $p = \alpha_1 \cap \alpha_2$; through p exactly one element π in $\mathcal{P}_1$ passes (see prop. IV). Obviously, $\alpha_1, \alpha_2 \in B(\pi)$ and the statement follows.

XI. $\pi \in \mathcal{P}_1$ implies $|\pi \cap V| = 1$. Furthermore, the lines in π which are the intersections of π by elements in $B(\pi)$ $(\subseteq \mathcal{P}_2)$ are precisely the lines in the pencil (in π) with centre at the point $v = \pi \cap V$. Thus, pairwise distinct points in $\pi \setminus \{v\}$ are collinear with v if they belong to the same element in $B(\pi)$.

Proof. Let α_1 and α_2 be two distinct elements of $B(\pi)$ (they do exist by prop. VIII) and set $v = \alpha_1 \cap \alpha_2$ (see axiom (1.2)); of course, v belongs to V. Assume $v \notin \pi$; through v a unique element π' of $\mathcal{P}_1$ passes other than π. Thus, the lines $l_1 = \alpha_1 \cap \pi'$ and $l_2 = \alpha_2 \cap \pi'$ are distinct, as $|\alpha_1 \cap \alpha_2| = 1$ and $|l_i| \geq 2$, $i = 1,2$, and belong to α_1 and α_2, resp. Since α_1 and α_2 belong to $B(\pi)$, the lines l_1 and l_2 meet π at the points q_1 and q_2, respectively, which are distinct, otherwise v would be coincident with $q_1 = q_2$, impossible as $v \notin \pi$. Hence, $\pi \cap \pi'$ contains the two points q_1 and q_2, a contradiction (see axiom (1.2)). Therefore v belongs to π and by prop. V $\pi \cap V = \{v\}$.

By axiom (1.5) any element in $B(\pi)$ contains the point $v = \alpha_1 \cap \alpha_2$; thus, it meets π at a line in $\mathcal{L}$ through v (see prop. VII: Since through every point in π there pass at least one element in $\mathcal{P}_2$, hence in $B(\pi)$, the lines in π, which are the sections of π by the elements in $B(\pi)$, are precisely the lines through v. The statement follows.

XII. $(\mathcal{P}_2, \mathcal{B})$ is an irreducible projective plane.

Proof. Take $\pi \in \mathcal{P}_1$ and $v = \pi \cap V$. A line l does exist in π not through v, as π is a proper subspace of $(\mathcal{P}, \mathcal{L})$. Since $(\mathcal{P}, \mathcal{L})$ is irreducible, $|l| \geq 3$. The lines joining v with points on l are distinct and contained in π. Thus, at least three lines exist in π through v. By prop. XI, on any line in $(\mathcal{P}_2, \mathcal{B})$ at least three points lie and the statement is proved.

4. THE PROOF OF PROPOSITION I

In this section prop. I will be proved.

Take $\alpha \in \mathcal{P}_2$ and $\mathcal{L}_\alpha$ be the set of all lines in $\mathcal{L}$ on α; consider the pencil $F(\alpha)$ of the lines in the projective plane $(\mathcal{P}_2, \mathcal{B})$ through the point $\alpha \in \mathcal{P}_2$; obviously,

$$F(\alpha) = \{B(\pi) \in \mathcal{B} : \alpha \cap \pi \in \mathcal{L}_\alpha\}$$

If $l \in \mathcal{L}_\alpha$, a correspondence $i_l : F(\alpha) \to F(\alpha)$ is defined as follows

$$(4.1) \qquad i_l(B(\pi)) \begin{cases} = B(\pi') \Leftrightarrow \pi \cap \pi' \in l, \ \pi \neq \pi', \\ = B(\pi) \Leftrightarrow \pi \cap V \in l. \end{cases}$$

By axiom (1.1), any element in $\mathcal{P}_1$ meeting α in a line shares a point with l. Moreover, the points on l not in V belong to A and (see prop. III) through each of them exactly two elements of $\mathcal{P}_1$ pass. Therefore, i_l is a bijection and

and involution of $F(\alpha)$ whose fixed lines are all the lines $B(\pi)$ in $F(\alpha)$ such that $\pi \cap V \in l$; furthermore, i_l interchanges the lines $B(\pi)$ and $B(\pi')$, $\pi \neq \pi'$, of $F(\alpha)$ if $\pi \cap \pi' \in l$. Thus, the next statement has been proved.

XIII. If $\alpha \in \mathcal{P}_2$ and $l \in \mathcal{L}_\alpha$, the bijection $i_l : F(\alpha) \to F(\alpha)$ defined by (4.1) is an involution.

If $\alpha \in \mathcal{P}_2$ then the family $\theta(\alpha) = \{i_l : l \in \mathcal{L}_\alpha\}$ of involutions of $F(\alpha)$ is defined.

XIV. If $\alpha \in \mathcal{P}_2$, then $|F(\alpha)| \geq 4$. Furthermore, the pair $(F(\alpha), \theta(\alpha))$ is a Buekenhout oval.

Proof. By prop. XIII, each bijection $i_l : F(\alpha) \to F(\alpha)$ $(l \in \mathcal{L}_\alpha)$ is an involution. Since $|l| \geq 3$, $|F(\alpha)| > 3$. Next, it will be shown that $(F(\alpha), \theta(\alpha))$ is a Buekenhout oval, i.e. that [2]

(i) every element of $\theta(\alpha)$ is an ivolution of $F(\alpha)$

(ii) for any two pairs $(B(\pi_1), B(\pi_2))$ and $(B(\pi'_1), B(\pi'_2))$, $\pi_i \neq \pi'_j$, $i,j = 1,2$ of lines in $F(\alpha)$ precisely one involution exists in $\theta(\alpha)$ interchanging $B(\pi_1)$ and $B(\pi_2)$ and $B(\pi'_1)$ with $B(\pi'_2)$.

From prop. XIII (i) follows. Thus, (ii) will be proved. If $B(\pi_1) = B(\pi_2)$ and $B(\pi'_1) = B(\pi'_2)$, then $\pi_1 = \pi_2$ and $\pi'_1 = \pi'_2$. A unique line l exists in α through the points $\pi_1 \cap V$ and $\pi'_1 \cap V$, both on α since $B(\pi_1)$, $B(\pi'_1) \in F(\alpha)$. If this occurs then i_l is the unique element in $\theta(\alpha)$ fixing both $B(\pi_1)$ and $B(\pi'_1)$.

On the other hand, if $B(\pi_1) \neq B(\pi_2)$ and $B(\pi'_1) \neq B(\pi'_2)$, then $\pi_1 \neq \pi_2$ and $\pi'_1 \neq \pi'_2$; the two points $\pi_1 \cap \pi_2$ and $\pi'_1 \cap \pi'_2$ in A belong to α and are distinct; hence, there is a unique line l in $\mathcal{L}_\alpha$ through both of them. Again, i_l is the unique element in $\theta(\alpha)$ interchanging $B(\pi_1)$ with $B(\pi_2)$ and $B(\pi'_1)$ with $B(\pi'_2)$.

A similar argument proves (ii) in the remaining cases. The statement follows.

Next, let $(\tilde{\mathcal{P}}, \tilde{\mathcal{L}}, \tilde{\mathcal{P}}_1, \tilde{\mathcal{P}}_2)$ be the Veronese space of $(\mathcal{P}_2, \mathcal{B})$ associated with the collection $\{(F(\alpha), \theta(\alpha)) : \alpha \in \tilde{\mathcal{P}}_2\}$ of Buekenhout ovals.

Consider the mapping $\phi : \tilde{\mathcal{P}} \to \mathcal{P}$ associating with every unordered pair $[B(\pi), B(\pi')]$ of distinct lines in $\mathcal{B}$ (i.e. $\pi \neq \pi'$) the point $\pi \cap \pi' \in \mathcal{P}$ and with the pair $[B(\pi), B(\pi)]$ of coincident lines in $\mathcal{B}$ the point $\pi \cap V$ in $\mathcal{P}$.

XV. The mapping ϕ is an isomorphism between $(\tilde{\mathcal{P}}, \tilde{\mathcal{L}}, \tilde{\mathcal{P}}_1, \tilde{\mathcal{P}}_2)$ and $(\mathcal{P}, \mathcal{L}, \mathcal{P}_1, \mathcal{P}_2)$.

Proof. By prop.s III, XI and axiom (1.2),

(4.2) ϕ is one-to-one and onto.

From the definition of $(F(\alpha),\theta(\alpha))$ the next statement follows.

(4.3) Any line in $\tilde{\mathcal{L}}$ on an element in $\tilde{\mathcal{P}}_2$ is mapped by ϕ onto a line in $\mathcal{L}$ on an element in $\mathcal{P}_2$; furhtermore, the inverse image under ϕ of any line in $\mathcal{L}$ on an element of $\mathcal{P}_2$ is a line in $\tilde{\mathcal{L}}$ on an element in $\tilde{\mathcal{P}}_2$.

By axiom (1.6), taking into account prop. XI;

(4.4) Any three collinear points of an element in $\tilde{\mathcal{P}}_1$ are mapped by ϕ onto three collinear points of an element in $\mathcal{P}_1$. From axiom (1.7) and prop. XI the next statement follows.

(4.5) Three collinear points on an element in $\mathcal{P}_1$ are mapped by ϕ^{-1} onto three collinear points on an element in $\tilde{\mathcal{P}}_1$. The statement follows from (4.2) - (4.5).

From prop.s X, XII, XIV, XV, prop. I follows.

5. THE PROOF OF PROP. II

Firstly, remark that all previous results but (4.5) were proved without the help of axiom (1.7). Thus, with the same notation as before, the next proposition can be stated.

XVI. Under the assumptions in prop. II for the mapping $\phi : \tilde{\mathcal{P}} \to \mathcal{P}$ (4.2), (4.3) and (4.4) hold.

Next, any $\pi \in \mathcal{P}_1$ determines both the line $B(\pi)$ belonging to $\mathcal{B}$ and the subspace $\tilde{\pi} = \pi_{B(\pi)} \in \tilde{\mathcal{P}}_1$ in $(\mathcal{P},\mathcal{L})$ defined by (see sect. 1):

$$\tilde{\pi} = \pi_{B(\pi)} = \{[B(\pi),B(\pi')]: \pi' \in \mathcal{P}_1\}.$$

Consider the mapping $\psi : \mathcal{B} \to \tilde{\pi}$ of the dual plane of $(\mathcal{P}_2,\mathcal{B})$ onto $\tilde{\pi}$ defined by

(5.1) $\psi(B(\pi')) = [B(\pi),B(\pi')]$.

Clearly (see sect. 1),

(5.2) ψ is an isomorphism between the dual plane of $(\mathcal{P}_2,\mathcal{B})$ and the subspace $\tilde{\pi}$ of $(\mathcal{P},\mathcal{L})$.

Define a mapping $\phi': \tilde{\pi} \to \pi$ by

(5.3) $\phi'([B(\pi),B(\pi')]) = \phi([B(\pi),B(\pi')])$;

obviously, ϕ' is one-to-one and onto. Since ϕ and ψ are bijections,

(5.4) the mapping $\phi'\psi: \mathcal{B} \to \pi$ is a bijection.

Next, assume $\mathcal{P}$ is finite. Then $\mathcal{P}_2$ is finite and so is the projective .

plane $(\mathcal{P}_2,\mathcal{B})$; if it is of order q, then $|\mathcal{B}| = q^2 + q + 1$; hence (see (5.4.));

(5.5) $|\pi| = q^2 + q + 1$.

By (5.2), (5.3), and (5.4)

(5.6) $\phi'\psi$ maps three concurrent lines of $\mathcal{B}$ onto three collinear points in π.

Furthermore, taking into account prop. XI, a line in π passes through the point $v = \pi \cap V$ if it is the image under $\phi'\psi$ of a pencil of lines in $(\mathcal{P}_2,\mathcal{B})$ through a point on $B(\pi)$. Consequently,

(5.7) There are precisely q + 1 lines in π through $v = \pi \cap V$ and on each of them q + 1 points lie.

Next, let l be a line in π not through v. Since every point on l is joined to v by a line, by (5.7) $|l| \leq q + 1$; hence

(5.8) Any line in π consists of q + 1 points at most.

The $q^2 + q + 1$ lines in $(\mathcal{P}_2,\mathcal{B})$ all have size q + 1 and each of them is mapped by $\phi'\psi$ onto a line in π (see (4.4) and (5.8)); therefore, on π at least $q^2 + q + 1$ lines lie each of them having size q + 1. Since π is a subspace of $(\mathcal{B},\mathcal{L})$, by (5.5),

(5,9) π is a projective plane of order q and the mapping $\phi'\psi$ is an isomorphism between the dual plane of $(\mathcal{P}_2,\mathcal{B})$ and π.

Let p_i, i = 1,2,3, be three points in π. Clearly,

$$\phi^{-1}(p_i) = (\phi')^{-1}(p_i) = (\psi\psi^{-1}(\phi')^{-1})(p_i) = \psi(\phi'\psi)^{-1}(p_i)$$

Since both ψ and $\phi'\psi$ are isomorphisms the three points $\phi^{-1}(p_i)$ are collinear iff the points p_i are collinear whence (4.5) follows.

By the previous argument and prop. XVI, under the assumptions in prop. II, ϕ: $\tilde{\mathcal{P}} \to \mathcal{P}$ satisfies (4.2) to (4.5) so that it is an isomorphism between $Q = (\tilde{\mathcal{P}}, \tilde{\mathcal{L}}, \tilde{\mathcal{P}}_1, \tilde{\mathcal{P}}_2)$ and $Q = (\mathcal{P}, \mathcal{L}, \mathcal{P}_1, \mathcal{P}_2)$. Since for the Veronese space Q axiom (1.7) holds, the same is true for Q and prop. II is proved.

REFERENCES

[1] E. Bertini, Introduzione alla geometria proiettiva degli iperspazi, Pisa, E. Spoerri (1907).

[2] F. Buekenhout, Etude intrinsêque des ovales, Rend. di Mat. V (1966) 333-393

[3] G. Tallini, Spazi parziali di rette, spazi polari. Geometrie subimmerse, Sem. Geom. Comb. Univ. Roma 14 (1979).

Annals of Discrete Mathematics 30 (1986) 69–84

S-PARTITIONS OF GROUPS AND STEINER SYSTEMS

Mauro Biliotti

Dipartimento di Matematica
Università di Lecce
Lecce - ITALIA

In this paper we investigate a special class of S-partitions of finite groups. These S-partitions are used for the construction of resolvable Steiner systems. Several classification theorems are also given.

The concept of S-partitions may be traced back to Lingenberg [13] although the actual introduction was made by Zappa [24] in 1964. Zappa developed some ideas of Lingenberg so as to provide a group-theoretical description of linear spaces with a group of automorphisms such that the stabilizer of a line acts transitively on the points of that line.

Afterward Zappa [26] and Scarselli [17] mainly investigated the following question: find conditions on a S-partition Σ of a group G relating the existence of Σ to that of a partition - in the usual group-theoretical sense - of a subgroup of G. In this case the linear space associated to Σ is simply the translation André structure associated to that partition [3]. From a geometrical point of view, the work of Zappa [25], Rosati [16] and Brenti [6] on the so-called Sylow S-partitions seems to be more interesting as Sylow S-partitions are useful in constructing some classes of Steiner systems. In this connection, another class of S-partitions is noteworthy. These S-partitions are those considered by Lingenberg [13] and later by Zappa [24]. We shall call these S-partitions "Lingenberg S-partitions".

Lingenberg S-partitions were inspired by a reconstruction method of the affine geometry $AG(n,K)$, K a field, by means of a special class of subgroups of $SL(n,K)$.

In this paper, we study Lingenberg S-partitions of finite groups. We mainly investigate "trivial intersection" S-partitions which we call type I S-partitions (see section 2). For type I S-partitions, we give a "geometric" characterization and somewhat determine the corresponding group structure and action. Also we obtain a classification theorem for Lingenberg S-partitions of doubly transitive permutation groups. We note that for some simple groups, Lingenberg S-partitions are useful in constructing resolvable Steiner systems. In these cases, the Steiner systems might be regarded as a natural affine geometry for the groups.

1. PRELIMINARIES

Groups and incidence structures considered here are always assumed to be finite. In general, we shall use standard notation. If G is a group and $H \leq G$, $K \trianglelefteq G$, then O(G) is the maximal normal subgroup of odd order of G, $S_p(G)$ is the set of all Sylow p-subgroups of G and HK/K is denoted by $\bar{H}$. If $H \cap K = \langle 1 \rangle$ then $K \rtimes H$ denotes the semidirect product of K by H. If G is a permutation group on a set Ω and $\Gamma \subseteq \Omega$ then G_Γ denotes the global stabilizer of Γ in G. A set Ω is a G-set if there is a homomorphism φ from G into the symmetric group on Ω. Usually we shall write G^Ω instead of $\varphi(G)$.

Let G be a group and S a subgroup of G with $S \neq G$. A set Σ of non-trivial subgroups of G such that $|\Sigma| \geq 2$ is said to be a (*regular*) S-*partition* of G if the following conditions are satisfied:

(i) $SH \cap SK = S$ for each $H,K \in \Sigma$ with $H \neq K$;

(ii) for each $g \in G$ there exists $H \in \Sigma$ such that $g \in SH$;

(iii) $H \in \Sigma$ implies $s^{-1}Hs \in \Sigma$ for each $s \in S$.

The above definition is due to Zappa [24]. Here we are interested in the following special class of S-partitions:

a S-partition Σ of a group G is said to be a *Lingenberg* S-*partition* with respect to the subgroup T of G if the following hold:

(j) $T \leq S < N_G(T) < G$;

(jj) $\langle T^x : x \in G \rangle = G$, $\bigcap_{x \in G} S^x = \langle 1 \rangle$;

(jjj) $\Sigma = \{T^x : x \in G - N_G(T)\} \cup \{N_G(T)\}$.

We point out that Σ is determined by the triple (G,S,T). Now we give some geometrical definitions.

As usual a *linear space* is a pair (Π, R), where Π is the set of *points* and R is a family of subsets of Π whose elements are called *lines*, such that two distinct points lie in exactly one line. Here we assume also that Π contains at least three non-collinear points.

An *André structure* (*A-structure*) is a triple $(\Pi, R, //)$, where (Π, R) is a linear space and "//" is an equivalence relation on R such that each equivalence class gives a set-theoretical partition of Π.

Let $\Psi = (\Pi, R, //)$ be an A-structure and let $\Pi_0 \subseteq \Pi$. Assume that whenever $P,Q,R \in \Pi_0$ with $P \neq Q$ then the line PQ and the parallel line through R to PQ are wholly contained in Π_0. If we set $R_0 = \{r : r \in R, r \subset \Pi_0\}$ then $(\Pi_0, R_0, //)$ is an A-structure (or a line) and it is said to be a *subspace* of Ψ. Isomorphisms and automorphisms of linear spaces and A-structures are defined in the usual manner with the requirement that the parallelism is preserved for A-structures.

A *resolvable Steiner system with parameters* (v,k) is an A-structure with v points such that each line contains exactly k points.

Following Zappa [24] a *quasi-translation A-structure* is an A-structure $\Phi=(\Pi,R,//)$ together with a set (or pencil) Θ of lines and a family $\vartheta = \{T(r) : r \in \Theta\}$ of automorphism groups of Φ satisfying the following conditions:

(Q_1) for each $P \in \Pi$ there exists exactly one line $r \in \Theta$ such that $P \in r$;

(Q_2) for each $s \in R$ there exists exactly one line $r \in \Theta$ such that $r//s$;

(Q_3) for each $r \in \Theta$ the group $T(r)$ fixes every point of r and every line parallel to r, leaves Θ invariant and acts transitively on the points of each line s such that $s//r$ and $s \neq r$.

In the following we shall say that Φ is a quasi-translation A-structure with respect to the pencil Θ and the family ϑ. It is very easy to see that, starting from Φ , a new A-structure $\overline{\Phi}$ may be obtained by adding to Φ a new point O incident with precisely the lines of Θ. $\overline{\Phi}$ will be called the *completion* of Φ. Any automorphism a of Φ extends to $\overline{\Phi}$ by setting $a(O) = O$.

Lingenberg S-partitions are essentially the geometrical counterpart of quasi-traslation A-structures. If Σ is a Lingenberg S-partition of a group G with respect to the subgroup T, define [G,S,T] as the triple given by

- the set of right cosets of S in G;
- the set $S(\Sigma)$ of the complexes of the form SXy with $X \in \Sigma$, $y \in G$;
- the following relation "//" on $S(\Sigma)$:

  if $r = SN_G(T)y = N_G(T)y$, then $r'//r$ if and only if either $r' = r$ or $r' = ST^xz$ with $T^x \neq T$ and $T^y = T^{xz}$;

  if $r = ST^xy$ with $T^x \neq T$, then $r'//r$ if and only if either $r' = N_G(T)z$ where $T^z = T^{xy}$ or $r' = ST^wz$ with $T^w \neq T$ and $T^{xy} = T^{wz}$.

The definition of the parallelism seems to be very involved, but it will make clear in the following.

PROPOSITION 1.1 (Zappa [24],2.2). *The triple* [G,S,T] *has the following properties:*

(1) *the mappings* $\tilde{y} : Sx \to Sxy$ *with* $y \in G$ *of the set of right cosets of* S *in* G *into itself form a point-transitive group* $\tilde{G}$ *of automorphisms of* [G,S,T] *which is isomorphic to* G;

(2) [G,S,T] *is a quasi-translation A-structure with respect to the pencil* Θ *whose elements are the lines* $N_G(T)x$ *with* $x \in G$, *and to the family* $\vartheta = \{\widetilde{T^x} : x \in G\}$.

Obviously in [G,S,T], we have that the point Sx belongs to the line SXy if and only if $Sx \subset SXy$. Also the converse holds. Indeed, we have:

PROPOSITION 1.2 (Zappa [24],2.1). *Let* Φ *be a quasi-translation A-structure with respect to the pencil* Θ *and to the family* $\vartheta = \{T(r) : r \in \Theta\}$. *Furthermore assume* ϑ *is a complete class of conjugate subgroups of* $G = \langle T(r) : r \in \Theta \rangle$. *If* P *is a point of* Φ, p *is the line of* Θ *through* P *and* $S = G_P$, *then* $\Sigma = \{T(r) : r \in \Theta - \{p\}\} \cup \cup \{N_G(T(p))\}$ *is a Lingenberg S-partition of* G *with respect to* $T(p)$ *and* $\Phi \simeq [G,S,T(p)]$.

Let Φ be a quasi-translation A-structure with respect to the pencil Θ and for each $r \in \Theta$ let $H(r)$ denote the group of all the automorphisms of Φ fixing every point of r and every line parallel to r and which leave Θ invariant. Then Φ is a quasi--translation A-structure with respect to Θ and to the family $\vartheta = \{H(r) : r \in \Theta\}$. Moreover, ϑ is a complete class of conjugate subgroups of $G = \langle H(r) : r \in \Theta \rangle$ so that Φ may always be represented in the form $[G, G_P, H(p)]$, where $P \in p \in \Theta$.

2. PROPERTIES AND EXAMPLES OF LINGENBERG S-PARTITIONS

LEMMA 2.1. *Let* (G,S,T) *determine a Lingenberg S-partition, then* $T^x \cap T^y = \langle 1 \rangle$ *for each* $x, y \in G$ *with* $T^x \neq T^y$.

Proof. Consider the A-structure $[G,S,T]$ and let $\tilde{z} \in \widetilde{T^x} \cap \widetilde{T^y}$. Each point P of $[G,S,T]$ is on a parallel line to $N_G(T)x$ and also on a parallel line to $N_G(T)y$. Since these lines are distinct and both are fixed by $\tilde{z}$ then $\tilde{z}(P) = P$ and therefore $\tilde{z} = I$.

Lingenberg S-partitions may be divided into two classes according to the following definition:

a Lingenberg S-partition Σ *of* G *with respect to the subgroup* T *is of type* I *if* $S \cap T^x = \langle 1 \rangle$ *for each* $x \in G$ *with* $T^x \neq T$. *If the S-partition is not of type* I, *we shall say that it is of type* II.

For Lingenberg S-partitions of type I we have:

PROPOSITION 2.2. (G,S,T) *determines a Lingenberg S-partition of type* I *if and only if the automorphism group* $\tilde{T}$ *of* $[G,S,T]$ *acts semiregularly on the lines of* Θ *distinct from* $N_G(T)$. *Furthermore, if* (G,S,T) *determines a Lingenberg S-partition of type* I *then the following hold:*

(1) $\tilde{T}$ *acts regularly on each line parallel to* $N_G(T)$ *and distinct from them,*

(2) $N_G(T) \cap T^x = \langle 1 \rangle$ *for each* $T^x \neq T$, *and*

(3) $[N_G(T):S] = |T|-1$.

Proof. Assume (G,S,T) determines a Lingenberg S-partition of type I and let $x \in N_G(T) \cap T^y$ with $T^y \neq T$. Then $\tilde{x}$ fixes both the line $N_G(T)$ and each line parallel to $N_G(T)y$, so that $N_G(T)$ is pointwise fixed by $\tilde{x}$. Therefore, $x \in S$ and from $S \cap T^y = \langle 1 \rangle$, it follows that $x = 1$ and (2) holds. Now let $z \in T$ and $N_G(T)wz = N_G(T)w$ for some $w \in G$ with $T^w \neq T$. Then $z \in N_G(T^w)$ and hence $z \in N_G(T^w) \cap T$. But, by (2), $N_G(T^w) \cap T = \langle 1 \rangle$ and so $z = 1$ and $\tilde{T}$ acts semiregularly on the lines of Θ distinct from $N_G(T)$. The argument may be reversed to prove the converse. Let r be a line parallel to $N_G(T)$ and distinct from them and assume $\tilde{z}(R) = R$ for some $z \in T$, $R \in r$. Then the line of Θ through R, which is distinct from $N_G(T)$, is fixed by $\tilde{z}$. Since $\tilde{T}$ acts semiregularly on $\Theta - \{N_G(T)\}$, it follows that $z = 1$ and (1) is proved. Now let $|G| = g$, $|N_G(T)| = n$, $|S| = s$ and $|T| = t$. In G, there exist $g/n - 1$ distinct complexes of the form ST^x with $T^x \neq T$ and each complex contains exactly st elements since $S \cap T^x = \langle 1 \rangle$. If we take account of the fact that $SN_G(T) = N_G(T)$ contains n elements of G and (G,S,T) determines a S-partition of G, then we must have $(st - s)(g/n - 1) + n = g$ so that (3) now follows.

COROLLARY 2.3. *If* (G,S,T) *determines a Lingenberg* S-*partition of type* I *then* $\overline{[G,S,T]}$ *is a resolvable Steiner system with parameters* (v,k) *where* $v = [G:S] + 1$ *and* $k = |T|$.

Proof. This is an immediate consequence of (1) and (3) of Proposition 2.2.

PROPOSITION 2.4. *Let* Φ *be a quasi-translation* A-*structure with respect to the pencil* Θ *and to the family* $\vartheta = \{T(r) : r \in \Theta\}$. *An automorphism* a *of* Φ *centralizes* $T(r)$ *for each* $r \in \Theta$ *if and only if it fixes every line of* Θ. *A non-identical automorphism of* Φ *fixing every line of* Θ *acts f.p.f. on the set of points of* Φ.

Proof. If a fixes every line of Θ then, by [24],3.1, a centralizes $T(r)$ for each $r \in \Theta$. Conversely, assume a centralizes $T(r)$ for each $r \in \Theta$ and there exists $s \in \Theta$ such that $a(s) \neq s$. $T(s)$ fixes every point of s and every line parallel to s. Likewise, $T(s) = a^{-1}T(s)a$ fixes every point of $a(s)$ and every line parallel to $a(s)$. Since $s \neq a(s)$, it is easy to see that this yields $T(s) = \langle 1 \rangle$, which is impossible. Now let a be an automorphism of Φ fixing every line of Θ and assume $a(P) = P$ for some point P. If r is a line through P and $s // r$, $s \in \Theta$ then $a(s) = s$ and so $a(r) // r$ which implies $a(r) = r$. Therefore, a fixes every line through P. Let Q be a point distinct from P and assume $PQ = q \notin \Theta$. If w denotes the line of Θ through Q, we have that $a(Q) = a(q \cap w) = a(q) \cap a(w) = q \cap w = Q$. If, on the contrary, $q \in \Theta$ then the relation $a(Q) = Q$ can be obtained by using the same argument as above by starting from a point $P' \notin q$. The thesis $a = I$ now follows.

Now we shall give some examples of Lingenberg S-partitions.

We assume the reader is acquainted with the structure of groups $SL(2,q)$; $PSU(3,q^2)$, $q=p^h$, p a prime; $Sz(2^{2n+1})$; $R(3^{2n+1})$, $n \geq 1$, and also with the elementary properties of linear groups. General references are in [11] and [12]. In particular, for Suzuki groups $Sz(2^{2n+1})$, Ree groups $R(3^{2n+1})$ and $PSU(3,q^2)$ see [20], [22] and [23], [7] respectively.

EXAMPLE I. $G \simeq SL(2,q)$, $q=p^h$, $q>2$. Let $P \in S_p(G)$ and assume $T = S = P$; then it is an easy exercise to show that (G,S,T) determines a Lingenberg S-partition of type I and that $\overline{[G,S,T]}$, the completion of $[G,S,T]$, is the affine plane over $GF(q)$. This is the classical example which inspired the work of Zappa [24]. It also explains the definition of the parallelism in [G,S,T] as given in section 1.

EXAMPLE II. $G \simeq Sz(q)$, $q=2^{2n+1}$, $n \geq 1$. Let $P \in S_2(G)$ and let Z(P) be the centre of P. If we assume $T = Z(P)$ and $S = P$ then (G,S,T) determines a Lingenberg S-partition of type I. Indeed, as it is well known, $N_G(T) = N_G(P)$ and if $x \notin N_G(T)$ then $N_G(T) \cap T^x = \langle 1 \rangle$ so that $N_G(T) \cap ST^x = S$. Now let $g \in ST^x \cap ST^y$ with $T \neq T^x \neq T^y \neq T$, then

$g = s_1 t_1^x = s_2 t_2^y$ with $s_1, s_2 \in S$, $t_1, t_2 \in T$ and hence $s_2^{-1}s_1 = t_2^y (t_1^{-1})^x$. If $t_1 \neq 1$, $t_2 \neq 1$ and G is regarded as acting in its usual doubly transitive representation of degree q^2+1 then the element $t_2^y(t_1^{-1})^x$, being the product of two involutions without common fixed points, fixes an even number of points. But $s_2^{-1}s_1$ lies in a Sylow 2-subgroup of G and hence it fixes exactly one point which is a contradiction. As we have previously shown, we cannot have $t_i=1$ for only one $i=1,2$ and so $t_i=1$ for $i=1,2$ and $g \in S$. This yields $ST^x \cap ST^y = S$. We still have $|T| = q$, $|S| = q^2$, $|N_G(T)| = q^2(q-1)$, $|G| = (q^2+1)q^2(q-1)$ and hence, if $T^{x_1},\dots,T^{x_{q^2}}$ are the q^2 subgroups of G which are conjugate to T and distinct from them, it is easily seen that the following relation holds:

$$\sum_{i=1}^{q^2} (|ST^{x_i}| - |S|) + |N_G(T)| = |G|.$$

This proves the assertion.

The completion of the A-structure [G,S,T] is a resolvable Steiner system with parameters $(q(q^2-q+1),q)$.

EXAMPLE III. $G \simeq PSU(3,q^2)$, $q=2^h$, $h>1$. Let $P \in S_2(G)$. It is well known that $N_G(P) = P \rtimes C$, where C il cyclic of order $(q^2-1)/d$ with $d=(3,q+1)$. Denote by C_1 the subgroup of C of order $(q+1)/d$ and set $T = Z(P)$, $S = P \rtimes C_1$. Then (G,S,T) determines a Lingenberg S-partition of type I. Indeed, we have again $N_G(T) = N_G(P)$ and, if $x \notin N_G(T)$, $N_G(T) \cap T^x = \langle 1 \rangle$, so that $N_G(T) \cap ST^x = S$. Now let T^x, T^y be such that $T \neq T^x \neq T^y \neq T$. By well known properties of G, we have that $M = \langle T^x, T^y \rangle \simeq SL(2,q)$ and $M \cap N_G(P) = \langle 1 \rangle$ or $Z(P) \rtimes C_2$, with C_2 cyclic of order $q-1$. Since q is even, we have also that $(q-1, q+1/d) = 1$ and so, if $S \cap M \neq \langle 1 \rangle$ then $S \cap M = Z(P) = T$. But, as we have seen in Example I, (M,T,T) determines a Lingenberg S-partition of type I and hence $T^xT^y \cap T = \langle 1 \rangle$. It follows that $T^xT^y \cap S = \langle 1 \rangle$ and therefore $ST^x \cap ST^y = S$. The thesis can now be achieved by a calculation similar to that carried out in Example II.

The completion of the A-structure [G,S,T] is a resolvable Steiner system with parameters $(q(q^3-q^2+1),q)$. In the case $q=2^h$, with h even, this Steiner system has been already obtained by Schulz [18].

EXAMPLE IV. $G \simeq R(q)$, $q=3^{2n+1}$, $n \geq 1$. We shall make use of the representation of G in $PG(6,q)$ due to Tits [22], §5. Let $x_1, x_2, \dots, x_7$ be a coordinate system for $PG(6,q)$ and let $\sigma \in Aut(GF(q))$, $\sigma : x \to x^{3^{n+1}}$. Furthermore, let I be the hyperplane of $PG(6,q)$ of equation $x_7=0$ and denote by A the affine space obtained from $PG(6,q)$ by assuming I as the ideal hyperplane. Then

$$x=x_1/x_7,\ y=x_2/x_7,\ z=x_3/x_7,\ u=x_4/x_7,\ v=x_5/x_7,\ w=x_6/x_7$$

is a non-homogeneous coordinate system for A. Finally, set $(\infty) \equiv (1,0,0,0,0,0,0)$ and denote by $\Gamma - \{(\infty)\}$ the set of points of A whose coordinates satisfy the equations

$$\begin{aligned} u &= x^2y - xz + y^\sigma - x^{\sigma+3} \\ v &= x^\sigma y^\sigma - z^\sigma + xy^2 + yz - x^{2\sigma+3} \qquad (1) \\ w &= xz^\sigma - x^{\sigma+1}y + x^{\sigma+3}y + x^2y^2 - y^{\sigma+1} - z^2 + x^{2\sigma+4}. \end{aligned}$$

Then $G \simeq PGL(6,q)_\Gamma$ and G acts on Γ in its usual doubly transitive representation of degree q^3+1. Let P be the unique Sylow 3-subgroup of G lying in $G_{(\infty)}$. By using the results of Tits [22],§5, about the representation of the elements of P as well as the fact that $|Z(P)| = q$, it is not hard to prove that the projectivities lying in Z(P) are exactly those of the form

$$t_c : (x_1,x_2,x_3,x_4,x_5,x_6,x_7) \to$$
$$\to (x_1,x_2,x_3+cx_7,-cx_1+x_4,cx_2+x_5-c^\sigma x_7,c^\sigma x_1-2cx_3+x_6-c^2x_7,x_7), \qquad c \in GF(q).$$

According to Tits [22],§5, we have also that

$$\omega : (x_1,x_2,x_3,x_4,x_5,x_6,x_7) \to (x_5,x_4,x_3,x_2,x_1,-x_7,-x_6)$$

is an involutorial projectivity of G which does not lie in $G_{(\infty)}$. Therefore, $\omega Z(P)\omega$ is the centre of a Sylow 3-subgroup Q of G which is distinct from P. Now it is our aim to prove that if $c,d \in GF(q)$, $c\neq 0$, $d\neq 0$, then $\omega t_c \omega t_d$ does not belong to any Sylow 3-subgroup of G. Since $N_G(P) \cap N_G(Q) = E$, where E is cyclic of order $q-1$, and Z(P)E is a Frobenius group with Frobenius kernel Z(P) (see [23],III.4), we can suppose, without loss of generality, $d=1$. We then have

$$\omega t_c \omega t_1 : (x_1,x_2,x_3,x_4,x_5,x_6,x_7)$$
$$(x_1+cx_4+c^\sigma x_6,x_2-cx_5,(1+2c)x_3-c^\sigma x_5-(c+c^2)x_6+x_7,$$
$$-x_1+(1-c)x_4-c^\sigma x_6,x_2-2cx_3+(1-c+c^\sigma)x_5+c^2x_6-x_7,$$
$$x_1-(2+2c)x_3+cx_4+c^\sigma x_5+(1+2c+c^2+c^\sigma)x_6-x_7,2cx_3-c^\sigma x_5-c^2x_6+x_7).$$

A straightforward calculation shows that $\omega t_c \omega t_1$ possesses the eigenvalue 1 whose eigenspace is generated by the vector $(0,1,1/2,-c^{\sigma-2},0,1/c,1)$. Now suppose $\omega t_c \omega t_1$ lies in a Sylow 3-subgroup of G, then the following hold:

- $\omega t_c \omega t_1$ does not have any eigenvalue different from 1, for we are in characteristic 3;
- $\omega t_c \omega t_1$ must fix a point of $\Gamma - \{(\infty)\}$.

From that which we have proved previously, we can infer that the fixed point of $\omega t_c \omega t_1$ on $\Gamma - \{(\infty)\}$ must have non-homogeneous coordinates $(0,1,1/2,-c^{\sigma-2},0,1/c)$. But these coordinates do not satisfy (1), a contradiction. Now we may argue as in the previous examples to show that if we set T = Z(P) and S = P then (G,S,T) determines a Lingenberg S-partition of type I.

The completion of the A-structure [G,S,T] is a resolvable Steiner system with parameters $(q(q^3-q^2+1),q)$.

EXAMPLE V. $G \simeq SL(n,q)$, $q=p^h$, $n\geq 3$. Let $K = GF(q)$, $V = K^n$ and U a 1-dimensional subspace of V. Denote by $T(\underline{a},\mu)$ the transvection $\underline{v} \to \underline{v}-\mu(\underline{v})\underline{a}$ where $\underline{a} \in V$ and $\mu \in Hom_K(V,K)$ with $\mu(\underline{a}) = \underline{0}$, $\mu\neq 0$. For a fixed non-zero vector $\underline{b}$ of U, set

$$T(U) = \{I,\ T(\underline{b},\mu) : 0\neq\mu \in Hom_K(V,K),\ \mu(\underline{b})=\underline{0}\}.$$

Then T(U) is a subgroup of G (see [11],II, Hilfssatz 6.5). Finally, denote by S(U) the subgroup of G fixing U pointwise. Then (G,S(U),T(U)) determines a Lingenberg S-partition of type II. It is indeed enough to observe that the A-structure $\dot{A}$ which is obtained from the affine space A associated to V by removing the origin $\underline{0}$ is a quasi-translation A-structure with respect to the pencil Θ of the lines through $\underline{O}$ (disregarding the point $\underline{O}$) and to the family ϑ of the subgroups of G which are conjugated to T(U). Furthermore, a transvection of T(U) with hyperplane H fixes all the lines of Θ lying in H, so that T(U) is not semiregular on Θ. The assertion now follows from Propositions 1.2 and 2.2.

3. FURTHER RESULTS ON LINGENBERG S-PARTITIONS OF TYPE I

We will require the following lemma.

LEMMA 3.1. *Let* G *be one of the following groups:*
$SL(2,q)$, $q=p^h$, *p prime*, $q\geq 4$;
$Sz(q)$, $q=p^{2n+1}$, $p=2$, $n\geq 1$;
$SU(3,q^2)$, $q=p^h$, *p prime*, $q>2$, $3|q+1$;
$PSU(3,q^2)$, $q=p^h$, *p prime*, $q>2$;
$R(q)$, $q=p^{2n+1}$, $p=3$, $n\geq 1$;
and let P *be a Sylow p-subgroup of* G. *If* T *is a normal subgroup of* $N_G(P)$ *satisfying conditions:*
(1) $|T|-1 \mid [N_G(P):T]$ *and*
(2) $T\cap Z(G) = \langle 1\rangle$,
then $T = Z(P)$.

Proof. We shall investigate the various cases separately.

Let $G = SL(2,q)$, $q\geq 4$. Assume q is odd. Then $|Z(G)| = 2$, $|P| = q$, $|N_G(P)| = q(q-1)$ and $\overline{N} = N_G(P)/Z(G)$ is a Frobenius group with Frobenius kernel $\overline{P}$. Since $\overline{T} \triangleleft \overline{N}$, then by [11],V, Satz 8.16, we have that either $\overline{T} < \overline{P}$ or $\overline{T} \geq \overline{P}$. In the first case, it follows that $T < PZ(G)$. But, $T\cap Z(G) = \langle 1\rangle$ and hence $T < P$, which implies $T = \langle 1\rangle$ since P is a minimal normal subgroup of $N_G(P)$. In the latter case, condition (1) yields $T = P$. Since P is elementary abelian, the proof is achieved. The case q even is similar.

Let $G = Sz(q)$. $N_G(P)$ is a Frobenius group with Frobenius kernel P, moreover $|N_G(P)| = q^2(q-1)$, $|P| = q^2$, $|Z(P)| = q$. We have that either $T \geq P$ or $T < P$. Condition (1) is unsatisfied when $T \geq P$. If $T < P$, then either $T \geq Z(P)$ or $T\cap Z(P) = \langle 1\rangle$ since Z(P) is a minimal normal subgroup of $N_G(P)$. As $T \triangleleft N_G(P)$, we have that $q-1 \mid |T|-1$. So, in the former case, it follows $T = Z(P)$ from condition (1). In the latter case we have $P = T \times Z(P)$. However, P/Z(P) is abelian and hence P must be abelian which is a contradiction.

Let $G = SU(3,q^2), 3|q+1$. We have $|Z(G)| = 3$ and $N_G(P) = P \rtimes C$, where $|P| = q^3$ and C is cyclic of order q^2-1 and contains Z(G). Moreover, $|Z(P)| = q$.
$\overline{N} = N_G(P)/Z(P)Z(G)$ is a Frobenius group with Frobenius kernel $\overline{P}$ and Frobenius complements isomorphic to $\overline{C}$. Since $\overline{T} \triangleleft \overline{N}$, we must have either $\overline{T} < \overline{P}$ or $\overline{T} \geq \overline{P}$. But, clearly $\overline{P}$ is a minimal normal subgroup of $\overline{N}$ and hence $\overline{T} = \langle 1\rangle$ in the first case. By condition (2) and since $(3,q)=1$, we then have that $T \leq Z(P)$ and hence $T = Z(P)$

because $Z(P)$ is a minimal normal subgroup of $N_G(P)$. In the latter case we cannot have $T \geqq P$ by condition (1), while $T \cap Z(P) = \langle 1 \rangle$ forces P to be $(T \cap P)Z(P)$, but as we have seen before then P must be abelian which cannot be the case.

Let $G = PSU(3,q^2)$. The proof is similar to the previous one.

Let $G = R(q)$. We have $N_G(P) = P \rtimes C$, where $|P| = q^3$ and C is cyclic of order $q-1$. Moreover, $Z(P) < P' = \Phi(P)$, $|Z(P)| = q$, $|P'| = q^2$. Since $\overline{N} = N_G(P)/P'$ is a Frobenius group with Frobenius kernel $\overline{P}$ (see [23],III.11), as in the previous cases, we have either $\overline{T} = \langle 1 \rangle$ or $\overline{T} \geqq \overline{P}$. In the first case $T \leqq P'$. If $T \cap Z(P) \neq \langle 1 \rangle$, then $T \geqq Z(P)$ since $Z(P)$ is a minimal normal subgroup of $N_G(P)$, but by [23],III.2, $T > Z(P)$ implies $T = P'$ and condition (1) is not satisfied. So $T = Z(P)$. We cannot have $T \cap Z(P) = \langle 1 \rangle$ since if $x \in P' - Z(P)$ then its centralizer in $N_G(P)$ has order $2q^2$ (see [23],III.2) and hence $|T| > q$, contrary to $T \leqq P'$. In the latter case, we cannot have $T \cap P' = \langle 1 \rangle$, since for each $x \in P - P'$ we have $o(x)=9$ and $x^3 \in Z(P) < P'$ (see [23], Theorem). Nevertheless, $T \cap P' \neq \langle 1 \rangle$ implies $|T| \geqq q^2$ (see [23],III.2) and again condition (1) is not satisfied. This completes the proof.

The following theorem is concerned with Lingenberg S-partitions of type I in the case of T being of even order.

THEOREM 3.2. *Let* (G,S,T) *determine a Lingenberg S-partition of type I. If T has even order then one of the following holds:*

(*a*) $G = O(G)T$ *and T is a Frobenius complement;*
(*b.1*) $G \simeq SL(2,q)$, $q=2^h$, $h\geq 2$; $T = S = P$ *with* $P \in S_2(G)$;
(*b.2*) $G \simeq Sz(q)$, $q=2^{2n+1}$, $n\geq 1$; $T = Z(P)$, $S = P$ *with* $P \in S_2(G)$;
(*b.3*) $G \simeq PSU(3,q^2)$, $q=2^h$, $h\geq 2$; $T = Z(P)$ *with* $P \in S_2(G)$ *and* $S = P \rtimes C_1$ *with* $|C_1| = (q+1)/d$, *where* $d=(3,q+1)$.

Proof. In the A-structure $[G,S,T]$, the pencil of lines Θ is a transitive $\tilde{G}$-set with $|\Theta|>1$. If $r = N_G(T) \in \Theta$ then $\tilde{G}_r = N_{\tilde{G}}(\tilde{T})$ and, by Proposition 2.2, $\tilde{T}$ acts semiregularly on $\Theta - \{r\}$. Then by [10], Theorem 2, either the case (a) occurs or $\tilde{G} \simeq SL(2,q)$, $Sz(q)$, $PSU(3,q^2)$, $SU(3,q^2)$ with $q=2^h$, $h>1$. In the latter case by [10], Lemma 3, $\tilde{G}$ acts on Θ in its usual doubly transitive representation of degree $q+1$, q^2+1, q^3+1, q^3+1 respectively. Then, it is well known that $\tilde{G}_r = N_{\tilde{G}}(\tilde{P})$ with $\tilde{P} \in S_2(\tilde{G})$ and hence $N_{\tilde{G}}(\tilde{T}) = N_{\tilde{G}}(\tilde{P})$. By taking account of Proposition 2.2, we see that $\tilde{T}$ satisfies conditions (1) and (2) of Lemma 3.1. Therefore $\tilde{T} = Z(\tilde{P})$. When $\tilde{G} \simeq SL(2,q)$, $Sz(q)$ or $PSU(3,q^2)$ we obtain (b.1) - (b.3) in view of Proposition 2.2,(3). If $\tilde{G} \simeq SU(3,q^2)$ with $3|q+1$ we have $|\tilde{T}| = |Z(\tilde{P})| = q$ and hence $|r|=q-1$. Since $|Z(G)|=3$, this implies that $3|q-1$ by Proposition 2.4, a contradiction. Therefore, the case $\tilde{G} \simeq SU(3,q^2)$, with $3|q+1$, cannot occur.

We point out that Examples I, II and III of section 2 show that the cases (b.1), (b.2) and (b.3) actually occur. On the contrary, it seems very difficult to achieve a complete classification of Lingenberg S-partitions of type I in the case (a).

In succession, we give some results and examples concerning this case.

PROPOSITION 3.3. *Let* (G,S,T) *determine a Lingenberg* S-*partition of type* I *and assume* $|T| \geq 3$. *If* $\tilde{G}$ *induces a Frobenius permutation group on the pencil* Θ *in the* A-*structure* [G,S,T], *then the following hold:*

(1) $G = M \rtimes T$, *where* M *is a nonabelian special p-group of order* q^{2m+1} *with* $q=p^h$, $m,h \geq 1$;

(2) $|Z(G)| = |Z(M)| = q$, $|T| = q+1$, $S = T$, $N_G(T) = TZ(M)$.

Proof. By Proposition 2.4, $Z(\tilde{G})$ is the kernel of the representation of $\tilde{G}$ on Θ. Therefore, $\tilde{G}^{\Theta} \simeq \tilde{G}/Z(\tilde{G})$ and $\tilde{G}^{\Theta}$ acts on Θ as a Frobenius group by our assumptions. Denote by $\overline{M}$ the Frobenius kernel of $\overline{G} = G/Z(G)$ and let $M \leq G$ such that $M/Z(G) = \overline{M}$. By Propositions 2.2 and 2.4, we have that $|Z(G)| \,\Big|\, |T|-1$ and hence $(|T|,|Z(G)|)=1$. Moreover, $(|\overline{T}|,|\overline{M}|)=1$ since $\overline{T}$ is contained in a Frobenius complement of $\overline{G}$. Therefore, $T \cap M = \langle 1 \rangle$ and $MT = M \rtimes T$. If $x \in G$ then, clearly, $T^x \subset MT$ and hence $G = MT$ and $\overline{G} = \overline{MT}$. We have $\overline{T} = N_{\overline{G}}(\overline{T}) \geq \overline{N_G(T)} \geq \overline{T}$, so that $\overline{T} = \overline{N_G(T)}$ and $N_G(T) = TZ(G)$. Set $|T| = t$, then $|Z(G)| = [N_G(T):T] \geq [N_G(T):S] = t-1$ by Proposition 2.2. Since, as we have previously seen, $|Z(G)| \,\Big|\, t-1$, it follows that $|Z(G)| = t-1$ and $S = T$. Note that since $\overline{M}$ is nilpotent so is M (see [11], V.8.7). Let P be a Sylow p-subgroup of M with $P \nleq Z(G)$ and let $N = PT$. Since (G,S,T) determines a Lingenberg S-partition of type I, the following relation holds

(2) $$(t^2 - t)(n/tc - 1) + tc \leq n ,$$

where $n = |N|$ and $c = |P \cap Z(G)|$. From (2), it follows that $t-1 \leq c$ since $n > tc$ and hence $c = t-1$ and $Z(G) < P$. This yields $M = P$. Consider the commutators of the form $[x,g]$ with $x \in T$ and $g \in M-Z(G)$. We have $[x,g] = x^{-1}(g^{-1}xg)$ and hence $[x,g] \in TT^g$. Each complex TT^g contains exactly $t-1$ non-identical distinct commutators of the form $[x,g]$. Moreover, if $T^g \neq T^f$ then $TT^g \cap TT^f = T$ and hence, the $t-1$ commutators lying in TT^g are distinct from those lying in TT^f. Since $|\{T^x : x \in G\}| = |\overline{M}|$, then by setting $|\overline{M}| = \overline{m}$ there exist at least $(t-1)(\overline{m}-1)+1$ distinct commutators lying in $[M,T]$. Since $[M,T] \leq M$ and $|M| = \overline{m}(t-1)$, it follows that $[M,T] = M$. Now suppose there exists a characteristic abelian subgroup A of M such that $A \nleq Z(G)$. Then the group AT contains exactly $a = [A:Z(G) \cap A]$ distinct conjugate elements of T. By using the same argument as before, we have

$$(t-1)a \geq |A| \geq |[A,T]| \geq (t-1)(a-1)+1 ,$$

where $a \geq 2$. From this it follows that $A = [A,T]$ and $|A| = (t-1)a$. The latter relation yields $Z(G) < A$, but this contradicts a result of Zassenhaus [11],III, Satz 13.4(b), since $|Z(G)| > 1$. Therefore, a characteristic abelian subgroup of M is central in G. In conclusion we have proved that:

(I) $(|M|,|T|) = 1$,

(II) $[M,T] = M$,

(III) T centralizes every characteristic abelian subgroup of M.

By a result of Thompson [11],III, Satz 13.6, we then have that M is a nonabelian special p-group. Moreover, since Z(M) is a characteristic abelian subgroup of M, we have that Z(M) = Z(G). Let $|Z(G)| = t-1 = p^h$, $h\geq 1$, and let $|M| = p^{h+n}$. Since a Frobenius complement of G/Z(G) has order $t=p^h+1$, it follows that $p^h+1 \mid p^n-1$. From this, we have that $p^h+1 \mid p^n+p^h=p^h(p^{n-h}+1)$ and hence $p^h+1 \mid p^{n-h}-p^h=p^h(p^{n-2h}-1)$. So $p^h+1 \mid p^{n-2h}-1$. Let $b \in \mathbb{Z}$ such that $bh\leq n\leq (b+1)h$. By iterating the above procedure, it is easy to prove that b must be even and $p^{n-bh}-1=0$. This completes the proof.

A Lingenberg S-partition of type I satisfying conditions (1) and (2) of Proposition 3.3 and its associated A-structure will be called *special*. An example is given below.

EXAMPLE VI. We assume the reader is familiar with [14],V,§32. Let π be the projective plane over GF(q^2), $q=p^h$, and let ρ be a hermitian polarity of π. It is well known that the absolute points and non-absolute lines of ρ make a Steiner system $\mathcal{U}$ with parameters $k=q+1$ and $v=q^3+1$, which is usually called the *classical unital*. Moreover the group P($\mathcal{U}$) consisting of the projectivities of π leaving $\mathcal{U}$ invariant is isomorphic to PGU($3,q^2$). According to Bose [5],§6, for each absolute line p of ρ, we may define a parallelism among the lines of $\mathcal{U}$ as follows: a class of parallel lines consists of a non-absolute line r through $\rho(p)$ and the non-absolute lines through $\rho(r)$. Note that $\rho(p)\in r$ implies $\rho(r)\in p$. Therefore, the group T(r) consisting of all $(\rho(r),r)$-homologies lying in P($\mathcal{U}$) preserves the parallelism just defined in $\mathcal{U}$, because it fixes the line p. The group T(r) fixes each line ℓ parallel to r and acts regularly on the points of ℓ lying in $\mathcal{U}$ because T(r) has order $q+1$. Moreover, there exists a unique Sylow p-subgroup M of P($\mathcal{U}$) which fixes $\rho(p)$ and so p itself and acts transitively on the q^2 non-absolute lines through $\rho(p)$. It follows that $\mathcal{U}$ - {P} is a quasi-translation A-structure with respect to the pencil Θ of non-absolute lines through $\rho(p)$ and to the family $\vartheta = \{T(r) : r\in\Theta\}$. It is easily seen that:

- $G = \langle T(r) : r\in\Theta\rangle = M \rtimes T(s)$, $s\in\Theta$;
- T(s) acts semiregularly on $\Theta - \{s\}$;
- $|Z(M)| = q$ and Z(M) consists of all $(\rho(p),p)$-elations lying in P($\mathcal{U}$) and therefore it fixes every line of Θ;
- $|M| = q^3$ and M/Z(M) acts regularly on Θ.

From this, it follows that Z(M) = Z(G) and G/Z(G) is a Frobenius group. Therefore, (G,T(s),T(s)) determines a special Lingenberg S-partition.

We now consider case (a) of Theorem 3.2. Assume,

(*i*) Σ *is a Lingenberg* S-*partition of* G *of type* I *with respect to the subgroup* T *with* $|T|\geq 3$ *and*

(*ii*) G = CT, *where* C *is solvable and* C $\triangleleft$ G.

Let Θ be the pencil of lines of [G,S,T]. By Propositions 2.2 and 2.4, we have that $(|\tilde{T}|,|Z(\tilde{G})|) = 1$ and hence $Z(\tilde{G}) < \tilde{C}$. Let $\tilde{L}/Z(\tilde{G})$ be a minimal normal subgroup of $\tilde{G}/Z(\tilde{G})$ contained in $\tilde{C}/Z(\tilde{G})$. Since $\tilde{L} \nleq Z(\tilde{G})$ and $\tilde{C}/Z(\tilde{G})$ acts transitively on Θ, then, if $r = N_G(T) \in \Theta$ and Ω is the orbit of r under $\tilde{L}$, we have that $|\Omega| > 1$. Moreover, since $\tilde{L}/Z(\tilde{G})$ is elementary abelian for $\tilde{C}$ is solvable, it follows that $\tilde{L}^{\Omega}$ is elementary abelian. It follows now that $\tilde{L}^{\Omega}$ acts regularly on Ω. Moreover, $\tilde{T}^{\Omega} \simeq \tilde{T}$ leaves Ω invariant and acts semiregularly on $\Omega - \{r\}$. From this, we infer that $\tilde{L}^{\Omega}\tilde{T}^{\Omega}$ is a Frobenius group. Now set $F = \langle T^x : x \in L\rangle$, $S_o = S \cap F$, $N_o = N_G(T) \cap F$, $|S_o| = s_o$, $|N_o| = n_o$, $|T| = t$ and $|F| = f$. Since Σ is a Lingenberg S-partition of G of type I, the following relation holds:

(3) $$(ts_o - s_o)(f/n_o - 1) + n_o \leq f.$$

From this, it follows that $n_o/s_o \geq t-1$. On the other hand we have that $[N_o:S_o] \leq$ $\leq [N:S] = t-1$. Therefore, $n_o/s_o = t-1$ and (3) holds as an equality. Using this, it is not difficult to see that if we set $R = \bigcap_{x\in F} S_o^x$, $S_1 = S_o/R$, $T_1 = TR/R$, $F_1 = F/R$, then $\Sigma_1 = \{T_1^x : x \in F_1\} \cup \{N_{F_1}(T_1)\}$ is a Lingenberg S-partition of F_1 of type I with respect to the subgroup T_1. But, $\tilde{F}_1^{\Omega} = \tilde{L}^{\Omega}\tilde{T}^{\Omega}$ and hence, by Proposition 3.3, Σ_1 is a *special Lingenberg* S-*partition.*

From a geometrical point of view, the previous result can be expressed as follows.

PROPOSITION 3.4. *Let* Σ *be a Lingenberg* S-*partition of* G *of type* I *with respect to the subgroup* T *with* $T \geq 3$. *Assume* G = CT, *where* C *is a solvable normal subgroup of* G. *Then the* A-*structure* [G,S,T] *contains a subspace which is a special* A-*structure.*

Proof. Let $[F_1,S_1,T_1]$ be the special A-structure related to the special S_1-partition Σ_1 of F_1 described above. If $S_1\bar{x}$ is a point of $[F_1,S_1,T_1]$ and $\bar{x} = Rx$, let η be the map from the set of points of $[F_1,S_1,T_1]$ into the set of points of [G,S,T] defined as follows

$$\eta : S_1\bar{x} \to Sx .$$

It is straightforward to show that η is well defined and gives an embedding of $[F_1,S_1,T_1]$ into [G,S,T].

4. LINGENBERG S-PARTITIONS OF DOUBLY TRANSITIVE PERMUTATION GROUPS

Assume (G,S,T) determines a Lingenberg S-partition. Since in Examples I-V we have that:

(a) *the group* $\tilde{G}$ *acts* 2-*transitively on the pencil of lines* Θ *of the* A-*structure* [G,S,T].

then the natural question arises whether it is possible to classify all the triples (G,S,T) which determine Lingenberg S-partitions satisfying condition (a). In the following, we shall prove that a rather satisfactory answer to this question may be given provided that the classification of doubly transitive permutation groups is assumed. As it is well known, such a classification follows from that of finite simple groups.

THEOREM 4.1. *Assume* (G,S,T) *determines a Lingenberg* S-*partition* Σ *satisfying condition* (a). *If* Σ *is of type* I *then one of the following holds:*

(1) $G \simeq SL(2,q)$, $q=p^h$, $p^h \geq 2$; $T = S = P$ *with* $P \in S_p(G)$;
(2) $G \simeq Sz(q)$, $q=2^{2n+1}$, $n \geq 1$; $T = Z(P)$, $S = P$ *with* $P \in S_2(G)$;
(3) $G \simeq PSU(3,q^2)$, $q=2^h$, $h \geq 2$; $T = Z(P)$ *with* $P \in S_2(G)$, $S = P \rtimes C_1$, *where* $|C_1| = (q+1)/d$, $d=(3,q+1)$;
(4) $G \simeq R(q)$, $q=3^{2n+1}$, $n \geq 1$; $T = Z(P)$, $S = P$ *with* $P \in S_3(G)$.

If Σ *is of type* II *then* $PSL(n,q) \leq G/Z(G) \leq P\Gamma L(n,q)$ *where* $n \geq 3$.

Proof. Assume (G,S,T) determines a Lingenberg S-partition of type I. By Proposition 2.2, the group $\tilde{G}^{\Theta} \simeq \tilde{G}/Z(\tilde{G})$ satisfies the following condition:

(h) for each $r \in \Theta$, the stabilizer of r in $\tilde{G}^{\Theta}$ contains a normal subgroup which acts semiregularly on $\Theta - \{r\}$.

From the classification theorem of finite doubly transitive permutation groups, we have that the so called "Hering conjecture" (see [4]) is actually a theorem (see [19],p.302) asserting that if $\tilde{G}^{\Theta}$ is 2-transitive on Θ and satisfies (h) then one of the following holds:

(j) $\tilde{G}^{\Theta}$ contains a regular normal subgroup,

(jj) $\tilde{G}^{\Theta} \simeq PSL(2,q)$, $q \geq 4$, $Sz(q)$, $PSU(3,q^2)$, $q>2$, or $R(q)$, $q>3$, and $\tilde{G}^{\Theta}$ acts on Θ in its usual doubly transitive representation.

We shall investigate these cases separately.

Case (j). Let $\tilde{N}^{\Theta}$ be the regular normal subgroup of $\tilde{G}^{\Theta}$. We have that $\tilde{G}^{\Theta} = \tilde{N}^{\Theta}\tilde{T}^{\Theta}$ and hence $\tilde{G}^{\Theta} \simeq \tilde{G}/Z(\tilde{G})$ is a Frobenius group. If $|T| = 2$ then $\tilde{G}^{\Theta} \simeq \tilde{G} \simeq SL(2,2)$. If $|T| \geq 3$ then by Proposition 3.3, we must have $|\tilde{N}^{\Theta}| = q^{2m}$, where q is a prime power and $m \geq 1$ Moreover, $|T| = q+1$. Since $\tilde{G}^{\Theta}$ is 2-transitive on Θ, it follows that $q^{2m}-1=q+1$. This implies $q=2$, $m=1$ and it is very easy to show that $G \simeq SL(2,3)$.

Case (jj). Note that if $L \leq G$ and $LZ(G) = G$ then $L = G$ since $(|T|,|Z(G)|) = 1$ and G is generated by T and its conjugates. Therefore, G is a central irreducible extension of $\tilde{G}^{\Theta}$. Since, in the case under consideration, G is a simple group (see [11],[12]) then there exists a unique representation group H of $\tilde{G}^{\Theta}$ and $G \simeq H/Z_0$ for some subgroup Z_0 of Z(H). Moreover, $Z(G) = Z(H)/Z_0$. If $M(\tilde{G}^{\Theta})$ denotes the Schur multiplier of $\tilde{G}^{\Theta}$ then $|Z(H)| = |M(\tilde{G}^{\Theta})|$ (see [21],Ch.2,§9). Since, by Proposition 2.2, we have that $|Z(G)| \mid |T| - 1$, it follows that $|Z(G)| \mid (|M(\tilde{G}^{\Theta})|,|T|-1)$. Now assume:

$\tilde{G}^{\Theta} \simeq PSL(2,q)$, $q=p^h$, $h \geq 2$. If p is odd then we cannot have $G \simeq PSL(2,q)$ since in this case every normal subgroup of $N_G(P)$, where $P \in S_p(G)$, contains P and hence relation (3) of Proposition 2.2 cannot be satisfied for $|P| = q$ and $|N_G(P)| = \frac{1}{2}q(q-1)$. Therefore, when $q \neq 4,9$ by [11],V,25.7, we must have $G = H \simeq SL(2,q)$ and the thesis follows from Lemma 3.1. If $q=4$ then T has even order and, by Theorem 3.2, $G \simeq SL(2,4)$. Let $q=9$. Since $N_G(T)/Z(G)$ is a Frobenius group of order $9 \cdot 4$ with Frobenius kernel of order 9, it follows that $|T| \geq 9$. On the other hand, $|T| \mid |\Theta|-1$ where $|\Theta| = 10$ and hence $|T| = 9$. By [8], Table 1, we have that $M(\tilde{G}^{\Theta}) \simeq Z_6$. Therefore, $|Z(G)| = 2$ and $G \simeq SL(2,9)$.

Let $\tilde{G}^{\Theta} \simeq Sz(q)$. In this case T has even order and (2) follows from Theorem 3.2.

Let $\tilde{G}^{\Theta} \simeq PSU(3,q^2)$, $q>2$. By [9], Theorem 2, we have either $G \simeq PSU(3,q^2)$ or $G \simeq SU(3,q^2)$. The case $G \simeq SU(3,q^2)$, $3|q+1$, may be excluded by arguing as in the proof of Theorem 3.2. Assume $G \simeq PSU(3,q^2)$, $q=p^h$, p an odd prime. By Lemma 3.1, we have that $T = Z(P)$ with $P \in S_p(G)$ and $S = P \rtimes C_1$, where C_1 is cyclic of order $q+1/d$, $d=(3,q+1)$. Moreover $C_1 < C$, where C is cyclic of order q^2-1/d and $C < N_G(P)$. Let C_2 be the subgroup of C of order $q-1$. As it is well known, there exists a subgroup M of G such that $M \cap N_G(P) = TC_2$ and $SL(2,q) \simeq M = \langle T,T^x \rangle$ for a suitable conjugate subgroup T^x of T, where T^x may be chosen in such a way that $C_2 < N_M(T^x)$. Let $b = ti$, where $1 \neq t \in T$ and i is the unique involution in C_2. As was shown in Example I, (M,T^x,T^x) determines a Lingenberg S-partition of type I. Since $N_M(T) \cap N_M(T^x) = C_2$, we have that $b \notin N_M(T^x)$. Therefore, there exists $T^y < M$ with $T \neq T^y \neq T^x$ such that $b \in T^xT^y$ and hence $T^xT^y \cap S \neq \langle 1 \rangle$ since $b \in S$. Note that $i \in C_1$ for q odd. It follows that (G,S,T) does not determine a Lingenberg S-partition. The thesis can now be achieved by using Lemma 3.1.

Let $\tilde{G}^{\Theta} \simeq R(q)$, $q>3$. By [1], Theorem 1, we have that $M(\tilde{G}^{\Theta}) = \langle 1 \rangle$ and (4) again follows from Lemma 3.1.

Now assume (G,S,T) determines a Lingenberg S-partition of type II. If $N_G(T)x$ is a line of Θ then the stabilizer $N_{\tilde{G}}(\tilde{T^x})^{\Theta}$ of this line in $\tilde{G}^{\Theta}$ contains the normal subgroup $\tilde{T}^{x\Theta}$ satisfying the condition

(1) $\tilde{T}^{x\Theta} \cap \tilde{T}^{y\Theta} = \langle 1 \rangle$ for each $\tilde{T}^{y\Theta}$ such that $\tilde{T}^{y\Theta} \neq \tilde{T}^{x\Theta}$.

Indeed, assume (1) does not hold. Then $T^xZ(G) \cap T^yZ(G) > Z(G)$ for some T^y with $T^x \neq T^y$. It follows that $T^xT^y \cap Z(G) \neq \langle 1 \rangle$ since $T^x \cap T^y = \langle 1 \rangle$ by Lemma 2.1. So there exists $z \in Z(G)$, $z \neq 1$, such that $z \in TT^{yx^{-1}} \subseteq ST^{yx^{-1}}$. Since $z \in N_G(T)$ and $ST^{yx^{-1}} \cap N_G(T) = S$ then $z \in S$, but this is a contradiction because $S \cap Z(G) = \langle 1 \rangle$. Moreover, by Proposition 2.2, we have:

(2) $\tilde{T}^{x\Theta}$ does not act semiregularly on $\Theta - \{N_G(T)x\}$.

By a well known result of O'Nan [15], Theorem A, conditions (1) and (2) imply $PSL(n,q) \leq \tilde{G}^{\Theta} \leq P\Gamma L(n,q)$ with $n \geq 3$. So the thesis follows from Proposition 2.4.

As a final remark, we note that the classification theorem of doubly transitive permutation groups is required only when (G,S,T) determines a Lingenberg S-partition of type I and T has odd order.

REFERENCES.

[1] Alperin, J.L. and Gorenstein, D., The multiplicators of certain simple groups, Proc. Am. Math. Soc. 17 (1966), 515-519.

[2] André, J., Über Parallelstrukturen, Teil I : Gundbegriffe, Math. Z. 76 (1961), 85-102.

[3] André, J., Über Parallelstrukturen, Teil II : Translationsstrukturen, Math. Z. 76 (1961), 155-163.
[4] Aschbacher, M., F-sets and permutation groups, J. Algebra 30 (1974), 400-416.
[5] Bose, R.C., On the application of finite projective geometry for deriving a certain series of balanced Kirkman arrangements, in: The Golden Jub. Comm., Calcutta Math. Soc. (1958-59), 341-354.
[6] Brenti, F., Sulle S-partizioni di Sylow in alcune classi di gruppi finiti, Boll. Un. Mat. It. (6) 3-B (1984), 665-685.
[7] Burkhardt, R., Über die Zerlegungszahlen der unitären Gruppen $PSU(3,2^{2f})$, J. Algebra 61 (1979), 548-581.
[8] Griess, R.L.,Jr., Schur Multipliers of the known finite simple groups, Bull. Am. Math. Soc. 78 (1972), 68-71.
[9] Griess, R.L.,Jr., Schur Multipliers of finite simple groups of Lie type, Trans. Am. Math. Soc. 183 (1973), 355-421.
[10] Hering, C., On subgroups with trivial normalizer intersection, J. Algebra 20 (1972), 622-629.
[11] Huppert, B., Endliche Gruppen I (Springer-Verlag, Berlin-Heidelberg-New York, 1979).
[12] Huppert, B. and Blackburn, N., Finite Groups III (Springer-Verlag, Berlin--Heidelberg-New York, 1982).
[13] Lingenberg, R., Über Gruppen projectiver Kollineationen, whelche eine perspective Dualität invariant lassen, Arch. Math. 13 (1962), 385-400.
[14] Luneburg, H., Translation Planes (Springer-Verlag, Berlin-Heidelberg-New York, 1980).
[15] O'Nan, M.E., Normal structure of the one-point stabilizer of a doubly-transitive permutation group. I, Trans. Am. Math. Soc. 214 (1975), 1-42.
[16] Rosati, L.A., Sulle S-partizioni nei gruppi non abeliani d'ordine pq, Rend. Sem. Mat. Univ. Padova 38 (1967), 108-117.
[17] Scarselli, A., Sulle S-partizioni regolari di un gruppo finito, Atti Acc. Naz. Lincei, Rend Cl. Sci. Fis. Mat. Nat. (8) 62 (1977), 300-304.
[18] Schulz, R.H., Zur Geometrie der $PSU(3,q^2)$, in: Beiträge zur Geometr. Algebra, Proc. Symp. Duisburg, 1976 (Birkhäuser, Basel, 1977), 293-298.
[19] Shult, E.E., Permutation groups with few fixed points, in: Geometry - von Staudt's Point of View, Proc. NATO Adv. Study Inst. Bad Windsheim, 1980 (D. Reidel P.C., Dordrecht, 1981), 275-311.
[20] Suzuki, M., On a class of doubly transitive groups, Ann. Math. 75 (1962), 104-145.
[21] Suzuki, M., Group Theory I (Springer-Verlag, Berlin-Heidelberg-New York, 1982)
[22] Tits, J., Les groupes simples de Suzuki et de Ree, in: Sem. Bourbaki, 13e année, 210 (1960/61), 1-18.
[23] Ward, H.N., On Ree's series of simple groups, Trans. Am. Math. Soc. 121 (1966), 62-89.
[24] Zappa, G., Sugli spazi generali quasi di traslazione, Le Matematiche (Catania) 19 (1964), 127-143.
[25] Zappa, G., Sulle S-partizioni di un gruppo finito, Ann. Mat. Pura Appl. (4) 74 (1966), 1-14.
[26] Zappa, G., Partizioni generalizzate nei gruppi, in: Coll. Int. Teorie Comb. 1973 (Acc. Naz. Lincei, Roma 1976), 433-437.

Annals of Discrete Mathematics 30 (1986) 85–98

COLLINEATION GROUPS STRONGLY IRREDUCIBLE ON AN OVAL

Mauro Biliotti

Dipartimento di Matematica
Università degli Studi di Lecce
via Arnesano, 73100-LECCE
Italy

Gabor Korchmaros

Istituto di Matematica
Università degli Studi della Basilicata
via N. Sauro, 85, 85100-POTENZA
Italy

In recent years, Hering has written several papers concerning the composition series of collineation groups of a finite projective plane. Prominent in his studies is the notion of *strongly irreducible collineation group on a projective plane,* one which does not leave invariant any point, line,triangle or proper subplane. There is a well developed theory of strongly irreducible collineation groups containing perspectivities, which has significant applications (see [4],[5],[15]). However, it should be noticed that only isolated results are known for such groups in the general case.

It should be interesting to investigate also "local" versions of the concept of irreducibility. In this connection, here we consider a finite projective plane π of even order with a collineation group Γ and a Γ-invariant oval Ω such that Γ does not leave invariant any point, chord or suboval of Ω. Here a suboval of Ω is a subset of points of Ω which is an oval in a proper subplane of π. We say that Γ is *strongly irreducible on the oval* Ω. Clearly Γ is not strongly irreducible on π since it fixes the knot K of Ω.

Our main result states that if Γ has even order then Γ contains some involutorial perspectivities, i.e. elations. The subgroup $\langle\Delta\rangle$ generated by all involutorial elations is essentially determined. If Γ has a fixed line then $\langle\Delta\rangle$ is the semidirect product of $O(\langle\Delta\rangle)$ with a subgroup of order two generated by an elation. If Γ has no fixed line then Γ acts as a "bewegend group" [6] on the dual affine plane of π with respect to the line at infinity K. From Hering's result [6] on bewegend groups containing involutorial elations, it then follows that $\langle\Delta\rangle$ is isomorphic to one of the simple groups: $SL(2,q)$, $Sz(q)$, $PSU(3,q^2)$, where q is a power of 2 and $q \geq 4$.

Clearly any collineation group of π mapping Ω onto itself and acting transitively on its points is strongly irreducible on Ω. As we shall prove in Section 5, such a group cannot involve $PSU(3,q^2)$. Hence, the only non-solvable collineation groups of π acting transitively on Ω are the groups Σ for which either $SL(2,q)\leq \Sigma \leq P\Gamma L(2,q)$ or $Sz(q)\leq \Sigma \leq Aut\, Sz(q)$. The groups are always 2-transitive on Ω. Furthermore, in the former case, π is a desarguesian plane of order q and Ω is a conic. For the latter, we may only assert that, at the present state of our knowledge, this situation occurs in the dual Lüneburg plane of order q^2 (see [11],[13],[14]).

2. NOTATION AND PRELIMINARY RESULTS

Fairly standard notation is used. A certain familiarity with finite projective planes as well as with finite groups is assumed. For the necessary background the reader is referred to [2],[9].

Throughout this paper, π denotes a projective plane of *even* order n containing an oval Ω. Here an oval is defined as a set of n+1 points no three of which are collinear.

The following elementary results are used in the proofs.

Through each point of Ω there exists exactly one tangent of Ω. The tangents are concurrent; their common point K is called the knot of Ω. Each line through K is a tangent of Ω. A line of π is an external line or a secant line of Ω according to whether $|r\cap\Omega|=0$ or 2. There are exactly $n(n-1)/2$ external lines and $n(n+1)/2$ secants of Ω in π. A chord of Ω is the pair of points which Ω has in common with a secant.

Let G be a collineation group of π mapping Ω onto itself. Then G fixes K. If G has no fixed point on Ω then it has no further fixed point in π. The only element of G with at least $\sqrt{n}+2$ fixed points on Ω is the identity collineation of π. The restriction map of G on Ω is a faithful representation.

Any non-trivial elation of G is involutorial. Its center does not belong to $\Omega\cup\{K\}$. Two distinct elations of G do not have the same center. The axis of any elation is a tangent of Ω. Any involutorial collineation of π is either an elation or a Baer-involution. The set of all fixed points and lines of a Baer-involution f is a subplane of order $\sqrt{n}$, called the Baer-subplane of f.

BAER INVOLUTIONS MAPPING Ω ONTO ITSELF

PROPOSITION 1. Suppose that a Baer-involution f *of* π *maps* Ω *onto itself and let*

F *denote the Baer-subplane of* f. *Then*

(1) $\Omega_F = F \cap \Omega$ *is a suboval of* F, *namely* Ω_F *is an oval of* F;

(2) *no line of* F *is an external line of* Ω.

Proof. Since n is even, Ω has an odd number of points. So, f has some fixed point on Ω. Given any fixed point P on Ω, the set of fixed points of f on Ω consists of P and all points Q for which the line PQ belongs to F. Other than the tangent of Ω at P, there are exactly $\sqrt{n}$ lines through P belonging to F. Therefore, $|\Omega_F| = \sqrt{n}+1$. This proves (1).

Let R be any point of $\Omega - \Omega_F$. There is a unique line r through R belonging to F. Moreover, f does not fix R and so r is not a tangent of Ω. Let $\{R,S\} = r \cap \Omega$. Then also $S \in \Omega - \Omega_F$. Therefore, r is an external line of Ω_F in the subplane F. Since $|\Omega - \Omega_F| = n - \sqrt{n}$, we obtain in this way each external line of Ω_F in the subplane. This proves (2).

PROPOSITION 2. If f *and* g *are two distinct Baer-involutions mapping* Ω *onto itself then their Baer-subplanes* F *and* G *are distinct.*

Proof. By way of contradiction, assume F=G. Choose a line r belonging to F which is an external line of Ω_F. By (2) of Prop. 1, $|r \cap \Omega| = 2$. Let $P \in r \cap \Omega$. Since $f(r)=g(r)$ and $f(P) \neq P$, $g(P) \neq P$, it follows that $f(P)=g(P)$. Hence $fg(P)=P$ with $P \notin F$. This implies that fg is the identity collineation of π which is a contradiction.

PROPOSITION 3. If f *and* g *are two distinct commuting Baer-involutions which map* Ω *onto itself, then* fg *is an elation.*

Proof. Let F (resp. G) be the Baer-subplane of f (resp. g). Since fg=gf, then f leaves G invariant. Let f' denote the involutorial collineation induced by f on G. Similarly, let g' denote the involutorial collineation induced by g on F. According to [2] 4.1.11, we have either

(i) f' *and* g' *are both Baer-involutions and* $F \cap G$ *is a subplane of order* $\sqrt[4]{n}$ *in both subplanes* F *and* G, *or*

(ii) f' *and* g' *are both elations; the point-set of* $F\cap G$ *consists of* $\sqrt{n}+1$ *collinear points lying on a tangent* r *of* Ω *which is the common axis of* f' *and* g'. $F\cap G$ *contains exactly* $\sqrt{n}+1$ *lines, each through the same point* $C\in r$ *which is the common center of* f' *and* g'.

We prove that the former possibility cannot occur. Suppose that $H=F\cap G$ is a subplane of order $\sqrt[4]{n}$. So $\Omega_H=H\cap\Omega$ is an oval of H by (1) of Prop. 1. Choose a line t of H such that $|t\cap\Omega_H|=0$. Applying (2) of Prop.1 to G, f' and Ω_G, we can infer that $|t\cap\Omega_G|=2$. Similarly, $|t\cap\Omega_F|=2$. This yields $|t\cap\Omega|\geq|t\cap\Omega_F|+|t\cap\Omega_G|+|t\cap\Omega_H|=$ $=4$. A contradiction, since Ω is an oval.
So we may assume that (ii) holds. In this case $\Omega_F\cap\Omega_G=\{r\cap\Omega\}$. The lines through C which are secants of either Ω_F or Ω_G belong to $F\cap G$. Since Ω is an oval, such lines are pairwise distinct. Thus

(4) $|t\cap(\Omega_F\cup\Omega_G)|=2$ *for any line* t *of* $F\cap G$ *distinct from* r.

Suppose there is a point $P\in\Omega-(\Omega\cap r)$ fixed by fg. Then $P\notin\Omega_F\cup\Omega_G$. Set $Q=f(P)=g(P)$. Again, $Q\notin\Omega_F\cup\Omega_G$. The line t joining P and Q meets $\Omega-(\Omega_F\cup\Omega_G)$ in two points. In particular, $t\neq r$. Both f and g leave t invariant. Thus, t belongs to H.
By (4), $|t\cap(\Omega_F\cup\Omega_G)|=2$. It follows that t has four common points with Ω. Since Ω is an oval, this is impossible.
Therefore, we have that fg has a unique fixed point on Ω. By (1) of Prof. 1, this implies that fg is an elation.

PROPOSITION 4. If g *is a collineation of order 4 mapping* Ω *onto itself and such that* g^2 *is a Baer-involution then* g *acts on the Baer-subplane* G *of* g^2 *as a Baer-involution.*

Proof. Let g' denote the involutorial collineation induced by g on G. By way of contradiction, assume that g' is either an elation or the identity. Choose an external line r of Ω_G in the subplane G such that r is fixed by g'. Applying (2) of Prop. 1 to G, g^2 and Ω_G, it follows that r meets $\Omega-\Omega_G$ in two points P and Q. As g leaves r invariant, then g^2 fixes P and Q. On the other hand g^2 fixes G pointwise. Since $P,Q\notin G$, it follows that g^2 is the identity collineation, contrary to our assumptions.

PROPOSITION 5. Let S be a 2-group of collineations of π mapping Ω onto itself. If S contains no elation then S is cyclic of order 2^t, where 2^{2^t} divides n.

Proof. By Prop. 3, S contains a unique involution. Then by [9], III. Satz 8.2, S is either a cyclic or a generalized quaternion group. We shall prove that the latter possibility cannot occur.

Denote by F the Baer-subplane of the unique involution f of S. Assume that S is a generalized quaternion group. Then the collineation group $\bar{S}$ induced by S on F admits an elementary Abelian subgroup $\bar{T}$ of order 4. By Prop. 4, each of the three involutions in $\bar{T}$ is a Baer-involution in F. By Prop. 2, their subplanes are pairwise distinct. But such a situation is excluded by applying Prop. 3 to F, Ω_F and any two involutions of $\bar{T}$. Finally, the statement concerning the order of S follows from [2] 4. 1.10.

PROPOSITION 6. Let Γ be a collineation group of π possessing a minimal normal solvable subgroup Ψ. If Γ is strongly irreducible on Ω then Γ contains no Baer-involution.

Proof. Let μ be the set of fixed points of Ψ on Ω. As Γ leaves invariant no point or chord of Ω we have that either $\mu=\emptyset$ or $|\mu|\geq 3$. If $|\mu|\geq 3$, Ψ fixes a quadrangle since the knot K of Ω is also fixed by Ψ. Thus, the fixed elements of Ψ in π form a subplane π' and $\mu=\Omega \cap \pi'$ is a suboval of Ω. Since Ψ is a normal subgroup of Γ, then Γ leaves μ invariant. As Γ is strongly irreducible on Ω, this is impossible. Thus μ is empty. As Ψ is an elementary Abelian p-group, this implies that p divides $|\Omega|$. Hence $p|n+1$.

Now we shall prove that Γ fixes exactly one line in the set E of all external lines of Ω. Since $|E|=n(n-1)/2$ and $(n+1,n(n-1)/2)=1$, then Ψ fixes at least one line of E. The common point of any two lines of E is distinct from the knot K of Ω. As μ is empty, it follows that Ψ cannot have further fixed lines in E.

Let r be the unique fixed line of Ψ in E. As Ψ is a normal subgroup of Γ, then Γ fixes r. But then, by (2) of Prop. 1, Γ has no Baer involution.

Since a finite group with cyclic Sylow 2-subgroups is solvable (see [9], IV. Satz 2.8), then Propositions 5 and 6 yield the following result:

THEOREM A. Any collineation group of even order of π which is strongly irreducible on Ω contains involutorial elations.

4. COLLINEATION GROUPS STRONGLY IRREDUCIBLE ON AN OVAL

THEOREM B. Let Γ be a collineation group of π which has even order and is strongly irreducible on Ω. Denote by Δ the set of all involutorial elations of Γ, and by $\langle\Delta\rangle$ the subgroup of Γ generated by Δ. Then, either $\langle\Delta\rangle$ is the semidirect product of $O(\langle\Delta\rangle)$ by a group of order two, and Γ contains no Baer-involution, or $\langle\Delta\rangle \simeq SL(2,q)$, $Sz(q)$, $PSU(3,q^2)$ where q is a power of 2 and $q \geq 4$. In any case $\Gamma \leq Aut\ \langle\Delta\rangle$.

Proof. We distinguish two cases according to whether Γ fixes exactly one line or it has no fixed line.

Assume Γ fixes a line r of π. Clearly, r is an external line of Ω. By (2) of Prop. 1, Γ contains no Baer-involution. Hence, any involution of Γ is an elation whose center belongs to r. But then two distinct involutions of Γ cannot commute since it is easily seen that their centers as well as their axes must be distinct. So any two distinct involutions in Γ generate a dihedral group with cyclic stem of odd order. By [3], Corollary 3, it follows that $\langle\Delta\rangle$ is the semidirect product of $O(\langle\Delta\rangle)$ by a group of order two generated by an involutorial elation.

Assume that Γ has no fixed line. By Theorem A, Δ is non-empty. So we can apply Hering's main theorem on bewegend groups [6]. As the knot K of Ω cannot be the center of any elation in Γ, our situation corresponds, up the duality, with that considered in Theorem 1 of [6]. It remains to exclude the possibility that $\langle\Delta\rangle \simeq SU(3,q^2)$, where q is a power of 2 and $q \geq 4$. In such a situation, $Z(\langle\Delta\rangle)$ has order 3 and fixes the axis of each elation in Δ. Thus, $Z(\langle\Delta\rangle)$ has some fixed points on Ω. Let μ be the set of all fixed points of $Z(\langle\Delta\rangle)$ on Ω. Γ leaves μ invariant. As Γ leaves invariant no point or chord of Ω then $|\mu| \geq 3$. But as we have shown in the proof of Prop. 6 $|\mu| \geq 3$ implies that μ is a suboval of Ω. Since Γ is strongly irreducible on Ω, this is impossible.

A similar argument shows that the centralizer of $\langle\Delta\rangle$ in Γ is trivial. Therefore, $\Gamma \leq Aut\ \langle\Delta\rangle$.

In the following, we shall be concerned with some geometrical properties of the set D of all points which are centers of involutorial elations of Δ. Also the set $D \cup S \cup \{K\}$ will be considered. Here S denotes the subset of Ω consisting of those points which are fixed by some involutorial elation of Δ. As we have seen in the proofs of Prop. 6 and Theorem B, the following statement holds:

PROPOSITION 7. D consists of collinear points if and only if Γ is solvable.

Now we shall prove

PROPOSITION 8. Let $\langle\Delta\rangle$ be isomorphic with one of the following simple groups: $SL(2,q)$, $Sz(q)$, $PSU(3,q^2)$, $q=2^\alpha$ and $q\geq 4$. Then $\langle\Delta\rangle$ acts on S as the corresponding simple group in its usual doubly transitive representation of degree $q+1$, q^2+1, q^3+1, respectively.

Proof. In our situation, $\langle\Delta\rangle$ has exactly one class of involutions. So all involutions in $\langle\Delta\rangle$ are elations.
Given any point $P\in S$, let σ denote an involutorial elation of Δ fixing P. Let $Z(\Sigma)$ be the center of the unique Sylow 2-subgroup Σ of $\langle\Delta\rangle$ containing σ. The involutions of $\langle\Delta\rangle$ commuting with σ are exactly those belonging to $Z(\Sigma)$. Each of them fixes P and has axis PK. Conversely, any two involutions of $\langle\Delta\rangle$ with the same fixed point P on S commute because they have the same axis PK. Thus, $\langle\Delta\rangle$ acts on S as the corresponding simple group acts on the set of its Sylow 2-subgroups. This completes our proof.
Assume $\langle\Delta\rangle\simeq SL(2,q)$, $q=2^\alpha$ and $q\geq 4$. By a result of Hering [7], Theorem 2.8.c, $D \cup S \cup \{K\}$ is a desarguesian subplane π' of order q of π and S is a suboval of Ω. Since Γ is strongly irreducible on Ω, then we must have $S=\Omega$. Hence, $\pi'=\pi$. Moreover, Ω is a conic. Therefore, we have

PROPOSITION 9. If $\langle\Delta\rangle\simeq SL(2,q)$, $q=2^\alpha$ and $q\geq 4$ then π is a desarguesian plane of order q and Ω is a conic. Moreover, $\langle\Delta\rangle$ acts on Ω as $SL(2,q)$ in its 3-transitive representation.

Assume either $\langle\Delta\rangle\simeq Sz(q)$, $q=2^\alpha$ $q>4$, or $\langle\Delta\rangle\simeq PSU(3,q^2)$, $q=2^\alpha$ $q\geq 4$. The present state of our knowledge does not allow us to determine the underlying plane π. For a geometrical approach to this essential question, it may be of interest to know the class of D as well as that of $D\cup S\cup\{K\}$. Here, a set U of points of a projective plane has class $[x_1,\ldots,x_K]$ when $|r\cap U|$ belongs to the integer set $\{x_1,\ldots,x_K\}$ for any line r in the plane. Results about sets with prescribed class are given in [8], [17], [18].

PROPOSITION 10. Let Γ *and* $\langle\Delta\rangle$ *as in Theorem B.*

(i) If $\langle\Delta\rangle\simeq Sz(q)$, $q=2^\alpha$ *and* $q>4$ *then* D *is of class* $[0,1,q-1,q\pm\sqrt{2q}+1]$, *and* $D\cup S\cup\{K\}$ *is of class* $[0,1,q+1,q\pm\sqrt{2q}+1]$.

(ii) If $\langle\Delta\rangle\simeq PSU(3,q^2)$, $q=2^\alpha$ *and* $q\geq 4$ *then* D *is of class* $[0,1,q-1,q+1]$, *and* $D\cup S\cup\{K\}$ *is of class* $[0,1,q]$.

Proof. Let σ_1 and σ_2 be any two distinct involutorial elations belonging to Δ. We shall denote their centers by R_i (i=1,2) and the line through them by r. We want to determine $|r\cap D|$.

We already remarked that each involution of $\langle\Delta\rangle$ is an elation. Moreover, as it was shown in the proof of Theorem B, $\langle\Delta\rangle$ does not leave r invariant. Hence $D\not\subset r\cap D$.

Assume first $\sigma_1\sigma_2=\sigma_2\sigma_1$. An argument similar to that used in the proof of Prop. 8 shows that r is a tangent of Ω such that $r\cap\Omega\in S$, and $r\cap D$ consists of the q-1 centers of the involutions in $Z(\Sigma)$, where Σ is the Sylow 2-subgroup of $\langle\Delta\rangle$ containing σ_1,σ_2.

Assume now $\sigma_1\sigma_2\neq\sigma_2\sigma_1$. Then r is not a tangent of Ω. By [14], Lemma 5.1, r is the unique fixed line of $\sigma_1\sigma_2$ which does not pass through K. Let Λ be any dihedral subgroup of $\langle\Delta\rangle$ which contains $\sigma_1\sigma_2$. Then Λ also leaves r invariant and fixes K. Thus, the centers of involutions of Λ belong to r, also. If Λ is such that no $\langle\Lambda,\sigma\rangle$ with $\sigma\in\Delta$, $\sigma\notin\Lambda$, leaves r invariant then $r\cap D$ consists of the centers of the involutions in Λ and $|r\cap D|=|\Lambda|/2$. We point out that this situation occurs when Λ is a maximal subgroup of $\langle\Delta\rangle$. We shall prove that such a dihedral group Λ exists in both of cases under consideration.

Assume $\langle\Delta\rangle\simeq Sz(q)$, $q=2^\alpha$ and q>4. Then $\langle\Delta\rangle$ admits exactly three coniugate classes of dihedral subgroups which are not properly contained in any other dihedral subgroup. They have orders 2(q-1) or $2(q\pm\sqrt{2q}+1)$. Those of order 2(q-1) are maximal subgroups of $\langle\Delta\rangle$. Moreover, each of those of order $2(q\pm\sqrt{2q}+1)$ is contained in exactly one ma-

ximal subgroup of order $4(q\pm\sqrt{2q}+1)$, which has cyclic Sylow 2-subgroups (see [16], Theorem 9). Thus, either $|r\cap D|=q-1$ or $|r\cap D|=q\pm\sqrt{2q}+1$.
Assume $<\Delta>\simeq PSU(3,q^2)$, $q=2^{\alpha}$ and $q\geq 4$. Then for any two distinct Sylow 2-subgroups Σ_1, Σ_2 we have $<Z(\Sigma_1),Z(\Sigma_2)>\simeq SL(2,q)$ (see [12], Satz 4.3.vii). Therefore, each dihedral subgroup of $<\Delta>$ is contained in some subgroups of $<\Delta>$ isomorphic to SL(2,q). By [9] II.8), there are exactly two coniugated classes of dihedral subgroups in SL(2,q) which are not properly contained in any other dihedral subgroup of SL(2,q). These groups have orders 2(q-1) or 2(q+1) according to whether $|r\cap S|=2$ or 0. In $PSU(3,q^2)$ the subgroups isomorphic to SL(2,q) are coniugate. Thus, the above assertion hold for $PSU(3,q^2)$, also. From these facts we can infer that $<\sigma_1,\sigma_2>$ is contained in a dihedral subgroup Λ of order 2(q-1) or 2(q+1) according to whether $|r\cap S|=2$ or 0.
Assume q=8. Let $|r\cap S|=2$. Let us consider the subgroup Ψ of $<\Delta>$ which leaves $r\cap S$ invariant. By [9], II.8, Ψ is the direct product of Λ with a cyclic group of order 3. Therefore the involutions of Ψ are exactly those of Λ. We may assume that r and S are disjoint. Let $M\simeq SL(2,8)$ be the subgroup of $<\Delta>$ containing Λ, Denote by Φ the subgroup of order 3 of $<\Delta>$ which centralizes M. If Ξ is the subgroup of order 9 of Λ then $\Phi\times\Xi$ is a Sylow 3-subgroup of $<\Delta>$. Since the centralizer of Ξ is contained in $\Phi\times M$ then $\Phi\times\Xi$ is the unique Sylow 3-subgroup of $<\Delta>$ which contains Ξ. Let R denote the set of all involutions in Δ whose centers lie in r. For any $\rho_1,\rho_2\in R$ with $\rho_1\neq\rho_2$, we have that $<\rho_1,\rho_2>$is contained in a dihedral subgroup of order $2\cdot 3^2$. Thus R is a full class of conjugate involutions in $<\Delta>$ and Glauberman's theorem (see [9] Cor. 3) may be applied. It follows that $|<R>|=2\cdot 3^{\beta}$ and $|<RR>|=3^{\beta}$. Moreover, $\Phi\times\Lambda$ has order $2\cdot 3^3$ and contains exactly 9 involutions, which are exactly those of Λ. Since Ξ is contained in $<RR>$, we have either $<RR>=\Xi$ or $<RR>=\Phi\times\Xi$. Thus, $<R>=\Lambda$ or $<R>=\Phi\times\Lambda$ holds. But the latter possibility cannot occur because all involutions of $\Phi\times\Lambda$ are in Λ.
Assume $q\neq 8$. Let $<\sigma_1,\sigma_2>$ be any dihedral subgroup of $<\Delta>$ of order 2(q-1) or 2(q+1). In order to prove that each involution $\sigma\in\Delta$, with center on r, belongs to Λ, we shall show that if we deny this then there exist two commuting involutions in $<\Delta>$ both leaving r invariant, which is a contradiction. In fact, such elations must have distinct centers, since they map Ω onto itself. Therefore, they must have the same axis t. But then, they both cannot leave r invariant since t, being a tangent of Ω, is distinct from r.
Let Σ_1 and Σ denote the Sylow 2-subgroups containing σ_1 and σ, respectively. If $\Sigma_1=$

$=\Sigma$, there is nothing to prove. Otherwise, assume first that $<Z(\Sigma_1),Z(\Sigma_2)>=<Z(\Sigma_1),Z(\Sigma)>$. If $\sigma\notin\Lambda$ then $<\sigma,\Lambda>\simeq SL(2,q)$, as Λ is a maximal subgroup of $<Z(\Sigma_1),Z(\Sigma)>\simeq SL(2,q)$ (see [9], II.8.27). Hence, $<\sigma,\Lambda>$ contains two commuting involutions. Assume now $<Z(\Sigma_1),Z(\Sigma_2)>\neq<Z(\Sigma_1),Z(\Sigma)>$. Then $N_{<\Delta>}(\Sigma_1)$ contains two distinct cyclic subgroups Θ and Θ_2 of order $(q+1)/d$ with $d=(3,q+1)$ which centralize $<Z(\Sigma_1),Z(\Sigma)>$ and $<Z(\Sigma_1),Z(\Sigma_2)>$, respectively (see [12], Satz 4.3.vii). These cyclic subgroups both leave r invariant, since the centers of σ, σ_1 and σ_2 lie on r. We prove that $<\Delta>_r$ admits two commuting involutions by showing that $|<\Theta,\Theta_2>\cap Z(\Sigma_1)|>2$. By [12], Satz 4.3.v-vii, $\Theta_2<\Psi=\Sigma_1\Theta$ and $\bar{\Psi}=\Sigma_1\Theta/Z(\Sigma_1)$ is a Frobenius group with Frobenius kernel $\bar{\Sigma}_1$. Moreover, $\bar{\Theta}$ and $\bar{\Theta}_2$ are Frobenius complements of Ψ. Since $|\Sigma_1|=q^2$, $|\Theta|=|\Theta_2|=(q+1)/d$, $q\neq 8$, it is not difficult to show that $<\bar{\Theta},\bar{\Theta}_2>=\bar{\Psi}$. If $<\Theta,\Theta_2>=\Sigma_0\Theta$ with $\Sigma_0<\Sigma_1$ then $|\Sigma_0\cap Z(\Sigma_1)|=2^\gamma$, where $\gamma\geq 1$ since each element of $\Sigma_1-Z(\Sigma_1)$ has order 4 and its square lies in $Z(\Sigma_1)$. Since the q-1 involutions in $Z(\Sigma_1)$ are conjugated under $N_{<\Delta>}(\Sigma_1)$ then each of them is the square of exactly q(q+1) elements of $\Sigma_1-Z(\Sigma_1)$. But, Σ_0 contains $2^\gamma\cdot(q^2-1)$ elements of $\Sigma_1-Z(\Sigma_1)$. Hence $|\Sigma_0\cap Z(\Sigma_1)|>2$.

From that we have seen so far, it follows that $|r\cap D|=q-1$ or $|r\cap D|=q+1$.

Now consider $D\cup S\cup\{K\}$. A line r through the knot K contains q-1 points of D and one point of S. Hence, $|r\cap(D\cup S\cup\{K\})|=q+1$. Assume $r\cap S=\{A,B\}$, with $A\neq B$. Since $<\Delta>$ acts on S in its usual doubly transitive representation, we have that $<\Delta>_{\{A,B\}}$ is a dihedral group with cyclic stem of order q-1 or $(q^2-1)/d$ according to whether $<\Delta>\simeq Sz(q)$ or $PSU(3,q^2)$ (see [16] and [12], Satz 4.3.vi). In the former case, $|r\cap D|=q-1$ In the latter, we have again $|r\cap D|=q-1$ since $<\Delta>_{\{A,B\}}$ contains exactly q-1 involutions, namely those lying in its dihedral subgroup of order 2(q-1).

Conversely, if $|r\cap D|=q-1$ then r is fixed by a dihedral group H of order 2(q-1) which, of course,does not fix any other line. But, H interchanges two points of S and hence it fixes the line through them. It follows that such a line must be r and so $|r\cap S|=2$. This completes the proof of Proposition 10.

5. COLLINEATION GROUPS OF EVEN ORDER WHICH ARE TRANSITIVE ON AN OVAL

THEOREM C. Let Γ be a collineation group of even order of π which maps Ω onto itself and acts transitively on Ω. If Δ is the set of all involutorial elations in Γ, and $<\Delta>$ denotes the subgroup generated by Δ then one of the following holds *(i)* $\langle\Delta\rangle$ *is the semidirect product of* $O(\langle\Delta\rangle)$ *with a group of order two and acts transitively on* Ω. $|\Delta|=n+1$ *and* Γ *contains no Baer-involution.*

(ii) $n=q$ *with* $q=2^{\alpha}$, $\alpha\geq 2$, $\langle\Delta\rangle\simeq SL(2,q)$, π *is a desarguesian plane of order* q *and* Ω *is a conic. Moreover,* $\langle\Delta\rangle$ *acts on* Ω *as* $SL(2,q)$ *in its 3-transitive representation.*

(iii) $n=q^2$ *with* $q=2^{2\beta+1}$, $\beta\geq 1$, $\langle\Delta\rangle\simeq Sz(q)$ *and* $\langle\Delta\rangle$ *acts on* Ω *as* $Sz(q)$ *in its usual 2-transitive representation.*

Proof. Clearly Γ is strongly irreducible on Ω. So we can apply Theorem B. Since Γ acts transitively on Ω, every point of Ω is fixed by an involution of Δ. But then, by Gleason's lemma (see [2], 4.3.15), $\langle\Delta\rangle$ also acts transitively on Ω. So, with the notation of Section 4, we have $S=\Omega$.

If Γ leaves a line r invariant then every point of r is the center of an involution of Δ and Propositions 6 and 7 yield (i).

If Γ does not leave any line invariant then it turns out that either (ii) or (iii) holds. In fact, actually Γ cannot involve $PSU(3,q^2)$. To see this, assume, by way of contradiction, that $\langle\Delta\rangle\simeq PSU(3,q^2)$ holds. Since $S=\Omega$, we have $n=q^3$ with $q=2^{\alpha}$, $\alpha\geq 2$. With the notation of Section 4, let P be any point P of $D\cup S\cup\{K\}$. By Prop. 10, for any line r through P, $|r\cap(D\cup S\cup\{K\})|\geq 2$ implies $|r\cap(D\cup S\cup\{K\})|=q+1$. Since $|D\cup S\cup\{K\}|=(q+1)(q+1)$, it follows that no line in the plane meets $D\cup S\cup\{K\}$ in a unique point. Hence, $D\cup S\cup\{K\}$ is actually of class $[0,q+1]$, i.e. it is a maximal $((q^3+1)q+1,q+1)$-arc. By [8] 12.2.1, this implies $(q+1)|(q^3+1)q+1$, a contradiction. Notice that the possibility $\langle\Delta\rangle\simeq PSU(3,q^2)$ can also be excluded by applying [1].

Research partially supported by G.N.S.A.G.A of C.N.R. and by M.P.I.

REFERENCES

[1] Biliotti, M. and Korchmaros, G., On the action of $PSU(3,q^2)$ on an affine plane of order q, Archiv Math. 44 (1985) 379-384.

[2] Dembowski, P., Finite Geometries (Springer Verlag, Berlin-Heidelberg-New York, 1968).

[3] Glauberman, G., Central elements in core-free groups, J. Algebra 4 (1966) 403-420.

[4] Hering, C., On the structure of finite collineation groups of projective planes, Abh. Math. Sem. Hamburg 49 (1979) 155-182.

[5] Hering, C., Finite collineation groups of projective planes containing nontrivial perspectivities, in: Finite Groups, Santa Cruz Conf. 1979, Proc. Symp. Pure Math. 37 (1980) 473-477.

[6] Hering C., On Beweglichkeit in affine planes, in: Finite geometries, Proc. Conf. Hon. T.G. Ostrom, Wash. Stat. Univ. 1981, Lect. Notes Pure Appl. Math. 82 (1983) 197-209.

[7] Hering, C., On projective planes of type VI, in: Colloq. int. Teorie comb., Roma 1973, Atti dei Convegni Lincei 17 Tomo II (1976) 29-53.

[8] Hirschfeld, J.W.P., Projective geometries over finite fields (Clarendon Press, Oxford, 1979).

[9] Huppert, B., Endliche Gruppen I (Springer Verlag, Berlin-Heidelberg-New York, 1967).

[10] Huppert, B. and Blackburn, N., Finite Groups III (Springer Verlag, Berlin-Heidelberg-New York, 1982).

[11] Kantor, W.M., Symplectic groups, symmetric designs and line ovals, J. Algebra 33 (1975) 43-58.

[12] Klemm, M., Charakterisierung der Gruppen $PSL(2,p^f)$ and $PSU(3,p^{2f})$ durch ihre Charactertafel, J. Algebra 24 (1973) 127-153.

[13] Korchmaros, G., Le ovali di linea del piano di Lüneburg d'ordine 2^{2r} che possono venir mutate in sé da un gruppo di collineazioni isomorfo al gruppo semplice $Sz(2^r)$, Atti Accad. Naz. Lincei, Memorie, Cl. Sci. Fis. Mat. Nat., (8) 15 (1979) 295-315.

[14] Lüneburg, H., Translation planes (Springer Verlag, Berlin-Heidelberg-New York, 1980).

[15] Stroth, G., On Chevalley-groups acting on a projective planes, J. Algebra 77 (1982) 360-381.

[16] Suzuki, M., On a class of doubly transitive groups, Ann. Math. 75 (1962) 105-145.

[17] Tallini, G., Problemi e risultati sulle geometrie di Galois, Relazione n.30 Ist. Matem. Univ. Napoli (1973).

[18] Tallini-Scafati, M., Sui (k,n)-archi di un piano grafico finito, con particolare riguardo a quelli a due caratteri, Atti dell'Accad. Naz. Lincei, Rendiconti, Cl. Fis. Mat. Nat. (8) 40 (1960) 812-818 and 1020-1025.

Annals of Discrete Mathematics 30 (1986) 99–104

ON SETS OF PLÜCKER CLASS TWO IN PG(3,q)

Paola Biondi and Nicola Melone (*)

Dipartimento di Matematica e Applicazioni
"R.Caccioppoli"
Università di Napoli
ITALY

In this paper the concept of the Plücker class of a k-set in PG(n,q) is introduced and certain k-sets of Plücker class two in PG(3,q) are characterized.

INTRODUCTION

The combinatorial characterization of geometric objects embedded in PG(n,q),the n-dimensional projective space over the Galois field GF(q),is one of the most interesting problems in combinatorial geometries.The theory of k-sets,i.e. the investigation of subsets of size k in PG(n,q) with respect to their possible intersections with all subspaces of a given dimension (see for instance [15], [16], [20]) turns out to be quite a powerful and useful tool in such characterizations.

A k-set K in PG(n,q) is said to be of <u>class</u> $[m_1,m_2,\dots,m_s]_d$ if for any d-subspace S_d we have $|K\cap S_d| \in \{m_1,m_2,\dots,m_s\}$,$(0\leq m_1< m_2< \dots m_s \leq \vartheta_d$, m_j a non-negative integer, $\vartheta_d = (q^{d+1}-1)/(q-1)$) .If $|K\cap S_d| = m_j$, S_d is called an <u>m_j-secant d-space</u>.If for all j=1,2,...,s some m_j-secant d-space exists,then K is of <u>type</u> $(m_1,m_2,\dots,m_s)_d$ .The k-set theory classifies all sets in a given class.

The first results from this point of view were obtained by B.Segre [10], [11], [12], whose characterization of conics in PG(2,q) (q odd),as k-sets of class $[0,1,2]_1$ is well known.

This result was generalized to quadrics by G.Tallini [13], [14] and later on by F.Buekenhout [3] ,as a result of the investigation of special sets of class $[0,1,2,q+1]_1$ in PG(n,q) .Furthermore,a thorough investigation of the class $[1,n,q+1]_1$ sets was carried out by M.Tallini Scafati [17] and J.W.P.Hirschfeld & J.A.Thas [7],[8] .

Whereas the literature is fairly abundant for d=1 , there are not so many results when d>1 .For instance,sets of type $(m,n)_d$ were investigated by A.Bichara [1] , M.J.de Resmini and M.de Finis [4],M.deFinis [5],G.Tallini [16], M.Tallini Scafati [18], [19], J.A.Thas [21]. Finally,results on more than two character sets,again when d>1 ,were obtained e.g. by A.Bichara [2], O.Ferri [6] and N.Melone [9].

A complete characterization of k-sets with more than two characters with respect to d-spaces, d>1 , seems to be extremely difficult.therefore we shall add some natural assumptions on the sets in order to classify them.Such assumptions may be of either arithmetic or structural nature.

(*) Work supported by Italian M.P.I.

From this point of view we define the Plücker class of a set in PG(n,q) as follows. Denote by K a k-set in PG(n,q). A hyperplane H is said to be tangent to K at its point p if any line in H through p is either a 1-secant or (q+1)-secant of K. For any (n-2)-space S_{n-2} in PG(n,q) we define $m(K,S_{n-2})$ as the number of tangent hyperplanes to K through S_{n-2}. If, for any S_{n-2} in PG(n,q), $m(K,S_{n-2}) \in \{0,1,\dots,m,q+1\}$ with $m<q+1$, then we say that K has Plücker class m.

In this paper we deal with k-sets in PG(3,q) of class $[1,q+1,2q+1]_2$ and Plücker class m=2. Taking into account the results in [15] and [21], the possible types for such a set are $(1,q+1,2q+1)_2$, $(1,q+1)_2$ and $(q+1,2q+1)_2$. Since the sets of type $(1,q+1)_2$ are the lines and the (q^2+1)-caps [21], it remains to consider the sets of types $(1,q+1,2q+1)_2$ and $(q+1,2q+1)_2$ and of Plücker class two. The next statement sums up the obtained results.

Theorem. Let K be a k-set in PG(3,q) of class $[1,q+1,2q+1]_2$ and Plücker class two containing ρ lines. Denote by t the number of tangent planes to K. Then one of the following occurs.

(i) If K is of type $(1,q+1,2q+1)_2$ with $\rho\geq 2$, then $\rho\geq 3$ and K is a cone projecting a (q+1)-arc of a plane from a point. Furthermore $\rho=3$ iff q=2.

(ii) If K is of type $(1,q+1)_2$, then K is either a line or a (q^2+1)-cap (i.e. an ovoid).

(iii) If K is of type $(q+1,2q+1)_2$ with $t>0$ and $\rho\geq 5$, then K consists of the points on q+1 pairwise skew lines, which either have one or two transversals or form a hyperbolic quadric.

1. k-SETS OF TYPE $(1,q+1,2q+1)_2$

In this section K is assumed to be a k-set in PG(3,q) of Plücker class two and type $(1,q+1,2q+1)_2$ with $\rho\geq 2$, i.e. in which there are at least two lines.

Consider a 1-secant plane H_o and let $p_o=K\cap H_o$. Since any line in K intersects H_o in a point, it must pass through p_o. Moreover, since any plane has at most 2q+1 points in common with K, no three lines contained in K can be coplanar. Consider first the case $\rho\geq 4$. Take four lines l_1,l_2,l_3,l_4 lying on K. Then the three distinct planes $\langle l_1,l_2\rangle$, $\langle l_1,l_3\rangle$, $\langle l_1,l_4\rangle$ are tangent to K. Since K is of Plücker class two, any plane through l_1 is tangent to K. Hence, for any point x on $K\setminus l_1$, the plane joining l_1 and x is tangent to K and meets K in two lines, namely l_1 and the line through p_o and x. Therefore, K is the union of ρ lines through p_o. Let H be a plane not through p_o. Then $K\cap H$ consists of the points in which the ρ lines on K through p_o meet H. Since no three of these lines are coplanar, $K\cap H$ is a ρ-arc. On the other hand, H is either a (q+1)-secant or a (2q+1)-secant plane of K; thus, $\rho=q+1$ and K is the cone projecting from p_o the (q+1)-arc $K\cap H$.

Next, we consider the case $\rho=2$. Let l,m be the two lines contained in K, and denote by H_1 the plane through l and m. Since $\rho=2$, no (2q+1)-secant plane different from H_1 is a tangent plane. Thus, any other possible tangent plane through l or m is a (q+1)-secant plane. We claim that neither l nor m is on three tangent planes; otherwise, any plane through l (or m, respectively) distinct from H_1 is a (q+1)-secant plane which is tangent to K. Thus, K consists only of the points on l and m, a contradiction since K is of type $(1,q+1,2q+1)_2$. Assume that there is a unique tangent plane $L\neq H_1$ through l. Then all other planes through l are (2q+1)-secant planes. Hence,

$$k = q^2+q+1 \qquad (1.1)$$

Consequently, by [15] eq. (18), the number t_{2q+1} of (2q+1)-secant planes of K equals

(1.2) $$t_{2q+1} = q(q+1)/2 \quad .$$

Next, take a 1-secant line r of K and denote by u,s,b the numbers of 1-secant,(q+1)-secant,(2q+1)-secant planes of K through r , respectively.Counting the points of K on the planes through r, we get in view of (1.1)

(1.3) $$b = u \quad .$$

Analogously,if r' is an n-secant of K , n 2 , and s' and b' denote the numbers of (q+1)-secant and (2q+1)-secant planes of K through r' ,respectively, then

(1.4) $$b' = n-1 \quad .$$

Each (2q+1)-secant plane other than H_1 meets H_1 in a line that, by (1.4) ,passes through the point p_o=l m . Consider now a 1-secant line r" through p_o in H_1 . Then,by (1.3),there is at least one 1-secant plane through r" . On the other hand, if there were two 1-secant planes through r" , then any plane through r" were a tangent plane and we would get a contradiction as above.Consequently, through any line through p_o in H_1 other than l and m there is a unique 1-secant plane. Hence,by (1.3) , t_{2q+1} = 2q+1 . Comparing this equality with (1.2) , we have q=2 ; moreover, K consists of seven points in PG(3,2) .Five one of these points are on the lines l and m, the remaining two are off the plane H_1 and non-collinear with p_o .Since this configuration is not type $(1,q+1,2q+1)_2$,we have a contradiction.

Now we assume that H_1 is the only tangrnt plane through l . Thus, each plane through l is a (2q+1)-secant plane and then

(1.5) $$k = (q+1)^2 \quad .$$

Now,equation (18) in 15 implies that there is no 1-secant plane, a contradiction.

Finally,we deal with the case =3. Let l,m and r be the three lines contained in K ; denote by H_1 , H_2 , H_3 the planes joining l and m , l and r and m and r ,respectively.Assume that through one of these lines, say l , there is a (q+1)-secant plane. Since such a plane is tangent to K , any plane through l distinct from H_1 and H_2 is a tangent (q+1)-secant plane. Thus, K is the union of three concurrent non-coplanar lines, and so q=2. On the other hand if any plane through l, m and r is a (2q+1)-secant plane, then $k=(q+1)^2$ and a counting argument shows that there are no 1-secant planes, a contradiction (compare 15 eq.s (18)).

Thus we have proved the following statement

<u>Theorem I.</u> Let K be a k-set of class $(1,q+1,2q+1)_2$ and Plücker class two in PG(3,q) such that 2 .Then 3 and K is the cone projecting from a point a (q+1)-arc.Furthermore, =3 iff q=2.

2. SETS OF TYPE $(q+1,2q+1)_2$.

In this section, K denotes a k-set of type $(q+1,2q+1)_2$ and Plücker class two in PG(3,q) satisfying O .

<u>Proposition 2.1</u> .We have $k=(q+1)^2$. Furthermore,each tangent plane is a (2q+1)-secant plane.

Proof. Since the tangent planes of K meet K in either one or two lines, it is sufficient to prove that the planes through a line of K are $(2q+1)$-secant planes. In our situation, the equation (22) of [15] reads

$$\vartheta_1 k^2-((m+n)\vartheta_2-q^2)k+mn\,\vartheta_3 = 0 .$$

For $m=q+1$, $n=2q+1$, this equation has the unique integral solution $k=(q+1)^2$. Counting the points of K on the planes through a line in K we get that all these planes are $(2q+1)$-secant.

Proposition 2.2. Through any n-secant line l there are precisely n $(2q+1)$-secant planes $(0\leq n\leq q+1)$.

Proof. Let b be the number of $(2q+1)$-secant planes through l. Counting the points of K on the planes through l, we get $k=q^2+(b+2-n)q+1$. It follows $b=n$, in view of proposition 2.1.

Proposition 2.3. Through any point on K there are at most two lines contained in K.

Proof. Assume that there are three concurrent lines l,m,r contained in K. Let p be their common point and denote by H the plane through l and m. Any line $r'\neq l,m$ through p in H is a 1-secant. Hence, by proposition 2.2, H is the unique $(2q+1)$-secant plane through it. Thus, the plane H' through r and r' is a $(q+1)$-secant plane. Considering the planes through r, we get therefore $k=3q+1$, which contradicts proposition 2.1.

Now we are ready to prove the following

Theorem II. Let K be a k-set in $PG(3,q)$ of type $(q+1,2q+1)_2$ and Plücker class two with $t>0$. Then $k=(q+1)^2$ and all the tangent planes are $(2q+1)$-secant planes. Furthermore, if $\rho\geq 5$ then either K consists of $q+1$ mutually skew lines with either one or two transversals, or K is a hyperbolic quadric.

Proof. Denote by D the set of points on K which are on two lines contained in K. Fix a tangent plane H_o which meets K at the lines l_o and m_o and let $p_o = l_o\cap m_o$. Set $D_o = D\cap H_o$ and $d_o = |D_o|$. Since $p_o\in D_o$, $d_o\geq 1$. Since $\rho\geq 5$, there are at least three lines l_1,l_2,l_3 different from l_o and m_o contained in K. These lines meet l_o or m_o; moreover, by proposition 2.3, the points of intersection are distinct and distinct from p_o. Consequently, $d_o\geq 4$. So, there are at least two points a,b in D_o, which are collinear with p_o, say $a,b\in l_o$. Since K is of Plücker class two, each plane through l_o is tangent. Hence, by proposition 2.3, $l_o\subseteq D_o$ which implies $d_o\geq q+1$. If $d_o=q+1$, then, by proposition 2.1, K consists of $q+1$ mutually skew lines with a unique transversal, namely l_o. If $d_o = q+2$, then D_o consists of the points on l_o and a unique point p_1 other than p_o on m_o. Denote by m the line through p_1 lying on K and other than m_o; l_o and m are skew and the $q+1$ lines of K on the planes through l_o meet both l_o and m. Consequently, K consists of $q+1$ pairwise skew lines having just two transversals, namely l_o and m. Finally, suppose $d_o\geq q+3$. Then, also $m_o\subseteq D_o$. Moreover, any line $l\neq l_o$ in K through a point of l_o meets any line $m\neq m_o$ in K through a point of m_o. Hence, K is a hyperbolic quadric.

REFERENCES

[1] Bichara, A., Sui k-insiemi di $S_{3,q}$ di tipo $((n-1)q+1,nq+1)_2$, Rend. Acc. Naz. Lincei, (8), 62 (1977) 480-488.

[2] Bichara, A., Sui k-insiemi di PG(r,q) di classe $[0,1,2,n]_2$, Rend. di Mat., Roma, 13(1980).

[3] Buekenhout, F., Ensembles quadratiques des espaces projectifs, Math. Z., 110 (1969) 306-318.

[4] de Finis,M. and de Resmini,M.J.,On a characterization of subgeometries PG(r, q) in PG(r,q),q a square,Europ.Jrnl.Comb.,3(1982) 319-328.

[5] de Finis,M.,On k-sets in PG(3,q) of type (m,n) with respect to planes,to appear.

[6] Ferri,O.,Su di una caratterizzazione grafica della superficie di Veronese di un $S_{5,q}$,Rend.Acc.Naz.Lincei,VIII,vol.LXI,6(1976)603-610.

[7] Hirschfeld,J.W.P. and Thas,J.A.,The characterization of projection of quadrics over finite fields of even order,Jrnl.London Math.Soc.,22(1980) 226-238.

[8] Hirschfeld,J.W.P. and Thas,J.A.,Sets of type (1,n,q+1) in PG(d,q),Proc.London Math.Soc.,41(1980)254-278.

[9] Melone,N.,The linear line geometry in PG(3,q) from a synthetic point of view, Pubbl.Ist.Mat."R.Caccioppoli",Univ.Napoli,38(1983)1-15.

[10] Segre,B.,Ovals in a finite projective plane,Canad.Jrnl.Math.,7(1955)414-416.

[11] Segre,B.,Curve razionali normali e k-archi negli spazi finiti,Ann.Mat.Pura e Appl.,(4),39(1955)357-379.

[12] Segre,B.,Le geometrie di Galois,Ann.Mat.,(4),48(1959)1-97.

[13] Tallini,G.,Sulle k-calotte di uno spazio lineare finito,Ann.Mat.(4),42(1956) 119-164.

[14] Tallini,G.,Caratterizzazione grafica delle quadriche ellittiche negli spazi finiti,Rend.Mat.,Roma,16(1957)328-351.

[15] Tallini,G.,Problemi e risultati sulle geometrie di Galois,Pubbl.Ist.Mat. "R.Caccioppoli",Univ.Napoli,30(1973).

[16] Tallini,G.,k-insiemi e blocking sets in PG(r,q) e in AG(r,q),Sem.Geom.Comb. Ist.Mat.Appl.Univ.Aquila,1(1982).

[17] Tallini Scafati,M.,Caratterizzazione grafica delle forme hermitiane di un $S_{r,q}$,Rend.Mat.,Roma,26(1976)273-303.

[18] Tallini Scafati,M.,Calotte di tipo (m,n) in uno spazio di Galois $S_{r,q}$, Rend.Acc.Naz.Lincei,(8),53(1973)71-81.

[19] Tallini Scafati,M.,Sui k-insiemi di uno spazio di Galois $S_{r,q}$ a due caratteri nella dimensione d ,Rend.Acc.Naz.Lincei,(8),40(1976)782-788.

[20] Tallini Scafati,M.,La teoria dei k-insiemi negli spazi di Galois,Sem.Geom. Comb.Univ.Roma,40(1982).

[21] Thas,J.A.,A combinatorial problem,Geom.Ded.,1(1973)236-240.

Annals of Discrete Mathematics 30 (1986) 105–106

A FREE EXTENSION PROCESS YIELDING A PROJECTIVE GEOMETRY

Flavio Bonetti
Dipartimento di Matematica - Via Machiavelli, 35 - 44100 Ferrara

Nino Civolani
Facoltà di Scienze - Università della Basilicata - 85100 Potenza

Summary. Presented is a free extension process, mainly based on the configuration of Veblen and yielding a projective geometry.

The basic definitions can be found e.g. in the following sources: partial plane, Desargues' condition (resp. configuration), projective plane, in [1]; (reducible) projective geometry, dimension, in [2]; free extension (resp. completion) process, free projective plane, in [3].

Let $\mathfrak{T} = (\mathfrak{P}, \mathfrak{L}, \mathrm{I})$ be a partial plane. The free projective plane $\mathfrak{F}(\mathfrak{T})$ can be associated to $\mathfrak{T}$ by the well-known free completion process $\mathfrak{F}$. We will describe another free extension process, $\mathfrak{F}^{PV}$, yielding a projective geometry $\mathfrak{F}^{PV}(\mathfrak{T})$, possibly reducible with dimension $\geqslant 3$, hence «mostly» different from the free projective plane $\mathfrak{F}(\mathfrak{T})$.

To this end we utilize the notion of *Veblen configuration* $\{c, a, a', b, b', A, B, C, C'\}$ (see Fig. 1), i.e. five distinct points $c, a, a', b, b' \in \mathfrak{P}$ and four distinct lines $A, B, C, C' \in \mathfrak{L}$ such that: c, a, a' I A; c, b, b' I B; a, b, I C; a', b' I C'. The two lines C, C' are the *entering* ones of the Veblen configuration, which is *closed* if they meet in a point.

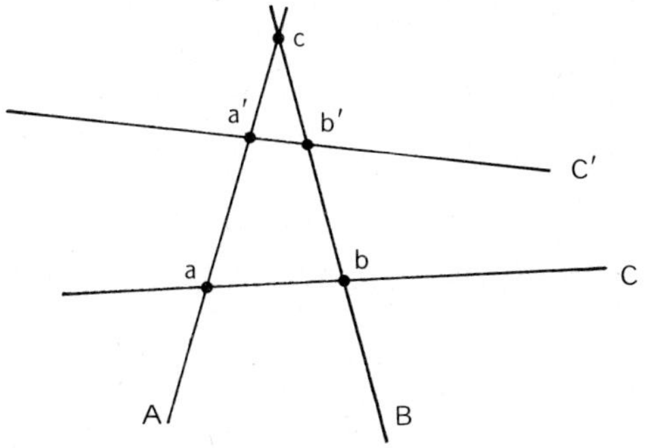

Fig. 1

Now we define the free extension process $\mathfrak{F}^{PV} = (\mathfrak{T}_n^{PV})_{n \geqslant 0}$:

$$\mathfrak{T}_0^{PV} = \mathfrak{T}$$

$$\mathfrak{T}_{n+1}^{PV} = (\mathfrak{P}_{n+1}^{PV}, \mathfrak{L}_{n+1}^{PV}, \mathrm{I}_{n+1}^{PV}), \text{ where}$$

$\mathfrak{P}_{n+1}^{PV} = \mathfrak{P}_n^{PV} \cup \{\{C, C'\} \mid C, C' \in \mathfrak{L}_n^{PV}$ are the two entering lines of a non-closed Veblen configuration of $\mathfrak{T}_n^{PV}\}$;

$\mathfrak{L}_{n+1}^{PV} = \mathfrak{L}_n^{PV} \cup \{\{p, p'\} \mid p, p' \in \mathfrak{P}_n^{PV}$ are distinct points not joined by any line of $\mathfrak{L}_n^{PV}\}$;

I^{PV}_{n+1} consists of the pairs of I^{PV}_n and also of those originated by the new elements of $\mathfrak{P}^{PV}_{n+1}$, $\mathfrak{L}^{PV}_{n+1}$, namely ({C, C'}, C), ({C, C'}, C'), (p, {p, p'}), (p',{p, p'}).

$$\mathfrak{F}^{PV}(\mathfrak{T}) = \left(\bigcup_{n \geqslant 0} \mathfrak{P}^{PV}_n, \bigcup_{n \geqslant 0} \mathfrak{L}^{PV}_n, \bigcup_{n \geqslant 0} I^{PV}_n \right)$$

Clearly $\mathfrak{F}^{PV}(\mathfrak{T})$ is a projective geometry, and $\mathfrak{F}^{PV}(\mathfrak{T}) = \mathfrak{T}$ if and only if $\mathfrak{T}$ is a projective geometry.

It is also easily seen that, if $\mathfrak{T}$ is a non-closed Veblen configuration, then at each stage $\mathfrak{T}_{2n-1}$ $(n \geqslant 1)$ there appears at least one new point.

The next results about this extension process are summarized by the following

THEOREM. *Let $\mathfrak{T}$ be a partial plane, with $\mathfrak{F}^{PV}(\mathfrak{T}) \neq \mathfrak{T}$. Then:*

i) $\mathfrak{F}^{PV}$ *has infinitely many distinct stages* $\mathfrak{T}^{PV}_n$;
ii) *dim* $\mathfrak{F}^{PV}(\mathfrak{T}) \geqslant 3 \Rightarrow \mathfrak{F}^{PV}(\mathfrak{T})$ *is reducible;*
iii) *dim* $\mathfrak{F}^{PV}(\mathfrak{T}) = 2 \Rightarrow \mathfrak{F}^{PV}(\mathfrak{T}) = \mathfrak{F}(\mathfrak{T})$.

Proofs are straightforward.

REFERENCES

[1] Dembowski,P., Finite Geometries (Springer, Berlin-Heidelber-New York; 1968).

[2] Dubreil-Jacotin, M.L., Lesieur, L. and Croisot, R., Leçons sur la Théorie des Treillis, des Structures algébriques ordonnées et des Treillis géométriques (Gauthier-Villars, Paris, 1953).

[3] Siebenmann, L.C., A Characterization of free projective Planes, Pacific J. Math. 15 (1965) 293 - 298.

Annals of Discrete Mathematics 30 (1986) 107–114

SYMMETRIC FUNCTIONS AND SYMMETRIC SPECIES

FLAVIO BONETTI — Università di Ferrara
GIAN-CARLO ROTA (*) — M.I.T. Boston
DOMENICO SENATO — Università di Napoli
ANTONIETTA M. VENEZIA — Università di Roma 1

INTRODUCTION

The idea of proving identities for symmetric functions by bijective arguments is quite old; it goes back to Lucas (Théorie des nombres, 1891) and probably earlier. To the best of our knowledge, the first glimmerings of a systematization of such bijective arguments goes back to one of the present author (cf. [9]); the idea was further developed by R.P. Stanley, wo gave a bijective proof of Waring's formula by Möbius inversion on the lattice of partition of a set, and later by Doubilet, who gaves bijective proofs of several identities in the theory of symmetric functions.

Joyal's theory of species led us to develop a systematic setting for such bijective proofs. We introduce here the notion of *symmetric species,* which can be viewed as a set-theoretic (a category-theoretic) counterpart of the notion of a symmetric function. To each of the classical classes of symmetric functions we associate a symmetric species. Operations on species, as introduced by Joyal, are generalized to symmetric species, and simple categorical operations yielded bijective proofs of all identities among elementary symmetric functions. By way of example, we give a bijective proof of Waring's formula, which we believe to be new, and dispenses altogether with Möbius inversion, as well as bijective proofs of several related identities.

This note is part of a communication presented in Bari at «Combinatorics 84».

1. DEFINITIONS AND PRELIMINARIES

We denote by Φ the category of finite sets and bijections, and we denote by:

$$I^* : \Phi \to \Phi$$

the *contravariant identity functor,* mapping every finite set E to itself, and such that, if:

$$u : E \to F$$

is a bijection, then $I^*[u] = u^{-1}$.

Recall that a species (Joyal) is a functor

$$M : \Phi \to \Phi .$$

We shall follow Joyal's terminology for species.

Let **X** be an infinite set, which will remain fixed throughout, whose elements will be called *variables.* To agree with current usage, we may occasionaly list the variables in linear order $x_1, x_2, \ldots$ through this listing is strictly speaking irrelevant.

Recall, that, if N* is a contravariant functor of Φ to Φ, the functor N from Φ to the category of sets whose objects are

(*) Research supported by N S F contract n. MCS 8104855.

$$N[E] = \mathrm{Hom}\,(N^*[E], \mathbf{X})$$

is covariant.

Indeed, if $u \in \mathrm{Hom}\,(E, F)$ and if $N^*[u] : N^*[F] \to N^*[E]$ then $N[u] : N[E] \to N[F]$ is defined by $N[u](f) = f \circ N^*[u]$.

Let M be a species, and let N^* be a contravariant functor of Φ to Φ. We denote by Pol (M, N^*) the functor defined as:

$$\mathrm{Pol}\,(M, N^*)\,[E] = M[E] \times \mathrm{Hom}\,(N^*[E], \mathbf{X})$$

and whose morphisms are, for $u : E \to F$,
$(M[u], N[u]) : M[E] \times \mathrm{Hom}\,(N^*[E], \mathbf{X}) \to M[F] \times \mathrm{Hom}\,(N^*[F], \mathbf{X})$.

A *polynomial spacies* **P** is a subfunctor of Pol (M, N^*), that is, for every object in Pol (M, N^*) it is a subset

$$\mathbf{P}[E] \subset M[E] \times \mathrm{Hom}\,(N^*[E], \mathbf{X})$$

such that if $u \in \mathrm{Hom}\,(E, F)$ and if $(s, f) \in \mathbf{P}[E]$ then

$$(M[u]\,(s),\ f \circ N^*[u]) \in \mathbf{P}\,[F].$$

In other words, the subset **P**[E] is functorially assigned.

When $N^* = \mathbf{I}^*$, we say that the polynomial species is *ordinary*.

Let $\varphi : \mathbf{X} \to \mathbf{X}$ be an isomorphism and let **P** be a polynomial species. We denote by $\mathbf{P}_\varphi$ the polynomial species defined by:

$$(s, f) \in \mathbf{P}_\varphi[E] \subset M[E] \times \mathrm{Hom}\,(N^*[E], \varphi(\mathbf{X})) \Longleftrightarrow (s,\ \varphi^{-1} \circ f) \in \mathbf{P}[E].$$

Clearly, $\mathbf{P} \to \mathbf{P}_\varphi$ is a natural trasformation of functors.

A polynomial species is a *symmetric species* when $\mathbf{P} = \mathbf{P}_\varphi$ for every isomorphism $\varphi : \mathbf{X} \to \mathbf{X}$.

EXAMPLE 1. The *elementary symmetric species* **E**.

E[E] = the set of all monomorphisms from E to $\mathbf{X} \subset$ Pol $(\mathbf{I}, \mathbf{I}^*)$

where I is the identity functor.

EXAMPLE 2. The *power-sum species* **S**

S[E] = the set of all functions from E to **X** of constant value $\subset$ Pol $(\mathbf{I}, \mathbf{I}^*)$.

EXAMPLE 3. The *disposition species* **H**.

Let S[E] be the set of all permutations on E. Take M = Exp (S) and let $r \in$ Exp (S) [E] (i.e. r consists of a partition π and a permutation on each block of π).

We let **H**[E] be the subset of

$$\mathrm{Exp}\,(S)\,[E] \times \mathrm{Hom}\,(E, \mathbf{X})$$

of all pairs (r, f) such that π is the kernel of the function $f : E \to \mathbf{X}$, that is, such that the blocks $B \in \pi$ are the sets $f^{-1}(x)$ whenever $f^{-1}(x)$ is non-empty, as x ranges on **X**.

This defines the species of dispositions.

EXAMPLE 4. $\mathbf{K}_\lambda$ = *the monomial elementary species of class* λ.

Let $\pi = \{B_1, \ldots, B_k\}$ be a partition of E, $|E| = n$, and let r_i $(1 \leqslant i \leqslant n)$ be the number of blocks of π with i elements. The class of π is the partition of the integer n defined by:

$$\mathrm{cl}(\pi) = (1^{r_1}, 2^{r_2}, \ldots, n^{r_n}) = (\lambda_1, \lambda_2 \ \ldots, \lambda_k)$$

where $\lambda_i = |B_i|$.

Let λ be a class (i.e. a partition of n), and let $M_\lambda[E]$ be the set of all partitions of class λ on set E.

We define $\mathbf{K}_\lambda[E]$ to be the symmetric species of all pairs (π, f) where π is the kernel of the function f. We call this the *monomial elementary species of class* λ.

EXAMPLE 5. The *cyclic species* C.

Let C[E] be the set of all cyclic permutations on E. We set C[E] to be the cyclic polynomial species on Pol (C, I*) of all pairs (μ, f) where $\mu \in C[E]$ and f is constant.

EXAMPLE 6. *The species* $\mathbf{H}_x$

$\mathbf{H}_x[E] = \{(\sigma, f) \mid \sigma \in S[E] \text{ and } f(e) = x \text{ for all } e \in E\} \subset S[E] \times \mathrm{Hom}(\mathbf{I}^*[E], \mathbf{X})$.

2. THE POLYNOMIAL OF POLYNOMIAL SPECIES

Let $\mathbf{Z}[\mathbf{X}]$ be the ring of all polynomials in the variables $x \in \mathbf{X}$.

We again denote by x the canonical image of x in the ring $\mathbf{Z}[\mathbf{X}]$.

Let C be a cofinite subset of $\mathbf{X}$. We write, if $p \in \mathbf{Z}[\mathbf{X}]$, $p/_{C=0}$ to denote the polynomial obtained from p by setting to zero all $x \in C$. If $p \in \mathbf{Z}[\mathbf{X}]$ and C is a cofinite subset of $\mathbf{X}$, we let

$$\mathbf{A}(p; C)$$

be the set of all $q \in \mathbf{Z}[\mathbf{X}]$ such that

$$p/_{C=0} = q/_{C=0}.$$

This defines a topology on the ring $\mathbf{Z}[\mathbf{X}]$. The completion of $\mathbf{Z}[\mathbf{X}]$ in this topology is denoted by $\mathbf{Z}[[\mathbf{X}]]$. Thus, an element $r \in \mathbf{Z}[[\mathbf{X}]]$ is an infinite sum of polynomials such that for every cofinite set C, $r/_{C=0}$ is an ordinary polynomial in $\mathbf{Z}[\mathbf{X}]$.

Let $\{p_C\}$ be a set of polynomials in $\mathbf{Z}[\mathbf{X}]$. We write

$$\lim_C p_C = r$$

when the set $\{p_C\}$ converges to r in $\mathbf{Z}[[\mathbf{X}]]$ along the filter of all cofinite sets.

A sufficient condition that ensures that $\{p_C\}$ converge is that, for $C' \subset C$

$$p_{C'}/_{C=0} = p_C.$$

The element of $\mathbf{Z}[[\mathbf{X}]]$ defined as

$$\mathrm{gen}(f) = \prod_{e \in E} \hat{f}(e) = \prod_{x \in \mathbf{X}} x^{|f^{-1}(x)|}$$

is called the *generating polynomial* of the function f (cf. [9]).

If $\mathbf{P}$ is a polynomial species on Pol (M, N*) we write:

$$\text{gen}(\mathbf{P}[E]) = \sum_{(s,f)\in \mathbf{P}[E]} \text{gen}(f).$$

Noting that gen ($\mathbf{P}[E]$) depends only on the cardinality n of E, we write

$$\text{gen}(\mathbf{P}, n) = \text{gen}(\mathbf{P}[E])$$

for any set E of cardinality n. We call this the *n-th polynomial coefficient* of the polynomial species **P**.

The generating function of a polynomial species **P** is the element

$$\text{Gen}(\mathbf{P}, z) = \sum_{n\geq 0} \text{gen}(\mathbf{P}, n) \frac{z^n}{n!}$$

of the algebra of formal power series in the variable z over $\mathbf{Z}[[\mathbf{X}]]$.

PROPOSITION 1. Let **P** be a symmetric species. Then the polynomial gen (**P**, n) is symmetric, in other words gen (**P**, n) is invariant for any bijection of **X**.

EXAMPLE 1. (cont. d.). The polynomial of the elementary symmetric species **E** is

$$\text{gen}(\mathbf{E}, n) = n! \sum_{i_1<\ldots<i_n} x_{i_1} x_{i_2} \ldots x_{i_n} = n!\, e_n(x_1, \ldots)$$

that is, except for the factor n!, it is the n-th elementary symmetric function. ■

EXAMPLE 2. (cont. d.). The polynomial of the power sum species **S** is:

$$\text{gen}(\mathbf{S}, n) = \sum_i x_i^n$$

that is the power-sum symmetric function s_n ■

EXAMPLE 3. (cont. d.). The polynomial of the disposition species **H** is:

$$\text{gen}(\mathbf{H}, n) = n!\, h_n(x_1, \ldots)$$

with

$$h_n(x_1, \ldots) = \sum_{i_1\leq\ldots\leq i_n} x_{i_1} \ldots x_{i_n}$$

that is h_n is the elementary homogeneous function of degree n ■

EXAMPLE 4. (cont. d.). The polynomial of the monomial elementary species $\mathbf{K}_\lambda$ of class λ is:

$$\text{gen}(\mathbf{K}_\lambda, n) = p(n, \lambda)\, r_1!\, r_2! \ldots k_\lambda$$

where p(n, λ) is the number of the partitions of class λ on E, k_λ is equal to $\Sigma n_{i_1}^{\lambda_1} \ldots n_{i_k}^{\lambda_k}$, and the sum ranges over all distinct monomials.

Except for a coefficient, gen ($\mathbf{K}_\lambda$, n) is a monomial elementary symmetric function ■

EXAMPLE 5. (cont. d.). The polynomial of the cyclic species **C** is

$$\text{gen}(\mathbf{C}, n) = (n-1)! \sum_i n_i^n$$

and the generating function of the polynomial species **C** is:

$$\text{Gen}(\mathbf{C}, z) = \sum_{n \geqslant 0} s_n \frac{z^n}{n} \blacksquare$$

EXAMPLE 6. (cont. d.) The polynomial of the species $\mathbf{H}_x$ is:

$$\text{gen}(\mathbf{H}_x, n) = n!\, x^n$$

Thus

$$\text{Gen}(\mathbf{H}_x\ z) = \frac{1}{1 - xz} \blacksquare$$

Sum, product and exponential of polynomial species are defined as in Joyal.

We recall the definition of product and exponential for ordinary species.

Let $\mathbf{P}_i$ be a species on Pol (M_i, I^*) $(i = 1, 2)$. We let *the product* $\mathbf{P}_1 \times \mathbf{P}_2$ be the polynomial species on Pol $(M_1 \times M_2, I^*)$ such that:

$$(\mathbf{P}_1 \times \mathbf{P}_2)[E] = \bigcup_{E_1 + E_2 = E} \Big\{ (s, f) \mid \exists (s_1, f_1) \in \mathbf{P}_1[E_1] \text{ and } (s_2, f_2) \in \mathbf{P}_2[E_2] \text{ such that}$$
$$s = (s_1, s_2) \text{ and } f|_{E_1} = f_1,\ f|_{E_2} = f_2 \Big\}$$

where $E_1 + E_2 = E$ ranges the set of all 2-scomposition of E.

PROPOSITION 2.

$$\text{Gen}(\mathbf{P}_1 \times \mathbf{P}_2, z) = \text{Gen}(\mathbf{P}_1, z) \cdot \text{Gen}(\mathbf{P}_2, z).$$

Let $\mathbf{P}$ be a polynomial species on Pol (M, I^*) without constant term (i.e. $\mathbf{P}[\phi] = \phi$) and let $\pi = \{B_1, \ldots, B_k\}$ be a partition of E in k blocks. An *assembly on* E *of order* k *of species* $\mathbf{P}$ is defined as the set of all pairs (s, f) such that, if s_i represents the structure induced by s on B_i, we have

$$(s_i, f/_{B_i}) \in \mathbf{P}[B_i], \text{ for } i = 1, \ldots, k.$$

The species of assemblies of species $\mathbf{P}$ *of order* k is the polynomial species $\text{Exp}_k(\mathbf{P})$ on Pol $(\text{Exp}_k(M), I^*)$ defined as follows:

$\text{Exp}_k(\mathbf{P})[E]$ = the set of all polynomial assemblies of species $\mathbf{P}$ on E of order k, where the partition ranges over the set of all partitions of E into k blocks.

The species of the assemblies of species $\mathbf{P}$ is the polynomial species on Pol (Exp(M), I*) defined by:

$$\text{Exp}(\mathbf{P})[E] = \bigcup_{k \geqslant 1} \text{Exp}_k(\mathbf{P})[E]$$

PROPOSITION 3. (Theorem of the assemblies).

$$\text{Gen}(\text{Exp}(\mathbf{P}), z) = e^{\text{Gen}(\mathbf{P}, z)}.$$

For present purposes, a family $\{\mathbf{P}_x\}_{x \in X}$ of polynomial species will be called *multipliable* if the following conditions are satisfied:

(i) $\mathbf{P}_x \subset \text{Pol}(M, I^*)$;

(ii) $\mathbf{P}_x[\phi] = \{(s, f : \phi \to \mathbf{X})\}, s \in M[\phi]$;

(iii) $(s, f) \in \mathbf{P}_x[E]$ iff $f(E) = \{x\}$.

If Comp (C) denotes the complement of the cofinite set C and $\mathbf{P}_C = \prod_{x \in \mathrm{Comp}(C)} \mathbf{P}_x$ we define *the product of the family* $\{\mathbf{P}_x\}_{x \in \mathbf{X}}$ as the $\lim_C \mathbf{P}_C$.

PROPOSITION 4.

$$\mathrm{Gen}\,(\lim_C \mathbf{P}_C, z) = \lim_C\,(\mathrm{Gen}\,(\mathbf{P}_C, z))$$

EXAMPLE. There exists a natural transformation from the species **H** of the disposition to the product species of the family $\{\mathbf{H}_x\}_{x \in \mathbf{X}}$.

As a consequence we obtain the classical identity:

$$\prod_{x \in X} \frac{1}{1 - xz} = \sum_{n \geqslant 0} h_n z^n .$$

3. A PROOF OF WARING'S FORMULA

PROPOSITION 5. There is a natural transformation between the species **H** of dispositions and the species Exp (C) of assemblies of cyclic species, i.e.

$$\mathbf{H} = \mathrm{Exp}\,(\mathrm{C})$$

Proof.

The following algorithm gives a canonical bijection between dispositions over a finite set E and the cyclic species assemblies on the same finite set.

STEP 1.

Let $(\{\sigma_i\}, f) \in \mathbf{H}[E]$ be a disposition on E. We shall write μ_i^j for to denote the cycles of σ_i and $\overline{\mu}_i^j$ the supports of these cycles. The assembly of associated cyclic species is obtained in the following way: the partition of E is that in which the blocks are the $\overline{\mu}_i^j$, on each block the cyclic permutation μ_i^j and the function $f/_{\overline{\mu}_i^j}$ with constant value x_i are defined.

STEP 2.

Given an assembly of cyclic species (s, f) relative to a partition $\pi = \{\overline{\nu}_1, \ldots, \overline{\nu}_k\}$. More explicitly, on each block $\overline{\nu}_j$ a cyclic permutation ν_j and a constant function are defined. The disposition associated with (s, f) is the fair $(\{\sigma_x\}, f)$ with σ_x a permutation whose cycles are those ν_j on the supported of which the function f assumes the constant value.

As a special case of propositions 3 and 5 we have the following:

THEOREM (Waring's formula)

$$\sum_{n \geqslant 0} h_n z^n = e^{s_1 z + s_2 z^2/2 + \ldots}$$

REFERENCES

[1] Aigner, M., Combinatorial theory, Springer-Verlag, New York (1979).

[2] Bourbaki, N., Element de Mathématique: Algèbre Commutative, Hermann, Paris (1965).

[3] Comtet, L., Advanced Combinatorics, Reidel, Dordrecht-Holland, Boston (1974).

[4] Doubilet, P., On the foundations of Combinatorial theory. VII: Symmetric functions through the theory of Distribution and Occupancy, Studies in App. Math. Vol. 51 (1972).

[5] Grätzer, G., Universal Algebra, D. Van Nostrand, Princeton, N.J. (1968).

[6] Joyal, A., Une théorie Combinatoire des séries formelles, Adv. in Math. v. 42 (1981).

[7] MacDonald, I.G., Symmetric functions and Hall polynomials, Clarendon Press, Oxford 1979.

[8] Metropolis, N., Rota, G.C., Witt vectors and the algebra of necklace, Adv. in Math. vol. 50 (1983).

[9] Rota, G.C., Baxter algebras and combinatorial identities I e II. Bull. Amer. Math. Soc. (1969).

[10] Rota, G.C., Finite operator Calculus, Academic Press, Inc. (1975).

Annals of Discrete Mathematics 30 (1986) 115–124

ON THICK (Q+2)-SETS

Rita Capodaglio Di Cocco

Universita' di Bologna

Summary: Un k-insieme K di un piano proiettivo finito viene detto denso se da ogni punto del piano esce almeno una s-secante di K, con $s>1$. Qui si studiano alcune proprieta' dei (q+2)-insiemi densi di un piano proiettivo d'ordine dispari.

INTRODUCTION

About forty years ago, Bose [5] and Qvist [20] introduced some subsets of a finite projective plane, called " non collinear systems of points". B. Segre and his school studied the same subsets, renamed k-arcs, and found very important results about them. In particular, if C is a k-arc of the plane PG(2,q) with q odd, we recall (see [28] p. 270-298) :

1) $k<q+2$
2) if $k=q+1$, then C is a conic
3) a q-arc cannot be complete (this proposition is still true if q is even [33], while it is wrong [2] in a non-desarguesian plane).
4) if $k=q-c$, where $c>0$, and is "large enough relative to c", then every k-arc is contained in a conic.
5) if C is complete, then its secants fill up the plane, i.e. each point of the plane is contained in at least one secant of C.

There arises the problem: for which values of k and q do complete k-arcs in PG(2,q) exist? As an answer, L. Lombardo Radice constructed complete $\left[(q+5)/2\right]$-arcs, where $q>5$, Korchmaros [17] constructed complete k-arcs with $\left[(q-1)/4\right]+3 \leqslant k \leqslant \left[(q-1)/2\right]+3$, and Pellegrino [19] showed that if q is a convenient odd number, then complete k-arcs with $k \leqslant (q+1)/2$ exist. Among the k-arcs which are not necessarly complete but are not contained in a conic, we recall the k-arcs contained in a cubic (see Di Comite [13], [14], [15] and Zirilli [37]).

It easy to see that the order of the known complete k-arcs is very "large" with regard to the theoretical valuation $k>(3+\sqrt{1+8q})/2$ (see [28]) and we are still a good way from finding a solution of the following problem: If q is fixed, what is the smallest number k, such that a complete k-arc of PG(2,q) exists?

Recently, in order to obtain new results, the definition of k-arc has been generalized as follows (see [3]).

DEFINITION 1: A k-set K of a finite projective plane $\mathscr{P}$ (not necessary desarguesian) is called thick if $\forall P \in \mathscr{P}$ there exists a s-secant of K passing through P with s>1.

DEFINITION 2: A thick k-set K is called minimal if every proper subset of K is not thick.

Obviously every complete k-arc is a minimal thick k-set, but the next examples show that the converse is wrong:
1) Let r and s be distinct lines of a projective plane of order $q>5$. If $P=r\cap s$, let $A_1, A_2, \ldots, A_{q-1}$ be points of r and B_1, B_2 be points of s, with $A_i \neq P \neq B_j$ $(i=1,2,\ldots,q-1; j=1,2)$. Then the set $K=\{A_1, A_2, \ldots, A_{q-1}, B_1, B_2\}$ is a minimal thick (q+1)-set.
2) Let r,s,t be three non-concurrent lines of PG(2,q). If we embed PG(2,q) in $PG(2,q^2)$, the set of the points of r,s,t is a minimal 3q-set of $PG(2,q^2)$. (E. Ughi's example).

In the sequel, we will be interested in the minimal thick k-sets for which k takes the maximal value, i.e. q+2 (see [3]).
We point out that minimal thick (q+2)-sets exist: a (q+2,q+1)-arc is a trivial example of such sets. Moreover we will show that a minimal thick (q+2)-set can be represented by a permutation polynomial, and so the study of these sets is connected to a subject to which many important papers have been devoted (see [6], [8], [9], [10], [11], [12]).

DEFINITION 3: A point N of a k-set K is called a nucleus of K if every line through N is a s-secant of K with s<3.

A (q+2)-set with a nucleus is obviously thick, but not necessarly minimal: for example let $q=p^h$, with $p\neq 2,3$ odd and $q\not\equiv 1 \pmod{3}$. In PG(2,q) assume Γ is an irreducible cubic with an isolated double point N. It is easy to see that Γ is a non-minimal (q+2)-set with nucleus N. Moreover if q=5, and F is a point of inflection of Γ, then $\Gamma - \{F\}$ is a minimal thick 6-set. Remark: Irreducible cubics with an isolated double point are used in [15] to construct (q+9)/2-arcs.
So it seems that the following problems are the most important in the theory of minimal thick (q+2)-sets:
I: Has every minimal thick (q+2)-set one and only one nucleus?
II: For which number n is a minimal thick (q+2)-set K a (q+2,n)-arc?
III: When is a (q+2)-set K with a nucleus minimal?
In the following we suppose that the order of the plane is odd and we give partial answers to these problems. In particular in:
problem I: we show that a minimal thick (q+2)-set has at most one nucleus. Moreover, if the plane is PG(2,q), we find conditions for the nucleus to exist.

problem II: we show that if K has a nucleus, then either $n=q+1$ or $n<q-1$. Moreover if the plane is $PG(2,q)$, then $4 \leq n \leq q/3$.
problem III: we find necessary conditions for K to be a minimal (q+2)-set of $PG(2,q)$.

I

Let $\mathscr{P}$ be any finite projective plane of odd order q.

DEFINITION 4: A point A of a thick k-set K is called essential if the set $K-\{A\}$ is not thick.

Obviously, a thick k-set K is minimal if and only if every point of K is essential. Suppose now that K is a minimal thick (q+2)-set, then for each $X \in K$ there esists at least a point P_X such that every line through P_X is an 1-secant for $K-\{X\}$. If X is not a nucleus of K, then $P_X \neq X$, whereas if K has a nucleus N and $X \neq N$, then any point P_X is on the line NX.
In this section we are interested in problem I, first we shall show that a minimal thick (q+2)-set has at most one nucleus.

THEOREM 1: If a (q+2)-set K has two distinct nuclei, then it is not minimal.
Proof: The plane $\mathscr{P}$ contains no (q+2)-arc, so K has at least a s-secant r with $s>2$. Then every point A of $K \cap r$ is not essential.

Corollary; If the minimal thick (q+2)-set K has a nucleus N and $X \in K$, $X \neq N$, then there exists at least one s-secant $r \ni X$ of K with $s>2$.

REMARK: We point out that some (q+2)-sets with two nuclei exist. In fact, if N_1 and N_2 are any two distinct points and r is the line N_1N_2, let ξ_i (i=1,2) be the set of the lines through N_i and different from r, moreover let $f: \xi_1 \rightarrow \xi_2$ be any bijection. It is easy to prove that $K=\{s \cap f(s);\ s \in \xi_1\} \cup \{N_1, N_2\}$ is a thick (q+2)-set and N_1, N_2 are nuclei of K. So the number of the (q+2)-sets with nuclei N_1, N_2 is equal to the number of the bijections from ξ_1 to ξ_2, i.e. q!. On the other hand, if $\mathscr{P}=$ $=PG(2,q)$, in [4] it is shown that a (q+2)-set can have more than two nuclei only if q is even.
Now we find conditions for a minimal thick (q+2)-set to have a nucleus.

THEOREM 2: Let K be any minimal thick (q+2)-set of $PG(2,q)$, with q odd; then K contains two points C and D, such that, if the frame is conveniently chosen, it is possible to represent the set $W=K-\{C,D\}$

by an equation $y=f(x)$, where $f(x)$ is a permutation polynomial with
1) the polynomial $f(x)-x$ has no root in $GF(q)$.
Moreover if K has a nucleus, we have
2) $\forall m \in GF(q)$, $m\neq 1$, the polynomial $f(x)-mx$ has only one root in $GF(q)$.
Proof : In the first place we suppose that K has no nucleus and we define an application $\tau: K \dashrightarrow PG(2,q)$ in the following way: we choose a point $X_1 \in K$ and we pose $\tau(X_1)=P_{X_1}$, where P_{X_1} is one of the above-stated points. The line $X_1\,\tau(X_1)$ intersects K at X_1 and at an other point, say X_2. We pose $\tau(X_2)=\tau(X_1)$. Then we choose a point $X_3\neq X_1, X_2$ and we call $\tau(X_3)$ one of the points P_{X_3}. The line $X_3\tau(X_3)$ intersects K at X_3 and at an other point, say X_4. If either $X_4=X_1$ or $X_4=X_2$, we have nothing to define; if $X_4\neq X_1, X_2$ we pose $\tau(X_4)=\tau(X_3)$ and so on. Since q+2 is odd, there must exist at least two points A and B of K such that the distinct lines $A\tau(A)$ and $B\tau(B)$ intersect K at the same point C, because otherwise the set K would have a partition in disjoint pairs. Obviously $\tau(A)$, $\tau(B)$ and C are not collinear, so we can choose $\tau(A)$ as the improper point of the axis x, $\tau(B)$ as the improper point of the axis y and C as the point (0,0). Let D be the only point of K on the improper line, then, using the terminology of [31], the set $W=K-\{C,D\}$ is a diagram relative both to $\tau(A)$ and to $\tau(B)$ and so it can be represented by an equation $y=f(x)$, where $f(x)$ is a permutation polynomial. If we choose the point (1,1) on the line CD, we obtain the cond. 1).
Now let K have a nucleus N and A,B be distinct points of K, with $A\neq N\neq B$. We choose a point P_A (resp. a point P_B) as the improper point of the axis x (resp. of the axis y) and we call D the only improper point of K. If we pose C=N, we can repeat the above proof. The cond. 2) is satisfied, because C is the nucleus of K.

THEOREM 3: In PG(2,q), with q odd, let W be the set represented by the equation $y=f(x)$, where $f(x)$ is a permutation polynomial with
1) the polynomial $f(x)-x$ has no root in $GF(q)$
2) $\forall m \in GF(q)$, $m\neq 1$, the polynomial $f(x)-mx$ has only one root in $GF(q)$.
If C is the point (0,0) and D is the improper point of the line $y=x$, then the set $K=W\cup\{C,D\}$ is a (q+2)-set with nucleus C.
Proof: Self evident.

II

Now we shall deal with problem II. In the plane $\mathscr{P}$ let K be a (q+2)-set with a nucleus N. In conformity with the terminology of [36], K is a (q+2,n)-arc for a convenient number n, and it has at least three characters, because it has s-secants for s=1,2,n. Let t be a line which intersects K exactly in the n points B_1, $B_2,\ldots,B_n$.

THEOREM 4: Suppose the minimal (q+2)-set K has a nucleus N, then either n=q+1 or n<(q-1) and so q>4.

Proof: If n=q+1, we have nothing to prove. Suppose n=q; if $C \in t$ is the point different from $B_1, B_2, \ldots, B_n$, we have $K=\{N, B_1, B_2, \ldots, B_n, A\}$, where A is a convenient point of the line NC; but this means that A is another nucleus for K, in contrast with th. 1. Suppose n=q-1; if C_1 and C_2 are the points of t different from $B_1, B_2, \ldots, B_n$, we have $K=\{N, B_1, B_2, \ldots, B_n, A_1, A_2\}$ where A_i is a convenient point of NC_i (i=1,2). Since A_1 cannot be a nucleus, the line A_1A_2 must pass through one of the points B_j; but this is impossible because this point would be not essential. So we have n<q-1.

If $\mathscr{P}$=PG(2,q), the above result is improved by the next theorem and its corollary.

THEOREM 5: Suppose K is a minimal thick set of PG(2,q) with a nucleus N. Then any line A_1A_2, $A_i \in K$, $A_i \neq N$, $A_1 \neq A_2$, is at least a s-secant of K with s>3.

Proof: Let $A_1 \neq N \neq A_2$ be two disinct points of K. Let P_{A_i} be the point of the line NA_i definied in the section I (i=1,2). For the sake of brevity we write P_i instead of P_{A_i}. After a B. Segre's scheme of proof (see [28]), we chose P_2, P_1, N as the fundamental triangle of a homogeneus coordinate system in PG(2,q). The line P_1P_2 contains only one point of K, say A_3. Let $U=P_2A_1 \cap NA_3$; we choose U as (1,1,1), so we have $A_1=(0,1,1)$, $A_3=(1,1,0)$ and $A_2=(1,0,a)$ with $a \neq 0$. For each point $C \in K$, $C \neq N, A_1, A_2, A_3$, the lines NC, P_1C, P_2C are represented respectively by the equations

$x_1 = m_2x_0$, $x_2 = m_1x_0$, $x_2 = m_0x_1$ with $m_1 = m_0 m_2$

If we consider all the points C of K, $C \neq N, A_1, A_2, A_3$, we obtain that m_2 and m_0 take all the values of GF(q) different from 0 and 1, while m_1 takes all the values of GF(q) different from 0 and a. Since the product of all non-zero elements of GF(q) is equal to -1, we have a=-1. This means that the points A_1, A_2, A_3 are collinear. Starting from the points A_1 and A_3 and putting $P_3=P_{A_3}$, by the above arguments, we have that the only point of K which is on the line P_1P_3 is also on the line A_1A_3. If this point is distinct from A_2, then the line A_1A_2 intersects K in at least four points; otherwise the point A_4, the intersection of K with the line P_2P_3, is certainly distinct from A_1 and lies on the line A_1A_2.

Corollary: Suppose the minimal thick set K of PG(2,q) has a nucleus N, then K is a (q+2,n)-arc with either n=q+1 or $4 \leqslant n \leqslant q/3$.

Proof: If n<q+1, let the line t intersect K exactly in the n points $B_1, B_2, \ldots, B_n$, and let $A \in K$, $A \neq N$, be not on t. By th. 5, each line AB_j (j=1,2,...,n) contains at least three points of K different from A and N. So $3n \leqslant q$.

III

We need a new definition. Let q be an odd prime power.
It is known [12] that a permutation polynomial for GF(q) is in reduced form if its degree is <q, otherwise it is called "crude". Let Φ be the set of all permutation polynomials for GF(q) in reduced form and let Ψ be the set of all permutations of GF(q). Then there exists a well-known bijection $\alpha: \Phi \longrightarrow \Psi$. For the sake of brevity we put $\alpha(f(x))=\psi_f$.

DEFINITION 5: A permutation polynomial is called pseudolinear if there exists a group Δ, subgroup of AGL(1,q), transitive on the elements of GF(q), except for at the most one element, and such that
$\forall \delta \in \Delta$ the permutation polynomial $\alpha^{-1}(\psi_f \, \delta \, \psi_f^{-1})$ is of the first degree.

All the permutation polynomials x^k, with GCD(k,q-1)=1, are pseudolinear because for them we can choose $\Delta = \{ x \longrightarrow ax;\ a \in GF(q),\ a \neq 0 \}$.
On the contrary, the polynomial corresponding to the permutation
$x \longrightarrow x$ if $x \neq 0,1$; $0 \longrightarrow 1; 1 \longrightarrow 0$
is not pseudolinear.

THEOREM 6: In PG(2,q) let H be the set represented by the equation y=f(x), where f(x) is a pseudolinear permutation polynomial and assume Π is the group of the affinities which map H onto itself. Then Π is transitive on H, except for at the most one point.
Proof : It is sufficient to point out that Π cointains all the affinities

$$\begin{cases} x = \alpha^{-1}(\delta)(x') \\ y = \alpha^{-1}(\psi_f \, \delta \, \psi_f^{-1})(y') \end{cases} \text{ where } \delta \in \Delta.$$

Now we treat the problem III. Suppose the (q+2)-set K of PG(2,q) has a nucleus N. We choose a coordinate system different from the one we used in the proof of th. 2. Let $A \neq N$ be a point of K, then we choose AN as the improper line and N as the improper point of the axis y. Since the set $H=K-\{N,A\}$ is a diagram relative to N, so it can be represented by an equation
(1) y=f(x)
where f(x) is a polynomial.

THEOREM 7: Suppose K is minimal, then it is possible to choose the improper point of the axis x such that f(x) is a permutation polynomial.
Proof: self-evident.

If in (1) $f(x)$ is a pseudolinear permutation polynomial, then, by th. 6, the group Π of the affinities that map H onto itself is not trivial. Now we suppose K is a $(q+2,n)$-arc with $n<q+1$ and we prove that if some conditions are satisfied, then K is not minimal.

THEOREM 8: Let $f(x)$ be a pseudolinear permutation polynomial. Assume Π contains a subgroup Π_1 of the group of the translations and Π_1 is transitive on H, then A and N are the only essential points of K.
Proof: Let B be any point of H and $t \in B$ be a s-secant of K with $s>1$; assume T is the improper point of t. Since $n<q+1$, then there exists $\sigma \in \Pi_1$ such that $\sigma(t)=t' \neq t$. Obviously t' passes through T and it is likewise a s-secant of K. On the other hand, since any affine line through T is either external or s-secant of K, then no 1-secant of K passes through T. Let now Q be any point of the line BN: since the line QT is not a 1-secant of $K-\{B\}$, then the point B is not essential.

THEOREM 9: Let $f(x)$ be a pseudolinear permutation polynomial. Assume any element of Π fixes the improper points of the axes and a proper point O (obviously O is on H); moreover assume Π is transitive on the points of H different from O, then O is not essential.
Proof: Let X be the improper point of the axis x; since the line XO is an 1-secant of K, then there exists at least a line $r \ni O$ that is a s-secant of K with $s>2$. The line r is a m-secant of H, where $m=s-1$ if A is on r, $m=s$ otherwise. Since Π is transitive on $H-\{O\}$, then the q points of H are shared among d lines which pass through O and are m-secant H, with obviously $q-1=d(m-1)$. It is easy to see that Π is transitive on the lines passing through X and different from XN and XO, and therefore also on the points of the line ON different from O and N. Suppose the line ON contains a point $Q \neq O,N$ such that any line $u \ni Q$ is t-secant H with $t<2$, then the same is true for every point R of ON ($R \neq O,N$). But evidently this is absurd. So the point O is not essential.

EXAMPLES :In PG(2,q), with $q=p^h$ odd, let N and A be respectively the improper point of the axis y and the improper point of the line $y=x$. Moreover let H be the set represented by the equation
$y=f(x)=c_0x+c_1x^p+\ldots+c_{h-}x^{p^{h-1}}$
where $c_i \in GF(p)$ and $GCD(c_0+c_1x+\ldots+c_{h-1}x^{h-1}, 1-x^h)=1$.
It is known [12] that $f(x)$ is a permutation polynomial; it is also pseudolinear because we can choose $\Delta=\{x \dashrightarrow x+b;\ b \in GF(q)\}$. Assume $K=H \cup \{A,N\}$. Since the conditions required by th. 8 are satisfied, no point of H is essential for K.
2) Let A,N,K have the same meaning as in 1), while H is the set represented by the equation
$y=x^k$
where $GCD(k,q-1)=1$. As we have already proved, x^k is a pseudolinear permutation polynomial. In this case the conditions

required by th. 9 are satisfied, so the point (0,0) of H is not essential for K.

LITERATURE

[1] A.Barlotti:Sui (k,n)-archi di un piano lineare finito. Boll. Un.Mat.Ital.11 (1956) 553-556.

[2] A.Barlotti: Un'osservaziome intorno ad un teorema di B. Segre sui q-archi. Matematiche (Catania) 21 (1966) 287-395.

[3] U.Bartocci: k-insiemi densi in piani di Galois;in corso di pubbl.sul Boll.Un.Mat.It.

[4] A.Bichara,G.Korchmaros: Note on (q+2)-sets in a Galois plane of order q. Ann. of discr.Math.14 (1982) 117-122

[5] R.C.Bose: Mathematical theory of the symmetrical factorial design. Sankhya 8 (1947) 323-338.

[6] A.Bruen: Permutation functions on a finite field. Canad. Math.Bull.15 (1972)

[7]:A.Bruen:The number of Lines Determined by n^2 Points.Journ.of Comb.Theory (A)15 (1973) 225-241.

[8] L.Carlitz:Permutations in a finite field. Proc.Amer.Math. Soc.4 (1953) 538.

[9] L.Carlitz: A Theorem on permutations in a finite field.Proc. Amer.Math.Soc.11 (1960) 456-459.

[10] L.Carlitz: A note on permutation functions over finite field. Duke Math.J.29 (1962) 325-332.

[11] L.Carlitz: Some theorems on permutation polynomials. Bull. Amer.Math. 68 (1962) 120-122.

[12].L.Carlitz: Permutations in finite field. Acta Sci.Math.Szeged. 24 (1963) 196-203.

[13] C. Di Comite: Sui k-archi deducibili da cubiche piane. Rend. Acc.Naz.Lincei (8) 33 (1962) 429-435.

[14] C. Di Comite: Sui k-archi contenuti in cubiche piane. Rend. Acc.Naz.Lincei (8) 35 (1963) 274-278.

[15] C. Di Comite: Intorno a certi (q+9)/2-archi di S(2,q). Rend. Acc.Naz.Lincei (8) 36 (1964) 819-824.

[16] F. Karteszi: Introduzione alle geometrie finite. Feltrinelli 1978 (traduzione italiana)

[17] G. Korchmaros: New examples of complete k-arcs in PG(2,q). Eur.J. of Comb.

[18] L.Lombardo Radice: Sul problema dei k-archi completi di S(2,q).Boll.Un.Mat.It.11 (1956) 178-181.

[19] G.Pellegrino: Sur les k-arcs complets des plans des Galois d'ordre impair.Ann. Discr. Math. 18 (1983) 667-694.

[20] B. Qvist: Some remarks concernin g curves of the second degree in a finite plane. Ann.Acad.Sci.Fenn.I 134 (1952)

[21] L. Redei: Uber eindeutig umkerbare Polinome in endliche Korpern. Acta Sci.Math.Szegd. 11 (1946-1948) 85-92

[22] M. Sce, L. Lunelli: Sulla ricerca dei k-archi completi mediante calcolatrice elettronica. Convegno reticoli e geometrie proiettive.(Palermo 1957) Roma Cremonese 81-86 (1958)

[23] B.Segre: Sulle ovali nei piani lineari finiti. Rend.Acc. Naz.Lincei (8) 17 (1954) 141-142.

[24] B.Segre: Curve razionali normali e k-archi negli spazi finiti. Ann. Mat.Pura Appl. 39 (1955) 357-379.

[25] B.Segre: Ovals in a finite projective plane. Canad. J. Math. 7 (1955) 414-416.

[26] B.Segre: Sui k-archi nei piani finiti di caratteristica 2.Rev. de Math.Pure et Appl. 2 (1957) 289-300.

[27] B.Segre: Le geometrie di Galois. Ann.Mat.Pura Appl. 48 (1959) 1-96

[28] B.Segre: Lectures on modern Geometry. Cremonese,Roma 1961

[29] B.Segre: Ovali e curve nei piani di Galois di caratteristica due. Rend.Acc.Naz.Lincei 32 (1962) 785-790

[30] B.Segre: Introduction to Galois geometries. Mem.Acc.Naz. Lincei, 8 (1967) 135-236.

[31] B.Segre,U.Bartocci: Ovali ed altre curve nei piani di Galois di caratteristica due. Acta Arithm. 18 (1971) 423-449

[32] B.Segre,G.Korchmaros: Una proprietà degli insiemi di punti di un piano di Galois caratterizzante quelli formati dalle singole rette esterne ad una conica,Rend.Acc.Naz.Lincei 62 (1977) 613-618.

[33] G.Tallini: Sui q-archi di un piano lineare finito di caratteristica p=2. Rend.Acc.Naz.Lincei 23 (1957) 242-245.

[34] G.Tallini: Le geometrie di Galois e le loro applicazioni alla statistica e alla teria del'informazione. Rend.Mat.e Appl. 19 (1960) 379-400.

[35] M.Tallini Scafati: Sui k,n -archi di un piano grafico finito. Atti Acc. Naz. Lincei.Rend. 40 (1966) 373-378.

[36] M.Tallini Scafati: k,n -archi di un piano grafico finito, con particolare riguardo a quelli con due caratteri (Note I e II) Atti Acc.Naz. Lincei Rend. 40 (1966) 812-818; 1020-1025.

[37] F.Zirilli: Su una classe di k-archi di un piano di Galois. Rend. Acc.Naz.Lincei 54 (1973) 393-397.

Annals of Discrete Mathematics 30 (1986) 125–136

ON A GENERALIZATION OF INJECTION GEOMETRIES

Pier Vittorio Ceccherini and Natalina Venanzangeli
Dipartimento di Matematica "G. Castelnuovo"
Università di Roma "La Sapienza"
Citta Universitaria, 00100 Roma, Italy

Injection geometries have been introduced in [3] as a generalization of permutation geometries studied in [1]. We present a generalization of injection geometries, namely $\mathcal{J}$-geometries, improving in some cases properties of injection geometries stated in [3].
$\mathcal{J}$-geometries have been introduced in [9], and also preannounced in [3], in a "Concluding remark" unknown to the authors.
$\mathcal{J}$-geometries have been recently considered also in [4], under the name of "squashed geometries" and in [6], [8].

1. INTRODUCTION

In what follows all sets will be finite. Let N be a non empty set and $N^2 = N \times N$. If $a \in N$, we get the generators $g_1(a) = \{(x,y) \in N^2 : x = a\}$ and $g_2(a) = \{(x,y) \in N^2 : y = a\}$; the first (resp. second) system of generators is $G_1 = \{g_1(a) : a \in N\}$ (resp. $G_2 = g_2(a) : a \in N$). A (partial) correspondence of N is any subset of N^2; <u>a partial application</u> (resp. <u>coapplication</u>) of N is any subset of N^2 which is 0- or 1-secant each $g_1 \in G_1$ (resp. $g_2 \in G_2$); a <u>subpermutation</u> F of N is any subset of N^2 which is 0- or 1-secant each generator; if dom F = N, then F is a permutation of N. In [7] sets and groups of permutations of N are studied (with special attention to transitivity properties) from this geometrical point of wiew. In [2] certain semigroups of subpermutations are characterized in the class of all semigroups. In [1] special sets of subpermutations of N (namely <u>permutation geometries</u>, cf. no. 2) are introduced by means of axioms similar to matroid axioms; more general results are obtained in a similar way in [5], where sets of partial applications (or coapplications) instead of subpermutations are considered. Other generalizations of permutation geometries are <u>injection geometries</u>, cf. [3]. If $d \geq 1$ is an integer, in the set N^d we have d systems of generators:

$G_i = \{g_i(a) : a \in N\}$, where $g_i(a) = \{(x_1,\dots,x_d) \in N^d : x_i = a\}$ $(i=1,\dots,d)$; a subset of N^d is called <u>injective</u> if it is 0- or 1-secant each generator; an injection geometry is a set of injective sets of N^d satisfying suitable axioms, which are similar to matroid axioms and which reduce to those for d = 1 and to permutation geometry axioms for d = 2.

This talk concerns a generalization of injection geometries (namely $\mathcal{I}$-geometries) given in [9] in a very abstract way which includes also partial application (or coapplication) geometries (cf. no. 2).

Prof. M. Deza informed us during this conference that the concept of $\mathcal{I}$-geometries is claimed in a concluding remark (added in proofs) in [3] and that it is going to be developed in [4], where our $\mathcal{I}$-geometries are called squashed geometries. A different and very elegant approach to squashed geometries as "bouquets of matroids" is given in [8], where instead of our set $\mathcal{I}$ an antichain C is considered satisfying suitable axioms (in our language C is the set of maximal elements of $\mathcal{A}$); that approach seems to be very efficient.

In what follows we sketch a theory of $\mathcal{I}$-geometries, improving in some cases properties of injection geometries stated in [3].

2. DEFINITION OF $\mathcal{I}$-GEOMETRIES

Let us start with the familiar concept of matroid.

<u>DEFINITION 2.1.</u> A matroid $M_r(X)$, or rank r on set X, is a pair $M_r(X) = (X,\mathcal{A})$ where $\mathcal{A}$ is a set of subsets of X, partitioned into $\mathcal{A} = \mathcal{A}_\circ \cup \dots \cup \mathcal{A}_r$ with $\mathcal{A}_r \neq \emptyset$, satisfying the following axioms (the elements $A_i \in \mathcal{A}_i$ are called the flats of rank i of $M_r(X)$, $0 < i < r$):

(1) $\mathcal{A}$ is closed under intersection;

(2) if $A_i \in \mathcal{A}_i$, $A_j \in \mathcal{A}_j$ and $A_i \subseteq A_j$ then $i \leq j$;

(3) if $A_i \in \mathcal{A}_i$, $i < r$ and $b \in X \setminus A_i$, then there exists an unique $A_{i+1} \in \mathcal{A}_{i+1}$ such that $A_{i+1} \supseteq A_i \cup \{b\}$; moreover A_{i+1} is included in each $A' \in \mathcal{A}$ such that $A' \supseteq A \cup \{b\}$.

We recall now the definitions of the structures mentioned in §1.

<u>DEFINITION 2.2</u> ([1]). Let N be a non empty set and $X \subseteq N \times N$. A <u>permutation</u>

geometry $P_r(X)$ of rank r on X, is a pair $P_r(X) = (X,\mathcal{A})$ where $\mathcal{A}$ is a set of subpermutations of N, partitioned into $\mathcal{A} = \mathcal{A}_\circ \cup \ldots \cup \mathcal{A}_r$ with $\mathcal{A}_r \neq \emptyset$, satisfying the axioms (1)-(3) with the restriction that axiom (3) holds for those $b \in X \setminus A_i$ such that $A_i \cup \{b\}$ is a subpermutation of N. (Note that the original definition given in [1] concerns the particular case $X = N^2$; i.e. our definition is slightly more general).

DEFINITION 2.3 ([5]). Let N be a non empty set and $X \subseteq N \times N$. A partial application (resp. coapplication) geometry of rank r on X is a pair $(X,\mathcal{A})$, where $\mathcal{A}$ is a set of partial applications (resp. coapplications) of N, partitioned into $\mathcal{A} = \mathcal{A}_\circ \cup \ldots \cup \mathcal{A}_r$ with $\mathcal{A}_r \neq \emptyset$, satisfying the axioms (1)-(3) with the restriction that axiom (3) holds for those $b \in X \setminus A_i$ such that $A_i \cup \{b\}$ is a partial application (resp. coapplication) of N.

DEFINITION 2.4 ([3]). Let $d > 1$ be an integer, N a non empty set and $X \subseteq N^d$. An injection geometry of rank r on X is a pair $I_r(X) = (X,\mathcal{A})$ where $\mathcal{A}$ is a set of injective subsets of N^d, partitioned into $\mathcal{A} = \mathcal{A}_\circ \cup \ldots \cup \mathcal{A}_r$ with $\mathcal{A}_r \neq \emptyset$, satisfying axioms (1)-(3) with the restriction that axiom (3) holds for those $b \in X \setminus A_i$ such that $A_i \cup \{b\}$ is an injective subset of N^d. The number d will be called the dimension of $I_r(X)$.

We give now the definition of $\mathcal{I}$-geometries.

DEFINITION 2.5 (cf. [9], and also [3], [4], [8] where the definition is given in a slightly different form). An $\mathcal{I}$-geometry of rank r on a set X, is a quadruple $G_r(X) = (S,\mathcal{I},X,\mathcal{A})$ where S is a non empty set, $\mathcal{I}$ is a simplicial complex of distinguished subsets of S (i.e. $Z \subset Z' \in \mathcal{I}$ implies $Z \in \mathcal{I}$), $\mathcal{A}$ is a subset of $\mathcal{I}$ partitioned into $\mathcal{A} = \mathcal{A}_\circ \cup \ldots \cup \mathcal{A}_r$ with $\mathcal{A}_r \neq \emptyset$ and $X = \bigcup_{A \in \mathcal{A}} A$, satisfying the axioms (1)-(3) with the restriction that axiom (3) holds for those $b \in X \setminus A_i$ such that $A_i \cup \{b\} \in \mathcal{I}$.

The elements $A_i \in \mathcal{A}_i$ are called the flats of rank i of the geometry $G_r(X)$.

We shall write $A_{i+1} = A_i \vee \{b\}$ for the set A_{i+1} mentioned in axiom (3).

A matroid $M_r(X)$ is a geometry $G_r(X) = (S,\mathcal{I},X,\mathcal{A})$ with $\mathcal{I} = 2^S$ and $X = S$, and conversely. A permutation geometry $P_r(X) = (X,\mathcal{A})$, with $X \subseteq N^2$, is a geometry $G_r(X) = (S,\mathcal{I},X,\mathcal{A})$ with $S = N^2$ and $\mathcal{I} = \{F \subseteq N^2 : F \text{ is a subpermutation of } N\}$, and conversely. Partial application (resp. coapplication) geometries can be easily characterized in a similar way between $\mathcal{I}$-geometries. An injection geometry $I_r(X) = (X,\mathcal{A})$ with $X \subseteq N^d$ is a geometry $G_r(X) = (S,\mathcal{I},X,\mathcal{A})$ with $S = N^d$ and $\mathcal{I} = \{F \supseteq N^d : F$ is injective$\}$, and conversely. Several examples of geometries $G_r(X)$ can then be deduced from [1], [5], [3] where examples of permutation geometries and of injection geometries are given. We now give some other examples.

EXAMPLE 2.6. Free $\mathcal{I}$-geometries. The free geometry $G_r(X) = (S,\mathcal{I},X,\mathcal{A})$ is defined by assuming $X = S$, $\mathcal{I}$ a simplicial complex of S, $\mathcal{A}_i = \{A \in \mathcal{I} : |A| = i\}$, $0 \leqslant i \leqslant r$, $r \geqslant 1$.

EXAMPLE 2.7. Star $\mathcal{I}$-geometries. A star geometry $G_1(X) = (S,\mathcal{I},X,\mathcal{A})$ with center C is defined by assuming $C \subset S$, $\mathcal{A}_\circ = \{C\}$, $\mathcal{A}_1 = \{D_1,\dots,D_t\}$ where $C \subset D_j \subset S$ and $D_i \cap D_j = C$ $(t \geqslant 2,\ i,j=1,\dots,t,\ i \neq j)$, $X = \overset{t}{\underset{1}{\cup}} D_i$, $\mathcal{I} = \overset{t}{\underset{i=1}{\cup}} 2^{D_i}$.

REMARK 2.8. Star $\mathcal{I}$-geometries are $\mathcal{I}$-geometries of rank 1 with $|\mathcal{A}_1| > 1$ and conversely (cf. Theorem 5.3).

EXAMPLE 2.9. Truncations of an $\mathcal{I}$-geometry. If $G_r(X) = (S,\mathcal{I},X,\mathcal{A})$ is an $\mathcal{I}$-geometry with $\mathcal{A} = \mathcal{A}_\circ \cup \dots \cup \mathcal{A}_r$ and if $1 \leqslant k < r$, we can consider $\mathcal{A}^{(k)} = \mathcal{A}_\circ \cup \dots \cup \mathcal{A}_k$. Then $G_r^{(k)} = (S,\mathcal{I},X,\mathcal{A}^{(k)})$ is an $\mathcal{I}$-geometry of rank k, called the k-truncation of G_r.

EXAMPLE 2.10. Cotruncation of an $\mathcal{I}$-geometry. If $G_r(X) = (S,\mathcal{I},X,\mathcal{A})$ is an $\mathcal{I}$-geometry and if $A_h \in \mathcal{A}_h$ $(0 \leqslant h < r)$ is included in some $A_r \in \mathcal{A}_r$, then we can consider $\bar{\mathcal{A}}_i = \bar{\mathcal{A}}_i(A_h) = \{A_{h+i} \in \mathcal{A}_{h+i} : A_h \subseteq A_{h+i},\ i=0,1,\dots,r-h\}$. Then $G_r(A_h) = (S,\mathcal{I},X,\mathcal{A})$ with $\bar{\mathcal{A}} = \bar{\mathcal{A}}_\circ \cup \dots \cup \bar{\mathcal{A}}_{r-h}$, is an $\mathcal{I}$-geometry of rank r-h. Note that $G_r(A_\circ) = G_r$, cf. 5.1 (a).

EXAMPLE 2.11. Bitruncations of an $\mathcal{I}$-geometry. If $G_r = (S,\mathcal{I},X,\mathcal{A})$ is an $\mathcal{I}$-geometry and if $A_h \in \mathcal{A}_h$, with $0 \leqslant h < k < r$ and $A_h \subset A_k$ for some $A_k \in \mathcal{A}_k$ then we can consider $\bar{\mathcal{A}}_i = \bar{\mathcal{A}}_i(A_h) = \{A_{h+i} \in \mathcal{A}_{h+i} : A_h \subset A_{h+i},\ i=0,\dots,k-h\}$. Then $G = (S,\mathcal{I},X,\mathcal{A})$ with $\bar{\mathcal{A}} = \bar{\mathcal{A}}_\circ \cup \dots \cup \bar{\mathcal{A}}_{k-h}$ is an $\mathcal{I}$-geometry of rank k-h.

EXAMPLE 2.12. Intervals of an $\mathcal{I}$-geometry. If $G_r = (S,\mathcal{I},X,\mathcal{A})$ is an $\mathcal{I}$-geometry and if $A_h \in \mathcal{A}_h$, $A_k \in \mathcal{A}_k$ with $A_h \subset A_k$, $0 \leqslant h < k < r$, then we can consider $\bar{\mathcal{A}}_i(A_h,A_k) = \{A_{h+i} \in \mathcal{A}_{h+i} : A_h \subseteq A_{h+i} \subseteq A_k,\ i=0,\dots,k-h\}$. Then $\bar{\mathcal{A}}_\circ \cup \dots \cup \bar{\mathcal{A}}_{k-h}$ gives a matroid of rank k-h on A_h.

EXAMPLE 2.13. Restrictions of $\mathcal{I}$-geometries. Let $G_r(X) = (S,\mathcal{I},X,\mathcal{A})$ be an $\mathcal{I}$-geometry and let be S' such that $A_1 \subseteq S' \subset X$ for some $A_1 \in \mathcal{A}_1$. Define $\mathcal{A}' = \{A \in \mathcal{A} : A \subset S'\}$, $X' = \bigcup_{A \in \mathcal{A}'} A$, $\mathcal{I}' = \bigcup_{A' \in \mathcal{A}'} 2^{A'}$, $r' = \max\{i : A_i \in \mathcal{A}'\}$. Then $G_{r'}(X') = (S',\mathcal{I}',X',\mathcal{A}')$ is an $\mathcal{I}$-geometry of rank r' on X'.

A special interesting case is obtained when $S' = X' = \bigcup_{A \in \mathcal{B}} A$ for some fixed $\mathcal{B} \subseteq \mathcal{A}$ such that $\mathcal{B} \cap \mathcal{A}_1 \neq \emptyset$; in this case $r' = \max\{i : \mathcal{A}_i \cap \mathcal{B} \neq \emptyset\}$, $\mathcal{A}' = \{A \in \mathcal{A} : A \subseteq B \text{ for some } B \subset \mathcal{B}\}$ and $\mathcal{I}' = \bigcup_{A \in \mathcal{B}} 2^A$.

3. DIRECT SUMS OF $\mathcal{I}$-GEOMETRIES

By starting from two given $\mathcal{I}$-geometries, it is possible to construct a new one (namely their direct sum), in the standard way described below.

DEFINITION 3.1. Direct sum $\oplus$ of $\mathcal{I}$-geometries. Let $G'_{r'} = (S',\mathcal{I}',X',\mathcal{A}')$ and $G''_{r''} = (S'',\mathcal{I}'',X'',\mathcal{A}'')$ be two $\mathcal{I}$-geometries; we can suppose without loss of generality that $S' \cap S'' = \emptyset$. Assume:

$$S = S' \cup S'',\quad X = X' \cup X'',\qquad \mathcal{I} = \{I' \cup I'' : I' \in \mathcal{I}',\ I'' \in \mathcal{I}''\},$$

$$\mathcal{A}_i = \{A'_{i'} \cup A''_{i''} : A'_{i'} \in \mathcal{A}'_{i'},\ A''_{i''} \in \mathcal{A}''_{i''},\ i' + i'' = i\},$$

$$\mathcal{A} = \mathcal{A}_\circ \cup \ldots \cup \mathcal{A}_r,\quad r = r' + r''.$$

Then $(S,\mathcal{I},X,\mathcal{A})$ will be called the direct sum $G'_{r'} \oplus G''_{r''}$ of $G'_{r'}$ and $G''_{r''}$. It is easy to prove the following:

THEOREM 3.2. The direct sum $G'_{r'} \oplus G''_{r''}$ of two $\mathcal{I}$-geometries is an $\mathcal{I}$-geometry of rank $r = r' + r''$. □

DEFINITION 3.3. Full direct sum $\bar{\oplus}$ of injection geometries. Let $G'_{r'} = (S',\mathcal{I}',X',\mathcal{A}')$ and $G''_{r''} = (S'',\mathcal{I}'',X'',\mathcal{A}'')$ be two injection geometries of the same dimension d. Actually $X' \subseteq S' = N'^d$ and $X'' \subseteq S'' = N''^d$. We can suppose without loss of generality that $N' \cap N'' = \emptyset$, so that $S' \cap S'' = \emptyset$. Assume:

$$N = N' \cup N'',\quad \bar{S} = N^d,\quad X = X' \cup X'',\quad \bar{\mathcal{I}} = \{I \in 2^{\bar{S}} : I \text{ is injective}\},$$

$$\mathcal{A}_i = \{A'_{i'} \cup A''_{i''} : A'_{i'} \in \mathcal{A}'_{i'},\ A''_{i''} \in \mathcal{A}''_{i''},\ i' + i'' = i\},$$

$$\mathcal{A} = \mathcal{A}_\circ \cup \ldots \cup \mathcal{A}_r,\quad r = r' + r''.$$

Then $(\bar{S},\bar{\mathcal{I}},X,\mathcal{A})$ will be called the full direct sum $\bar{G}_r = G'_{r'} \bar{\oplus} G''_{r''}$ of $G'_{r'}$ and $G''_{r''}$.

THEOREM 3.4. The full direct sum $G'_{r'} \bar{\oplus} G''_{r''}$ of two injection geometries of dimension d is an injection geometry of rank $r = r' + r''$ and dimension d.

Proof. Let $G_r = (S,\mathcal{I},X,\mathcal{A}) = G'_{r'} \oplus G''_{r''}$ be the direct sum of $G'_{r'}$ and $G''_{r''}$ considered as $\mathcal{I}$-geometries (cf. Theorem 3.2). Actually $S = S' \cup S'' = N'^d \cup N''^d \subset N^d = \bar{S}$ and $\mathcal{I} \subset \bar{\mathcal{I}}$. It follows immediately that $\bar{G}_r$ is an injection geometry of rank $r = r' + r''$ and dimension d. □

For d = 2 we have

COROLLARY 3.5. The full direct sum $G'_{r'} \bar{\oplus} G''_{r''}$ of two permutation geometries (considered as injection geometries of dimension d=2) is an injection geometry of dimension two, i.e. it is a permutation geometry. □

REMARK 3.6. Let $G'_{r'}$ and $G''_{r''}$ be two "permutation geometries" in the restricted meaning of [1]. Then the full direct sum $G'_{r'} \bar{\oplus} G''_{r''}$ is a permutation geometry (in our meaning), but it is not a "permutation geometry" in the meaning of [1], because (with the notation of Definition 2.2) $X \subset N^2$.

4. REGULAR $\mathcal{I}$-GEOMETRIES

A geometry $G_r(X) = (W,\mathcal{I},X,\mathcal{A})$ is called regular if each $A \in \mathcal{A}$ is included in some $A_r \in \mathcal{A}_r$. Every star $\mathcal{I}$-geometry is regular. The following $G_2(X) = (S,\mathcal{I},X,\mathcal{A})$ is not regular:

$$X = S = \{a,b,c,d\},\ \mathcal{A}_o = \{A_o = \{a\}\},\quad \mathcal{A}_1 = \{A_1 = \{a,b\},\ A'_1 = \{a,c\},\ A''_1 = \{a,d\}\},$$

$$\mathcal{A}_2 = \{A_2 = \{a,b,c\}\},\quad \mathcal{A} = \mathcal{A}_o \cup \mathcal{A}_1 \cup \mathcal{A}_2,$$

$\mathcal{I} = \{A_o, \{b\}, \{c\}, \{d\}, A_1, A'_1, A''_1, \{b,c\}, A_2\}$. We note that A''_1 is not included in A_2.

The same example shows that if $G_r(X) = (S,\mathcal{I},X,\mathcal{A})$ is a geometry, then $\mathcal{I}$ is not necessarily the family of the independent sets of a matroid on S. It is also easy to give examples of regular $G_r(X)$ with the same property ([9]).

PROPOSITION 4.1. A geometry $G_r(X) = (W,\mathcal{I},X,\mathcal{A})$ is regular if and only if for each $A_i \in \mathcal{A}_i$, with $i < r$, there exists $b \in X \setminus A_i$ such that $A_i \cup \{b\} \in \mathcal{I}$.

Proof. Suppose $G_r(X)$ regular, $A_i \in \mathcal{A}_i$, $i < r$. Let $A_i \subseteq A_r \in \mathcal{A}_r$. We have $A_i \subset A_r$ since $i < r$ (cf. Prop. 5.1. (e)). If $b \in A_r \setminus A_i$, then $b \in X \setminus A_i$ and $A_i \cup \{b\} \in \mathcal{I}$, since $A_i \cup \{b\} \subseteq A_r \in \mathcal{I}$ and $\mathcal{I}$ is a simplicial complex.

Conversely, if for each $A_i \in \mathcal{A}_i$, with $i < r$, there exists $b \in X \setminus A_i$ such that $A_i \cup \{b\} \in \mathcal{I}$, then A_i is included in some $A_r \in \mathcal{A}_r$: this follows by applying (3) r-i times. □

PROPOSITION 4.2. A geometry $G_r(X)$ is regular with $|\mathcal{A}_r| = 1$ iff $\mathcal{A}_r = \{X\}$. (In this case $G_r(X)$ is exactly a matroid).

Proof. If $\mathcal{A}_r = \{X\}$, then each $A_i \in \mathcal{A}_i$ with $i < r$ is included in $A_r = X = \bigcup_{A \in \mathcal{A}} A$ so that G_r is regular with $|\mathcal{A}_r| = 1$. Conversely let $G_r(X)$ be regular with $\mathcal{A}_r = \{A_r\}$. If $x \in X = \bigcup_{A \in \mathcal{A}} A$, then $x \in A_i$ for some $A_i \in \mathcal{A}_i$, so that $x \in A_i \subseteq A_r$ by regularity. Therefore $X \subseteq A_r$, i.e. $X = A_r$. □

5. FIRST PROPERTIES OF $\mathcal{I}$-GEOMETRIES

PROPOSITION 5.1. Let $G_r = (S, \mathcal{I}, X, \mathcal{A})$ be an $\mathcal{I}$-geometry.

(a) $|\mathcal{A}_0| = 1$ and the element $A_0 \in \mathcal{A}_0$ is the minimum of $\mathcal{A}$.

(b) If $A_h \in \mathcal{A}_h$, $A_k \in \mathcal{A}_k$ with $A_h \subset A_k$ and $0 \leqslant h < k \leqslant r$, then there exists a chain $A_h \subset A_{h+1} \subset \ldots \subset A_k$ with $A_s \in \mathcal{A}_s$ $(s = h, h+1, \ldots, k)$.

(c) Each $A_i \in \mathcal{A}_i$ $(0 < i \leqslant r)$ is included in a chain $A_0 \subset A_1 \subset \ldots \subset A_i$ with $A_s \in \mathcal{A}_s$ $(s = 0, 1, \ldots, i)$.

(d) If G_r is regular, each $A_i \in \mathcal{A}_i$ $(0 < i \leqslant r)$ is included in a chain $A_0 \subset \ldots \subset A_i \subset \ldots \subset A_r$ with $A_s \in \mathcal{A}_s$ $(s = 0, \ldots, r)$.

(e) If $A_i \in \mathcal{A}_i$, $A_j \in \mathcal{A}_j$ (with $i, j = 0, 1, \ldots, r$), and if $A_i \subseteq A_j$, then $i \leqslant j$, where $A_i = A_j$ if and only if $i = j$.

Proof. (a) Let $\bigcap_{A \in \mathcal{A}} A = A_i \in \mathcal{A}_i$. From (2) it follows that $i = 0$. If $A_0' \in \mathcal{A}_0$, then $A_0 \subseteq A_0'$. Then $A_0 = A_0'$; otherwise if $b \in A_0' \setminus A_0$, then from (3) it follows that $A_0 \vee \{b\} = A_1 \subseteq A_0'$ with $A_1 \in \mathcal{A}_1$, contradicting (2).

(b) Let $b \in A_k \setminus A_h$. From (3), $A_h \vee \{b\} = A_{h+1} \subseteq A_k$ with $A_{h+1} \in \mathcal{A}_{h+1}$. If $h+1 = k$, then $A_{h+1} = A_k$ by (3). If $h+1 < k$, then $A_{h+1} \subset A_k$ and iteration of the same argument leads to a chain $A_h \subset \ldots \subset A_{k-1} \subset A_k' \subseteq A_k$ with $A_s \in \mathcal{A}_s$ $(h \leqslant s \leqslant k)$ and $A_k' \in \mathcal{A}_k$. Actually $k-1 < r$, so that (3) implies $A_k' = A_k$, and $A_h \subset \ldots \subset A_{k-1} \subset A_k$ is a chain as required.

(c) Take $h = 0$ and $k = i$ in (b).

(d) If $i = r$, (d) follows from (c). If $i < r$, then $A_i \subset A_r$ for some $A_r \in \mathcal{A}_r$ by regularity. By (c) we have a chain $A_0 \subset \ldots \subset A_i$ and by (b) we have a chain $A_i \subset \ldots \subset A_{r-1} \subset A_r$, so that we get a chain $A_0 \subset \ldots \subset A_i \subset \ldots \subset A_r$ $(A_s \in \mathcal{A}_s$, $s = 0, \ldots, r)$.

(e) We have $i \leqslant j$ by (2). From $A_i = A_j$ it follows obviously that $i = j$. Suppose now $A_i \subseteq A'_i$ with $A_i, A'_i \in \mathcal{A}_i$. If $i=0$, then $A_\circ = A'_\circ$ by (a). If $i \neq 0$, then by (c) we get a chain $A_\circ \subset \ldots \subset A_{i-1} \subset A_i \subseteq A'_i$, so that condition (3) implies $A_i = A_{i-1} \vee \{b\} = A'_i$ where $b \in A_i \setminus A_{i-1}$. So $A_i \subseteq A'_i$ implies $A_i = A'_i$. □

From 5.1 (e) follows

COROLLARY 5.2. Let G_r be an $\mathcal{I}$-geometry. If $A_i, A'_i \in \mathcal{A}_i$ with $A_i \neq A'_i (0 < i \leqslant r)$, then $A_i \cap A'_i \in \mathcal{A}_j$ with $0 \leqslant j \leqslant i-1$. □

THEOREM 5.3. Let $G_r = (S, \mathcal{I}, X, \mathcal{A})$ be an $\mathcal{I}$-geometry.

(a) Each $x \in X \setminus A_\circ$ is included in a unique $A_1 \in \mathcal{A}_1$. If x is included in some $A_j (j > 0)$, then x is included in some $A_i \in \mathcal{A}_i$ for every $0 < i \leqslant j$. In particular if G_r is regular, then each x is included in some $A_i \in \mathcal{A}_i$ for every $0 < i \leqslant r$.

(b) If $0 < i < j \leqslant r$, each $A_j \in \mathcal{A}_j$ is the union of the elements of $\mathcal{A}_i$ which are included in A_j.

(c) If $A_i \in \mathcal{A}_i (0 \leqslant i < r)$ is included in two distinct element $A'_{i+1}, A''_{i+1} \in \mathcal{A}_{i+1}$, then $A_i = A'_{i+1} \cap A''_{i+1} = \bigcap_{A_i \subseteq A \in \mathcal{A}_{i+1}} A$.

(d) The following conditions are equivalent:

(d_1) for all $0 \leqslant i < r$, each $A_i \in \mathcal{A}_i$ is included in two distinct elements of $\mathcal{A}_{i+1}$;

(d_2) for all $0 \leqslant i < r$, each $A_i \in \mathcal{A}_i$ is included in two distinct elements of $\mathcal{A}_r$;

(d_3) for all $0 \leqslant i < r$, each $A_i \in \mathcal{A}_i$ is the intersection of elements of $\mathcal{A}_r$;

(d_4) for all $0 \leqslant i < j \leqslant r$, each $A_i \in \mathcal{A}_i$ is the intersection of the elements of $\mathcal{A}_j$ containing A_i.

Proof. (a) Suppose $x \in X \setminus A_\circ$. Since $X = \bigcup_{A \in \mathcal{A}} A$, we have $x \in A_j$ for some $A_j \in \mathcal{A}_j$. Then $x \in A_\circ \vee \{x\} = A_1 \subseteq A_j$, and A_1 is the unique element of $\mathcal{A}_1$ including x. If $x \in A_j$ and $0 < i < j$, there is a chain $A_\circ \subset A_1 \subset \ldots \subset A_i \subset \ldots \subset A_j$ $(A_s \in \mathcal{A}_s, 0 \leqslant s \leqslant j)$, with $x \in A_1 = A_\circ \vee \{x\}$, so that $x \in A_1 \subset A_i$.

(b) It is enough to prove that if $x \in A_j \setminus A_\circ$, there exists some $A_i \in \mathcal{A}_i$ such that $x \in A_i \subset A_j$. This follows from (a).

(c) $A_i \subseteq A'_{i+1} \cap A''_{i+1} = A_j \in \mathcal{A}_j$, with $i \leqslant j \leqslant (i+1)-1$ by Corollary 5.2, so that $j = i$ and $A_i = A'_{i+1} \cap A''_{i+1}$ by Proposition 5.1 (e).

(d) (d_1) ⇒ (d_2). From (d_1) it follows that each $A_i \in \mathcal{A}_i (0 \leqslant i < r)$ is included in a chain $A_i \subset \ldots \subset A_{r-1}$ and A_{r-1} is included in two distinct elements of $\mathcal{A}_r$.

$(d_2) \Rightarrow (d_1)$. From (d_2) we have that $A_i \subseteq A'_r \cap A''_r$ with A'_r, A''_r distinct elements of $\mathcal{A}_r$. If $x' \in A'_r \setminus A''_r$ and $x'' \in A''_r \setminus A'_r$, then $A_i \subseteq A'_{i+1} \cap A''_{i+1}$ with $A'_{i+1} = A_i \vee \{x'\}$ and $A''_{i+1} = A_i \vee \{x''\}$, $A'_{i+1} \neq A''_{i+1}$.

$(d_3) \Rightarrow (d_2)$. Obviously.

$(d_1) \Rightarrow (d_3)$. From (d_1) we have $A_i \subseteq A^1_{i+1} \cap A^2_{i+1}$ with $A^1_{i+1} \neq A^2_{i+1}$ and from (a) we get $A_i = A^1_{i+1} \cap A^2_{i+1}$. If $i+1 = r$ then (d_3) is proved. If $i+1 < r$ then from (d_1) and from (a) we get $A^1_{i+1} = A^{11}_{i+2} \cap A^{12}_{i+2}$ and $A^2_{i+2} = A^{21}_{i+2} \cap A^{22}_{i+2}$ so that

$A_i = \bigcap_{1 \leq h,\, k \leq 2} A^{hk}_{i+2}$, and so on.

$(d_4) \Rightarrow (d_3)$. Obviously.

$(d_1) \Rightarrow (d_4)$. Induction on j-i. If $j-i = 1$, then, by (a), (d_1) implies (d_4). Suppose that each $A_i \in \mathcal{A}_i$ is the intersection of the elements of $\mathcal{A}_{j-1}$ including A_i. By the previous argument each such A_{j-1} is the intersection of the elements of $\mathcal{A}_j$ including A_{j-1}; all such A_j are precisely the elements of $\mathcal{A}_j$ including A_i (because if $A_i \subset \bar{A}_j$ for some $\bar{A}_j \in \mathcal{A}_j$, then there exists $A_{j-1} \in \mathcal{A}_{j-1}$ with $A_i \subset A_{j-1} \subset \bar{A}_j$, by Proposition 5.1 (b)). □

If $A_h \in \mathcal{A}_h$ and $A_k \in \mathcal{A}_k$, we say that A_h and A_k are <u>joinable</u> if they are included in some $A \in \mathcal{A}$. If A_h and A_k are joinable, we define $A_h \vee A_k = \bigcap_{\substack{A \in \mathcal{A} \\ A \supset A_h \cup A_k}} A$; then $A_h \vee A_k = A_c \in \mathcal{A}_c$ for some $c \geq \max\{h,k\}$.

<u>PROPOSITION 5.4.</u> If $A_h \in \mathcal{A}_h$, $A_k \in \mathcal{A}_k$ are joinable with $A_h \vee A_k = A_c \in \mathcal{A}_c$ and $A_h \cap A_k = A_i \in \mathcal{A}_i$ then $h+k \geq i+c$.

<u>Proof.</u> Induction on $k-i \geq 0$. If $k=i$, then $A_k = A_i \subseteq A_h$ and $A_c = A_h \vee A_k = A_h \vee A_i = A_h$; it follows $h+k = c+i$.

If $k \geq i+1$, let be $A_{k-1} \in \mathcal{A}_{k-1}$ with $A_i \subseteq A_{k-1} \subset A_k$. Then $A_{c'} = A_h \vee A_{k-1} \subseteq A_h \vee A_k = A_c$. Then $c' \leq c$. If $c' < c$ then $A_k \not\subset A_{c'}$; if $x \in A_k \setminus A_{c'}$, then $x \notin A_{k-1}$ so that $A_c = A_h \vee A_k = A_h \vee (A_{k-1} \vee \{x\}) = (A_h \vee A_{k-1}) \vee \{x\} = A_{c'} \vee \{x\} = A_{c'+1}$. In conclusion $c'+1 \geq c$. By the induction hypothesis $h + k - 1 \geq i + c'$, so that $h + k \geq i + c' + 1 \geq i + c$. □

<u>THEOREM 5.5.</u> Let $G_r(X) = (S, \mathcal{I}, X, \mathcal{A})$ be a geometry of rank $r \leq 3$. Then $G_r(X)$ satisfies the following condition:

(*) if $A_h \in \mathcal{A}_h$, $A_k \in \mathcal{A}_k$, $A_h \cap A_k = A_i$ with $h + k \leq i + r$ and if $A_h \cup A_k \in \mathcal{I}$, then A_h and A_k are joinable.

Proof. We can suppose without loss of generality that $h \leqslant k$. The theorem is true for $r = 1$.

If $r = 2$, then actually $h + k - i \leqslant 2$. We can obviously assume that A_h and A_k are not in inclusion. In this case, the possible values of h, k and i are: $i=0$, $h=k=1$. Let A_1, $A_1' \in \mathcal{A}_1$ be such that $A_1 \cup A_1' \in \mathcal{I}$ with $A_1 \cap A_1' = A_o \in \mathcal{A}_o$. Then $A_1 \neq A_1'$, and if $x \in A_1' \setminus A_1$ it is easy to prove that $A_1 \vee \{x\}$ is an element of $\mathcal{A}_2$ including A_1 and A_1'.

If $r = 3$ then actually $h + k - i \leqslant 3$. We can obviously assume that A_h and A_k are not in inclusion. In this case, the possible values of h, k and i, which are distinct from the choice $i=0$, $h=k=1$ already considered, are $\{i=0, h=1, k=2\}$ and $\{i=1, h=k=2\}$.

When $i=0$, $h=1$ and $k=2$, we have $A_1 \in \mathcal{A}_1$, $A_2 \in \mathcal{A}_2$, $A_1 \cap A_2 = A_o$, $A_1 \cup A_2 \in \mathcal{I}$. It is easy to check that the element $A_3 \in \mathcal{A}_3$ defined by $A_2 \vee \{x\}$ where $x \in A_1 \setminus A_o$, contains A_1 and A_2.

When $i=1$, $h=k=2$, we have A_2, $A_2' \in \mathcal{A}_2$, $A_2 \cap A_2' = A_1 \in \mathcal{A}_1$ and $A_2 \cup A_2' \in \mathcal{I}$. Then it is easy to prove that the element $A_3 \in \mathcal{A}_3$, defined by $A_3 = A_2 \vee \{x\}$ where $x \in A_2' \setminus A_2 = A_2' \setminus A_1$, contains A_2 and A_2'. □

REMARK 5.6. When $r \geqslant 4$, condition (*) is not necessarily satisfied (cf. Theorem 5.8). It can be proved that the first values of i, h and k, for which the previous argument fails, are $\{i=0, h=k=2\}$.

LEMMA 5.7. There exists an $\mathcal{I}$-geometry $G_4(X) = (S,\mathcal{I},X,\mathcal{A})$, namely a permutation geometry of rank 4, for which condition (*) is not satisfied.

Proof. Let be $N = \{1,2,\dots,6\}$. An element $(h,k) \in N^2$ will be written hk. Let $X = \{11, 22, 33, 44, 45, 35, 26, 15\}$, $A_o = \emptyset$, $\mathcal{A}_o = \{A_o\}$,

$\mathcal{A}_1$ = {the set of singletons of X} = $\{\{xy\} : xy \in X\}$,

$\mathcal{A}_2$ = {the set of all the injective pairs of X}, $\mathcal{A}_3 = \bigcup_{i=1}^{10} A_3^{(i)}$ with:

$A_3^{(1)} = \{11, 22, 33, 45\}$, $A_3^{(2)} = \{11, 22, 44, 35\}$,

$A_3^{(3)} = \{33, 44, 11, 26\}$, $A_3^{(4)} = \{33, 44, 22, 15\}$,

$A_3^{(5)} = \{11, 45, 26\}$, $A_3^{(6)} = \{11, 35, 26\}$, $A_3^{(7)} = \{33, 45, 26\}$,

$A_3^{(8)} = \{33, 26, 15\}$, $A_3^{(9)} = \{44, 35, 26\}$, $A_3^{(10)} = \{44, 26, 15\}$,

$\mathcal{A}_4 = \bigcup_{i=1}^{3} A_4^{(i)}$ with: $A_4^{(1)} = \{11, 33, 45, 26\}$,

$A_4^{(2)} = \{11, 44, 35, 26\}$, $A_4^{(3)} = \{33, 44, 26, 15\}$. Let be

$\mathcal{I} = \{I \subseteq N^2 : I \text{ is injective}\}$ and $\mathcal{A} = \bigcup_{i=0}^{4} \mathcal{A}_i$.

It is easy to check that $(S, \mathcal{I}, X, \mathcal{A})$ is a permutation geometry $G_4(X)$ of rank 4 on X.

Let us consider the elements of $\mathcal{A}_2$: $A_2 = \{11, 22\}$ and $A_2' = \{33, 44\}$. Actually $A_2 \cap A_2' = \emptyset = A_\circ$, $2 + 2 - 0 \leqslant 4$, $A_2 \cup A_2' \in \mathcal{I}$, but A_2 and A_2' are not joinable, i.e. $G_4(X)$ does not satisfy condition (*). □

THEOREM 5.8. For each $r \geqslant 4$ there exists an $\mathcal{I}$-geometry (resp. a permutation geometry) $G_r(X)$ of rank r, such that condition (*) is not satisfied.

Proof. Induction on $r \geqslant 4$. For $r = 4$, the theorem reduces to Lemma 5.7.

Suppose that there exists an $\mathcal{I}$-geometry (resp. a permutation geometry) $G'_{r-1} = (S', \mathcal{I}', X', \mathcal{A}')$ of rank $r-1 \geqslant 4$, for which condition (*) is not satisfied; in other words, there exist $A_h' \in \mathcal{A}_h'$, $A_k' \in \mathcal{A}_k'$ such that: $A_h' \cap A_k' = A_i'$, $h + k - i \leqslant r - 1$, $A_h' \cup A_k' \in \mathcal{I}'$ and A_h' and A_k' are not joinable.

Let $G_1'' = (S'', \mathcal{I}'', X'', \mathcal{A}'')$ be an $\mathcal{I}$-geometry (resp. a permutation geometry) of rank 1 with $\mathcal{A}_\circ'' = \{A_\circ''\}$. Then we claim that $G_r = G'_{r-1} \oplus G_1'' = (A, \mathcal{I}, X, \mathcal{A})$ (resp. $G_r = G'_{r-1} \,\overline{\oplus}\, G_1''$) is an $\mathcal{I}$-geometry (resp. a permutation geometry) of rank r such that condition (*) is not satisfied. Indeed

$A_h = A_h' \cup A_\circ'' \in \mathcal{A}_h$, $A_k = A_k' \cup A_\circ'' \in \mathcal{A}_k$ are such that:

$A_h \cap A_k = (A_h' \cap A_k') \cup A_\circ'' = A_i' \cup A_\circ'' \in \mathcal{A}_i$, $h + k - i \leqslant r - 1 < r$,

$A_h \cup A_k = (A_h' \cup A_k') \cup A_\circ'' \in \mathcal{I}$ (resp. $\in \overline{\mathcal{I}}$), but A_h and A_k are not joinable, because if $A_h \cup A_k \subseteq (A' \cup A'') \in \mathcal{A}$ then $A_h' \cup A_k' \subseteq A' \in \mathcal{A}'$, which is impossible.

□

REMARK 5.9. From Theorem 5.8. we get a counterexample for Proposition 2.3. of [3] for all $r \geqslant 4$. The proof of Proposition 2.3. is incorrect because the flats "A" and "B'" are not necessarily such that $A \cup B'$ is an injective set.

We note that Proposition 2.3. of [3] is true for $r \leqslant 3$, by Theorem 5.6.

Other properties of $\mathcal{I}$-geometries, stated in [9], will be developed in another paper.

ACKNOWLEDGEMENT. This research was partially supported by GNSAGA of CNR and by MPI.

BIBLIOGRAPHY

[1] P.J. Cameron and M. Deza, On permutation geometries, J. London Math. Soc. (2) 20 (1979) 373-386.

[2] P.V. Ceccherini and G. Ghera, A Vagner-Preston type theorem for semigroups with right identities, Quad. Sem. Geom. Comb. Ist. Mat. Appl. Univ. L'Aquila 4, (1984).

[3] M. Deza and P. Frankl, Injection geometries, J. Comb. Theory (B) 37 (1984) 31-40.

[4] M. Deza and P. Frankl, On squashed designs, (to appear).

[5] G. Ghera, Algebra e geometria delle corrispondenze parziali di un insieme in sé, Tesi, Roma, Ist. Mat. "G. Castelnuovo", luglio 1983.

[6] M. Laurent, Geometries laminées: aspects algebriques et algorithmiques, These Univ. Paris VII (to appear).

[7] B. Segre, Istituzioni di geometria superiore (a.a. 1963-64), Appunti di P.V. Ceccherini Vol. III: Complessi, reti, disegni, Roma, Ist. Mat. "G. Castelnuovo", 1965.

[8] M.C. Schilling, Gêometries laminées et bouquets de matroids, These, Univ. Paris VI (to appear).

[9] N. Venanzangeli, Geometrie di permutazioni, geometrie iniettive e loro generalizzazione, Tesi, Roma, Ist. Mat. "G. Castelnuovo", luglio 1984.

Annals of Discrete Mathematics 30 (1986) 137–142

A NEW CHARACTERIZATION OF HYPERCUBES

Pier Vittorio Ceccherini and Anna Sappa
Dipartimento di Matematica "G. Castelnuovo"
Università di Roma "La Sapienza"
Città Universitaria, 00100 Roma, Italy

By using a theorem of S. Foldes [2], we prove that a finite graph G is a hypercube iff it is connected, bipartite, and the number of geodesics between any two vertices of the graph $G \times K_2$ depends only on their distance. Graphs of the type $G \times K_m$ are also considered.

1. INTRODUCTION

In what follows, all graphs will be finite without loops or multiple edges. If $G = (V,E)$ is a graph and if two vertices $x,y \in V$ are joined by a path, the distance $d(x,y)$ is defined as the number of edges in a geodesic (shortest path) between x and y. We denote by $\gamma(x,y) = \gamma_G(x,y)$ the number of distinct geodesics of G between x and y. If x=y, we put $d(x,y)=0$ and $\gamma(x,y)=1$.

We say that a connected graph G is a graph with a geodetic function if there exists a map $F:\{0,1,.., \text{diam } G\} \to N$ such that $\gamma(x,y) = F(d(x,y))$; in this case we shall say also that G is F-geodetic. A study of F-geodetic graphs is developed in [8] in a more general context.

For each positive integer n, the n-cube Q_n is defined (uniquely up to isomorphism) as the graph whose vertices are the subsets of a set S with n elements and two vertices are joined by an edge if and only if they differ for exactly one element. In other words the vertex set of Q_n is the set of ordered n-ples of 0 and 1; two vertices are joined by an edge if and only if they differ for exactly one digit. A hypercube is a graph isomorphic to some Q_n.

Several characterizations of hypercubes were given in [1] -[5]; in [5] the previous characterizations are also summarized.

We mention here the following characterization given by S. Foldes.

THEOREM 1.1 (S. Foldes [2]). A graph G is a hypercube if and only if

(1) G is connected and bipartite,

(2) G is F-geodetic with $F(k)=k!$.

Theorem 1.1 has been extended in [6] to a q-analogous result. A q-hypercube Q_n^q is defined as the graph whose vertices are the subspaces of a graphic space S of order q and dimension n-1, and two vertices are joined by an edge iff one of them is covered by the other in the lattice of all subspaces. For q=1, we have $Q_n^1 = Q_n$.

In this paper we give a characterization of hypercubes which is based on theorem 1.1 and on the concept of the translation graph TG' of a graph G' = (V',E'). The graph TG' is the permutation graph $(G',id_{V'})$, i.e. $TG' = G'xK_2$, where K_2 is the complete graph on two vertices (1 and 2), (cf. also [7]).

The description of TG' = (V,E) as the permutation graph $(G',id_{V'})$ is the following. Let G" = (V",E") be a copy of G' = (V',E') and let $x''\in V''$ denote the copy of $x'\in V'$; more generally, if $A'\subseteq V'$, then $A''\subseteq V''$ will denote the copy of A', i.e. $A'' = \{x''\in V'': x'\in A'\}$. Then TG' = (V,E) where $V = V'\cup V''$ and
$E = E'\cup E''\cup\{(x',x'')\in V'xV'': x'\in V'\}$.

The description of TG' = (V,E) as the graph $G'xK_2$ is the following:

$V = V'x\{1,2\} = \{(x',h) : x'\in V', h=1,2\}$,

$E = \{((x',h),(y',k))\in VxV : [(x'=y' \text{ and } (h,k)\in E(K_2)) \text{ or } ((x',y')\in E' \text{ and } h=k)]\}$.

We shall prove the following

THEOREM 1.2 A graph G' is a hypercube if and only if

(1) G' is connected and bipartite,

(2) $G' \times K_2$ is F-geodetic, for some F.

2. PROOF OF THEOREM 1.2.

We shall start with the following lemma.

LEMMA 2.1. Let G' be a connected graph and let $G = G'xK_2 = TG'$ be the translation graph of G'. The following conditions are equivalent

(a) G' is F'-geodetic with $F'(k) = k!$,

(b) G is F-geodetic with $F(k) = k!$,

(c) G is F-geodetic for some F.

Proof. Obviously (b) $\Rightarrow$ (a),(c). We use the description of $G = G'xK_2 = (V,E)$ as the permutation graph $(G', id_{V'})$ of $G' = (V',E')$ (§ 1).

(a) $\Rightarrow$ (b): let $x,y \in V$. It is enough to suppose $d(x,y) \geq 2$. If $x,y \in V'$, then $\gamma_G(x,y) = \gamma_{G'}(x,y) = d_{G'}(x,y)! = d_G(x,y)!$. Similarly if $x,y \in V''$, then $\gamma_G(x,y) = d_G(x,y)!$.

If $x=x' \in V'$ and $y=y'' \in V''$, let $x'' \in V''$ and $y' \in V'$ be the copies of x' and y'' respectively. We have $d_G(x,y) = d_G(x',y'') = d_{G'}(x',y') + d_G(y',y'') = d_G(x',y')+1$. Moreover each geodesic of G between $x=x'$ and $y=y''$ is of the form $g(x,y) = (x',\dots,z',z'',\dots,y'')$ where $g'(x',y') = (x',\dots,z',\dots,y')$ is a geodesic of G' between x' and y' and its copy $g''(x'',y'') = (x'',\dots,z'',\dots,y'')$ is a geodesic of G'' between x'' and y''. (The case $z'=x'$, i.e. $z''=x''$, and the case $z'=y'$, i.e. $z''=y''$, are not excluded). In other words, each geodesic $g(x,y)$ is obtained exactly once from a geodesic $g'(x',y')$ by choosing in $g'(x',y')$ a vertex z' (as the basis of the bridge (z',z'') between V' and V''). The number of such z' is $d_{G'}(x',y')+1$. So

$$\gamma_G(x,y) = \gamma_{G'}(x',y') \cdot (d_{G'}(x',y')+1) = d_{G'}(x',y')!\,(d_{G'}(x',y')+1) =$$

$$= (d_{G'}(x',y')+1)! = d_G(x,y)!.$$

(c) $\Rightarrow$ (b): actually $F(0) = F(1) = 1$. When $x,y \in V$ with $d_G(x,y) = k \geq 2$, there exist $x' \in V'$ and $y'' \in V''$ such that $d_G(x',y'') = k$. If $y' \in V'$ is the copy of y'', we have $d_{G'}(x',y') = d_G(x',y') = k - 1 \geq 1$. With the same argument used for proving (a) $\Rightarrow$ (b) we obtain from (c):

$$F(k) = \gamma_G(x',y'') = \gamma_{G'}(x',y') \cdot (d_{G'}(x',y')+1) =$$

$$= \gamma_G(x',y') \cdot (d_G(x',y')+1) = F(k-1) \cdot k.$$

From $F(1) = 1$ and $F(k) = k \cdot F(k-1)$, it follows $F(k) = k!$. □

By using Theorem 1.1 we can now give a new proof of the following well known result.

LEMMA 2.2. $Q_{n+1} = Q_n \times K_2$.

Proof. We can use the description of $Q_n \times K_2 = TQ_n$ as the permutation graph $(Q_n, id_{V'})$ of $Q_n = (V',E')$ (§ 1).

The graph Q_n is connected and bipartite (theorem 1.1), so that TQ_n is also connected and bipartite: if $A' \cup B'$ is a partition of the vertex set V' of Q_n then $(A' \cup B'') \cup (A'' \cup B')$ is a partition of the vertex set $V = V' \cup V''$ of TQ_n, where A'' and B'' are the copies of A' and B' respectively.

By theorem 1.1 the graph Q_n is F'-geodetic with $F'(k) = k!$, so that (Lemma 2.1) $Q_n x K_2$ is F-geodetic with $F(k) = k!$. The graph TQ_n is consequently a hypercube by theorem 1.1, and the lemma is proved since $\text{diam}(TQ_n) = \text{diam} Q_n + 1 = n+1$. □

<u>COROLLARY 2.3.</u> If T^n denotes the n-th power of the operator T (translation of a graph), we have $Q_{n+1} = T^n K_2$. □

We give now the

<u>Proof of theorem 1.2.</u> Let G' a hypercube. By theorem 1.1, condition (1) is satisfied. Moreover $G = TG' = G' x K_2$ is a hypercube by lemma 2.2, so that (by theorem 1.1 again) $G' x K_2$ is F-geodetic with $F(k) = k!$, $(0 \leqslant k \leqslant \text{diam } G)$, and condition (2') is satisfied.

Conversely if G' is a graph satisfying (1) and (2'), then, by lemma 2.1, G' satisfies condition (1) and (2) of theorem 1.1, so that G' is a hypercube. □

3. A GENERALIZATION OF LEMMA 2.1.

If $G' = (V', E')$ is any graph and K_m is the complete graph on m vertices $(m \geqslant 2)$, the graph $G = G' x K_m = (V, E)$ is defined by

$$V = V' x \{1,2,\dots,m\} = \{(x',h) : x' \in V',\ h = 1,2,\dots,m\},$$

$$E = \{((x',h),(y',k) \in VxV : [(x'=y' \text{ and } (h,k) \in E(K_m)) \text{ or } ((x',y') \in E' \text{ and } h=k)]\}.$$

An other description of the same graph $G = G' x K_m = (V,E)$ is the following. Let $G^i = (V^i, E^i)$ be a copy of $G' = (V', E')$ and let $x^i \in V^i$ denote the copy of $x' \in V'$, $i=1,2,\dots,m$, where we assume $G^1 = G'$, $x^1 = x'$. Then

$$V = \bigcup_{i=1}^{m} V^i,\ E = \bigcup_{i=1}^{m} E^i \cup \{(x^i, x^j) \in V^i x V^j : x' \in V',\ i,j=1,2,\dots,m; i \neq j\}.$$

Lemma 2.1 is included for m=2 in the following

<u>PROPOSITION 3.1.</u> Let $G' = (V', E')$ be a connected graph and let be $G = G' x K_m$ $(m \geqslant 2)$. The following conditions are equivalent

(a) G' is F'-geodetic with $F'(k) = k!$

(b) G is F-geodetic with $F(k) = k!$

(c) G is F-geodetic for some F.

<u>Proof.</u> We can use the same argument as for Lemma 2.1. Indeed, if $x,y \in V$ with $x=x^i \in V^i$ and $y=y^j \in V^j$ $(i \neq j)$, then the only essential point is that $G^i x \{i,j\} = TG^i \simeq TG'$. □

COROLLARY 3.2. A graph G' is a hypercube if and only if

(1) G' is connected and bipartite,

(2") $G'xK_m$ is F-geodetic for some F and some m. □

REMARK 3.3. Proposition 3.1 easily yields a family of connected graphs which are F-geodetic with F(k) = k! and which are not hypercubes (because they are not bipartite).

ACKNOWLEDGEMENTS. This research was partially supported by GNSAGA of CNR and by MPI.

BIBLIOGRAPHY

[1] L.R. Alvarez, Undirected graphs realizable as graphs of modular lattices, Can. J. Math. 17 (1965) 923-932.

[2] S. Foldes, A characterization of hypercubes, Discrete Math. 17 (1977) 155-159.

[3] J.M. Laborde, Characterization locale du graphe du n-cube, Journée Combinatoires, Grenoble (1978).

[4] J.D. McFall, Hypercubes and their characterizations, University of Waterloo Dept. of Combinatorics and Optimization Research Report CORR 78-26 (1978).

[5] J.D. McFall, Characterizing hypercubes, Annals Discrete Math. 9 (1980) 237-241.

[6] P.V. Ceccherini, A q-analogous of the characterization of hypercubes as graphs, J. Geometry 22 (1984) 57-74.

[7] A. Sappa, Caratterizzazione di grafi tramite geodetiche, Tesi, Univ. di Roma, Dipartimento di Matematica (1984).

[8] P.V. Ceccherini and A. Sappa, F-binomial coefficients and related combinatorial topics: perfect matroid designs, partially ordered sets of full binomial type and F-graphs, Annals Discrete Math. (this volume).

Annals of Discrete Mathematics 30 (1986) 143–158

F-BINOMIAL COEFFICIENTS AND RELATED COMBINATORIAL TOPICS: PERFECT MATROID DESIGNS, POSETS OF FULL BINOMIAL TYPE AND F-GEODETIC GRAPHS

Pier Vittorio Ceccherini and Anna Sappa
Dipartimento di Matematica "G. Castelnuovo"
Università di Roma "La Sapienza"
Città Universitaria, 00100 Roma, Italy

We introduce F-binomial coefficients as a natural generalization of binomial and q-binomial coefficients. A general calculus with these numbers leads to unify the arithmetical properties of (finite) projective and affine spaces and of Steiner systems S(t,k,v) into those of perfect matroid designs (§ 2). Partially ordered set of full binomial type (§ 3) and graphs such that the number of geodesics between any two vertices depends only on their distance (§ 4) are also studied by means of this formal calculus.

1. F-BINOMIAL COEFFICIENTS

Let N (resp. Q) denote the set of non negative integers (resp. rational numbers) and let be $N^* = N\setminus\{0\}$, $Q^* = Q\setminus\{0\}$.

Let $F: N \to Q^*$ be any function such that $F(0) = F(1) = 1$ and let $f: N \to Q$ be any function such that $f(0) = 0$, $f(1) = 1$ and $f(N^*) \subseteq Q^*$.

In what follows, F and f <u>will be functions satisfying the above conditions</u>. Given an F, the <u>associated</u> f will be defined by:

$f(0) = 0,\ f(n) = F(n)/F(n-1)$ for $n \geq 1$.

Given an f, the <u>associated</u> F will be defined by:

$F(0) = 1,\ F(n) = f(n)F(n-1)$ i.e. $F(n) = f(n)f(n-1)\ldots f(1)$ for $n \geq 1$.

Two such functions F and f are then <u>mutually associated</u>, and sometimes below they will be used interchangeably.

<u>DEFINITION 1.1.</u> Given a pair (F,f) of mutually associated functions and given any integers k,n with $0 \leq k \leq n$, we shall define the <u>F-binomial coefficient</u> (or <u>f-binomial coefficient</u>) $\left\{ {n \atop k} \right\}_F$ as the rational number

$$\left\{ {n \atop 0} \right\}_F = \left\{ {n \atop 0} \right\}_f := 1,\quad \left\{ {n \atop k} \right\}_F := \frac{F(n)}{F(k)F(n-k)} = \frac{f(n)f(n-1)\ldots f(n-k+1)}{f(k)f(k-1)\ldots f(1)} := \left\{ {n \atop k} \right\}_f \quad (1 \leq k \leq n).$$

These numbers turn out to be positive integers in the following examples (in which f and F also take integer values).

EXAMPLE 1.2. Binomial coefficients. Consider the pair of mutually associated functions

$$f_1(t) = t \quad \text{and} \quad F_1(t) = t! \quad \text{for all } t \in N.$$

Then

$$\{{n \atop k}\}_{F_1} = \{{n \atop k}\}_{f_1} = \{{n \atop k}\} \text{ is the usual binomial coefficient } (0 \leqslant k \leqslant n).$$

EXAMPLE 1.3. Gaussian numbers. Given an integer $q \geqslant 1$, consider the pair of mutually associated functions

$$f_q(t) = [t]_q \quad \text{and} \quad F_q(t) = [t]_q!$$

where $[\ ]_q$ and $[\ ]_q!$ are defined by

$$[0]_q = 0, \quad [1]_q = 1, \quad [t]_q = q^{t-1} + \dots + q+1,$$

$$[0]_q! = [1]_q! = 1, \quad [t]_q! = [t]_q\,[t-1]_q \dots [1]_q, \qquad t \geqslant 2.$$

Then

$$\{{n \atop k}\}_{F_q} = \{{n \atop k}\}_{f_q} = [{n \atop k}]_q \text{ is the q-binomial coefficient (gaussian number, cf. [4]).}$$

EXAMPLE 1.4. Constant coefficients. Given an integer $a \geqslant 1$, consider the pair of mutually associated functions

$$f(0) = 0, \quad f(1) = 1, \quad f(t) = a; \quad F(0) = F(1) = 1, \quad F(t) = a^{t-1} \qquad (t \geqslant 2).$$

Then

$$\{{n \atop n}\}_F = \{{n \atop 0}\}_F = 1, \quad \text{and } \{{n \atop k}\}_F = a \quad \text{for} \quad 0 < k < n.$$

Usual binomial identities and q-binomial identities (cf. [4]) are particular cases of F-binomial identities (obtained for $F = F_1$ and for $F = F_q$ resp.). We give now some of them, written as f-binomial identities.

PROPOSITION 1.5. The following f-binomial identities hold:

$$\left\{{n \atop 0}\right\}_f = \left\{{n \atop n}\right\}_f = 1, \quad \left\{{n \atop 1}\right\}_f = f(n), \quad \left\{{n \atop k}\right\}_f = \left\{{n \atop n-k}\right\}_f ,$$

$$\left\{{n \atop k}\right\}_f = \frac{f(n)}{f(n-k)}\left\{{n-1 \atop k}\right\}_f = \frac{f(n)}{f(k)}\left\{{n-1 \atop k-1}\right\}_f = \frac{f(n-k+1)}{f(k)}\left\{{n \atop k-1}\right\}_f \qquad (0< k <n),$$

$$\left\{{n \atop k}\right\}_f = \left\{{n-1 \atop k}\right\}_f + \frac{f(n)-f(n-k)}{f(k)}\left\{{n-1 \atop k-1}\right\}_f \qquad (0 < k < n),$$

$$\left\{{n \atop s}\right\}_f \left\{{s \atop k}\right\}_f = \left\{{n \atop k}\right\}_f \left\{{n-k \atop s-k}\right\}_f = \left\{{n \atop k}\right\}_f \left\{{n-k \atop n-s}\right\}_f \qquad (k \leqslant s \leqslant n). \quad \square$$

These formulas suggest the following

DEFINITION 1.6. Given a function f and any integers k,n with $0 < k < n$, define

$$\Delta^f_{n,k} = \Delta_{n,k} = \frac{f(n)-f(k)}{f(n-k)} .$$

PROPOSITIONS 1.7. Let $\Delta^f_{n,k}$ be as Def. 1.6. The following conditions are equivalent:

(a) $\Delta^f_{n,k}$ is independent of n;

(b) $f = f_q$ for some $q \in Q^*$ (where f_q is formally defined as in Ex. 1.3).

(c) $\Delta^f_{n,k} = q^k$ for some $q \in Q^*$.

Proof. Obviously (c) $\Rightarrow$ (a). We have (b) $\Rightarrow$ (c) because $\Delta^f_{n,k} = \Delta^{f_q}_{n,k} = \frac{f_q(n)-f_q(k)}{f_q(n-k)} = \frac{q^{n-1} + \ldots + q^k}{q^{n-k-1}+\ldots+q+1} = q^k$. We prove now that (a) $\Rightarrow$ (b). Let us put $\Delta^f_{n,1} = q$ for all n. Then $f(1) = 1$ and for all integers $n \geqslant 2$ we get $f(n) = \Delta^f_{n,1} f(n-1)+f(1)$, so that the equality $f(n) = q^{n-1} + \ldots + q+1$ follows by induction. $\square$

2. PERFECT MATROID DESIGNS

A combinatorial geometry (or simple matroid) M on a finite set S is a matroid on S such that $\emptyset$, S and every singleton of S are flats of M (cf. [2], [9]).

A <u>perfect matroid design</u> (PMD) is a combinatorial geometry M such that every k-flat (flat of rank k) has the same number f(k) of points, $k = 0,1,\dots,\mathrm{rk}\,M$ (cf. [10], [12], [14]). We shall say that f is the <u>size-function</u> of M. Obviously $f(0) = 0$ and $f(1) = 1$, so that f-binomial coefficients $\left\{ {n \atop k} \right\}_f$ can be considered ($0 \leqslant k \leqslant n \leqslant \mathrm{rk}\,M$).

PMDs include boolean sets 2^X, projective spaces, affine spaces, t-(v,k,1) designs (cf. [14]).

<u>PROPOSITION 2.1.</u> Let M be a PMD on a finite set S with size-function f.

(a) M is the matroid of all subsets of S if and only if $f = f_1$. In this case $f(n) = n$, $\left\{ {n \atop k} \right\}_f = \binom{n}{k}$ and $\Delta^f_{n,k} = 1$.

(b) M is the matroid of flats of a projective space of order q if and only if $f = f_q$ (with $q \geqslant 2$). In this case $f(n) = q^{n-1} + \dots + q+1$, $\left\{ {n \atop k} \right\}_f = \left[{n \atop k} \right]_q$ and $\Delta^f_{n,k} = q^k$. (Conversely, each of these equalities implies $f = f_q$).

(c) If M is the matroid of flats of an affine space of order q, then $f(n) = q^{n-1}$; the converse is true when $q \geqslant 4$. In this case $\left\{ {n \atop k} \right\}_f = q^{k(n-k)}$ and $\Delta^f_{n,k} = q^k - q^{2k-n}$. (Conversely, each of these equalities implies $f(n) = q^{n-1}$).

<u>Proof.</u> (a) is obvious. For (b) (resp. (c)), we have to prove only the "if" part. A simple counting argument shows that every 3-flat is a projective (resp. an affine) plane of order q, so that the result follows from a well known characterization of projective (resp. affine) spaces by means of planes, cf. [2] (resp. [1]). □

With an argument which is standard for finite projective spaces (cf. [14]) one can prove the following

<u>PROPOSITION 2.2.</u> Let M be a PMD with size-function f.

(a) The number of k-flats included in a r-flat (with $0 < k \leqslant r$) is given by

$$\alpha(k,r) = \frac{f(r)(f(r)-f(1))\dots(f(r)-f(k-1))}{f(k)(f(k)-f(1))\dots(f(k)-f(k-1))} =$$

$$= \frac{\Delta_{r,1}\dots\Delta_{r,k-1}}{\Delta_{k,1}\dots\Delta_{k,k-1}} \; \frac{f(r)\dots f(r-k+1)}{f(k)\dots f(1)} = \frac{\Delta_{r,1}\dots\Delta_{r,k-1}}{\Delta_{k,1}\dots\Delta_{k,k-1}} \left\{ {r \atop k} \right\}_f .$$

(b) When a k-flat is included in a r-flat, the number of j-flats between them ($0 \leqslant k < j < r$) is given by

$$\beta(k,j,r) = \prod_{s=k}^{j-1} \frac{f(r)-f(s)}{f(j)-f(s)} = \frac{\Delta_{r,k}\cdots\Delta_{r,j-1}}{\Delta_{j,k}\cdots\Delta_{j,j-1}} \left\{{r-k \atop j-k}\right\}_f .$$

In particular $\beta(0,j,r) = \alpha(j,r)$ and

$$\beta(k,k+1,r) = \frac{\Delta_{r,k}}{\Delta_{k+1,k}} f(r-k) .$$

(c) When a k-flat A is included in a r-flat B, the number of maximal chains of flats between them is given by

$$\vec{\gamma}(k,r) = \beta(k,k+1,r)\ \beta(k+1,k+2,r)\ldots\beta(r-2,r-1,r) = \frac{\Delta_{r,k}\cdots\Delta_{r,r-2}}{\Delta_{k+1,k}\cdots\Delta_{r-1,r-2}} F(r-k)$$

where F is associated to f. □

PROPOSITIONS 2.3. Let M be a PMD on a finite set S with size-function f (and associated function F) and let $\alpha(k,r)$, $\beta(k,j,r)$, $\vec{\gamma}(k,r)$ be defined as in proposition 2.2. The following conditions are equivalent:

(1) M is the PMD of a graphic space S of order q (i.e. a projective space of order q (possibly a "line"), when $q \geq 2$, or the boolean set 2^S, when $q = 1$);

(1a) $\alpha(k,r) = \left\{{r \atop k}\right\}_f$ for all $0 < k \leq r \leq \text{rk } M$;

(1b) $\beta(k,j,r) = \left\{{r-k \atop j-k}\right\}_f$ for all $0 \leq k < j < r \leq \text{rk } M$;

(1b') $\beta(k,k+1,r) = f(r-k)$ for all $0 \leq k < r \leq \text{rk } M$;

(1c) $\vec{\gamma}(k,r) = F(r-k)$ for all $0 \leq k < r \leq \text{rk } M$.

Moreover, if condition (1) is satisfied, then

(2) $f = f_q$, $\alpha(k,r) = \left[{r \atop k}\right]_q$, $\beta(k,j,r) = \left[{r-k \atop j-k}\right]_q$, $\vec{\gamma}(k,r) = [r-k]!$

Proof. It is well known (cf. [3], [16]) that (1) implies (1a)-(1c) and (2).

(1a) ⇒ (1). By Prop. 2.2, we get $\alpha(2,r) = \frac{\Delta_{r,1}}{\Delta_{2,1}} \cdot \left\{{r \atop 2}\right\}_f = \left\{{r \atop 2}\right\}_f \Rightarrow$

$\Delta_{r,1} = \Delta_{2,1} = f(2)-f(1) = q\ (\in \mathbb{N}^*)$, say $(r=2,3,\ldots,\text{rk } M)$. From $f(n) = \Delta_{n,1} f(n-1)+f(1)$ it follows that $f(n) = q^{n-1}+\ldots+q+1 = f_q(n)$ by induction on n. Since $f = f_q$, we get (1) by Prop. 2.1.

(1b') ⇒ (1). By Prop. 2.2, we get $\beta(k,k+1,r) = \frac{\Delta_{r,k}}{\Delta_{k+1,k}} f(r-k) = f(r-k) \Rightarrow$ $\Delta_{r,k} = \Delta_{k+1,k}$; for k=1 we obtain $\Delta_{r,1} = \Delta_{2,1}$ and (1) follows as above.

(1b) ⇒ (1a). $\alpha(k,r) = \beta(0,k,r) = \{{r-0 \atop k-0}\}_f = \{{r \atop k}\}_f$.

(1c) ⇒ (1b). In each maximal chain of flats between a k-flat A and a r-flat B there exists exactly one j-flat C ($0 \leq k < j < r \leq rk\ M$); given such a C, the joining of a maximal chain between A and C and of a maximal chain between C and B is a maximal chain between A and B. The double counting argument, applied to the set $\{(C,\phi): C$ is a j-flat belonging to a chain ϕ between A and B$\}$, gives:

$$F(r-k) = \beta(k,j,r)\, F(j-k)F(r-j), \text{ i.e. } \beta(k,j,r) = \frac{F(r-k)}{F(r-j)F(j-k)} = \{{r-k \atop j-k}\}_f. \quad \square$$

<u>REMARK 2.4.</u> Let M be the matroid of flats of an affine space of order q and define $\alpha(k,r)$, $\beta(k,j,r)$ and $\vec{\gamma}(k,r)$ as in Prop. 2.2. From (2) of Prop. 2.3, it follows that $f(0) = 0$, $f(t) = q^{t-1}$, $F(t) = q^{t(t-1)/2}$ and for $k > 0$

$$\alpha(k,r) = q^{r-k} \left[{r-1 \atop k-1}\right]_q, \quad \beta(k,j,r) = \left[{r-k \atop j-k}\right]_q, \quad \vec{\gamma}(k,r) = [r-k]_q!.$$

<u>REMARK 2.5.</u> We give now an other example of a PMD M with size-function f such that $\vec{\gamma}(k,r) \neq F(r-k)$. Let S be a Steiner system S(2,k,n) (e.g. the Steiner system whose blocks are the lines of a projective (resp. affine) space S of order q with k = q+1 (resp. k=q)) and let M be the PMD (of rank 3) whose flats are the empty set, the single points, the blocks and the full set of points. The size function f is such that $f(0) = 0$, $f(1) = 1$, $f(2) = k$, $f(3) = n$; the associated function F is such that $F(0) = F(1) = 1$, $F(2) = k$, $F(3) = nk$; moreover $\Delta_{2,0} = \Delta_{1,0} = 1$, $\Delta_{2,1} = k-1$, $\Delta_{3,1} = \frac{n-1}{k}$. Therefore

$$\vec{\gamma}(1,3) = \frac{\Delta_{3,1}}{\Delta_{2,1}} F(3-1) = \frac{n-1}{k(k-1)} F(2) = \frac{n-1}{k(k-1)} k = \frac{n-1}{k-1} \neq k = F(3-1)$$

whenever $n \neq k^2-k+1$ (in the previous examples, whenever the space S is not a projective plane).

3. POSETS OF FULL BINOMIAL TYPE

These structures have been introduced in [11], (cf. also [7]). We shall use the notation of §1.

Let $(\mathcal{P}, \leqslant)$ be a partially ordered set (poset) with minimum 0, and let a,b be elements of $\mathcal{P}$. If $a \leqslant b$, the interval between them is defined by $I(a,b) = \{c \in \mathcal{P} : a \leqslant c \leqslant b\}$. If $a < b$ a maximal chain between a and b is a chain $a = a_0 < a_1 < \ldots < a_n = b$ of I(a,b) which is not included in a longer chain of I(a,b); the number n is called the length of the chain. We shall denote by $\vec{\gamma}(a,b)$ the number of maximal chains of I(a,b), and we shall assume $\vec{\gamma}(a,b) = 1$ when a=b.

We shall say that $\mathcal{P}$ is a JD-poset if it satisfies the following condition (JD) (Jordan-Dedeking chain condition): if $a,b \in \mathcal{P}$ with $a < b$, all the maximal chains of I(a,b) have one and the same length, denoted by d(a,b) and called the distance between a and b (or the length of the interval (I(a,b)). If a=b, we put d(a,b) = 0.

If $\mathcal{P}$ is a JD-poset with minimum 0, then the rank of an element $a \in \mathcal{P}$ is defined by rk a=d(0,a), and the rank of $\mathcal{P}$ is defined by $\mathrm{rk}\ \mathcal{P} = \max\{\mathrm{rk}\ a : a \in \mathcal{P}\}$. We note that if $a,b \in \mathcal{P}$ with $a \leqslant b$, then $d(a,b) \leqslant \mathrm{rk}\ \mathcal{P}$.

We shall say that a JD-poset $\mathcal{P}$ (with minimum 0) has a chain function if there exists a function $F : \{0,1,\ldots,\mathrm{rk}\ \mathcal{P}\} \to N$ such that, for all $a,b \in \mathcal{P}$ with $a < b$, the number $\vec{\gamma}(a,b)$ of maximal chains between them is F(d(a,b)), i.e. every interval of length n of $\mathcal{P}$ has F(n) maximal chains. In this case we shall say that $\mathcal{P}$ is an F-chain poset.

DEFINITION 3.1. A poset $\mathcal{P}$ is called a poset of full binomial type if

(1) $\mathcal{P}$ has minimum 0,

(2) $\mathcal{P}$ is a JD-poset,

(3) $\mathcal{P}$ is a F-chain poset, for some F.

If $\mathcal{P}$ is a JD-poset with minimum 0, for each interval I(a,b) of $\mathcal{P}$ and for each integer k, with $0 \leqslant k \leqslant d(a,b)$, we shall put

$$I_k(a,b) = \{c \in I(a,b) : d(a,c) = k\}.$$

We shall say that $\mathcal{P}$ has a size function if there exists a fucntion f, $f : \{0,1,\ldots,\mathrm{rk}\ \mathcal{P}\} \to N$ such that, for all intervals I(a,b) of $\mathcal{P}$,

$$|I_1(a,b)| = f(d(a,b)).$$

In this case we shall say that $\mathcal{P}$ is an f-size poset. The following proposition yields other equivalent definitions of poset of full binomial type.

PROPOSITION 3.2. Let $\mathcal{P}$ be a JD-poset with minimum 0; the following conditions are equivalent, where f and F are mutually associated functions:

(a) $\mathcal{P}$ is an F-chain poset (i.e. $\mathcal{P}$ is a poset of full binomial type).

(b) $|I_k(a,b)| = \{{d(a,b) \atop k}\}_f \qquad (0 \leqslant k \leqslant d(a,b))$.

(c) $\mathcal{P}$ is an f-size poset.

Proof. Let $a,b \in \mathcal{P}$, $a \leqslant b$.

(a) $\Rightarrow$ (b). Let be $0 \leqslant k \leqslant d(a,b)$. In each maximal chain of $I(a,b)$ there exists exactly one $c \in \mathcal{P}$ such that $d(a,c) = k$. If c is such an element of $I(a,b)$, the joining of a maximal chain of $I(a,c)$ and of a maximal chain of $I(c,b)$ is a maximal chain of $I(a,b)$. The double counting argument applied to the set $\{(c,\alpha) : c \in \alpha,$ α is a maximal chain of $I(a,b)$, $d(a,c) = k\}$ gives:

$$F(d(a,b)) = |I_k(a,b)|\ F(k)\ F(d(a,b)-k),$$

so that (b) follows.

(b) $\Rightarrow$ (c). For k=1 we obtain $|I_1(a,b)| = \{{d(a,b) \atop 1}\}_f = f(d(a,b)$.

(c) $\Rightarrow$ (a). Actually $|I_1(a,b)| = f(d(a,b))$. We now apply induction on $d(a,b)$; if $d(a,b) = 0,1$, then $1 = \vec{\gamma}(a,b) = F(d(a,b))$. If $d(a,b) \geqslant 2$, apply the double counting argument to the set $\{(c,\alpha) : c \in \alpha, \alpha$ is a maximal chain of $I(a,b)$, $d(a,c)=1\}$. By the previous argument and by the induction hypothesis, we have

$$\vec{\gamma}(a,b) = |I_1(a,b)|\ F(1)\ F(d(a,b)-1) = f(d(a,b))\ F(d(a,b)-1) = F(d(a,b)). \ \square$$

EXAMPLE 3.3. If X is a finite set and if $\mathcal{P} = 2^X$ is the set of all subsets of X ordered by inclusion, then $\mathcal{P}$ is a poset of full binomial type, with size function $f(t) = t$ and chain function $F(t) = t!$ $(0 \leqslant t \leqslant |X|)$ (cf. [13]). Conversely any such functions f and F (with dom f = dom F = $\{0,1,\ldots,n\}$), (cf. example 1.2), can be considered as size function and chain function of a boolean set $\mathcal{P} = 2^X$ with $|X| = n$. We note that, if $\mathcal{P} = \{X\} \cup \binom{X}{0} \cup \binom{X}{1} \cup \ldots \cup \binom{X}{k}$ with $2 \leqslant k < |X| - 1$, then the poset $\mathcal{P}$ is not of full binomial type: if $Y \in \binom{X}{2}$ and $Z \in \binom{X}{k-1}$, then $d(\emptyset,Y) = d(Z,X) = 2$, but there are 2 chains from $\emptyset$ to Y and $|X| - (k-1) > 2$ chains from Z to X.

EXAMPLE 3.4. If X is a finite projective space PG(n-1,q) of order q and dimension $n-1 \geq 2$ and if $\mathcal{P}$ is the set of flats of X ordered by inclusion, then $\mathcal{P}$ is a poset of full binomial type with size function $f(t) = [t]_q$ and chain function $F(t) = [t]_q!$. (cf. [3], prop. 3.3.). Conversely, any such functions $f = [\ ]_q$ and $F = [\ ]_q!$ with $q = p^h$ ($h \geq 1$, p prime ≥ 2) and dom f = dom F = $\{0,1,\dots,n\}$ (cf. example 1.3) can be considered as the size and the chain functions of the poset of the subspaces of a projective space X = PG(n-1,q) of order q and dimension n-1.

We note that, if $\mathcal{P}$ is the set of all the i-dimensional flats of X = PG(n-1,q), with $i = -1,0,\dots,k,n-1$ where $1 \leq k < n-2$, then the poset $\mathcal{P}$ is not of full binomial type: if Y is a line and Z is a (k-1)-dimensional flat, then $d(\emptyset,Y) = d(Z,X) = 2$ but there are q+1 chains from $\emptyset$ to Y and $q^{n-k-1} + \dots + q+1 > q+1$ chains from Z to X.

EXAMPLE 3.5. If $\mathcal{P} = \{x_0, x_{11},\dots,x_{1a},\dots,x_{n1},\dots,x_{na}\}$ is any set with an+1 elements ($a,n \geq 1$), ordered by assuming x_0 as minimum and $x_{ij} < x_{hk}$ iff $i < h$, then $\mathcal{P}$ is a poset of full binomial type with size function $f(0)=0$, $f(1)=1$, $f(t)=a$ and with chain function $F(0) = F(1) = 1$, $F(t) = a^{t-1}$ ($2 \leq t \leq n$). Conversely, any such functions f and F (cf. example 1.3.) can be considered as the size and the chain functions of a poset of full binomial type $\mathcal{P}$ as above.

REMARK 3.6. Let M be the PMD of the flats of an affine space X = AG(n-1,q) of dimension $n-1 \geq 2$, and let $\mathcal{P}$ be the poset of the flats of X. The poset $\mathcal{P}$ is not of full binomial type: if Y is a line and Z is an (n-3)-dimensional flat, then $d(\emptyset,Y) = d(Z,X) = 2$, but there are q chains from $\emptyset$ to Y and q+1 chains from Z to X.

Therefore, if M is a PMD on a finite set X and if $\mathcal{P}$ is the set of the flats of M ordered by inclusion, then $\mathcal{P}$ is not necessarily a poset of full binomial type: the number of maximal chains in an interval I(a,b) does not depend only on the length of the interval, but also on the ranks of a and b (cf. also Remark 2.5, and Examples 3.3, 3.4). The case when $\mathcal{P}$ is a poset of full binomial type is characterized by the following proposition.

PROPOSITION 3.7. Let M be a PMD on a finite set X and let $\mathcal{P}$ be the poset of the flats of M ordered by inclusion. The poset $\mathcal{P}$ is of full binomial type if and only if M is one of the following PMD's:

(a) M is the trivial PMD of rank 1 or the trivial PMD of rank 2 on X (i.e. M is a trivial graphic space of dimension 0 on 1 on X resp.);

(b) M is the matroid 2^X of all subsets of X, i.e. M is the graphic space of order 1 and dimension $|X|-1$;

(c) M is the matroid $M_{n-1,q}(X)$ of all flats of a projective space, of dimension n-1 and order $q \geq 2$, having X as set of points.

Proof. If M is a PMD as in (a)-(c), then the poset $\mathcal{P}$ of its flats is a poset of full binomial type (cf. Ex. 3.3 and Ex. 3.4).

Let M be a PMD on X such that the poset $\mathcal{P}$ of its flats is a poset of full binomial type. If rk $M \leq 2$, then we are in the case (a). If rk $M > 2$, let f denote the size function of M.

For any flat Y of rank 2, we have that $|I_1(\emptyset,Y)| = |Y| = f(2) = f(d(\emptyset,Y))$. In other words, the poset $\mathcal{P}$ has the same size function f than the PMD M. Thus condition (b) of Prop. 3.2 holds; it means that condition (1b) of Prop. 2.3 holds. So condition (1) of Prop. 2.3 holds too, and the result (b)-(c) is proved. □

4. F-GEODETIC GRAPHS

In what follows, all graphs will be finite without loops or multiple edges, and all directed graphs will be without directed circuits.

Any directed graph $\vec{G} = (V,\vec{E})$ is obviously the Hasse diagram of a poset $\mathcal{P} = \mathcal{P}(\vec{G}) = (V,\leq)$ where $x < y$ if and only if there exists a directed path from x to y; conversely the Hasse diagram of a poset $\mathcal{P} = (V,\leq)$ is a directed graph $\vec{G} = \vec{G}(\mathcal{P}) = (V,\vec{E})$, where $(x,y) \in \vec{E}$ if and only if x is covered by y. We shall say that $\vec{G}$ and $\mathcal{P}(\vec{G})$ are mutually associated.

If $G = (V,E)$ (resp. $\vec{G} = (V,\vec{E})$) is a graph (resp. directed graph) and if two vertices $x,y \in V$ are joined by a path (resp. directed path), the distance $d(x,y)$ (resp. $\vec{d}(x,y)$) is defined as the number of edges in a geodesic i.e. in a shortest path (resp. directed shortest path) between x and y. We denote by $\Gamma(x,y) = \Gamma_G(x,y)$ (resp. by $\vec{\Gamma}(x,y) = \vec{\Gamma}_{\vec{G}}(x,y)$) the set of distinct geodesics of G (resp. of $\vec{G}$) between x and y; we put:

$$\gamma(x,y) = |\Gamma(x,y)| \text{ and } \vec{\gamma}(x,y) = |\vec{\Gamma}(x,y)| \text{ if } x \neq y;$$

$$\vec{\gamma}(x,x) = \gamma(x,x) = 1 \text{ and } d(x,x) = \vec{d}(x,x) = 0;$$

$$\text{diam } G = \max\{d(x,y) : x,y \in V\}, \quad \text{diam } \vec{G} = \max\{\vec{d}(x,y) : x,y \in V\}.$$

We shall say that a connected graph $G = (V,E)$ (resp. a directed graph $\vec{G} = (V,\vec{E})$)

is a graph with a geodetic function if there exists a map $F : \{0,1,\dots,\text{diam } G\} \to N$ (resp. $F : \{0,1,\dots,\text{diam } \vec{G}\} \to N$) such that $\gamma(x,y) = F(d(x,y))$ (resp. $\vec{\gamma}(x,y) = F(\vec{d}(x,y))$ for all $x,y \in V$. In this case we shall also say that G (resp. $\vec{G}$) is F-geodetic. Note that $F(0) = 1$ and that $F(n) \neq 0$ for all $n \in \text{dom } F$, so that the associated function f can be considered as in §1.

Let $\vec{G} = (V,\vec{E})$ be a directed graph and let $x,y \in V$ be such that $x \leqslant y$. Then the interval $I(x,y)$ is defined by

$$I(x,y) := \{z \in V: x \leqslant z \leqslant y\}, \text{ i.e. } I(x,y) \text{ is defined as in } \mathcal{P}(\vec{G}),$$

and the geodetic interval $I^g(x,y)$ is defined by

$$I^g(x,y) := \{z \in V: z \in \vec{g}(x,y) \text{ for some } \vec{g} \in \Gamma(x,y)\}.$$

Note that $I^g(x,y) \subseteq I(x,y)$ and that $I^g(x,y) = \{z \in I(x,y): \vec{d}(x,z) = \vec{d}(x,y) - \vec{d}(y,z)\}$. For any $1 \leqslant k \leqslant \vec{d}(x,y)$, let

$$I^g_k(x,y) := \{z \in I^g(x,y): \vec{d}(x,z) = k\} = \{z \in I^g(x,y): \vec{d}(z,y) = \vec{d}(x,y) - k\}.$$

We say that $\vec{G} = (V,\vec{E})$ has a source $0 \in V$ if for any $x \in V \setminus \{0\}$ there exists a directed path from 0 to x.

A graph $\vec{G} = (V,\vec{E})$ will be called a directed graph of full binomial type (with geodetic function F) if:

(a) $\vec{G}$ has a source 0,

(b) for all $x,y \in V$ with $x < y$: $I^g(x,y) = I(x,y)$,

(c) G is F-geodetic for some F.

PROPOSITION 4.1. Let $\vec{G} = (V,\vec{E})$ be a directed graph and let $\mathcal{P} = \mathcal{P}(\vec{G}) = (V,<)$ be the poset associated with $\vec{G}$ (so that $\vec{G} = \vec{G}(\mathcal{P})$). Then

(1) $\mathcal{P}$ has a minimum 0 if and only if $\vec{G}$ has source 0;

(2) $\mathcal{P}$ is a JD-poset if and only if $I^g(x,y) = I(x,y)$, for all $x,y \in V$ with $x < y$;

(3) $\mathcal{P}$ is a poset of full binomial type with chain function F if and only if $\vec{G}$ is a directed graph of full binomial type with geodetic function F.

Proof. (1) is obvious. (2): $\mathcal{P}$ is a JD-poset $\Leftrightarrow$ for all $x,y \in V$ with $x < y$ the set $M(x,y)$ of maximal chains of $\mathcal{P}$ in $I(x,y)$ is the set $\vec{\Gamma}(x,y)$ of the geodesics $\vec{g}(x,y)$ of $\vec{G}$ $\Leftrightarrow$ for all $x,y \in V$ with $x < y$ $I(x,y) = I^g(x,y)$. (3): $\mathcal{P}$ is a poset of full binomial type with chain function $F \Leftrightarrow \mathcal{P}$ has minimum 0, $\mathcal{P}$ is a JD-poset and $|M(x,y)| = F(\vec{d}(x,y))$ for all $x<y$ in $V \Leftrightarrow \vec{G}$ has source 0, for all $x<y$ in V $I^g(x,y) = I(x,y)$ and $|\vec{r}(x,y)| = F(\vec{d}(x,y)) \Leftrightarrow \vec{G}$ is a directed graph of full binomial type with geodetic function F. □

The poset $\mathcal{P} = (V,\leqslant)$, where $V = \{0,x,y,z,t\}$ and $0<x<z<y$, $0<t<y$, is not a JD-poset, but the graph $\vec{G}(\mathcal{P})$ is F-geodetic (with F=1); note that $I^g(0,y) = \{0,t,y\} \subset I(0,y) = V$.

PROPOSITION 4.2. Let $\vec{G} = (V,\vec{E})$ be a directed graph. The following conditions are equivalent (where F and f are mutually associated):

(a) $\vec{G}$ is F-geodetic for some F,

(b) for all $x<y$ in V and for all $1 \leqslant k \leqslant \vec{d}(x,y)$: $|I^g_k(x,y)| = \left\{{\vec{d}(x,y) \atop k}\right\}_F$,

(c) for all $x<y$ in V: $|I^g_1| = f(\vec{d}(x,y))$.

Proof. Let x,y be elements of V with $x<y$. (a) ⇒ (b). The double counting argument applied to the set $Z_k = (z,\vec{g}(x,y))$: $z \in \vec{g}(x,y)$, $\vec{g} \in \vec{r}(x,y)$, $\vec{d}(x,z) = k\}$ gives: $F(d(x,y)) = |I^g_k(x,y)|\, F(k)\, F(\vec{d}(x,y)-k)$, so that $I^g_k(x,y) = \left\{{\vec{d}(x,y) \atop k}\right\}_F$.

(b) ⇒ (c) for $k = 1$.

(c) ⇒ (a). We apply induction on $\vec{d}(x,y) \geqslant 1$. If $\vec{d}(x,y) = 1$, then $|\vec{r}(x,y)| = 1 = F(1)$. Suppose $\vec{d}(x,y) \geqslant 2$; by the induction hypothesis, when $z \in Z_1$ we have $|\vec{r}(z,y)| = F(\vec{d}(z,y)) = F(\vec{d}(x,y)-1)$. Therefore the double counting argument applied to the set Z_1 gives: $|\vec{r}(x,y)| = |I^g_1(x,y)|F(1)F(\vec{d}(x,y)-1) = f(\vec{d}(x,y))\, F(\vec{d}(x,y)-1) = F(\vec{d}(x,y))$. □

If $\vec{G}$ is a directed graph, we shall denote by G the undirected graph associated to $\vec{G}$.

PROPOSITION 4.3. Let $\vec{G} = (V,\vec{E})$ be a directed graph with source $0 \in V$. Suppose that for each $x \in V$ the parity of the length of any directed path from 0 to x depends only on x; write $p(x) = 0$ if it is even and $p(x) = 1$ if it is odd. Then the undirected graph G associated to $\vec{G}$ is connected and bipartite.

Proof. G is connected since 0 is a source. We verify that G is bipartite by assuming $V = V^0 \cup V^1$ where $V^i = \{x \in V\colon p(x) = i\}$, $i = 0,1$. We have to prove that if $(x,y) \in E$ then $p(x) \neq p(y)$. We can suppose without loss of generality that $(x,y) \in \vec{E}$. If $\vec{g}(0,x)$ and $\vec{g}(0,y)$ are geodesics from 0 to x and to y resp., then

$\vec{g}(0,y)$ and $\vec{g}(0,x) \cup (x,y)$ are both directed paths from 0 to y, so that their lengths have the same parity p(y). It follows that $p(x) \neq p(y)$. □

PROPOSITION 4.4. Let G = (V,E) be a connected bipartite graph and let 0 be any vertex of G. Then by starting from 0, a natural orientation can be defined on E, in such a way that $\vec{G} = (V,\vec{E})$ is a directed graph with source 0 (and without directed circuits). It follows that $\mathcal{P} = \mathcal{P}(\vec{G}) = (V,\leqslant)$ is a poset with minimum 0, where $x < y$ if and only if there is in $\vec{G}$ a directed path from x to y.

Proof. If $x,y \in V$ with $d(0,x) = d(0,y)$, then x and y cannot be adjacent because G is bipartite. Let (x,y) be an edge of G. We have either $d(0,y) = d(0,x)+1$ or $d(0,x) = d(0,y)+1$. Orient the edge from x to y in the first case, from y to x in the second. It is easy to show that, whenever there is an edge $(x,y) \in \vec{E}$, then (x,y) is the only directed path from x to y, and that whenever there exists a directed path from x to y then there is no directed path from y to x; indeed (by induction on i), if $(x_0,x_1,\ldots,x_i)$ is a directed path, then we have $(x_0,x_1),\ldots,(x_{i-1},x_i) \in \vec{E}$ and $d(0,x_i) = d(0,x_0) + i$. □

THEOREM 4.5. Let G = (V,E) be a connected bipartite graph and let $\vec{G} = (V,\vec{E})$ be the directed graph obtained by starting from a given vertex $0 \in V$ as in Prop. 4.4. Then, whenever x and y are elements of V with $x < y$, the following conditions are equivalent:

(a) p(x,y) is a directed path from x to y in $\vec{G}$,

(b) p(x,y) is a geodesic from x to y in G,

(c) p(x,y) is a geodesic from x to y in $\vec{G}$.

Proof. (a) ⇒ (b). Assume x = 0 first. Let $p(0,y) = (0=x_0,\ldots,x_i=y)$ be any directed path of lenght i from x to y in $\vec{G}$. We have $d(0,y)=i$, by the induction argument sketched at the end of the proof of Prop. 4.4. Therefore p(0,y) = p(x,y) is a geodesic in G.

Assume now $0 < x < y$. Let p(0,x) and p(x,y) be any directed paths in $\vec{G}$ from 0 to x and from x to y resp. By glueing p(0,x) and p(x,y) we get a directed path p(0,y) in $\vec{G}$. For the previous case p(0,y) is a geodesic in G. Thus its subpath p(x,y) must also be a geodesic in G.

(b) ⇒ (a). Induction on $i=d(x,y)$. When $i=1$, we have $p(x,y) = (x,y) \in \vec{E}$ since $x < y$, so that p(x,y) is a directed path in $\vec{G}$. Assume now $i \geqslant 2$ and suppose that any geodesic of G of length j with $1 \leqslant j < i$ is a directed path in $\vec{G}$. The geodesic $p(x,y) = (x_0=x, x_1, \ldots, x_{i-1}, x_i=y)$ is obtained by glueing the geodesics $p(x,x_{i-1})$ and (x_{i-1}, x_i) of G. These are both directed paths in $\vec{G}$ by the induction hypothesis. Then p(x,y) is a directed path in $\vec{G}$.

(c) ⇒ (a) obviously.

(b) ⇒ (c). If $p(x,y) = (x_0=x, x_1, \ldots, x_i=y)$ is a geodesic in G, then p(x,y) is a directed path in $\vec{G}$ (because (b) ⇒ (a)), and it must be a geodesic in $\vec{G}$, because any shorter directed path p'(x,y) in $\vec{G}$ would be a path of G shorter than p(x,y), which is impossible since p(x,y) is a geodesic in G. □

COROLLARY 4.6. Let G = (V,E) be a connected bipartite graph and let $\vec{G}$ be the directed graph obtained by starting from a given vertex $0 \in V$ as in Prop. 4.4. Then

(1) whenever $x,y \in V$ with $x \leqslant y$, we have $d(x,y) = \vec{d}(x,y)$, $\Gamma(x,y) = \vec{\Gamma}(x,y)$ so that $\gamma(x,y) = \vec{\gamma}(x,y)$;

(2) G is F-geodetic if and only if $\vec{G}$ is F-geodetic for all $0 \in V$.

Proof. (1) is obvious. For (2) it is enough to note that $\vec{\Gamma}(x,y) = \Gamma(x,y)$ when we assume 0 = x. □

COROLLARY 4.7. Let G = (V,E) be a connected bipartite graph. Then the following conditions are equivalent, where F and f are mutually associated functions:

(a) G is F-geodetic,

(b) $x,y \in V$, $0 \leqslant k \leqslant d(x,y) \Rightarrow |\{z \in V: d(x,z) = k,\ d(z,y) = d(x,y)-k\}| = \{{d(x,y) \atop k}\}_f$,

(c) $x,y \in V \Rightarrow |\{z \in V: d(x,z) = 1,\ d(z,y) = d(x,y)-1\}| = f(d(x,y))$. □

REMARK 4.8. The statement of Corollary 4.7 also holds if G is not bipartite; it can be proved by a direct argument (cf. [16]). This can also be obtained from the proof of Prop. 4.2, by replacing $\vec{g}(x,y)$, $\vec{\Gamma}(x,y)$, $\vec{d}(x,y)$ resp. with g(x,y), Γ(x,y), d(x,y) (and by exchanging consequently the set $I_k^g(x,y)$ and the set $\{z \in V: d(x,z) = k,\ d(z,y) = d(x,y)-k\})$.

EXAMPLE 4.9. The complete graph K_n, a tree G, the (2k+1)-circuit G are F-geodetic graphs with F=1. An F-geodetic graph with F=1 is called geodetic-graph (cf. [16], [18]). The 2k-circuit G is F-geodetic with F(t)=1, for $t < k$, and with F(2)=2 otherwise.

EXAMPLE 4.10. The complete bipartite graph $K_{n,n}=(V,E)$ with $V = V' \cup V''$,

$|V'| = |V''| = n$ is F-geodetic with $F(0) = F(1) = 1$, $F(2) = n$.

EXAMPLE 4.11. A hypercube is an F-graph with $F(t) = t!$. Conversely any connected bipartite F-graph with $F(t) = t!$ is a hypercube (cf. [13]).

EXAMPLE 4.12. The graph $K_n \times K_m$ is F-geodetic with $F(t) = t$ $(0 \leq t \leq 2 \leq n,m)$ (cf. [5], [15]).

EXAMPLE 4.13. If Q_n is the n-cube, the graph $Q_n \times K_m$ is F-geodetic with $F(t) = t!$ $(0 \leq t \leq n+1)$ (cf. [5], Prop. 3.1).

EXAMPLE 4.14. If G is a connected graph and if $G x K_m$ is F-geodetic for some F and some $m \geq 2$, then $F(t) = t!$. Moreover, if G is bipartite, then G is a hypercube (and $G x K_m$ is also a hypercube if and only if $m = 2$) (cf. [5], Prop. 3.1, Cor. 3.2)

EXAMPLE 4.15. Let $\vec{G}$ be the directed graph whose vertices are the subspaces of a graphic space of dimension n and of order $q \geq 1$ and (x,y) is an edge if the flat x is covered by the flat y. Then $\vec{G}$ is F-geodesic with $F(t) = [t]_q!$ (cf. [3], Prop. 3.3).

REMARK 4.16. Let G be the undirected graph associated to the directed graph $\vec{G}$ considered in the Ex. 4.15. In other words, G is the q-analogue of Q_{n-1}. Note that when $q \geq 2$, G is not F-geodesic: if x,y are two points and z is a plane containing x, then $d(x,y) = d(x,z) = 2$, but $2 = \gamma(x,y) \neq \gamma(x,z) = q+1$. Note also that the directed graph $\vec{G}$ obtained from G by starting from the empty flat is the graph of Example 4.15; when we start from a flat $0 \neq \emptyset$, then $\vec{G}$ is F-geodetic if and only if $q=1$.

EXAMPLE 4.17. Let $\vec{G} = \vec{G}(\mathcal{P})$ be the directed graph associated to the poset $\mathcal{P}$ of Ex. 3.5. Then $\vec{G}$ is F-geodetic accordingly to Prop. 4.1 (but G is not F-geodetic).

ACKNOWLEDGEMENT. This research was partially supported by GNSAGA of CNR and by MPI.

BIBLIOGRAPHY

[1] F. Buekenhout, Une charactērization des espaces affins basée sur la notion de droite, Math. Z. 111 (1969) 367-371.

[2] P.V. Ceccherini, Sulla nozione di spazio grafico, Rend. Mat. (5) 6 (1967) 78-98.

[3] P.V. Ceccherini, A q-analogous of the characterization of hypercubes as graphs, J. Geometry 22 (1984) 57-74.

[4] P.V. Ceccherini, A. Dragomir, Combinazioni generalizzate, q-coefficienti binomiali e spazi grafici, Atti Convegno Geometria Combinatoria e sue applicazioni (Perugia, Settembre 1970) 137-158.

[5] P.V. Ceccherini, A. Sappa, A new characterization of hypercubes, Annals Discrete Math. (this volume).

[6] L. Cerlienco, F. Piras, Coefficienti binomiali generalizzati, Rend. Sem. Fac. Sci. Cagliari 52 (1982) 47-56.

[7] L. Cerlienco, F. Piras, G-R-Sequences and incidence coalgebras of posets of full binomial type, (to appear).

[8] R.J. Cook, D.G. Pryce, Uniformly geodetic graphs, to appear.

[9] H. Crapo, G.C. Rota, Combinatorial geometries, MIT Press, Cambridge (1970).

[10] M. Deza, N.M. Singhi, Some properties of perfect matroid designs, Annals Discrete Math. 6 (1980) 57-76.

[11] P. Doubilet, G.C. Rota, R.P. Stanley, On the foundations of Combinatorial theory VI: the idea of generating function, in G.C. Rota (ed.), Finite Operator Calculus, Academic Press, New York (1975) 83-134.

[12] J. Edmonds, U.S.R. Murti, P. Young, Equicardinal matroids and matroid designs, in "Combinatorial Mathematics and its Applications", Second Chapel Hill Conference (1970).

[13] S. Foldes, A characterization of hypercubes, Discrete Math. 17 (1977) 155-159.

[14] B.L. Rothschild, N.M. Singhi, Characterizing k-flats in geometric designs, J. Comb. Theory A 20 (1976), 398-403.

[15] A. Sappa, Caratterizzazione di grafi tramite geodetiche, Tesi, Univ. di Roma, Dipartimento di Matematica (1984).

[16] R. Scapellato, On geodetic graphs of diameter two and some related structures (to appear).

[17] B. Segre, Lectures on modern geometry. With an Appendix by L. Lombardo-Radice, Cremonese, Roma (1961).

[18] J.C. Stempe, Geodetic graphs of diameter two, J. Comb. Theory B 17 (1974) 266-280.

Annals of Discrete Mathematics 30 (1986) 159–170

POLYNOMIAL SEQUENCES ASSOCIATED WITH A CLASS OF INCIDENCE COALGEBRAS.†

Luigi Cerlienco
Dip. di Matematica
Univ. di Cagliari
09100 Cagliari
Italy

Giorgio Nicoletti
Dip. di Matematica
Univ. di Bologna
40127 Bologna
Italy

Francesco Piras
Dip. di Matematica
Univ. di Cagliari
09100 Cagliari
Italy

A few special sequences of polynomials associated with both automorphisms and hemimorphisms of a particular class of coalgebras as well as their links with locally finite posets of binomial type are analysed.

§0.

The purpose of this paper is to briefly study a class of coalgebras, which we call *coalgebras of binomial type*. These are the coalgebras C having a countable basis (b_i) such that

$$\Delta b_i = \sum_{j=0}^{i} h_j^i \, b_j \otimes b_{i-j}, \qquad \varepsilon b_i = \delta_i^0$$

where h_j^i are integers and $h_0^0=1$.

Coalgebras of binomial type are the background for some recent work in combinatorics beginning with polynomials of binomial type [17]. We show that, under mild conditions, the structure constants h_j^i behave much like binomial coefficients, and that the analog of sequence of polynomials of binomial type is obtained in any coalgebra of binomial type from coalgebra morphism.

The one new coalgebraic notion introduced in this work is the notion of *hemimorphism*, namely, a linear map f on C into itself such that

$$\Delta \circ f = (I \otimes f) \circ \Delta .$$

The image of the distinguished basis (b_i) under a hemimorphism generalizes sequences of polynomials introduced by Goldman and Rota [14].

We believe the general setting of coalgebras (and Hopf algebras) to be suited to a variety of combinatorial problems which we propose to study in future publications.

§1.

1.1. Let $C:=(V,\Delta,\varepsilon)$ be a coalgebra over a field K of characteristic zero. Here V is a K-vector space and $\varepsilon: V \longrightarrow K$ *(counit)* and $\Delta: V \longrightarrow V \otimes V$ *(comultiplication or diagonalization)* are linear maps such that the following diagrams commute:

† Research partially supported by "Fondi Ministeriali per la Ricerca 40% e 60%".

(1) $$\begin{array}{ccc} V & \xrightarrow{\quad\Delta\quad} & V\otimes V \\ {\scriptstyle\Delta}\downarrow & & \downarrow{\scriptstyle\Delta\otimes I} \\ V\otimes V & \xrightarrow{\quad I\otimes\Delta\quad} & V\otimes V\otimes V \end{array}$$ (coassociativity)

(2) $$\begin{array}{ccccc} K\otimes V & \xleftarrow{\ \varepsilon\otimes I\ } & V\otimes V & \xrightarrow{\ I\otimes\varepsilon\ } & V\otimes K \\ & {\scriptstyle\psi}\nwarrow & \uparrow{\scriptstyle\Delta} & \nearrow{\scriptstyle\psi} & \\ & & V & & \end{array}$$ (counitary property)

(ψ is the canonical isomorphism and I the identical map).

If, moreover, we have $\Delta=T\circ\Delta$, where $T:V\otimes V\longrightarrow V\otimes V$, $v\otimes w\longmapsto w\otimes v$ (twist--operator), C is said to be *cocommutative*.
As usual, we shall denote by $C^*=(V^*,m,u)$ the dual K-algebra of C; we have $u=\varepsilon^*$ and $m=\Delta^*\circ j$, where $j:V^*\otimes V^*\longrightarrow(V\otimes V)^*$ is the canonical embedding.
A linear map $f:V\longrightarrow V$ is an *endomorphism* of C if

(3) $$\Delta\circ f = (f\otimes f)\circ\Delta$$

(4) $$\varepsilon\circ f = \varepsilon .$$

A linear map $f:V\longrightarrow V$ is said to be a *right hemimorphism* of C if

(5) $$\Delta\circ f = (I\otimes f)\circ\Delta .$$

Left hemimorphisms are defined in a similar way. If C is cocommutative, each right hemimorphism is also a left hemimorphism, and conversely. Let us denote by *Hem(C)* the set of all right hemimorphisms of the coalgebra C. Hem(C) is closed under linear combination and functional composition. Hence Hem(C) is an algebra. Moreover, if f^* is the dual map of $f\varepsilon\mathrm{Hem}(C)$, then for every $\alpha\varepsilon V$ we have:

$$f^*(\alpha) = f^*(m(\alpha\otimes 1)) = (f^*\circ m)(\alpha\otimes 1) = (f^*\circ(\Delta^*\circ j))(\alpha\otimes 1) =$$

$$= ((\Delta\circ f)^*\circ j)(\alpha\otimes 1) = (((I\otimes f)\circ\Delta)^*\circ j)(\alpha\otimes 1) = (\Delta^*\circ(I\otimes f)^*\circ j)(\alpha\otimes 1) =$$

(because of $(f\otimes g)^*\circ j = j\circ(f^*\otimes g^*)$)

$$= ((\Delta^*\circ j)\circ(I^*\otimes f^*))(\alpha\otimes 1) = (\Delta^*\circ j)(\alpha\otimes f^*(1)) = m(\alpha\otimes f^*(1)),$$

i.e., using $\alpha\beta$ instead of $m(\alpha\otimes\beta)$:

(6) $$f^*(\alpha) = \alpha\cdot f^*(1) \qquad \text{with } 1=u(1_K)\varepsilon V,$$

which becomes $f^*(\alpha) = f^*(1)\cdot\alpha$ in the case of left hemimorphisms.

The element $f^*(1)$ will be called the *indicator* of f and denoted by ind(f):

(7) $$\mathrm{ind}(f) := f^*(1) .$$

As a consequence of (6) we obtain:

<u>*Prop.1.*</u> *The map*

(7') $$ind : Hem(C) \longrightarrow C^*$$
$$f \longmapsto ind(f)$$

is a monomorphism of algebras.

(Notice, however, that (7') is an antimonomorphism in the case of left hemimorphisms).

Proof. It remains to prove that ind(f∘g) = ind(f)·ind(g); in fact, we have

$$(f\circ g)^*(1) = (g^*\circ f^*)(1) = g^*(f^*(1)) = f^*(1)\cdot g^*(1).$$ □

1.2. In the following we shall always assume that $C=(V,\Delta,\varepsilon)$ has a countable basis $(b_i)_{i\in N}$ such that

$$\Delta b_i = \sum_{j=o}^{i} h_j^i\, b_j \otimes b_{i-j}, \qquad h_o^o = 1 \tag{8}$$

$$\varepsilon b_i = \delta_i^o. \tag{9}$$

If this is the case, C is said to be a *coalgebra of binomial type*. It is then a strictly graded coalgebra. In particular, b_o is the unique group-like element (i.e. $\Delta b_o = b_o \otimes b_o$) and b_1 is a primitive element (i.e. $\Delta b_1 = b_1 \otimes b_o + b_o \otimes b_1$).

The dual algebra C^* is naturally endowed with a topological structure (the so-called *finite topology*) assuming the family

$$U^i = \{\beta\varepsilon V \mid (\forall j)(j\leq i \Rightarrow \beta(b_j)=0\}, \qquad i\varepsilon N$$

as a base for a system of neighbourhoods of zero. With this topology, each element α of C^* can be represented as follows:

$$\alpha = \sum_{i\geqslant o} \alpha_i b^i := \lim_{n\to\infty} \sum_{i=o}^{n} \alpha_i b^i$$

where

$$\alpha_i := \langle\alpha | b_i\rangle := \alpha(b_i) \;\varepsilon\, K$$

and

$$b^i : V \longrightarrow K$$
$$b_j \longmapsto \delta_j^i$$

This is sometimes expressed by saying that $(b^i)_{i\varepsilon N}$ is a *pseudo-basis* of V^*.

With reference to a fixed basis (b_i) and its dual pseudo-basis (b^i), we can represent each couple of elements

$$a = \sum_{i=o}^{n} a^i b_i \;\varepsilon\, C, \qquad \alpha = \sum_{i\geqslant o} \alpha_i b^i \;\varepsilon\, C^*$$

respectively as a column-vector (a^i) of entries a^i and as a row-vector (α_i) of entries α_i. Thus, we may write

$$\alpha(a) =: \langle\alpha | a\rangle = (\alpha_i)\times(a^i).$$

Moreover, for any given linear map $f: V \longrightarrow V$ let us define the *representing matrix M(f)*, in the same way as in the finite-dimensional case, to be the $(N\times N)$-matrix whose (r,s)-entry is given by

$$\langle r|f|s\rangle := \langle b^r | f(b_s)\rangle = \langle f^*(b^r) | b_s\rangle .$$

Of course the same matrix also represents the dual map f^*: $M(f)=M(f^*)$.

Notice that the i-th column $f(b_i)$ of M(f) has a finite support.

Using the above notational conveniences, we may easily deduce the following:

Prop. 2. *A linear map* $g: C^* \longrightarrow C^*$ *is a continuous (relative to the* *finite topology) if and only if there exists a linear map $f:C\to C$ such that $g=f^*$.*

Proof. In fact g is continuous if and only if its representing matrix M(g) has columns with finite support. □

If $\alpha=\Sigma\alpha_i b^i$, $\beta=\Sigma\beta_i b^i$ and $\gamma=\Sigma\gamma_i b^i:=m(\alpha\otimes\beta)$, we have

(10) $$\gamma_i = \sum_{j=0}^{i} h_j^i \alpha_j \beta_{i-j}$$

so that h_j^i's are also the structure constants of the algebra C^* relative to the pseudo-basis (b^i). This, together with Prop.1 and 2, implies:

Prop. 3. If C is a coalgebra of binomial type, then map (7') is an isomorphism of algebras.

Proof. In fact, for every $\beta\in C^*$ the representing matrix of the linear map $m(-\otimes\beta)$, because of (10), has columns with finite support. □

1.3. Commutativity of diagrams (1) and (2) gives rise to the following identities relative to structure constants h_j^i:

(11) $$h_j^i h_r^j = h_r^i h_{j-r}^{i-r}$$

(12) $$h_o^i = 1 = h_i^i .$$

Obviously, C is cocommutative if and only if

(13) $$h_j^i = h_{i-j}^i .$$

The structure constants h_j^i of a coalgebra of binomial type can all be expressed in terms of h_1^i alone, which will be more simply denoted by $\eta_i:=h_1^i$. In fact, we have the following:

Prop. 4. If $\eta_j\neq 0$ for $1\leq j\leq s-1$ but $\eta_s=0$, then for every i and every $j\in\{1,2,\ldots,s-1\}$ we have:

a) $$h_j^i = \frac{\eta_i\eta_{i-1}\cdots\eta_{i-j+1}}{\eta_1\eta_2\cdots\eta_j} =: \left[{i \atop j}\right]_\eta ;$$

b) if $\eta_p=0$ for some $p\geq s$, then $h_j^p=h_j^{p+1}=\ldots=h_j^{p+j-1}=0$;

c) $$h_j^i\cdot h_j^{i+1}\cdots\cdot h_j^{i+s-j} = 0.$$

If instead $\eta_j\neq 0$ for every j, then $h_j^i = \left[{i \atop j}\right]_\eta = \dfrac{\eta_i!}{\eta_j!\,\eta_{i-j}!}$ (where $\eta_o!:=1$ and $\eta_i!:=\eta_{i-1}!\,\eta_i$) for every i and every j. In such a case, C is cocommutative.

Proof. For r=1, (11) gives $h_j^i = h_{j-1}^{i-1}\,\eta_i/\eta_j$. From this, we deduce by recurrence both a) and b).

Let us first prove c) when j=1. From (11) and a) we get

$$h_s^{i+s-1}\eta_s = \eta_{i+s-1}\, h_{s-1}^{i+s-2} = \eta_{i+s-1}\,\frac{\eta_{i+s-2}\cdots\eta_i}{\eta_{s-1}\cdots\eta_1} ,$$

and then, since $\eta_s=0$, $\eta_i\eta_{i+1}\cdots\eta_{i+s-1}=h^i_1\cdot h^{i+1}_1\cdots h^{i+s-1}_1=0$. When $j>1$, c) follows from the case $j=0$ and from b). The second part of the statement is now trivial. □

Thus, each coalgebra of binomial type C is characterized, relative to the basis (b_i), by the sequence $\eta=(\eta_n)_{n\in N}$. Accordingly, we shall write $C_\eta=(V,\Delta_\eta,\varepsilon)$ instead of $C=(V,\Delta,\varepsilon)$. Furthermore, C_η is said to be of *full binomial type* if $\eta_1=1$ and $\eta_i\neq 0$ for every $i>0$.

<u>*Prop. 5*</u>. *Any two coalgebras of full binomial type, say C_η and C_λ, are isomorphic as coalgebras:*

$$C_\eta \xrightarrow{\sim} C_\lambda \qquad\qquad (14)$$

$$b_i \longmapsto \eta_i!/\lambda_i!\ b_i.$$

<u>Proof</u>. Trivial. □

Here are some examples:

1) Coalgebra of polynomials: $C_\eta=K[x]$, $b_i=x^i$, $h^i_j=\binom{i}{j}$, $\eta_n=n$. C^*_η is the algebra of divided power series. In the following we shall denote this coalgebra with C_N.

2) Coalgebra of divided powers: $C_\eta=K[x]$, $h^i_j=\eta_i=1$. C^*_η is the algebra of formal power series.

3) q-eulerian coalgebra: $C_\eta=K[x]$, $h^i_j=\binom{i}{j}_q=\dfrac{[i]_q!}{[j]_q!\,[i-j]_q!}$ (Gaussian coefficients) and $\eta_i=[i]_q:=1+q+q^2+\ldots+q^{i-1}$. C^*_η is said to be the algebra of formal eulerian series.

Coalgebras like these have a significant combinatorial counter-part. Let $\mathcal{P}$ be a locally finite partially ordered set (for short, l.f. poset) that satisfies the following further conditions:

a) all maximal chains in a given interval $[x,y]$ of $\mathcal{P}$ have the same cardinality (equal to "1+length$[x,y]$") (Jordan-Dedekind chain condition);

b) all intervals of length n in $\mathcal{P}$ possess the same number, say B_n, of maximal chains;

c) there exists in $\mathcal{P}$ only one minimal element.

After [12], these posets are said to be *l.f. posets of full binomial type*. With every l.f. poset of full binomial type of infinite length one can associate a coalgebra of full binomial type $C_\eta=(K[x],\Delta_\eta,\varepsilon)$ - the so-called maximally reduced incidence coalgebra of $\mathcal{P}$ - by denoting with b_i the residual class of all intervals of the same length i in $\mathcal{P}$ and assuming $\eta_i=B_i$. Thus, each structure constant h^i_j gives the number $h^i_j = B_i/B_jB_{i-j}$ of elements of rank j in any interval of length i. In this way, the coalgebras considered above correspond respectively to the following posets: a) the lattice of all finite subsets of a countable set; b) the countable chain; c) the lattice of all finite-dimensional subspaces of a vector space of dimension ω over GF(q).

§2.

In this section we shall show how both automorphisms and hemimorphisms of a coalgebra of full binomial type C_η are associated with special sequences of polynomials, whose great interest is well-known (at least in the particular case of the coalgebra of polynomials).

2.1. Let us begin by generalizing the notion of polynomial sequence of binomial type (see [17]).
In order to study analitically a coalgebra $C=(V,\Delta,\varepsilon)$ given in some intrinsic way, it is clear that we may arbitrarily choose any basis (v_i) of V. Then, all we have to know is the value of structure constants τ_i^{jr}, ε_i occurring in $\Delta v_i = \sum_{j,r} \tau_i^{jr} v_j \otimes v_r$ and $\varepsilon(b_i)=\varepsilon_i$. However, the chosen basis (v_i) is nothing but a useful tool. Thus, it may happen that the analyses regarding C carried out using two different bases (v_i), $(v_i^!)$ cannot be compared to each other by means of the map $v_i \longmapsto v_i^!$. This remark justifies the following definition.
Let $C_\eta=(V,\Delta_\eta,\varepsilon)$ be a coalgebra of (full) binomial type and let (b_i) be a basis fixed on it. A new basis $(b_i^!)$ of V is said to be an η-*basis* of C_η if the the map

(15) $$f: C_\eta \longrightarrow C_\eta \qquad b_i \longmapsto b_i^!$$

is an automorphism of coalgebras, that is

(16) $$\Delta_\eta b_i^! = \sum_{j=o}^{i} \begin{bmatrix} i \\ j \end{bmatrix}_\eta b_j^! \otimes b_{i-j}^! .$$

Consider the isomorphism

(17) $$\phi: C_\eta \longrightarrow C_N \qquad b_i \longmapsto \eta_i!/i!\; x^i$$

from C_η to the coalgebra of polynomials C_N. We shall say that a sequence $p_i(x)$ of polynomials is η-*nomial* if there exists a η-basis $(b_i^!)$ in C_η such that $p_i(x)=\phi(b_i^!)$. It is simple to prove that:

<u>Prop. 6</u>. *A polynomial sequence $p_i(x) \varepsilon K[x]$, $i \varepsilon N$, is η-nomial if and only if the following statements hold:*

1) $deg(p_i) = i$;

2) $p_o(x) = 1$;

3) $p_i(0) = 0$ for every $i \neq 0$;

4) $p_i(x+y) = \sum_{j=o}^{i} \begin{bmatrix} i \\ j \end{bmatrix}_\eta p_j(x)\, p_{i-j}(y)$. □

The interest in η-nomial sequences of polynomials is due to the fact that they enable us to carry out η-analog of umbral calculus along the lines followed by Rota and others [17], [18] (see also [8], [9], [10]).

The following propositions provide us with a useful tool in ordet to get η-nomial sequences.

<u>Prop. 7</u>. *Let C_η be a coalgebra of full binomial type and let $f: C_\eta \longrightarrow C_\eta$ be a morphism of coalgebras. Then the representative matrix $M(f)$ is completely determined by $f^*(b^1)$:*

(18) $$f^*(b^i) = (f^*(b^1))^i/\eta_i!$$

where the i-th power is calculated in C_η.

Proof. With a straightforward calculation, from (3) we get

$$<r+s|f|t>\cdot\left[{r+s \atop t}\right]_\eta = \sum_{j=0}^{t}\left[{t \atop j}\right]_\eta <r|f|j><s|f|t-j>; \qquad <0|f|t>=\delta_t^0$$

which imply (18). □

If $\alpha=\sum_{i\geq 0}\alpha_i b^i$, $\beta=\sum_{i\geq 1}\beta_i b^i \in C_\eta^*$, the element $\sum_{i\geq 0}(\alpha_i/\eta_i!)\beta^i \in C_\eta^*$ is said to be the *composition of* α *and* β and denoted by $\alpha\circ\beta$.

Prop. 8. The map

(19) $$\begin{array}{ccc} Aut(C_\eta) & \longrightarrow & \hat{C}_\eta^* \\ f & \longmapsto & f^*(b^1) \end{array}$$

is an isomorphism of the group $Aut(C_\eta)$ *of the automorphisms of the coalgebra* C_η *on the compositional group* $(\hat{C}_\eta^*,\circ)$ *of the elements* $\alpha=\Sigma\alpha_i b^i \in C_\eta^*$ *such that* $\alpha_0=0\neq\alpha_1$.

Proof. Because of (18), map (19) is a bijection. Moreover, from $M(f\circ g)=M(f)\times M(g)$ it follows:

$$(f\circ g)^*(b^1) = \sum_{i\geq 0} <f^*(b^1)|b_i> \frac{(g^*(b^1))^i}{\eta_i!} = f^*(b^1)\circ g^*(b^1). \quad \square$$

2.2. We come now to sequences of polynomials associated with hemimorphisms.

For the sake of simplicity, in the remainder of this section we assume that the underlying vector space of the coalgebra of binomial type (not necessarily of full binomial type) C_η is $V=K[x]$ with its canonical basis $b_i=x^i$:

$$\Delta_\eta x^i = \sum_{j=0}^{i} h_j^i\, x^j\otimes x^{i-j}.$$

Moreover, let us identify the linear dual of $K[x]$ with $K[[x]]$ and denote also b^i by x^i; thus, in general, the "series" x^i is not the i-th power of x^1 in C_η^*.

Consider in C_η^* the element $\zeta=\sum_{i\geq 0} x^i$ (zeta-function) and let $\mu=\sum_{i\geq 0}\mu_i x^i$ (Möbius function) be its multiplicative inverse: $\zeta\mu=1$. We have:

Prop. 9. Let $f\colon C_\eta \to C_\eta$ *be a hemimorphism such that* $ind(f)=f^*(1)=\mu$ *and let*

$$p_n(x) := f(x^n) = \sum_{i=0}^{n} h_i^n\,\mu_{n-i}\,x^i \quad \in C_\eta.$$

Then the sequence $(P_n(x,y))_{n\in N}$ *of homogeneous polynomials defined by* $P_n(x,1)=p_n(x)$ *satisfies the identities:*

(20) $$P_n(x,y) = \sum_{k=0}^{n} h_k^n\,P_k(x,z)\,P_{n-k}(z,y)$$

(21) $$P_n(1,0) = 1.$$

More generally:

Prop. 10. *If* $\sigma=\sum_{i\geq o}\sigma_i x^i$ *is an element of* C_η^*, $g: C_\eta \longrightarrow C_\eta$ *is the hemimorphism of indicator* σ, $s_n(x)$ *is the polynomial* $g(x^n)=\sum_{i=o}^{n} h_i^n \sigma_{n-i} x^i$ *and* $S_n(x,y)$ *the homogeneous polynomial such that* $S_n(x,1)=s_n(x)$, *then we have*

$$S_n(x,y) = \sum_{k=o}^{n} h_k^n\, P_k(x,z)\, S_{n-k}(z,y) \tag{22}$$

where $P_k(x,y)$ *is as in Prop.9.*

In order to prove the previous propositions, we need the following four lemmas.

Lemma 1. *For arbitrarily given polynomials* $P_n(x,y)=\sum_{r=o}^{n} p_n^r x^r y^{n-r}$ *and* $S_n(x,y)=\sum_{r=o}^{n} s_n^r x^r y^{n-r}$, *the identity*

$$\sum_{k=r}^{q} h_k^n\, p_k^r\, s_{n-k}^{q-k} = \delta_r^q\, s_n^q, \qquad q\geq r \tag{23}$$

is equivalent to (22).

Proof. Consider the identity

$$\sum_{k=o}^{n}\sum_{r=o}^{k}\sum_{t=o}^{n-k} A(k,r,t) = \sum_{t=o}^{n}\sum_{r=o}^{n-t}\sum_{k=r}^{t+r} A(k,r,r+t-k) \tag{24}$$

Formula (22) becomes:

$$S_n(x,y) - \sum_{k=o}^{n} h_k^n\, P_k(x,z)\, S_{n-k}(z,y) =$$

$$= \sum_{r=o}^{n} s_n^r x^r y^{n-r} - \sum_{k=o}^{n}\sum_{r=o}^{k}\sum_{t=o}^{n-k} h_k^n\, p_k^r\, s_{n-k}^t\, x^r y^{n-k-t} z^{k-r+t} =$$

(because of (24))

$$= \sum_{r=o}^{n} s_n^r x^r y^{n-r} - \sum_{t=o}^{n}\sum_{r=o}^{n-t}\sum_{k=r}^{t+r} h_k^n\, p_k^r\, s_{n-k}^{r+t-k}\, x^r y^{n-r-t} z^t =$$

$$= \sum_{r=o}^{n} \{s_n^r - h_r^n\, p_r^r\, s_{n-r}^o\} x^r y^{n-r} - \sum_{t=1}^{n}\sum_{r=o}^{n-t}\{\sum_{k=r}^{t+r} h_k^n\, p_k^r\, s_{n-k}^{r+t-k}\} x^r y^{n-r-t} z^t = 0.$$

Putting r+t=q, this is equivalent to (23). □

Lemma 2. *Let* $P_n(x,y)=\sum_{r=o}^{n} p_n^r x^r y^{n-r}$ *be a sequence of polynomials satisfying (20) and (21). Let us consider the matrices* $P=(p_n^r)$ *and* $H=(h_k^n)$ *(with* $h_k^n=o$ *for* $n<k$*). Then*

$$^tH\cdot P = P\cdot {}^tH = I. \tag{25}$$

Thus, the sequence $P_n(x,y)$ *is completely determined by the coefficients* h_k^n.

Proof. From (20), when y=o and z=1, we get

$$x^n = \sum_{k=o}^{n} h_k^n\, P_k(x,1).$$

This, together with

$$P_n(x,1) = \sum_{k=o}^{n} p_n^k\, x^k,$$

expresses the change of bases $x^n \longleftrightarrow P_n(x,1)$ in K[x], that is (25) □

Lemma 3. In the hypotheses of Lemma 2., the identity

$$s_n^q\, h_t^q = h_t^n\, s_{n-t}^{q-t} \tag{26}$$

is equivalent to (23), and then to (22).

Proof. (23) implies (26). In fact, if we multiply (23) by h_t^r and sum with respect to the index r, then we get:

$$s_n^q\, h_t^q = \sum_{r=t}^{q} \sum_{k=r}^{q} h_k^n\, p_k^r\, s_{n-k}^{q-k}\, h_t^r = \sum_{k=t}^{q} h_k^n\, s_{n-k}^{q-k} \sum_{r=t}^{k} p_k^r\, h_t^r =$$

(because of (25))

$$= \sum_{k=t}^{q} h_k^n\, s_{n-k}^{q-k}\, \delta_t^k = h_t^n\, s_{n-t}^{q-t}.$$

Conversely, (26) implies (23):

$$\delta_r^q\, s_n^q = \text{(because of (25))} = s_n^q \sum_{k=r}^{q} h_k^q\, p_k^r = \sum_{k=r}^{q} h_k^n\, s_{n-k}^{q-k}\, p_k^r. \qquad \square$$

Lemma 4. Let $P_n(x,y)=\sum_{k=0}^{n} p_n^k\, x^k y^{n-k}$ be an arbitrary sequence of polynomials. Then, (20) $\Longleftrightarrow$ (23) $\Longleftrightarrow$ ((25) and (26)) with $s_n^k=p_n^k$.

Proof. As in Lemma 1. and Lemma 3. $\square$

Proof of Prop.9. Consider the coefficients $p_n^i=h_i^n\cdot\mu_{n-i}$. Observe that the hypothesis $\zeta\mu=1$ is expressed by (25). Therefore, owing to Lemma 4., it is sufficient to prove formula (26). In fact we have:

$$h_t^n\, p_{n-t}^{q-t} = h_t^n\, h_{q-t}^{n-t}\, \mu_{n-q} = \text{(because of (11))} = h_q^n\, h_t^q\, \mu_{n-q} = h_t^q\, p_n^q. \qquad \square$$

By a similar argument, one can prove Prop. 10.

Conversely, if we call any sequence of homogeneous polynomials $P_n(x,y)$ (respectively, $S_n(x,y)$) satisfying (20) and (21) (respectively, (22)) relative to suitable coefficients h_k^n a *Goldman-Rota-sequence* (respectively *Goldman-Rota-Sheffer-sequence*), it is possible to prove that:

Prop. 11. The coefficients h_k^n ($h_o^n\neq 0$) associated with any Goldman-Rota-sequence can be assumed as structure costants of a coalgebra of binomial type.

Proof. We have to prove that (20) implies both (11) and (12). We have

$$h_r^i\, h_{j-r}^{i-r} = \sum_{k=r}^{i} \delta_k^i\, h_r^k\, h_{j-r}^{k-r} = \text{(because of Lemma 2.)}$$

$$= \sum_{k=r}^{i} \{\sum_{n=k}^{i} p_n^k\, h_n^i\} h_r^k\, h_{j-r}^{k-r} = \text{(because of (26))}$$

$$= \sum_{k=r}^{i} \sum_{n=k}^{i} h_n^i\, h_{j-r}^{k-r}\, h_r^n\, p_{n-r}^{k-r} = \sum_{n=r}^{i} h_n^i\, h_r^n \sum_{k=r}^{n} h_{j-r}^{k-r}\, p_{n-r}^{k-r} =$$

(because of Lemma 2.)

$$= \sum_{n=r}^{i} h_n^i\, h_r^n\, \delta_{j-r}^{n-r} = h_j^i\, h_r^j\,,$$

that is formula (11). Moreover, (12) is a straightforward consequence of (21) and Lemma 2. $\square$

In conclusion, we give a combinatorial interpretation to both Goldman-Rota-sequences and Goldman-Rota-Sheffer-sequences.

Prop. 12. *Let C_η be the maximally reduced incidence coalgebra of a l.f. poset of full binomial type $\mathcal{P}$. In respect of this coalgebra C_η, let us consider the Goldman-Rota-sequence $P_n(x,y)$ and the Goldman--Rota-Sheffer-sequence associated with the series $\zeta=\sum_{n\geq 0} x^n$. Furthermore, let $p_n(x)=P_n(x,1)$ and $s_n(x)=S_n(x,1)$. Then each interval of length n in $\mathcal{P}$ has $p_n(x)$ as its characteristic polynomial and $s_n(x)$ as its level number indicator (that is, the coefficients of $s_n(x)$ are the level numbers of second kind of the given interval).*

Proof. See [7]. □

REFERENCES:

[1] Abe, E., Hopf Algebras (Cambridge Univ. Press, Cambridge, 1980).

[2] Aigner, M., Combinatorial Theory (Springer-Verlag, New York, 1979).

[3] Allaway, W.R., A Comparison of two Umbral Algebras, J.Math. Anal.Appl. 85 (1982) 197-235.

[4] Andrews, G.E., On the Foundations of Combinatorial Theory V: Eulerian Differential Operators, Studies in Appl.Math. 50 (1971) 345-375.

[5] Cerlienco, L. and Piras, F., Coalgebre graduate e sequenze di Goldman-Rota, Actes du Séminaire Lotharingien de Combinatoire, Publ. de l'IRMA, Strasbourg, 230/S-09 (1984) 113-125.

[6] Cerlienco, L. and Piras, F., Aspetti coalgebrici del calcolo umbrale, Atti del Convegno "Geometria combinatoria e d'incidenza: fondamenti e applicazioni", Rend.Sem.Mat. Brescia 7 (1984) 205-217.

[7] Cerlienco, L. and Piras, F., G-R-sequences and incidence coalgebras of posets of full binomial type, J.Math.Anal.Appl. (to appear).

[8] Cerlienco, L., Nicoletti, G. and Piras, F., Automorphisms of graded coalgebras and analogs of the umbral calculus, Actes du Séminaire Lotharingien de Combinatoire, Publ. de l'IRMA, Strasbourg, 230/S-09 (1984) 126-132.

[9] Cerlienco, L., Nicoletti, G. and Piras, F., Coalgebre e Calcolo Umbrale, Rend.Sem.Mat.Fis. Milano (to appear).

[10] Cerlienco, L., Nicoletti, G. and Piras, F., Umbral Calculus, Actes du Séminaire Lotharingien de Combinatoire, Publ. de l'IRMA, Strasbourg, 266/S-11 (1985) 1-27.

[11] Comtet, L., Advanced Combinatorics (Reidel P.C., Boston, 1974).

[12] Doubilet, P., Rota, G.-C. and Stanley, R.P., On the Foundations of Combinatorial Theory VI: The Idea of Generating Function, in: Rota, G.-C. (Ed.), Finite Operator Calculus (Academic Press, New York,1975) 83-134.

[13] Garsia, A.M. and Joni, S.A., Composition Sequences, Comm. Algebra 8 (1980) 1195-1266.

[14] Goldman, G.R. and Rota, G.-C., On the Foundations of Combinatorial Theory IV: Finite Vector Spaces and Eulerian Generating Functions, Studies in Appl.Math. 49 (1970) 239-258.

[15] Ihrig, E.C. and Ismail, M.E., A q-umbral Calculus, J.Math.Anal. Appl. 84 (1981) 178-207.

[16] Joni, S.A. and Rota, G.-C., Coalgebras and Bialgebras in Combinatorics, Studies in Appl.Math. 61 (1979) 93-139.

[17] Mullin, R. and Rota, G.-C., On the Foundations of Combinatorial Theory III: Theory of Binomial Enumeration, in: Harris (Ed), Graph Theory and its Applications (Academic Press, New York, 1970) 167-213.

[18] Rota, G.-C., Kahaner, D. and Odlyzko, A., On the Foundations of Combinatorial Theory VIII: Finite Operator Calculus, J.Math. Anal.Appl. 42 (1973) 684-760.

[19] Taft, E.J., Non-cocommutative Sequences of Divided Powers, Lecture Notes in Math. 933 (Springer-Verlag, New York, 1980) 203-209.

Annals of Discrete Mathematics 30 (1986) 171–184

$\mathcal{R}$-REGULARITY AND CHARACTERIZATIONS OF THE GENERALIZED QUADRANGLE P(W(s),(∞))

M. De Soete and J.A. Thas

Seminar of Geometry
State University of Ghent
Krijgslaan 281
B-9000 Gent-Belgium

In a generalized quadrangle of order (s,s+2), s ≠ 1, regular points cannot occur. Therefore we introduce the notion of $\mathcal{R}$-regularity for points and lines in those generalized quadrangles of order (s,s+2) which contain a spread $\mathcal{R}$. Using these new concepts we give three characterization theorems for P(W(s),(∞)).

I. *INTRODUCTION*

1. DEFINITIONS

A finite *generalized quadrangle* is an incidence structure S = (P,B,I) where P and B are sets of elements called *points* and *lines* resp., with a symmetric incidence relation I which satisfies the following axioms :

(i) each point is incident with 1+t lines ($t \geq 1$) and two distinct points are incident with at most one line;

(ii) each line is incident with 1+s points ($s \geq 1$) and two distinct lines are incident with at most one point;

(iii) for each non-incident point-line pair (x,L), there exists a unique pair $(y,M) \in P \times B$ such that x I M I y I L.

We call s and t the parameters of the generalized quadrangle, and (s,t) (or s if s = t) is the *order* of S. There holds $|P| = v = (1+s)(1+st)$, $|B| = b = (1+t)(1+st)$ and $s+t \mid st(s+1)(t+1)$ [12]. Moreover there is a *point-line duality* for generalized quadrangles of order (s,t); this means that in any definition or theorem the words point and line and the parameters s and t may be interchanged.

If the points x,y (resp. lines L,M) are collinear (resp. concurrent) we write $x \sim y$ (resp. $L \sim M$). The line defined by distinct collinear points x,y is denoted by xy; the point defined by distinct concurrent lines L,M is denoted by LM or $L \cap M$.

Let S be a generalized quadrangle of order (s,t). For $x \in P$, we define the *star* of x as $x^{\perp} = \{z \in P \parallel z \sim x\}$; remark that $x \in x^{\perp}$. The *trace* of two distinct points (x,y) is the set $x^{\perp} \cap y^{\perp}$ and is denoted by $\{x,y\}^{\perp}$. There holds $|\{x,y\}^{\perp}| = s+1$ if $x \sim y$ and $|\{x,y\}^{\perp}| = t+1$ if $x \not\sim y$. More generally, for $A \subset P$ we define $A^{\perp} = \cap\{x^{\perp} \parallel x \in A\}$. For $x \neq y$ we define the *span* of (x,y) as the set $\{x,y\}^{\perp\perp} = \{u \in P \parallel u \in z^{\perp}, \forall z \in \{x,y\}^{\perp}\}$. If $x \not\sim y$, the set $\{x,y\}^{\perp\perp}$ is also called the *hyperbolic line* defined by x and y. We have $|\{x,y\}^{\perp\perp}| = s+1$ if $x \sim y$ and $|\{x,y\}^{\perp\perp}| \leqslant t+1$ if $x \not\sim y$. A pair of distinct points (x,y) is *regular* provided $x \sim y$ or $x \not\sim y$ and $|\{x,y\}^{\perp\perp}| = t+1$. A point x is *regular* provided (x,y) is regular for all $y \in P$, $y \neq x$. If S has a regular pair of non-collinear points, then $s = 1$ or $s \geqslant t$ [9]. A *triad* of points is a triple of pairwise non-collinear points. A point u is called a *center* of the triad (x,y,z) iff $u \in x^{\perp} \cap y^{\perp} \cap z^{\perp}$. A *spread* of S is a subset $\mathcal{R}$ of B such that each point of S is incident with exactly one line of $\mathcal{R}$. There holds $|\mathcal{R}| = st+1$.

2. THE MODELS W(q), $T_2(O')$, $T_2^*(O)$ AND P(S,x)

(a) The points of PG(3,q) together with the totally isotropic lines with respect to a symplectic polarity, define a generalized quadrangle W(q) with parameters $s = t = q$, $v = b = (q+1)(q^2+1)$. We remark that the lines of W(q) are the elements of a linear line complex in PG(3,q) [10].

All points of W(q) are regular; the lines of W(q) are regular iff q is even [9]. We note also that W(q) is self-dual iff q is even [9].

(b) Let O' be an oval [4] of PG(2,q) = H, where H is embedded in PG(3,q) = P. Define points as (i) the points of $P \setminus H$, (ii) the planes X of P with $|X \cap O'| = 1$, (iii) a new symbol (∞). The lines are (a) the lines of P which are not contained in H and meet O', and (b) the points of O'. The incidence is defined as follows. A point of type (i) is only incident with lines of type (a) and the incidence is that of P. A point of type (ii) is incident with the lines of type (a) and (b) contained in it. The point (∞) is incident with no line of type (a) but with all lines of type (b). The obtained structure is a generalized quadrangle with parameters $s = t = q$, $v = b = (q+1)(q^2+1)$ and is denoted by $T_2(O')$.

These quadrangles are due to J. Tits [4]. The quadrangle $T_2(O')$ is isomorphic to W(q) iff q is even and O' is an irreducible conic [9]. Further, all points of type (b) are regular, and the point (∞) is regular iff q is even [9].

(c) Let O be a complete oval (i.e. a (q+2)-arc [10]) of the plane at infinity of AG(3,q), q even. The points of the structure $T_2^*(O)$ are the points of AG(3,q); the lines of $T_2^*(O)$ are the lines of AG(3,q) intersecting the plane at infinity in the points of O; the incidence is that of AG(3,q). Then $T_2^*(O)$ is a generalized quadrangle of order (q-1,q+1). It was first discovered by R.W. Ahrens and G. Szekeres [1] and independently by M. Hall Jr. [5].

A pair of non-concurrent lines (L,M) of $T_2^*(O)$ is regular iff the lines are parallel [9].

(d) In [7] S.E. Payne gives an important construction of generalized quadrangles of order (s-1,s+1). Consider a generalized quadrangle S = (P,B,I) of order s,s > 1, with a regular point x. Define P' as the set $P \setminus x^\perp$. In B' there are two types of elements : the elements of type (a) are the lines of B which are not incident with x, the elements of type (b) are the hyperbolic lines $\{x,y\}^{\perp\perp}$, $y \not\sim x$. Now we define the incidence relation. If $y \in P'$, $L \in B'$ with L a line of type (a), then y I' L iff y I L; if $y \in P'$ and $L \in B'$ with L a line of type (b) then y I' L iff $y \in L$. Then the structure S' = (P',B',I') is a generalized quadrangle of order (s-1,s+1) and is denoted by P(S,x).

In the even case the generalized quadrangle P(W(q),x), x a point of W(q), is isomorphic to a $T_2^*(O)$ (here O is an irreducible conic together with its nucleus) [9]. The generalized quadrangle $P(T_2(O'),(\infty))$, with $T_2(O')$ as in (b) and q even, is isomorphic to $T_2^*(O)$ where $O = O' \cup \{n\}$ with n the nucleus of O' [9].

In P(W(q),x), q odd, a pair of non-concurrent lines (L,M) is regular iff one of the following cases occur : (i) L and M are lines of type (b), (ii) L and M are concurrent lines of W(q) (but are not concurrent in P(W(q),x)); in P(W(q),x), q even, a pair of non-concurrent lines (L,M) is regular iff one of the following cases occurs : (i) L and M are lines of type (b), (ii) in W(q) some line of $\{L,M\}^\perp$ is incident with x.

II. R-REGULARITY OF POINTS AND LINES

1. DEFINITIONS

Consider a generalized quadrangle S = (P,B,I) of order (s,s+2), s > 1. Since 1 < s < t regular points cannot occur [9]. Moreover, in the known examples also regular lines do not occur. Therefore we introduce the concept of R-regularity.

In what follows we always assume that the generalized quadrangle $S = (P,B,I)$ of order $(s,s+2)$ contains a spread R ($|R| = (s+1)^2$). For $x \in P$, we define $x^{\perp *} = \{z \in P \parallel z \sim x,\ z \neq x,\ zx \notin R\} \cup \{x\}$. For a pair of distinct points x,y we denote the set $x^{\perp *} \cap y^{\perp *}$ as $\{x,y\}^{\perp *}$. If $x \not\sim y$ or $x \sim y$ but $xy \notin R$, there holds $|\{x,y\}^{\perp *}| = s+1$. More generally, for $A \subset P$ we define $A^{\perp *} = \cap\{x^{\perp *} \parallel x \in A\}$. So for a pair of non-collinear points x,y we have $\{x,y\}^{\perp * \perp *} = \{u \in P \parallel u \sim z,$ $uz \notin R,\ \forall z \in \{x,y\}^{\perp *}\}$. So we obtain $|\{x,y\}^{\perp * \perp *}| \leqslant s+1$. If $x \sim y$ but $xy \notin R$, then clearly $\{x,y\}^{\perp * \perp *} = xy$ and so $|\{x,y\}^{\perp * \perp *}| = s+1$.

A pair of distinct points x,y is called *R-regular* provided $x \sim y$ and $xy \notin R$, or $x \not\sim y$ and $|\{x,y\}^{\perp * \perp *}| = s+1$. A point x is *R-regular* provided (x,y) is R-regular for all $y \in P$, $y \not\sim x$.

A *R-grid* in S is a substructure $S' = (P',B',I')$ of S defined as follows :

$P' = \{x_{ij} \in P \parallel i=1,\ldots,s+2,\ j=1,\ldots,s+2, \text{ and } i \neq j\}$,

$B' = \{L_1,\ldots,L_{s+2}, M_1,\ldots,M_{s+2}\} \subset B \setminus R$, and

$I' = I \cap ((P' \times B') \cup (B' \times P'))$, with $L_i \not\sim L_j$, $M_i \not\sim M_j$,
$L_i \cap M_j = x_{ij}$ if $i \neq j$, $L_i \not\sim M_i$, $x_{ij} \sim x_{ji}$, and
$x_{ij}x_{ji} = R_{ij} = R_{ji} \in R$ for $1 \leqslant i, j \leqslant s+2$.

We denote the set $\{L_1,\ldots,L_{s+2}\}$ (resp. $\{M_1,\ldots,M_{s+2}\}$) by $\{L_i,L_j\}^{\perp\perp *}$ or $\{M_i,M_j\}^{\perp *}$ (resp. $\{M_i,M_j\}^{\perp\perp *}$ or $\{L_i,L_j\}^{\perp *}$) for any $i \neq j$.

If $L_1,L_2 \in B \setminus R$, $L_1 \not\sim L_2$, then by definition the pair (L_1,L_2) is *R-regular* iff (L_1,L_2) belongs to a R-grid. In such a case there exists a unique $R \in R$ for which $L_1 \sim R \sim L_2$. A line $L \in B \setminus R$ is *weak R-regular* iff (L,M) is R-regular for all $M \in B \setminus R$ with $L \not\sim M$ and $|\{L,M\}^{\perp} \cap R| = 1$. A line $L \in B \setminus R$ is *R-regular* iff L is weak R-regular and for all $M \in B \setminus R$, $L \not\sim M$, with $|\{L,M\}^{\perp} \cap R| \neq 1$, the pair (L,M) is regular.

Finally, notice that R-regularity for lines is not the dual of R-regularity for points.

2. EXAMPLES

2.1. Theorem. *Consider* $P(W(q),x) = (P',B',I')$ *and let* R *be the set of all lines of type (b) in* B' *(see* I.2.(d)). *Then each point is* R*-regular. Each line of* $B' \setminus R$ *is* R*-regular iff* q *is even.*

Proof. Let $W(q) = (P,B,I)$. Choose a point x in $W(q)$. It is obvious that the set $\{\{x,y\}^{\perp\perp} \parallel y \in P,\ x \not\sim y\}$ defines a spread R in $P(W(q),x)$.

Let $y,z \in P'$, $y \not\sim' z$. Then it follows that $y \not\sim z$. We know that (y,z) is regular in $W(q)$. Let $z_0, z_0' \in x^\perp$ be defined by $\{y,z\}^\perp \cap x^\perp = \{z_0\}$ and $\{y,z\}^{\perp\perp} \cap x^\perp = \{z_0'\}$. From II.1. and the definition of R we obtain that $|\{y,z\}^{\perp'*}| = |\{y,z\}^\perp \setminus \{z_0\}| = q$ and $|\{y,z\}^{\perp'*\perp'*}| = |\{y,z\}^{\perp\perp} \setminus \{z_0'\}| = q$. Hence each point of $P(W(q),x)$ is R-regular.

Let $L_1,L_2 \in B' \setminus R$, $L_1 \not\sim' L_2$. If $L_1 \sim L_2$, (L_1,L_2) is a regular pair in $P(W(q),x)$ and $\{L_1,L_2\}^{\perp'} \subset R$. If $L_1 \not\sim L_2$ but L_1,L_2 are both concurrent with a same line of $W(q)$ through x, then (L_1,L_2) is regular (both in $W(q)$ and in $P(W(q),x)$) iff q is even. Here we have $\{L_1,L_2\}^{\perp'} \cap R = \phi$. Finally, suppose that $L_1 \not\sim L_2$ and L_1,L_2 are not concurrent with a same line through x in $W(q)$. Then $|\{L_1,L_2\}^{\perp'} \cap R| = 1$. If q is even, (L_1,L_2) is regular in $W(q)$. In this case, let $\{L_1,L_2\}^\perp = \{M_1,\ldots,M_{q+1}\}$, $\{L_1,L_2\}^{\perp\perp} = \{L_1,\ldots,L_{q+1}\}$ and let $L_i \mathrm{I}\, x_i \mathrm{I}\, M_i$, $x_i \in x^\perp$. Then the lines L_i,M_i, $1 \leq i \leq q+1$, define a R-grid in $P(W(q),x)$. Indeed, $L_i \sim' M_j$ iff $i \neq j$, and since for any $i \neq j$ $\{x_i,x_j\} \subset \{x_i,x_{ij}\}^\perp$, $x_{ji} \sim x_i$, $x_{ji} \sim x_j$ (with $x_{ij} = L_i \cap M_j$ and $x_{ji} = L_j \cap M_i$) and hence $x_{ji} \in \{x,x_{ij}\}^{\perp\perp}$, we have $x_{ij} \sim' x_{ji}$ with $x_{ij}x_{ji} \in R$. We conclude that each line of $B' \setminus R$ is R-regular iff q is even. □

2.2. <u>Theorem</u>. *Let R be the set of all lines through a point x of O in $T_2^*(O)$. A point y, resp. a line $L \notin R$, is R-regular if and only if $O \setminus \{x\}$ is a conic.*

Proof. It is immediate that R is a spread in $T_2^*(O)$. If $O \setminus \{x\}$ is a conic there holds $T_2^*(O) \cong P(W(q),y)$, q even. Applying the foregoing theorem there follows the R-regularity.

To prove the converse, we remark that $T_2^*(O) \cong P(T_2(O'),(\infty))$, $O' = O \setminus \{x\}$ (I.2.(d)). Assume that $T_2^*(O) = (P,B,I)$ and $T_2(O') = (P',B',I')$. Consider a point $y_1 \in P$ such that y_1 is R-regular. Then y_1 is a point of type (i) in $T_2(O')$. We prove that y_1 is regular in $T_2(O')$.

Let $y_2 \in P'$, $y_2 \not\sim' y_1$, with y_2 a point of type (i). If $y_1 \not\sim y_2$, the pair (y_1,y_2) is R-regular in $T_2^*(O)$. Let $\{y_1,y_2\}^{\perp*\perp*} = \{y_1,y_2,\ldots,y_q\}$. Since $\{y_1,y_2\}^{\perp*} \subset \{y_1,y_2\}^{\perp'}$ there results $y_i \sim' z_j$, $\forall y_i \in \{y_1,y_2\}^{\perp*\perp*}$, $\forall z_j \in \{y_1,y_2\}^{\perp'}$ (see 1.3.4. in [9]). Consequently $\{y_1,y_2\}^{\perp*\perp*} \subset \{y_1,y_2\}^{\perp'\perp'}$. Using again 1.3.4. in [9] we obtain that (y_1,y_2) is regular in $T_2(O')$. Now suppose that $y_2 \sim y_1$, i.e. the line $y_1y_2 \in R$. Denote by $x_1,\ldots,x_{q+1}$ the points of $O \setminus \{x\}$, and by $W_1,\ldots,W_{q+1}$ the planes defined by y_1,y_2,x_i, $i = 1,\ldots,q+1$, which meet $O \setminus \{x\} = O'$ in a unique point. Then $\{y_1,y_2\}^{\perp'} = \{W_1,\ldots,W_{q+1}\}$

and $\{y_1,y_2\}^{\perp'\perp'} = \{z \parallel z \text{ I } y_1y_2\} \cup \{(\infty)\}$ since $y_1y_2 \subset W_i$, $\forall\, i = 1,\ldots,q+1$. Again (y_1,y_2) is a regular pair in $T_2(O')$. Next, let $X \in P'$ be a point of type (ii) with $y_1 \not\sim' X$. The trace $\{y_1,X\}^{\perp'}$ contains at least two points x_1,x_2 of type (i) with $x_1 \not\sim' x_2$. On the other hand $\{x_1,x_2\}^{\perp'}$ contains, besides y_1 and X, at least one point y_2 of type (i). Clearly $y_2 \not\sim' y_1$. From the foregoing it follows that (y_1,y_2) is regular in $T_2(O')$. Since $\{x_1,x_2\} \subset \{y_1,y_2\}^{\perp'}$ and $\{y_1,X\} \subset \{x_1,x_2\}^{\perp'}$, the pair (y_1,X) is also regular in $T_2(O')$. Finally, from the regularity of (∞) in $T_2(O')$, q even, there results that $(y_1,(\infty))$ is regular. We conclude that y_1 is a regular point in $T_2(O')$ and hence O' is a conic [9].

Next we suppose that $L_1 \in B \setminus R$ is R-regular and that (L_1,L_2) is a R-regular pair in $T_2^*(O)$. We use the notations of II.1. with $q = s+1$. For each pair (L_i,M_i), $1 \leqslant i \leqslant q+1$, there holds $\{L_i,M_i\}^{\perp} = \{R_{i1},\ldots,R_{i,s+1}\} \subset R$. Each pair of lines of R is regular (they define the same point x of O (I.2.(c)). Hence (L_i,M_i) is a regular pair of $T_2^*(O)$, and so defines a point x_i of O, $i = 1,\ldots,q+1$ (I.2.(c)). There follows that the sets $\{L_1,L_2\}^{\perp'\perp'} = \{L_1,\ldots,L_{q+1}\}$ and $\{L_1,L_2\}^{\perp'} = \{M_1,\ldots,M_{q+1}\}$ are the two sets of generators of a hyperbolic quadric Q in PG(3,q). The oval O' is a plane intersection of Q. Hence O' is a conic. □

3. R-REGULARITY AND AFFINE PLANES

3.1. Theorem. *Let* x *be a R-regular point of the generalized quadrangle* S *of order* (s,s+2) *with spread R. Then the incidence structure* (P^*,B^*,I^*) *where* $P^* = x^{\perp^*}$, $B^* = \{L \in B \parallel x \text{ I } L \text{ and } L \notin R\} \cup \{\{y,z\}^{\perp^*\perp^*} \parallel y,z \in x^{\perp^*}, z \not\sim y\}$ *and* I^* *the natural incidence, is a* $2\text{-}((s+1)^2,s+1,1)$ *design i.e. an affine plane of order* s+1.

Proof. Immediate. □

3.2. Theorem. *Let* S *be a generalized quadrangle of order* (s,s+2) *which contains a spread R and a R-regular point* x *with* $x \text{ I } L \in R$. *Then each pair* (L,M), $M \in R$, *is regular and* $\{L,M\}^{\perp\perp} \subset R$.

Proof. Let $x \in L \in R$ and $M \in R \setminus \{L\}$. Choose a point $y \in M$ such that $x \not\sim y$. Then (x,y) is a R-regular pair. Let $\{x,y\}^{\perp^*\perp^*} = \{z_i, 1 \leqslant i \leqslant s+1\}$ with $x = z_1$, $y = z_{s+1}$. Each point z_i is incident with a unique line $R_i \in R$ with $R_1 = L$, $R_{s+1} = M$, and with a unique line L_i which does not contain a point of $\{x,y\}^{\perp^*}$, for all $1 \leqslant i \leqslant s+1$. Since $z_i \not\!\text{I}\ R_j$, $i \neq j$, $1 \leqslant i,j \leqslant s+1$, we obtain, applying axiom (iii) of I.1. w.r.t. z_i and R_j, that $L_i \sim R_j$, for all $j \neq i$, $1 \leq i,j \leq s+1$. So (L,M) is regular and moreover $\{L,M\}^{\perp} = \{L_i, 1 \leq i \leq s+1\}$ and $\{L,M\}^{\perp\perp} = \{R_i, 1 \leq i \leq s+1\}$. □

3.3. Corollary. Let L be a set of non-concurrent lines of a generalized quadrangle of order (s,t). Then L is *normal* (see [7]) provided each pair of lines of L is regular and their span is a subset of L.

The foregoing theorem shows that in a generalized quadrangle S of order $(s,s+2)$ with a spread R in which all points are R-regular, the spread R is a normal set. Then an affine plane of order $s+1$ can be defined in the following way. Let $P^* = R$, $B^* = \{\{M,N\}^{\perp\perp}, M,N \in R\}$ and I^* the natural incidence. Then the structure (P^*,B^*,I^*) is an affine plane A_R of order $s+1$.

3.4. Theorem. *Let* S *be a generalized quadrangle of order* $(s,s+2)$ *with a spread* R. *Let* $R, R' \in R$, $R \neq R'$ *and assume that* (R,R') *is regular with* $\{R,R'\}^{\perp\perp} = \{R_1 = R, R_2,\ldots,R_{s+1} = R'\} \subset R$. *If all pairs of points incident with different lines of* $\{R,R'\}^{\perp\perp}$ *are* R-*regular, we can define an affine plane* (P',B',I') *of order* $s+1$ *with*
$P' = \{z \in P \parallel z \text{ I } R_i, R_i \in \{R,R'\}^{\perp\perp}, 1 \leq i \leq s+1\}$,
$B' = \{R,R'\}^{\perp} \cup \{R,R'\}^{\perp\perp} \cup \{\{z,z'\}^{\perp*\perp*} \parallel z,z' \in P', z \not\sim z'\}$ *and* I' *the natural incidence.*

Proof. First we prove that for $z,z' \in P'$, $z \not\sim z'$, the set $\{z,z'\}^{\perp*\perp*} \subset P'$. Let $z \text{ I } R_i$, $z' \text{ I } R_j$, $R_i, R_j \in \{R,R'\}^{\perp\perp}$. From the proof of II.3.2. it follows that for any $z_k \in \{z,z'\}^{\perp*\perp*}$ the line $R_k \in R$, $z_k \in R_k$, belongs to $\{R_i,R_j\}^{\perp\perp}$. Hence $z_k \in P'$. Now it is immediate that (P',B',I') is a 2-$((s+1)^2,s+1,1)$ design. □

3.5. Theorem. *Let* S *be a generalized quadrangle of order* $(s,s+2)$ *with a spread* R *and a* R-*regular line* L. *If* $R,R' \in L^{\perp} \cap R$, $R \neq R'$, *then* (R,R') *is regular and* $\{R,R'\}^{\perp\perp} = L^{\perp} \cap R$.

Proof. Let $R,R' \in L^{\perp} \cap R$, $R \neq R'$. Consider an arbitrary point $x \text{ I } R$, $x \not\text{I}\, L$. The point x is incident with $s+2$ lines of $B \setminus R$. Since $|L^{\perp} \cap R| = s+1$ there exists at least one line $L' \in B \setminus R$, $x \text{ I } L'$, such that $\{L,L'\}^{\perp} \cap R = \{R\}$. In view of the R-regularity of L, (L,L') belongs to a R-grid. Using the notations of II.1. we put $L = L_1$, $L' = L_2$. Then $\{L_1,M_1\}^{\perp} = \{R_{1i}, i = 2,\ldots,s+2\} = L^{\perp} \cap R$ with $R = R_{12}$, $R' = R_{1k}$, $3 \leq k \leq s+2$. Again by the R-regularity of L (L_1,M_1) is a regular pair. Hence (R,R') is regular and $\{R,R'\}^{\perp\perp} = L'^{\perp} \cap R$. □

3.6. Theorem. *Let* S *be a generalized quadrangle of order* (s,s+2) *with a spread* R. *If* S *contains a* R-*regular line* L, *then the incidence structure* $(P',B',\in)$ *with* $P' = L^{\perp} \setminus (\{L\} \cup R)$,
$B' = \{\{K,K'\}^{\perp\perp *} \cap P' \parallel K,K' \in P',\ K \not\sim K'$ *and* $\{K,K'\}$ *belongs to a* R-*grid*$\} \cup \{\{K,K'\}^{\perp\perp} \parallel K,K' \in P',\ K \not\sim K'$ *and* $\{K,K'\}$ *does not belong to a* R-*grid*$\} \cup \{x_B \parallel x \text{ I } L\}$ *with* x_B *the set of all lines of* P' *incident with* x, *is an affine plane of order* s+1.

Proof. Denote $L^{\perp} \cap R$ by $\{R_1,\ldots,R_{s+1}\}$. Each pair of different lines of this set is regular (II.3.5.). Let $K,K' \in P'$, $K \not\sim K'$ and let $L_1 \in \{K,K'\}^{\perp}$, $L \neq L_1$. From the R-regularity of L there results if $|\{L,L_1\}^{\perp} \cap R| \neq 1$ that (L,L_1) is a regular pair. Hence (K,K') is also regular with $|\{K,K'\}^{\perp\perp}| = s+1$. It is obvious that in this case $\{K,K'\}^{\perp\perp}$ is the unique element of B' through K and K'. Now suppose that $|\{L_1,L\}^{\perp} \cap R| = 1$. In view of the R-regularity of L the pair (L,L_1) defines a R-grid. This R-grid contains K and K'. Hence $|\{K,K'\}^{\perp\perp *} \cap P'| = s+1$. Clearly $\{K,K'\}^{\perp\perp *}$ is the unique element of B' through K and K'. Finally, if $K,K' \in P'$ with K I x I K' then x_B is the unique line of B' through K and K'. We have again $|x_B| = s+1$.

Hence, we conclude that $(P',B',\in)$ is a 2-$((s+1)^2,s+1,1)$ design. □

3.7. Theorem. *Let* S *be a generalized quadrangle of order* (s,s+2) *which contains a spread* R. *If* x *is a* R-*regular point and* (L_1,L_2) *a* R-*regular pair of lines such that* x *is incident with no line* $x_{ij}x_{ji}$ *of* R *(with the notations of* II.1.), *then* s *is odd.*

Proof. Clearly x is incident with no line of $\{L_1,L_2\}^{\perp *}$. Let $\{L_1,L_2\}^{\perp\perp *} = \{L_1,\ldots,L_{s+2}\}$ and let y_i be defined by $x \sim y_i$, y_i I L_i, $1 \leq i \leq s+2$. The points $x,y_1,\ldots,y_{s+2}$ are s+3 points of the affine plane of order s+1 defined by x (II.3.1.). Each line through x in that plane has a unique point in common with the set $\{y_1,\ldots,y_{s+2}\}$. Suppose that y_i,y_j,y_k, $i \neq j \neq k \neq i$, $i,j,k \in \{1,\ldots,s+2\}$, are collinear in that affine plane. If this is the case, there holds $|\{y_i,y_j,y_k\}^{\perp *}| = s+1$. Suppose that one of the points y_i,y_j,y_k is incident with a line $M \in \{L_1,L_2\}^{\perp *}$ which is concurrent with L_i,L_j,L_k, e.g. y_i I M. Let u_j (resp. u_k) be the intersection of M with L_j (resp. L_k). Then $u_k \in \{y_i,y_k\}^{\perp *}$. Hence $y_j \sim u_k$, with $u_k y_j \notin R$, and a triangle $u_j u_k y_j$ arises, a contradiction.

There follows that there is a line of $\{L_1,L_2\}^{\perp *}$ which is incident with y_i,y_j or y_k and concurrent with eactly two of the lines L_i,L_j or L_k. Suppose e.g. that y_i I M_j, $M_j \in \{L_1,L_2\}^{\perp *}$, and M_j I u_k I L_k. From the definition of a R-grid it follows that L_j I x_{jk} I M_k and

$x_{jk}u_k \in R$. Hence $y_j = x_{jk}$. On the other hand $u_k \in \{y_i, y_k\}^{\perp *}$ and hence $u_k \sim y_j$, $u_k y_j \notin R$, a contradiction. So the points y_i, y_j, y_k are not collinear in the affine plane of order s+1. Hence $\{y_1, \ldots, y_{s+2}\}$ defines an oval in the corresponding projective plane. The s+2 tangent lines of the oval are concurrent at x. So we conclude that the order s+1 of the plane is even. □

3.8. <u>Corollary</u>. *Let* S *be a generalized quadrangle of order* (s,s+2) *which contains a spread* R. *If* x *is a* R*-regular point and* L *a* R*-regular line such that* $x \not\mathrm{I}\, R$, $\forall R \in L^{\perp} \cap R$, *then* s *is odd.*
Proof. Let $x \mathrm{I} R_x$, $R_x \in R$. Then $R_x \not\sim L$. Put $\{L, R_x\}^{\perp} = \{K_1, \ldots, K_{s+1}\}$ and $L^{\perp} \cap R = \{R_1, \ldots, R_{s+1}\}$. If $M \in \{R_1, R_2\}^{\perp}$, $M \neq L$, then $M \sim R_i$, for all $1 \leqslant i \leqslant s+1$ (II.3.5.). Hence there exists at least one line L', L' ≠ M, incident with y, where $y = M \cap R_j$, $j \in \{1, \ldots, s+1\}$, such that $L' \not\sim K_i$, $1 \leqslant i \leqslant s+1$. Then $\{L, L'\}^{\perp} \cap R = \{R_j\}$ and so (L,L') is a R-regular pair. Let $\{L, L'\}^{\perp\perp *} = \{L_1, \ldots, L_{s+2}\}$. If R_x is of the form $x_{ij}x_{ji}$ (w.r.t. the R-grid defined by L and L'), then $\{L_i, L_j\} \cap \{L, L'\} = \phi$ and so $M_i \sim L$, $M_i \sim L'$, $M_j \sim L$, $M_j \sim L'$. Hence one of the lines K_i is concurrent with L', a contradiction. □

III. CHARACTERIZATIONS OF P(W(s+1),(∞))

1. THEOREM

Let S = (P,B,I) *be a generalized quadrangle of order* (s,s+2) *which contains a spread* R. *If all points are* R*-regular, then* S *is isomorphic to* P(W(s+1)),(∞)) *with* (∞) *an arbitrary point of* W(s+1).
Proof. We define an incidence structure S' = (P',B',I') as follows.

P' contains three types of points :

(i) the points $x \in P$;

(ii) the sets $\{R_1, R_2\}^{\perp}$, $R_1, R_2 \in R$;

(iii) a unique point (∞).

B' contains two types of lines :

(a) the lines $L \in B \setminus R$;

(b) sets, each being the union of all traces of pairs of elements of a same parallelclass in A_R (cf. II.3.3.).

I' is defined as :

- a point of type (i) is incident with a line of type (a) iff they are incident in S; it is incident with no line of type (b);

- a point $\{R_1,R_2\}^{\perp}$ of type (ii) is incident with each line of type (a) which belongs to $\{R_1,R_2\}^{\perp}$ and with the unique line of type (b) from which it is a subset.
- the symbol (∞) is incident with all lines of type (b) but with no line of type (a).

It is easy to check that S' is a generalized quadrangle of order s+1 (see also [7]). We prove now that moreover $S' \cong W(s+1)$.

<u>1st proof</u>. Applying the theorem of C.T. Benson [2] we only have to check that each point of S' is regular, i.e. $|\{x,y\}^{\perp'\perp'}| = s+2$ for all pairs (x,y) in S' with $x \not\sim' y$.

Suppose that x and y are both points of type (i) with $x \not\sim' y$. Let $x \not\sim y$. Since all points of S are R-regular, we have $|\{x,y\}^{\perp*\perp*}| = s+1$. Let $x \text{ I } R_x \in R$, $y \text{ I } R_y \in R$. Then $\{x,y\}^{\perp'} = \{x,y\}^{\perp*} \cup \{\{R_x,R_y\}^{\perp}\}$. From the proof of II.3.4. it follows that each point of $\{x,y\}^{\perp*\perp*}$ is incident with a line of $\{R_x,R_y\}^{\perp}$. Hence $\{x,y\}^{\perp*\perp*} \subset \{x,y\}^{\perp'\perp'}$. Applying 1.3.4. of [9] we obtain $|\{x,y\}^{\perp'\perp'}| = s+2$. If $x \text{ I } R \text{ I } y$, $R \in R$, then $\{x,y\}^{\perp'}$ consists of the s+2 points $\{R,R'\}^{\perp}$, $R' \in R$ and $R \neq R'$, of type (ii). Clearly $\{x,y\}^{\perp'\perp'} \supset \{z \parallel z \text{ I } R\} \cup \{(\infty)\}$, and so $|\{x,y\}^{\perp'\perp'}| = s+2$. Hence (x,y) is regular in S'.

Let x be a point of type (i) and $y = \{R,R'\}^{\perp}$ a point of type (ii), with $x \not\sim' y$. If $x_1,x_2 \in \{x,y\}^{\perp'}$, with x_1,x_2 points of type (i), the pair (x_1,x_2) is regular in S'. Hence there follows the regularity of (x,y) in S'.

Next we suppose that x is a point of type (i) and $y = (\infty)$. If $x \text{ I } R \in R$, then $\{x,(\infty)\}^{\perp'}$ consists of the s+2 points $\{R,R'\}$, $R' \in R$ and $R \neq R'$, in S'. Clearly $\{x,(\infty)\}^{\perp'\perp'} \supset \{y \parallel y \text{ I } R\} \cup \{(\infty)\}$, and so $|\{x,(\infty)\}^{\perp'\perp'}| = s+2$. Hence $(x,(\infty))$ is regular in S'.

Finally, let $x = \{R_x,R'_x\}^{\perp}$, $y = \{R_y,R'_y\}^{\perp}$, $R_x,R'_x,R_y,R'_y \in R$, be two points of type (ii), $\{R_x,R'_x\}^{\perp} \not\sim' \{R_y,R'_y\}^{\perp}$. Choose $x_1,x_2 \in \{x,y\}^{\perp'}$ such that x_1,x_2 are two points of type (i). From the regularity of (x_1,x_2) in S', there follows that (x,y) is regular in S'.

Since these are the only possible combinations for a pair of distinct points (x,y), $x \not\sim' y$, we conclude that $S' \cong W(s+1)$. Hence the quadrangle S is isomorphic to $P(W(s+1),(\infty))$. The lines of the spread R in S can be identified with the sets $\{(\infty),x\}^{\perp'\perp'}$, $x \not\sim' (\infty)$.

<u>2nd proof</u>. In [11] C. Somma defines the Steiner system $\mathcal{D} = (P_1,B_1,I_1)$ with $P_1 = P$, $B_1 = B \cup \{\{x,y\}^{\perp*\perp*} \parallel x,y \in P, x \not\sim y\}$ and I_1 the natural incidence. Now we prove that each substructure of $\mathcal{D}$

generated by an arbitrary point $x \in P_1$, and an arbitrary line $L \in B_1$, with $x \not{I}_1 L$, is an affine plane of order s+1.

(i) Suppose $x \in P_1$, $L \in B_1$, with $L = \{y,z\}^{\perp*\perp*}$ and $x \not{I}_1 L$.

(a) If $x \in \{y,z\}^{\perp*}$ then $xw \in B \setminus \mathcal{R}$, $\forall w \in \{y,z\}^{\perp*\perp*}$. The affine plane defined by x (cf. II.3.1.) is a substructure of $\mathcal{D}$ which contains x and L. Hence x and L generate an affine plane of order s+1.

(b) If $x^{\perp*} \cap \{y,z\}^{\perp*} = \{r\}$, $r \neq x$, then the affine plane defined by r (cf. II.3.1.) is a substructure of $\mathcal{D}$ which contains x and L. So in this case, x and L define again an affine plane of order s+1.

(c) Suppose $x^{\perp*} \cap \{y,z\}^{\perp*} = \phi$ and $x \sim w$ with $w \in \{y,z\}^{\perp*\perp*}$, $xw \notin \mathcal{R}$. If $y \mathrm{I} R_y$, $z \mathrm{I} R_z$, $R_y, R_z \in \mathcal{R}$, then we consider the affine plane of order s+1 defined by $\{R_y,R_z\}$ (see II.3.4.). The line $xw \in \{R_y,R_z\}^{\perp}$ (see II.3.2.). Hence this affine plane is also generated by x and L.

(d) Let $x^{\perp*} \cap \{y,z\}^{\perp*} = \phi$ and $x \sim w$ with $w \in \{y,z\}^{\perp*\perp*}$, $xw \in \mathcal{R}$. If $y \mathrm{I} R_y$, $z \mathrm{I} R_z$, $R_y,R_z \in \mathcal{R}$, then we consider the affine plane of order s+1 defined by $\{R_y,R_z\}$ (see 3.4.). From II.3.2. it follows that $xw \in \{R_y,R_z\}^{\perp\perp}$. Again x and L generate an affine plane of order s+1.

(e) Suppose that $x^{\perp} \cap \{y,z\}^{\perp*\perp*} = \phi$ and $x^{\perp*} \cap \{y,z\}^{\perp*} = \phi$. Counting the number of points u in S such that $u^{\perp} \cap \{y,z\}^{\perp*\perp*} \neq \phi$ or $u^{\perp*} \cap \{y,z\}^{\perp*} \neq \phi$ we obtain $(s+1)^2(s-1)+2(s+1)^2 = (s+1)^3 = v$. Hence this case cannot occur.

(ii) Suppose $x \in P_1$, $L \in B$ with $L \notin \mathcal{R}$ and $x \not{I}_1 L$. Since S is a generalized quadrangle there exists a pair $(y,M) \in P \times B$ such that $x \mathrm{I} M \mathrm{I} y \mathrm{I} L$.

(a) If $M \in \mathcal{R}$ we consider the affine plane defined by $\{M,M'\}$ (cf. 3.4.) with $M \sim L$, $M' \in \mathcal{R}$. This plane of order s+1 is also generated by x and L.

(b) If $M \notin \mathcal{R}$ we construct the affine plane of roder s+1 defined by y (cf. II.3.1.). This plane is generated by x and L.

(iii) Let $x \in P_1$, $L \in B$ with $L \in \mathcal{R}$ and $x \not{I}_1 L$. Since S is a generalized quadrangle there exists a pair $(y,M) \in P \times B$ such that $x \mathrm{I} M \mathrm{I} y \mathrm{I} L$ with $M \notin \mathcal{R}$. Let $L' \sim M$, $L' \in \mathcal{R}$ and $L' \neq L$. Then we consider the affine plane defined by $\{L,L'\}$ (cf. II.3.4.). Again this plane is generated by x and L.

In each of the cases the plane generated by x and L is an affine plane of order s+1. If $s \geqslant 3$ we can apply the theorem of F. Buekenhout [3]. Then $\mathcal{D}$ is a threedimensional affine space of order s+1. By the embedding theorem of J.A. Thas [13] and since $s \geq 3$, we have $S \cong P(W(s+1),(\infty))$, or $S \cong T_2^*(O)$ with O a complete oval of PG(2,s+1) with s odd. Suppose we are in the second case. The spread R of $T_2^*(O)$ is a normal set (see II.3.3.). Hence all lines of R are concurrent in a same point x of O (II.1.(c)). Since all points are R-regular, $O \setminus \{x\}$ is a conic (II.2.2.). So in this case we also have $T_2^*(O) \cong P(W(s+1),(\infty))$ (cf. II.1.(d)). If s = 2 (resp. s = 1) there is, up to isomorphism, only one generalized quadrangle of order (2,4) (resp. (1,3)) [9], and again the result follows. □

2. THEOREM

Let $S = (P,B,I)$ *be a generalized quadrangle of order* (s,s+2) *containing a normal spread* R . *If each line of* $B \setminus R$ *is weak* R*-regular, then* $S \cong P(W(s+1),(\infty))$, s *odd, with* (∞) *an arbitrary point of* W(s+1).

Proof. Analogously as in III.1. we construct the generalized quadrangle S' of order s+1. Now we prove that $S' \cong W(s+1)$, s odd, using the dual of Theorem 5.2.6. in [9]. Assume that the lines L_1,L_2 of S define a R-grid of S. From the definition of a R-grid it follows that each line of $\{L_1,L_2\}^{\perp *}$, as well as each line of $\{L_1,L_2\}^{\perp\perp *}$ is concurrent (in S') with a unique line of type (b), and that different lines from $\{L_1,L_2\}^{\perp *}$ or $\{L_1,L_2\}^{\perp\perp *}$ are never concurrent (in S') with a common line of type (b). Now we notice that L_i,M_i with $L_i \in \{L_1,L_2\}^{\perp\perp *}$, $M_i \in \{L_1,L_2\}^{\perp *}$, $1 \leq i \leq s+2$, are concurrent with a same line of type (b) and moreover are incident with a same point of that line of type (b). Indeed, for the points x_{ki} I M_i, $k \neq i$, x_{ik} I L_i, $k \neq i$, there holds $x_{ik}x_{ki} \in R$. There easily follows that (L_1,L_2) is a regular pair in S'.

Consider the point (∞). Analogously as in the proof of III.1. we show that (∞) is regular in S'. Let x be an arbitrary point of type (i) and put $\{x,(\infty)\}^{\perp'} = \{\{R,R_i\}^{\perp}, x \text{ I } R \in R, R_i \in R\}$. Let (L_1,L_2,L_3) be a triad of lines of S' which are incident with points of $\{x,(\infty)\}^{\perp'}$. Then the following two cases may occur. The three lines are of type (a), or two lines are of type (a) and one is of type (b). Suppose that L_1,L_2 are of type (a). In S the lines of R which are concurrent with L_1 (resp. L_2) form a line L_1 (resp. L_2) of the affine plane A_R. Since in S' the lines L_1,L_2 are concurrent with different lines of type (b), the lines L_1, L_2 of A_R are not parallel. If $R' \in R$ is the common element of L_1 and L_2, then clearly R' is the unique element of R which is concurrent (in S) with L_1

and L_2. Hence the pair $\{L_1,L_2\}$ belongs to a *R*-grid in S, and by the previous paragraph it is a regular pair in S'. By 1.3.6. (ii) in [9] the triad (L_1,L_2,L_3) has at least one center. Using the dual of 5.2.6. in [9], there results that S' is isomorphic to W(s+1), s odd. Hence the quadrangle S is isomorphic to P(W(s+1),(∞)). The lines of the spread *R* in S can be identified with the sets $\{(\infty),x)\}^{\perp'\perp'}$, $x \not\sim' (\infty)$. □

3. THEOREM

Let S *be a generalized quadrangle of order* (s,s+2) *containing a spread* R. *If each line of* B \ *R is R-regular then* S ≅ P(W(s+1),(∞)) s *odd, with* (∞) *an arbitrary point of* W(s+1).

Proof. This is immediate from II.3.5. and Theorem II.2. □

Taking account of III.1., III.2. and III.3., we can formulate the next theorem.

4. THEOREM

Let S *be a generalized quadrangle of order* (s,s+2) *which contains a spread* R. *Then the following statements are equivalent :*

(i) *each point is R-regular and* s *is odd;*

(ii) *each line is weak R-regular and R is a normal set;*

(iii) *each line is R-regular;*

(iv) S ≅ P(W(s+1), (∞)), s *odd, with* (∞) *an arbitrary point of* W(s+1).

REFERENCES

[1] Ahrens,R.W. and Szekeres,G., On a combinatorial generalization of 27 lines associated with a cubic surface, J. Austr. Math. Soc. 10 (1969) 485-492.

[2] Benson,C.T., On the structure of generalized quadrangles, J. Algebra 15 (1970) 443-454.

[3] Buekenhout,F., Une caractérisation des espaces affins basée sur la notion de droite, Math. Z. 111 (1969) 367-371.

[4] Dembowski,P., Finite geometries (Springer-Verlag,1968).

[5] Hall,M. Jr., Affine generalized quadrilaterals, Studies in Pure Math. (ed. L. Mirsky), Academic Press (1971) 113-116.

[6] Payne,S.E., The equivalence of certain generalized quadrangles, J. Comb. Th. 10 (1971) 284-289.

[7] Payne,S.E., Quadrangles of order (s-1,s+1), J. Algebra 22 (1972) 97-110.

[8] Payne,S.E. and Thas,J.A., Generalized quadrangles with symmetry, Part II, Simon Stevin 49 (1976) 81-103.

[9] Payne,S.E. and Thas,J.A., Finite generalized quadrangles, Research Notes in Mathematics #110 (Pitman Publ. Inc. 1984).

[10] Segre,B., Lectures on modern geometry (Ed. Cremonese Roma 1961).

[11] Somma,C., Generalized quadrangles with parallelism, Annals of Discrete Math. 14 (1982) 265-282.

[12] Thas,J.A., Combinatorics of partial geometries and generalized quadrangles, in : Higher combinatorics (ed. M. Aigner), Nato Advanced Study Institute Series, Reidel Publ. Comp. (1976) 183-199.

[13] Thas,J.A., Partial geometries in finite affine spaces, Math. Z. 158 (1978) 1-13.

Annals of Discrete Mathematics 30 (1986) 185–202

ON PERMUTATION ARRAYS, TRANSVERSAL SEMINETS AND RELATED STRUCTURES

Michel Deza and Thomas Ihringer

Université Paris VII, U.E.R. de Math., Paris, France

Technische Hochschule, Fachbereich Mathematik, Darmstadt, Federal Republic of Germany

Exploiting some ideas of [4], this paper is focused on the equivalence between sets of mutually orthogonal permutation arrays and a special class of seminets (the so-called transversal seminets). Besides this equivalence, Section 2 contains a construction method for transversal seminets using groups. Nonsolvable permutation groups of prime degree and the projective special linear groups $PSL(2,2^m)$ yield examples for this method. In Section 3 some upper bounds are proved for the number of mutually orthogonal permutation arrays, depending on the intersection structure of these arrays. With the results of Sections 4 it is shown that all examples of Section 2 are row-extendible. Section 5 deals with several incidence structures associated to transversal seminets. The consequences are investigated when these incidence structures have special properties (for instance, when they are pairwise balanced designs). Section 6 discusses briefly the relations of transversal seminets with other mathematical structures, e.g. with transversal packings, generalized orthogonal arrays, and sets of mutually orthogonal partial quasigroups.

1. INTRODUCTION

A $v\times r$ matrix $A = (a_{ij})$ with entries a_{ij} from the set $\{1,2,...,r\}$ is called a *permutation array* if each row of A contains each of the elements $1,2,...,r$ exactly once, i.e. if the rows of A represent permutations of $\{1,2,...,r\}$. The *intersection structure* of A is defined as the $v\times v$ matrix $F(A) = (F_{ii'}(A))$ with $F_{ii'}(A) := \{j \mid 1\le j\le r,\ a_{ij} = a_{i'j}\}$. The $v\times r$ permutation arrays $A = (a_{ij})$ and $B = (b_{ij})$ are called *similar* if $F(A) = F(B)$. Clearly, A and B are similar if and only if, for all indices i,i',j,

$$a_{ij} = a_{i'j} \iff b_{ij} = b_{i'j} .$$

The permutation arrays A and B are called *orthogonal* if they are similar and if, for all i,i',j,j',

$$a_{ij} = a_{i'j'} \text{ and } b_{ij} = b_{i'j'} \implies j = j' .$$

This concept of orthogonality generalizes the well-known idea of orthogonal latin rectangles. It has been defined and investigated by Bonisoli and Deza in [4].

(See also [7] for closely related considerations.)

Similarly as for latin rectangles and latin squares, one will be interested in sets $\mathcal{A} = \{A_1, A_2, \ldots, A_t\}$ of t mutually orthogonal permutation arrays, with t possibly greater than 2. In this case the *intersection structure* $F(\mathcal{A}) = (F_{i\,i'}(\mathcal{A}))$ of $\mathcal{A}$ is defined as the common intersection structure of the A_k, i.e. $F(\mathcal{A}) := F(A_1) = F(A_2) = \ldots = F(A_t)$.

A permutation array $A = (a_{ij})$ is called *standardized* if $a_{1j} = j$ for all j.

Without loss of generality, it will be assumed in this paper that each permutation array is standardized, and that it satisfies the following nontriviality conditions (C_1) and (C_2).

(C_1) The permutation array A has no constant column, i.e. each column of A contains at least two distinct values,

(C_2) any two rows of A are distinct.

Let X be a nonempty set, and let $L_0, L_1, \ldots, L_t$ (with $t \geq 1$) be mutually disjoint sets of nonempty subsets of X. The elements of X will be called *points* and the elements of $\bigcup_{k=0,1,\ldots,t} L_k$ *lines*. Then $\mathcal{S} := (X; L_0, L_1, \ldots, L_t)$ is called a *seminet* or (more precisely) a (t+1)-*seminet* if

(S_1) any two distinct lines intersect in at most one point,

(S_2) each class L_i partitions the point set X.

Condition (S_2) justifies the term *parallel class* for each of the lines L_i. The notion of a seminet generalizes such well-known structures like affine planes, nets and (more generally) the parallel structures of André [2]. A subset of X is called a *transversal* of the seminet $\mathcal{S}$ if it intersects each line of $\mathcal{S}$ in exactly one point. If $\mathcal{S}$ has a transversal consisting of r points, then each parallel class L_i contains exactly r lines (and hence each further transversal of the seminet consists also of exactly r points). If $T_1, T_2, \ldots, T_v$ are transversals of $\mathcal{S}$ then $\mathcal{T} := (X; L_0, L_1, \ldots, L_t; T_1, T_2, \ldots, T_v)$ is called a *transversal seminet* (or *transversal* (t+1,r)-*seminet* if each transversal consists of r points). Bonisoli and Deza [4] pointed out that there is a close relationship between sets of mutually orthogonal permutation arrays and other mathematical structures. For instance, they proved that each set of t mutually orthogonal $v \times r$ permutation arrays is equivalent to a 1-design with v treatments, replication number r and $t+1$ mutually orthogonal resolutions (see Section 5). Moreover, it was shown that any of these are equivalent to a transversal (t+1,r)-seminet with v transversals. Therefore many of the examples and results in this paper can be translated into analogous statements on combinatorial designs with mutually ortogonal resolutions.

2. AN EQUIVALENCE AND A CONSTRUCTION METHOD

Let $\mathcal{T} := (X;L_0,L_1,\ldots,L_t;T_1,T_2,\ldots,T_v)$ be a transversal seminet with $L_0 := \{l_0^1,l_0^2,\ldots,l_0^r\}$. (In fact, throughout this paper the set of parallel classes, the set of transversals and the set of lines of L_0 of any transversal seminet are assumed to be linearly ordered, by the numbering of their elements.) One can now define t mutually orthogonal $v\times r$ permutation arrays $A_k = (a_{ij}^k)$, $k=1,2,\ldots,t$, in the following way: For $i \in \{1,2,\ldots,v\}$, $j \in \{1,2,\ldots,r\}$, $k \in \{1,2,\ldots,t\}$ let x be the unique point contained in $T_i \cap l_0^j$, and let l be the L_k-line through x. Let y be the unique point with $y \in T_1 \cap l$, and let l_0^c be the L_0-line through y. Finally, define $a_{ij}^k := c$. From the properties of transversal seminets one can conclude that $\mathcal{A}(\mathcal{T}) := \{A_1,A_2,\ldots,A_t\}$ is, in fact, a set of mutually orthogonal permutation arrays.

The conditions (C_1) and (C_2) of Section 1 can be paraphrased in terms of transversal seminets as follows:

(D_1) There is no point of the seminet which is contained in all transversals, i.e. $\bigcap_{i=1,2,\ldots,v} T_i = \emptyset$. (This implies, in particular, $|l| \geq 2$ for each line l, and $v \geq 2$.)

(D_2) Any two transversals are distinct, i.e. $T_i \neq T_j$ for $i \neq j$.

In the construction procedure for $\mathcal{A}(\mathcal{T})$ only those points of X have been used which are contained in one of the transversals T_i. Therefore one can always restrict oneself to the *reduced* transversal seminet $(X';L_0',L_1',\ldots,L_t';T_1,T_2,\ldots,T_v)$ defined by $X' := \bigcup_{i=1,2,\ldots,v} T_i$, $L_k' := \{l \cap X' \mid l \in L_k\}$.

In the rest of this paper all transversal seminets are assumed to be reduced and to satisfy the conditions (D_1) and (D_2).

The process of constructing mutually orthogonal permutation arrays from transversal seminets can be reversed: Let $\mathcal{A}$ be a set of t mutually orthogonal $v\times r$ permutation arrays $A_k = (a_{ij}^k)$, $k=1,2,\ldots,t$. Define $Y := \{1,2,\ldots,v\} \times \{1,2,\ldots,r\}$, and let ψ be the equivalence relation on Y with $(i,j)\ \psi\ (i',j')$ if and only if $j = j'$ and $j \in F_{ii'}(\mathcal{A})$. Define the point set X of the seminet as the set of equivalence classes of ψ, i.e. $X := Y/\psi = \{[(i,j)]\psi \mid (i,j)\in Y\}$. For $c = 1,2,\ldots,r$ and $k = 1,2,\ldots,t$ let $l_0^c := \{[(i,j)]\psi \mid j=c,\ i=1,2,\ldots,v\}$ and $l_k^c := \{[(i,j)]\psi \mid a_{ij}^k = c\}$, and define $L_0 := \{l_0^1,l_0^2,\ldots,l_0^r\}$, $L_k := \{l_k^1,l_k^2,\ldots,l_k^r\}$. Finally, let $T_i := \{[(i,j)]\psi \mid j=1,2,\ldots,r\}$, $i=1,2,\ldots,v$. Then $\mathcal{T}(\mathcal{A}) := (X;L_0,L_1,\ldots,L_t;T_1,T_2,\ldots,T_v)$ is a (reduced) transversal seminet with $\mathcal{A}(\mathcal{T}(\mathcal{A})) = \mathcal{A}$. Summarizing, one obtains

2.1. PROPOSITION. *The existence of a set of* t *mutually orthogonal* $v\times r$ *permutation arrays is equivalent to the existence of a transversal* $(t+1,r)$*-seminet with* v *transversals.*

This equivalence was already observed in [4]. One can show even a little more. Let $\mathcal{T} = (X;L_0,L_1,\ldots,L_t;T_1,T_2,\ldots,T_v)$ and $\mathcal{U} = (Y;M_0,M_1,\ldots,M_t;U_1,U_2,\ldots,U_v)$ be reduced transversal seminets with $L_0 = \{l_0^1,l_0^2,\ldots,l_0^r\}$ and $M_0 = \{m_0^1,m_0^2,\ldots,m_0^r\}$, and assume $\mathcal{A}(\mathcal{T}) = \mathcal{A}(\mathcal{U})$. Define a mapping $\phi\colon X \longrightarrow Y$ as follows. For $x\in T_i\cap l_0^j$ let $\phi(x)$ be the unique element contained in $U_i\cap m_0^j$. Then ϕ is an isomorphism of $\mathcal{T}$ and $\mathcal{U}$, i.e. ϕ is a bijection which maps parallel lines onto parallel lines (in fact, ϕ satisfies $\phi L_i = M_i$, $\phi T_i = U_i$ and $\phi l_0^i = m_0^i$ for all i). This yields

2.2. PROPOSITION. *If* $\mathcal{T}$ *and* $\mathcal{U}$ *are reduced transversal seminets with* $\mathcal{A}(\mathcal{T}) = \mathcal{A}(\mathcal{U})$ *then* $\mathcal{T}$ *and* $\mathcal{U}$ *are isomorphic.*

The transversal seminet $\mathcal{T} = (X;L_0,L_1,\ldots,L_t;T_1,T_2,\ldots,T_v)$ corresponds to a set $\mathcal{A}(\mathcal{T}) = \{A_1,A_2,\ldots,A_t\}$ of mutually orthogonal *latin rectangles* exactly if $T_i\cap T_j = \emptyset$ for all i,j, $i\neq j$. The arrays $A_1,A_2,\ldots,A_t$ form a set of mutually orthogonal *latin squares* if and only if $(X;L_0,L_1,\ldots,L_t,L_{t+1})$, with $L_{t+1} := \{T_1,T_2,\ldots,T_v\}$, is a net. In other words, the equivalence of Proposition 2.1 specializes to the classical correspondence of mutually orthogonal latin squares with nets.

The following theorem provides a construction method of sets of mutually orthogonal permutation arrays via seminets, using groups.

2.3. THEOREM. *Let* G *be a finite group with neutral element* e. *Let* t *and* s *be positive integers, and let* $S_0,S_1,\ldots,S_t$ *and* $F_1,F_2,\ldots,F_s$ *be nontrivial subgroups of* G *such that the following conditions are satisfied for all* $i,j\in\{0,1,\ldots,t\}$, $k,l\in\{1,2,\ldots,s\}$:

(1) $i\neq j \Longrightarrow S_i\cap S_j = \{e\}$,

(2) $S_i\cap F_k = \{e\}$,

(3) $k\neq l \Longrightarrow F_k\neq F_l$,

(4) $|F_k| = [G:S_i]$.

Then there exists a set of t *mutually orthogonal* $v\times r$ *permutation arrays, with* $r := |F_1|$ *and* $v := \frac{s}{r}\cdot|G|$.

Proof. For each $i\in\{0,1,\ldots,t\}$ let L_i consist of the right cosets of S_i, i.e. $L_i = \{S_ig \mid g\in G\}$. Then $(G;L_0,L_1,\ldots,L_t)$ is a $(t+1)$-seminet: Condition (S_1)

is trivially satisfied while (S_2) is a consequence of (1). Each right coset $F_k h$ of one of the subgroups F_k is a transversal of $(G;L_o,L_1,\dots,L_t)$: It has to be shown that $|S_i g \cap F_k h| = 1$ for all $g,h \in G$. As a consequence of (2) one obtains $|S_i g \cap F_k h| \le 1$. Assumption (4) then implies $\bigcup_{f \in F_k} S_i f = G$, and hence $|S_i g \cap F_k h| \ge 1$. Each transversal has $r = |F_1|$ elements, and each F_k has $\frac{|G|}{r}$ distinct right cosets. Thus there are $v = \frac{s}{r} \cdot |G|$ distinct transversals of the form $F_k h$, with $k \in \{1,2,\dots,s\}$ and $h \in G$. Finally, the nontriviality conditions (D_1) and (D_2) are a consequence of the nontriviality of the subgroups F_k and of (3), respectively. By Proposition 2.1, the proof is complete. □

The seminet $(G;L_o,L_1,\dots,L_t)$ of the above proof is, in fact, a *translation seminet*, i.e. it has a translation group operating regularly on its points: In the right regular representation of G each mapping $x \mapsto xg$, $g \in G$, maps every line onto a parallel line. On the other hand, each translation seminet can be obtained in this way from a group G and subgroups $S_o,S_1,\dots,S_t$ satisfying condition (1). Analogous group theoretic characterizations have been given, for instance, for translation planes, translation nets, translation structures and translation group divisible designs (see e.g. [1], [15], [22], [3] and [20]). Marchi [18] uses similar ideas for his characterization of regular affine parallel structures by partition loops. Probably one can formulate an analogue of Theorem 2.3 using loops instaed of groups. The problem would be to find examples for such a generalization. The rest of this section yields two classes of examples for Theorem 2.3. Cf. Huppert [14] and Wielandt [24] for the group theoretic notations.

2.4. EXAMPLE. Let G be a nonsolvable transitive permutation group of prime degree p. Let $v := p^2$, $r := \frac{|G|}{p}$, and let d be the positive integer with $d \le p-1$ and $d \equiv r \pmod p$. Then one can construct a set of $t := \frac{r}{d} - 1$ mutually orthogonal $v \times r$ permutation arrays: Assume G to operate on $\{a_1,a_2,\dots,a_p\}$. For each $k \in \{1,2,\dots,p\}$ define F_k to be the stabilizer of a_k in G, i.e. $F_k := G_{a_k}$. Then $F_k \ne F_l$ for $k \ne l$ since G is doubly transitive (cf. Theorem 11.7 of [24]). Let $S_o,S_1,\dots,S_{t'}$ be the Sylow p-subgroups of G (with $t' \ge 1$ because G is nonsolvable). Obviously, these subgroups satisfy the assumptions of Theorem 2.3. Hence there are t' mutually orthogonal $v \times r$ permutation arrays. It remains to show $t' = t$ or, equivalently, that G has exactly $\frac{r}{d}$ Sylow p-subgroups. Let P be a Sylow p-subgroup of G. The only Sylow p-subgroup of the normaliser $N_G(P)$ of P is P itself. Hence $N_G(P)$ is solvable and thus of order $p \cdot d'$ with $d' | p-1$ (cf. [14], Satz II.3.6). Therefore the number n of Sylow p-subgroups satisfies $n = [G:N_G(P)] = \frac{G}{p \cdot d'} = \frac{r}{d'}$. From $n \equiv 1 \pmod p$ one obtains $d' \equiv r \pmod p$, i.e. $d = d'$ and $n = \frac{r}{d}$.

The nonsolvable transitive permutation groups of prime degree have been completely determined, due to the classification of finite simple groups (see

Corollary 4.2 of Feit [11]).

Notice that *solvable* transitive permutation groups of prime degree p cannot be used in the above construction: These groups have exactly one Sylow p-subgroup, which would imply $t = 0$.

2.5. EXAMPLE. For each integer $m \geq 2$ one can construct a set of t mutually orthogonal $v \times r$ permutation arrays, with $t := 2^{m-1}(2^m-1)-1$, $v := (2^m+1)^2$ and $r := 2^m(2^m-1)$: Regard the projective special linear group $G = PSL(2,q)$, with $q = 2^m$, as a permutation group operating canonically on the $q+1$ points $\{a_1, a_2, \ldots, a_{q+1}\}$ of the projective line over the q-element field. For each $k \in \{1,2,\ldots,q+1\}$ define F_k to be the stabilizer of a_k in G, i.e. $F_k := G_{a_k}$. Let $S_0, S_1, \ldots, S_{t'}$ be the (mutually conjugate) cyclic subgroups of G of order $q+1$. By the results in [14], pp. 191-193, these subgroups satisfy the assumptions of Theorem 2.3. For the number t' of conjugates of S_0 one obtains $t'+1 = [G:N_G(S_0)] = \frac{(q-1)q(q+1)}{2(q+1)} = \frac{q(q-1)}{2} = t+1$, i.e. $t' = t$.

Notice that Hartman [12] used some of the groups PSL(2,q) in order to construct designs with mutually orthogonal resolutions. For instance, for each $q \in \{19,31,43\}$ there exists a design with $v = q+1$ treatments, $r = \frac{q(q-1)}{6}$ replications and $t+1 = \frac{q-1}{6}$ mutually orthogonal resolutions.

3. BOUNDS FOR THE NUMBER OF MUTUALLY ORTHOGONAL PERMUTATION ARRAYS

A set $\{A_1, A_2, \ldots, A_t\}$ of mutually orthogonal permutation arrays is called *maximal* if there exists no permutation array A_{t+1} which is orthogonal to all A_k, $k = 1,2,\ldots,t$. A transversal seminet $(X; L_0, L_1, \ldots, L_t; T_1, T_2, \ldots, T_v)$ is called *L-maximal* if there exists no additional parallel class L_{t+1} such that $(X; L_0, L_1, \ldots, L_t, L_{t+1}; T_1, T_2, \ldots, T_v)$ is again a transversal seminet. The following lemma is obvious.

3.1. LEMMA. *A set $\mathcal{A}$ of mutually orthogonal permutation arrays is maximal if and only if the associated transversal seminet $\mathcal{T}(\mathcal{A})$ is L-maximal.*

3.2. PROPOSITION. *Each set $\mathcal{A}$ of mutually orthogonal permutation arrays of Example 2.5 is maximal.*

Proof. Let $G = PSL(2,2^m)$ be the group used for the construction of $\mathcal{A}$. The associated transversal seminet $\mathcal{T}$ has G as point set. The subgroups $S_0, S_1, \ldots, S_t$ are exactly the lines through the neutral element e of G, and the subgroups $F_1, F_2, \ldots, F_{q+1}$ are exactly the transversals through e (cf. the proof of Theorem 2.3). By Satz II.8.5 of Huppert [14], these subgroups cover G. Hence there cannot be any additional line through e, and $\mathcal{T}$ is therefore L-maximal. By Lemma 3.1 the proof is complete, since Proposition 2.2 yields $\mathcal{T} \cong \mathcal{T}(\mathcal{A})$. □

Actually, the assertion of Proposition 3.2 depends only on the intersection structure $F(\mathcal{A})$ of $\mathcal{A}$: Let $\mathcal{B} = \{B_1, B_2, \ldots, B_{t'}\}$ be a set of mutually orthogonal permutation arrays with $F(\mathcal{B}) = F(\mathcal{A})$. The construction procedure of Section 2 shows that the transversal seminets $\mathcal{T}(\mathcal{B})$ and $\mathcal{T}(\mathcal{B})$ have the same point sets and the same transversals. The transversal seminet $\mathcal{T}$ of the proof of Proposition 3.2 contains $n := |S_0| = [G:F_1]$ pairwise disjoint transversals $F_1 s$, $s \in S_0$, with $\bigcup_{s \in S_0} F_1 s = G$. Therefore each line of $\mathcal{T}$ contains exactly n points, and $\mathcal{T} \cong \mathcal{T}(\mathcal{A})$ implies that the same is true for $\mathcal{T}(\mathcal{A})$ and also for $\mathcal{T}(\mathcal{B})$. As a consequence, there is a point x of $\mathcal{T}(\mathcal{B})$ such that the number $t'+1$ of lines of $\mathcal{T}(\mathcal{B})$ through x cannot exceed the number $t+1$ of lines of $\mathcal{T}$ through e (in fact, this is true for *each* point x of $\mathcal{T}(\mathcal{B})$). Therefore Proposition 3.2 can be improved as follows.

3.3. PROPOSITION. *Let* $\mathcal{A} = \{A_1, A_2, \ldots, A_t\}$ *be one of the sets of mutually orthogonal permutation arrays of Example 2.5. Let* $\mathcal{B} = \{B_1, B_2, \ldots, B_{t'}\}$ *be a set of mutually orthogonal permutation arrays with* $F(\mathcal{B}) = F(\mathcal{A})$. *Then* $t' \le t$.

The next lemma gives an upper bound for the number of mutually orthogonal permutation arrays depending on the intersection structure of the arrays. The proof of this lemma is a unified version of the proofs of several related results in [4] and [7].

3.4. LEMMA. *Let* $\mathcal{A} = \{A_1, A_2, \ldots, A_t\}$ *be a set of mutually orthogonal* $v \times r$ *permutation arrays, and let* $I \subseteq \{1,2,\ldots,v\}$, $i_0 \in \{1,2,\ldots,v\}$, $J \subseteq \{1,2,\ldots,r\}$, $j_0 \in \{1,2,\ldots,r\}$ *satisfy*

a) $I \neq \emptyset$,

b) $\forall i_1, i_2 \in I$: $j_0 \in F_{i_1 i_2}(\mathcal{A})$,

c) $\forall i \in I$: $j_0 \notin F_{i i_0}(\mathcal{A})$,

d) $\forall j \in J \;\; \exists i \in I$: $j \in F_{i i_0}(\mathcal{A})$.

Then $t \le r - |J| - 1$.

<u>Proof</u>. Let $A = (a_{ij})$ be a $v \times r$ permutation array with $F(A) = F(\mathcal{A})$. By a) there exists an element $i_1 \in I$. Define $c := a_{i_1 j_0}$. From b) one obtains $a_{ij_0} = c$ for all $i \in I$, and c) implies $a_{i_0 j_0} \neq c$. For each $j \in J$ there exists an $i \in I$ with $a_{i_0 j} = a_{ij}$, by d). Thus $j \neq j_0$ and $a_{ij} \neq a_{ij_0} = c$. Hence $a_{i_0 j} \neq c$. There-

fore $a_{i_0 j} = c$ for exactly one element j of the $(r-|J|-1)$-element set $\{1,2,\dots,r\} \setminus (J \cup \{j_0\})$. In particular, this is true for each of the permutation arrays $A_k \in \mathcal{A}$ with c and j replaced by c_k and j_k. By orthogonality, the mapping $k \mapsto j_k$ is injective. This implies $t \le r - |J| - 1$. □

The following corollary translates Lemma 3.4 into the language of transversal seminets. Notice that this corollary could have been used in order to prove the Propositions 3.2 and 3.3.

3.5. COROLLARY. *Let* $\mathcal{T} = (X;L_0,L_1,\dots,L_t;T_1,T_2,\dots,T_v)$ *be a transversal seminet, and let* $I \subseteq \{1,2,\dots,v\}$, $i_0 \in \{1,2,\dots,v\}$ *and* $x \in X$ *satisfy*

a) $I \neq \emptyset$,

b) $x \in \bigcap_{i \in I} T_i$,

c) $x \notin T_{i_0}$.

Then $t \le r - \delta - 1$, *with* $r := |T_1|$ *and* $\delta := |T_{i_0} \cap (\bigcup_{i \in I} T_i)|$.

3.6. COROLLARY. *Let* $\mathcal{A} = \{A_1,A_2,\dots,A_t\}$ *be a set of mutually orthogonal* $v \times r$ *permutation arrays, and let* $\mu := \max\{|F_{ii'}(\mathcal{A})| \mid i,i'=1,2,\dots,v,\ i \neq i'\}$. *Then* $t \le r - \mu - 1$.

<u>Proof</u>. Choose $i,i_0 \in \{1,2,\dots,v\}$ and $j_0 \in \{1,2,\dots,r\}$, with $|F_{i i_0}(\mathcal{A})| = \mu$ and $j_0 \notin F_{i i_0}(\mathcal{A})$. Define $I := \{i\}$ and $J := F_{i i_0}(\mathcal{A})$. Then I, i_0, J, j_0 satisfy the assumptions of Lemma 3.4. □

3.7. COROLLARY. *Let* $\mathcal{A} = \{A_1,A_2,\dots,A_t\}$ *be a set of mutually orthogonal* $v \times r$ *permutation arrays. Let* $\lambda := \min\{|F_{ii'}(\mathcal{A})| \mid i,i'=1,2,\dots,v\}$, *and assume* $\lambda \ge 1$. *Then* $t \le r - \lambda - 2$.

<u>Proof</u>. Choose $i_0,i_1,i_2 \in \{1,2,\dots,v\}$, $i_1 \neq i_2$, and $j_0 \in \{1,2,\dots,r\}$ with $j_0 \in F_{i_1 i_2}(\mathcal{A})$, $j_0 \notin F_{i_1 i_0}(\mathcal{A})$ and $j_0 \notin F_{i_2 i_0}(\mathcal{A})$. Define $I := \{i_1,i_2\}$, $J := F_{i_1 i_0}(\mathcal{A}) \cup F_{i_2 i_0}(\mathcal{A})$. Then I, i_0, J, j_0 satisfy the assumptions of Lemma 3.4. Hence the proof is complete in the case $|J| \ge \lambda+1$. Assume now $|J| = \lambda$. Then $\lambda = |F_{i_1 i_0}(\mathcal{A})| = |F_{i_2 i_0}(\mathcal{A})|$, and thus $|F_{i_1 i_2}(\mathcal{A})| = |J \cup \{j_0\}| = \lambda+1$. Therefore the case $|J| = \lambda$ is settled by Corollary 3.6. □

The Corollaries 3.5, 3.6 and 3.7 yield slight generalizations for some of the results in [4], [7] and [17] (which are formulated in terms of designs with mutually orthogonal resolutions).

A $v \times r$ permutation array A is called *row-transitive* if the rows of A form a set of permutations operating transitively on the set $\{1,2,\dots,r\}$.

3.8. PROPOSITION. *Let* $\mathcal{A} = \{A_1, A_2, \ldots, A_t\}$ *be a set of mutually orthogonal* $v \times r$ *permutation arrays. Assume one of these arrays (and hence all of them) to be row-transitive. Then* $t \le \min\{r-1, m+1\}$, *with* m *the largest number of mutually orthogonal latin squares of order* r.

Proof. Row-transitivity implies each line of the associated transversal semi-net $\mathcal{T}(\mathcal{A}) = (X; L_0, L_1, \ldots, L_t; T_1, T_2, \ldots, T_v)$ to have exactly r points, since any line $l \in L_i$, $i \ge 1$, intersects each line of L_0 exactly once. Thus $(X; L_0, L_1, \ldots, L_t)$ is a $(t+1)$-net of order r. The existence of such a net is equivalent to the existence of $t-1$ mutually orthogonal latin squares of order r. Hence $t-1 \le m$. A complete $(t+1)$-net of order r has exactly $r+1$ parallel classes. But a net with a transversal cannot be complete. Therefore $t+1 < r+1$. □

4. EXTENSION BY ROWS

A set $\mathcal{A} = \{A_1, A_2, \ldots, A_t\}$ of mutually orthogonal $v \times r$ permutation arrays is called *row-extendible* if it is possible to adjoin a new row to each of the arrays such that the resulting $(v+1) \times r$ permutation arrays are again mutually orthogonal.

A transversal seminet $(X; L_0, L_1, \ldots, L_t; T_1, T_2, \ldots, T_v)$ is called *transversal-extendible* if there exists a transversal seminet $(Y; M_0, M_1, \ldots, M_t; T_1, T_2, \ldots, T_v, T_{v+1})$ with $Y \supseteq X$ and $L_i = \{m \cap X \mid m \in M_i\}$. As transversal-extendibility is the obvious translation of row-extendability into the language of transversal seminets, one obtains

4.1. LEMMA. *A set* $\mathcal{A}$ *of mutually orthogonal permutation arrays is row-extendible if and only if the associated transversal seminet* $\mathcal{T}(\mathcal{A})$ *is transversal-extendible.*

The above definition of row-extendability is strictly stronger than the one given in [4] where the resulting arrays were only assumed to be similar. Both definitions coincide if $Y = X$ in the transversal seminets used in the definition of transversal-extendibility. In particular, the definitions coincide if $(X; L_0, L_1, \ldots, L_t)$ is a net (cf. Proposition 4.4 of [4]). In the case of mutually orthogonal latin squares, row-extendibility is equivalent to the existence of a common *column-transversal* (i.e. a usual transversal of latin squares with the condition "no two cells are on the same row" replaced by the weaker condition "not all cells are on the same row"); see [4], Proposition 4.2. Let $(X; L_0, L_1, \ldots, L_t; T_1, T_2, \ldots, T_v)$ be the transversal seminet associated to the latin squares $A_1, A_2, \ldots, A_t$.

Clearly, the latin squares have a common column-transversal exactly if there exists a transversal T_{v+1} of $(X;L_o,L_1,\ldots,L_t)$ with $T_{v+1} \notin L_{t+1} := \{T_1,T_2,\ldots,T_v\}$. There is no such T_{v+1} if the net $(X;L_o,L_1,\ldots,L_t,L_{t+1})$ is an affine plane (i.e. if $A_1,A_2,\ldots,A_t$ form a complete set of mutually orthogonal latin squares). In general, it is an open question whether such a transversal exists if $(X;L_o,L_1,\ldots,L_{t+1})$ is a *transversal-free* net in the sense of Dow [10].

Although the next proposition is easy to prove, the result is quite surprising.

4.2. PROPOSITION. *All sets* $\mathcal{A}$ *of mutually orthogonal permutation arrays of the Examples 2.4 and 2.5 are row-extendible.*

Proof. Let G be the group used in Example 2.4 or 2.5 for the construction of $\mathcal{A}$, and let $\mathcal{T} = (X;L_o,L_1,\ldots,L_t;T_1,T_2,\ldots,T_v)$ be the transversal seminet associated to G and the subgroups $S_o,S_1,\ldots,S_t$ and $F_1,F_2,\ldots,F_s$ of G. Recall from the proof of Theorem 2.3 that $X = G$, $L_i = \{S_i g \mid g\in G\} = \{S_i f \mid f\in F_1\}$ for all $i \in \{0,1,\ldots,t\}$, and $\{T_1,T_2,\ldots,T_v\} = \{F_k g \mid g\in G,\ k=1,2,\ldots,s\}$. By Lemma 4.1 and because of $\mathcal{T} \cong \mathcal{T}(\mathcal{A})$, it is sufficient to show that the transversal seminet $\mathcal{T}$ is transversal-extendible. Let $\overline{F}_1 = \{\overline{f} \mid f\in F_1\}$ be a copy of F_1 with $\overline{F}_1 \cap X = \emptyset$, and define $Y := X \cup \overline{F}_1$. As all subgroups S_i, $i=0,1,\ldots,t$, are conjugate to each other, there exist $a_o,a_1,\ldots,a_t \in G$ with $S_i = a_i S_o a_i^{-1}$. Let $M_i := \{S_i a_i f \cup \{\overline{f}\} \mid f\in F_1\}$, $i=0,1,\ldots,t$, and define $T_{v+1} := \overline{F}_1$. Trivially then $Y \supseteq X$ and $L_i \supseteq \{m\cap X \mid m\in M_i\}$. In order to show $L_i = \{m\cap X \mid m\in M_i\}$, assume $f_1,f_2 \in F_1$ and $S_i a_i f_1 = S_i a_i f_2$. Then $(a_i S_o a_i^{-1})a_i f_1 = (a_i S_o a_i^{-1})a_i f_2$ and thus $f_1 f_2^{-1} \in S_o$. From $S_o \cap F_1 = \{e\}$ one obtains $f_1 = f_2$. This implies $|L_i| = |\{m\cap X \mid m\in M_i\}|$, and therefore $L_i = \{m\cap X \mid m\in M_i\}$. The same argument implies condition (S_2) for $(Y;M_o,M_1,\ldots,M_t)$. It remains to show (S_1). Let $i,j \in \{0,1,\ldots,t\}$, $f_1,f_2 \in F_1$ and $x,y \in X$ satisfy $i \neq j$, $x \neq y$ and $x,y \in (S_i a_i f_1 \cup \{\overline{f}_1\}) \cap (S_j a_j f_2 \cup \{\overline{f}_2\})$. As $(X;L_o,L_1,\ldots,L_t)$ is a seminet, either x or y cannot be contained in X. Assume $y \notin X$. Then $y = \overline{f}_1 = \overline{f}_2$ and $x \in S_i a_i f_1 \cap S_j a_j f_2 = (a_i S_o a_i^{-1})a_i f_1 \cap (a_j S_o a_j^{-1})a_j f_1 = a_i S_o f_1 \cap a_j S_o f_1$. This implies $x f_1^{-1} \in a_i S_o \cap a_j S_o$. Hence $a_i S_o = a_j S_o$ and thus $S_i = S_j$, contradicting $i \neq j$. □

The alternating group A_5 yields an example for some of the results of the preceding sections. Since A_5 and $PSL(2,2^2)$ are isomorphic as permutation groups, both of the Examples 2.4 and 2.5 imply the existence of a set $\mathcal{A}$ of 5 mutually orthogonal 25×12 permutation arrays. By Proposition 3.2, the set $\mathcal{A}$ is maximal, and by Proposition 4.2 $\mathcal{A}$ is row-extendible, i.e. there exists a set $\mathcal{A}'$ of 5 mutually orthogonal 26×12 permutation arrays. It is unknown whether $\mathcal{A}'$ is row-extendible or not.

5. TRANSVERSAL SEMINETS CARRYING DESIGNS

The concept of transversal seminets comprises a large variety of different mathematical structures. For detailed investigations one has therefore to add further restrictions. Sections 2 and 4 and parts of Section 3 treated transversal seminets with a group of translations operating regularly on the points. In this section additional assumptions will be imposed on the following incidence structures which are associated to each transversal seminet $\mathcal{T} = (X;L_0,L_1,\ldots,L_t;T_1,T_2,\ldots,T_v)$ (let $L := L_0 \cup L_1 \cup \ldots \cup L_t$ and $T := \{T_1,T_2,\ldots,T_v\}$, i.e. L is the set of lines and T the set of transversals of $\mathcal{T}$):

$I(X,L)$ — the treatments of this incidence structure are the points and the blocks are the lines of $\mathcal{T}$,

$I(T,X)$ — the treatments are the transversals and the blocks are the points of $\mathcal{T}$,

$I(X,L\cup T)$ — the treatments are the points and the blocks are the lines *and* the transversals of $\mathcal{T}$.

The incidence structure $I(X,L)$ is essentially the seminet of $\mathcal{T}$. In all three cases the incidence is defined naturally. For example, $x \in X$ and $T_i \in T$ are incident in $I(T,X)$ exactly if $x \in T_i$. $I(T,X)$ is the design with mutually orthogonal resolutions mentioned in Section 1: The blocks $x,y \in X$ are defined to be parallel in the i th parallel class if and only if there exists a line $l \in L_i$ with $x,y \in l$. Two interesting cases discussed later in this section occur when $I(T,X)$ or $I(X,L\cup T)$ are PBD's (pairwise balanced designs). For instance, $I(X,L\cup T)$ has this property if the associated set of permutation arrays is a complete set of mutually orthogonal latin rectangles (the examples after Proposition 5.3 show that the converse of this statement is not true). Before going into detail, some definitions are necessary.

The seminet $\mathcal{S} = (X;L_0,L_1,\ldots,L_t)$ is called n-*regular* if each line contains exactly n points. In this case $\mathcal{S}$ is a $(t+1,r)$-seminet with $r = |X|/n$ (i.e. $|L_i| = r$ for all i), and $\mathcal{S}$ is also called an (r,n)-*Mano configuration*. One has $n \leq r$, with equality if and only if $\mathcal{S}$ is a net or, equivalently, if the associated permutation arrays are row-transitive (see Section 3). An n-regular transversal seminet satisfies $n \leq v$ (because $rv = \Sigma_{i=1,2,\ldots,v}|T_i| \geq |X| = rn$), with equality if and only if the T_i's are pairwise disjoint. In this situation the set of transversals can be considered as a new parallel class L_{t+1}, and the associated permutation arrays are latin rectangles. Therefore, if $n = r = v$, then $I(X,L\cup T)$ is a $(t+2,r)$-net, and the associated permutation arrays are latin squares.

The transversal seminet $\mathcal{T} = (X;L_0,L_1,\ldots,L_t;T_1,T_2,\ldots,T_v)$ is called q- *uniform* if each point is contained in exactly q transversals. A q-uniform transversal seminet is n-regular with $n = v/q$. Each column of each of the associated permutation arrays contains each element $i \in \{1,2,\ldots,r\}$ either q or 0 times. An example of an n-regular (with n=4) but *not* q-uniform transversal seminet is provided by the orthogonal permutation arrays

$$A_1 = \begin{pmatrix} 1 & 2 & 3 & 4 \\ 2 & 1 & 4 & 3 \\ 3 & 4 & 1 & 2 \\ 4 & 3 & 2 & 1 \\ 1 & 3 & 4 & 2 \end{pmatrix}, \qquad A_2 = \begin{pmatrix} 1 & 2 & 3 & 4 \\ 4 & 3 & 2 & 1 \\ 2 & 1 & 4 & 3 \\ 3 & 4 & 1 & 2 \\ 1 & 4 & 2 & 3 \end{pmatrix}.$$

(The set $\{A_1,A_2\}$ is a row-extension of two orthogonal latin squares. It is easy to check that A_1,A_2 have no orthogonal mate.)

For each point x of a transversal seminet, let $\tilde{x}$ denote the number of transversals through x, and define $\tilde{X} := \{\tilde{x} \mid x \in X\}$. For each transversal seminet the incidence structure $I(T,X)$ is a 1-design $S_r(1,\tilde{X},v)$, i.e. each of the v treatments $T_i \in T$ is incident with exactly r blocks, and each block x is incident with $\tilde{x} \in \tilde{X}$ treatments. Notice that $I(T,X)$ may have repeated blocks. The blocks of $I(T,X)$ have $t+1$ mutually orthogonal resolutions (i.e. any two parallel classes of distinct resolutions coincide in at most one block). The resolutions are given by $L_0,L_1,\ldots,L_t$.

If there is a nonnegative integer λ with $|T_i \cap T_j| = \lambda$ for all i,j with $i \neq j$ (i.e. if the transversals form an (r,λ)-equidistant code), then $I(T,X)$ becomes a PBD with any two distinct treatments contained in exactly λ blocks. Equivalently, the associated permutation arrays are *equidistant* with Hamming-distance $r-\lambda$, i.e. any two rows of each of the permutation arrays coincide in exactly λ positions. If, moreover, the transversal seminet is q-uniform, then $I(T,X)$ is a 2-design $S_\lambda(2,q,v)$. As a consequence, $r(q-1) = \lambda(v-1)$. Analogously, if $I(T,X)$ is an $S_{\lambda'}(t',q,v)$, with $t' \geq 2$, then $r\binom{q-1}{t'-1} = \lambda'\binom{v-1}{t'-1}$. In this case, any t' distinct rows of each of the permutation arrays coincide in exactly λ' positions.

The transversal seminets constructed in the proof of Theorem 2.3 are n-regular and q-uniform, with $n = \frac{|G|}{r}$ and $q = s$. For none of these examples $I(T,X)$ is a PBD.

Many examples of transversal seminets are given in [17], with $I(T,X)$ an $S_{\lambda'}(t',q,v)$ or a group-divisible design (in this case $|T_i \cap T_j| \leq 1$ for all distinct i,j). For instance, by Theorem 1.5 of [17] there is a positive integer v_1 such that, for $v \geq v_1$, the condition $v \equiv 3 \pmod{12}$ is equivalent to the existence of a transversal seminet with $t = 2$ and the property that $I(T,X)$ is an $S(2,3,v)$. There are similar results of other authors; some of them are listed here:

1) Dinitz [8]. For each prime power of the form $q = 2^k c + 1$, with k a positive integer and $c>1$ an odd integer, there exists a transversal seminet such that $t = c-1$ and $I(T,X)$ is an $S(2,2,q+1)$.

2) Kramer et al. [16]. There exists a transversal semient such that $t = 12$ and $I(T,X)$ is an $S(5,8,24)$.

3) Hartman [12]. There exist transversal seminets such that $I(T,X)$ is an $S(3,4,v)$ for $(v,t) = (20,3),(32,5),(44,7)$.

Transversal seminets can be represented by the following (t+1)-dimensional array of side r: the cell $(i_0,i_1,\dots,i_t)$ contains the set $\{T_i \in T \mid x \in T_i\}$ if $l_0^{i_0} \cap l_1^{i_1} \cap \dots \cap l_t^{i_t} = \{x\}$, and it is empty otherwise (the lines of each parallel class are assumed to be linearly ordered, i.e. $L_k = \{l_k^1, l_k^2, \dots, l_k^r\}$). If $I(T,X)$ is an $S(2,2,v)$, then this array is called a Room (t+1)-cube of side $r = v-1$. Each such array is equivalent to $t+1$ pairwise orthogonal symmetric latin squares of size $(v-1)\times(v-1)$ (see [9]). If $I(T,X)$ is an $S_{\lambda'}(t',q,v)$, then the above (t+1)-dimensional array is a (t+1)-dimensional Room design; for instance, $S(5,8,24)$ yields a 13-dimensional Room design of side 253, see [16].

The incidence structure $I(X,L)$ of the seminet $\mathcal{S} = (X;L_0,L_1,\dots,L_t)$ is a 1-design $S_{t+1}(1,\{|l| \mid l \in L\},|X|)$ without repeated blocks. If $\mathcal{S}$ is n-regular, then $I(X,L)$ becomes an $S_{t+1}(1,n,rn)$. $\mathcal{S}$ is called an *André seminet* (or *affine parallel structure*) if $I(X,L)$ is a linear space, i.e. if any two distinct treatments of $I(X,L)$ are incident with exactly one block. Obviously, André seminets do not admit transversals. A transversal seminet is called *almost-André* (or *complete*) if any two distinct points are either connected by a line or by a transversal. As each almost-André transversal seminet is L-maximal, Lemma 3.1 yields

5.1. REMARK. *Let the transversal seminet* $\mathcal{T}$ *be almost-André. Then the associated set* $\mathcal{A}(\mathcal{T})$ *of mutually orthogonal permutation arrays is maximal.*

Actually, Proposition 3.2 was proved essentially by showing that the involved transversal seminets are almost-André.

Recall that a set of t mutually orthogonal latin rectangles of size $v\times r$ is called *complete* if $t = r-1$.

5.2. PROPOSITION. *Let* $\mathcal{T}$ *be a transversal* (t+1,r)*-seminet with* v *transversals. Then* $t \le r-1$, *with equality if and only if the associated set* $\mathcal{A}(\mathcal{T})$ *of permutation arrays is a complete set of mutually orthogonal latin rectangles. In this case* $I(X,L\cup T)$ *is an* $S(2,\{r,v\},rv)$.

Proof. The inequality $t \le r-1$ is a trivial consequence of Corollary 3.5. Assume now $t = r-1$. By Corollary 3.6 then $\mu = 0$ for $\mathcal{A}(\mathcal{T})$ or, equivalently, $\mathcal{A}(\mathcal{T})$ is a set of latin rectangles. The completeness of $\mathcal{A}(\mathcal{T})$ implies $\mathcal{T}$ to be complete, i.e. any two elements of X are connected by a block from $L \cup T$. As $T_i \cap T_j = \emptyset$ for all i,j with $i \ne j$, one obtains that $I(X, L \cup T)$ is an $S(2,\{r,v\},rv)$. □

Notice that complete sets of mutually orthogonal latin rectangles have been constructed in Quattrocchi, Pellegrino [19] for all v not exceeding the smallest prime divisor of r.

For a transversal seminet $(X;L_0,L_1,\ldots,L_t;T_1,T_2,\ldots,T_v)$ and $M \subseteq \{1,2,\ldots,v\}$, $M \ne \emptyset$, let $t_M := |\bigcap_{i\in M} T_i|$, and define $d := \Sigma_{M\subseteq\{1,2,\ldots,v\},M\ne\emptyset} (-1)^{|M|-1}\binom{t_M}{2}$. Then

$$(1) \qquad d + \Sigma_{l\in L}\binom{|l|}{2} \le \binom{|X|}{2},$$

with equality if the transversal seminet is almost-André.

5.3. PROPOSITION. *Let* $\mathcal{T}$ *be a transversal* $(t+1,r)$-*seminet with* v *transversals. Let* $\mathcal{T}$ *be* n-*regular, and assume* $I(X,L\cup T)$ *to be an* $S(2,\{r,n\},rn)$. *Then* $t = \frac{(v-n)(r-1)}{n(n-1)} - 1$, *and there are exactly* $\frac{v}{n}$ *transversals through each* $x \in X$.

Proof. As the transversal seminet is almost-André, (1) is valid with equality. Since $|T_i \cap T_j| \le 1$ for $i \ne j$, one obtains $d = \Sigma_{1\le i\le v}\binom{|T_i|}{2} = v\binom{r}{2}$. With $|l| = n$ for all $l \in L$ and $|X| = rn$, (1) turns into

$$v\binom{r}{2} + (t+1)r\binom{n}{2} = \binom{rn}{2}.$$

This can easily be transformed into the claimed equality for t. Let α be the number of transversals through some $x \in X$. Then $v-1 = rn-1 = \alpha(r-1) + (t+1)(n-1)$ together with the equality just proved yields $\alpha = \frac{v}{n}$. □

A class of examples satisfying the assumptions of Proposition 5.3 can be obtained as follows. Let $(X;L_0,L_1,\ldots,L_r)$ be an affine plane of order r. For $k \ge 2$, consider the transversal seminet $\mathcal{T} = (X;L_0,L_1,\ldots,L_{r-k};T_1,T_2,\ldots,T_{rk})$ where $T_1,T_2,\ldots,T_{rk}$ are the lines contained in $\bigcup_{r-k<i\le r} L_i$. Then $\mathcal{T}$ satisfies the assumptions of Proposition 5.3, with $n = r$ and $v = rk$. In this situation, Corollary 3.6 yields the bound $t \le r-2$ (since $\mu=1$) while the exact value is $t = r-k$.

However, there also exist transversal seminets for which $I(X,L\cup T)$ is a linear space $S(2,\{r,n\},rn)$ with $r \ne n$. In terms of [13] these transversal seminets correspond exactly to the *partially resolvable* 2-*partitions* PRP 2-$(n,r,v;t+1)$

with the additional property that $v = nr$. Together with Proposition 5.3, the results of [13] and [5] on these designs imply

5.4. PROPOSITION. *An* n-*regular transversal* $(t+1,r)$-*seminet with* $I(X,L\cup T)$ *an* $S(2,\{r,n\},rn)$ *exists*

a) in the case $n = 2$, $r \geq 3$ *if and only if eihter* $t = 0$, $r = 3$ *or* $t = r-1$,

b) in the case $n = 3$, $r = 2$ *if and only if* $t = 0$,

c) in the case $n = 3$, $r = 4$ *if and only if* $t = 0$ *or* $t = 3$.

Notice that all seminets in the above proposition are either complete sets of mutually orthogonal latin rectangles ($t = r-1$) or transversal designs ($t=0$).

6. SOME STRUCTURES RELATED TO TRANSVERSAL SEMINETS

It is well-known that nets are equivalent to sets of mutually orthogonal latin squares, to transversal designs (via duality), to orthogonal arrays, to optimal codes, etc. Proposition 2.1 is a generalization of the first of these equivalences. Next, the second equivalence is generalized to transversal seminets.

Each transversal seminet $\mathcal{T} = (X;L_0,L_1,\ldots,L_t;T_1,T_2,\ldots,T_v)$ is equivalent to a *transversal packing*, via the incidence structure $I(L,X)$. Each block $x\in X$ hits each of the *groups* L_i in exactly one treatment $l\in L_i$, two treatments from distinct groups L_i,L_j are joined by at most one block, and there is no such block if $L_i = L_j$. Moreover, there are additional *main treatments* T_i; each T_i partitions the treatments of $I(L,X)$ into disjoint blocks.

Suppose now X and each L_i to be linearly ordered, i.e. $X = \{x_1,x_2,\ldots\}$ and $L_i = \{l_i^1,l_i^2,\ldots,l_i^r\}$. Let the matrix $B = (b_{ij})$ of size $(t+1)\times|X|$ be defined by $b_{ij} = k :\Longleftrightarrow x_j \in l_{i-1}^k$. This matrix is an OA (*orthogonal array*) if the seminet is a net (see e.g. [9]). In the general case, the set of all $|X|$ columns of B forms a *code* of *length* $t+1$ over the alphabet $\{1,2,\ldots,r\}$ with $|X|$ *words* and *minimal distance* t, since any two distinct columns coincide in at most one position. Each transversal T_j corresponds to a family of codewords (columns of B) such that, for all $k \in \{1,2,\ldots,r\}$ and all $i \in \{0,1,\ldots,t\}$, there is exactly one codeword in this family with value k in row i. The transversal T_j can also be regarded as an *injective diagonal* subset of $\{1,2,\ldots,r\}^{t+1}$ (i.e. as a set of r words of length $t+1$ over $\{1,2,\ldots,r\}$ which differ in any coordinate; see [6]).

Let the associated permutation arrays $A_1,A_2,\ldots,A_t$ now be latin rectangles or, equivalently, let $T_i \cap T_j = \emptyset$ for all i,j, $i\neq j$. Assume the numbering of the points $x\in X$ and of the lines $l\in L_i$ $(i\geq 1)$ now be more special than above: Let

$l = l_i^k$ if $l \cap T_1 \subseteq l_o^k$, and let $x = x_j$ with $j = (i-1)r + k$ if $x \in T_i \cap l_o^k$. Under these assumptions the first row of B satisfies $b_{1j} = k$ if $j \equiv k \pmod r$ and, for $m>1$, the m th row of B becomes a "linearization" of the latin rectangle A_{m-1}, i.e. $(b_{m,(i-1)r+1}, b_{m,(i-1)r+2}, \ldots, b_{m,ir})$ is the i th row of A_{m-1}. Notice that the transversal T_i corresponds now to the set of columns $(i-1)r+1$, $(i-1)r+2, \ldots, ir$ of B. For example, considering the simple complete set $A_1 = \begin{pmatrix} 1 & 2 & 3 \\ 2 & 3 & 1 \end{pmatrix}$, $A_2 = \begin{pmatrix} 1 & 2 & 3 \\ 3 & 1 & 2 \end{pmatrix}$ of mutually orthogonal latin rectangles, one obtains

$$B = \begin{pmatrix} 1 & 2 & 3 & 1 & 2 & 3 \\ 1 & 2 & 3 & 2 & 3 & 1 \\ 1 & 2 & 3 & 3 & 1 & 2 \end{pmatrix}.$$

The permutation arrays A and A' are similar exactly if, for all j, column j of A' can be obtained from column j of A by renaming the symbols. Let $Q(A,A')$ be the $r \times r$ matrix (q_{ij}) defined by $q_{ij} = k$ if the i's in column j of A become k's in column j of A', and $q_{ij} = *$ otherwise. For example,

$$Q\left(\begin{pmatrix} 1 & 2 & 3 \\ 2 & 3 & 1 \end{pmatrix}, \begin{pmatrix} 1 & 2 & 3 \\ 3 & 1 & 2 \end{pmatrix}\right) = \begin{pmatrix} 1 & * & 2 \\ 3 & 2 & * \\ * & 1 & 3 \end{pmatrix}.$$

As all permutation arrays are assumed to be standardized, $q_{ii} = i$ for all i in $Q(A,A')$. In no column of this matrix a symbol appears twice. Therefore $Q(A,A')$ can be considered as (the multiplication table of) a partial left-cancellative groupoid defined on $\{1,2,\ldots,r\}$. The permutation arrays A,A' are orthogonal if and only if this groupoid also is right-cancellative, and $Q(A,A')$ then becomes a partial quasigroup. Moreover, $Q(A,A')$ is a complete quasigroup if and only if the permutation arrays are row-transitive. Let $A_1,A_2,\ldots,A_t$ be similar permutation arrays. By Proposition 1.2 of [4], these arrays are mutually orthogonal exactly if $Q(A_1,A_2), Q(A_1,A_3), \ldots, Q(A_1,A_t)$ are mutually orthogonal partial quasigroups. Notice that the associated transversal seminet is n-regular if and only if each $Q(A_i,A_j)$, $i \neq j$, has the following properties: Each $i \in \{1,2,\ldots,r\}$ appears exactly n times, and there are exactly n distinct symbols in each row and in each column.

There are many incidence structures built from nets. Examples are transversal geometries [6], d-dimensional nets [21], and extensions of dual affine planes [23]. It will be interesting to study similar structures with nets replaced by transversal seminets.

REFERENCES

[1] J. André, Über nicht-Desarguessche Ebenen mit transitiver Translationsgruppe. *Math. Z.* 60 (1954) 156-186.

[2] J. André, Über Parallelstrukturen. Teil I: Grundbegriffe. *Math. Z.* 76 (1961) 85-102.

[3] J. André, Über Parallelstrukturen. Teil II: Translationsstrukturen. *Math. Z.* 76 (1961) 155-163.

[4] A. Bonisoli and M. Deza, Orthogonal permutation arrays and related structures. *Acta Univ. Carolinae* 24 (1983) 23-38.

[5] A.E. Brouwer, A. Schrijver and H. Hanani, Group divisible designs with block size four. *Discrete Mathematics* 20 (1977) 1-10.

[6] M. Deza and P. Frankl, Injection geometries. *J. Comb. Theory* (B) 37 (1984) 31-40.

[7] M. Deza, R.C. Mullin and S.A. Vanstone, Orthogonal systems. *Aeq. Math.* 17 (1978) 322-330.

[8] J. Dinitz, New lower bounds for the number of pairwise orthogonal symmetric latin squares. *Congressus Numerantium* 22 (1979) 393-398.

[9] J. Dinitz and D.R. Stinson, The spectrum of Room cubes. *Europ. J. Combinatorics* 2 (1981) 221-230.

[10] S. Dow, Transversal-free nets of small deficiency. *Arch. Math.* 41 (1983) 472-474.

[11] W. Feit, Some consequences of the classification of finite simple groups. *Proceedings of Symposia in Pure Mathematics* 37 (1980) 175-181.

[12] A. Hartman, Doubly and orthogonally resolvable quadruple systems. In: R.W. Robinson et al. (eds.), *Combinatorial Mathematics VII*. Lecture Notes in Mathematics 829 (Springer, Berlin Heidelberg New York, 1980).

[13] Ch. Huang, E. Mendelsohn and A. Rosa, On partially resolvable t-partitions. *Annals of Discrete Mathematics* 12 (1982) 169-183.

[14] B. Huppert, *Endliche Gruppen I* (Springer, Berlin Heidelberg New York, 1967).

[15] D. Jungnickel, Existence results for translation nets. In: P.J. Cameron et al. (eds.), *Finite geometries and designs*. Lecture Notes London Math. Soc. 49 (Cambridge Universitiy Press, Cambridge New York, 1981).

[16] E.S. Kramer, S.S. Magliveras and D.M. Mesner, Some resolutions of S(5,8,24). *J. Comb. Theory* (A) 29 (1980) 166-173.

[17] E.R. Lamken and S.A. Vanstone, Designs with mutually orthogonal resolutions. *Europ. J. Combinatorics* (to appear).

[18] M. Marchi, Partition loops and affine geometries. In: P.J. Cameron et al. (eds.), *Finite geometries and designs*. Lecture Notes London Math. Soc. 49 (Cambridge University Press, Cambridge New York, 1981).

[19] P. Quattrocchi and C. Pellegrino, Rettangoli latini e "transversal designs" con parallelismo. *Atti Sem. Mat. Fis. Univ. Modena* 28 (1980), 441-449.

[20] R. H. Schulz, On the classification of translation group divisible designs. *Europ. J. Combinatorics* (to appear).

[21] A.P. Sprague, Incidence Structures whose planes are nets. *Europ. J. Combinatorics* 2 (1981) 193-204.

[22] A.P. Sprague, Translation nets. *Mitt. math. Sem. Gießen* 157(1982) 46-68.

[23] A.P. Sprague, Extended dual affine planes. *Geom. Dedic.* 16 (1984) 107-124.

[24] H. Wielandt, *Finite permutation groups* (Academic Press, New York London, second printing 1968).

Annals of Discrete Mathematics 30 (1986) 203–216

PASCALIAN CONFIGURATIONS IN PROJECTIVE PLANES

Giorgio Faina

Dipartimento di Matematica
Università di Perugia
06100 Perugia
ITALIA

INTRODUCTION

Following Tits [19], let Ω be a projective oval of a projective plane Π and let $P(\Omega)$ be the figure formed by all of the secants or tangents to Ω which are pascalian lines with respect to Ω (see [4,p. 370]). The figures $P(\Omega)$ are called Ω-pascalian configurations of Π.

The problem of determining the configuration $P(\Omega)$ was introduced by F. Buekenhout in [4] and, by using these configurations, it is possible to produce an interesting classification for the projective ovals (see, also, [7]). The purpose of this paper is to prove the following result:

THEOREM. *Let* $P(\Omega)$ *be a* Ω*-pascalian configuration of one of the four known projective plane of order* 9*: Desargues, Hughes, and two Hall planes. Then* $P(\Omega)$ *coincides with one of the following sets, respectively:*

1) *all the tangents and all the secants;*
2) *the empty set;*
3) *all the tangents and all the secants through a unique point of* $\Pi \setminus \Omega$.

It seems extremely interesting to mention that in a Galois plane $PG(2,q)$, $q=p^h$, the problem of determining the non-empty Ω-pascalian configurations is completely resolved. If q is odd, by the theorem of Segre [18], the ovals are conics, for which each line is pascalian (see [4, p. 372]). If instead q is even then there is, other than the class of conics, another class of ovals (called *translation ovals* [10]) which have a single non-exterior pascalian line and such a line is a tangent (see [10]).

We also note that there are projective ovals having non-exterior pascalian lines in non-desarguesian planes, but in this setting the problem is very far from being resolved (see [4, p. 380],[21]). At present two other Ω-pascalian configurations were found:

a) If Ω is the Wagner's oval [21] then the Ω-pascalian configuration contains a unique line (see [7]); Buekenhout also discovered that this line is a tangent to Ω (see [4]).

b) If Ω is the Tits *ovoïde à translation* [19], then $P(\Omega)$ coincides with the set of tangents to Ω (see [4, p. 382] and [7]).

So we see that in *2) and 3)* of our Theorem we have exhibited two new types of Ω-pascalian configurations.

1. DEFINITIONS AND PRELIMINARY RESULTS

For definitions of the terms projective plane, collineation, translation plane, desarguesian plane and near-field see, for example, Segre [18].

A *projective oval* is defined (see [19]) as a set of points Ω of a projective plane Π such that no three are collinear and through each there passes one and only one line (the tangent) that contains no other points of Ω.

A *hexagon* of Ω is a 6-ple $(a_1,a_2,a_3,b_1,b_2,b_3)$ of points of Ω, not necessarily distinct, such that : $\overline{a_i,b_j} \neq \overline{a_j,b_i}$, $a_i \neq a_j$ and $b_i \neq b_j$ for $i\neq j$ and $a_i \neq b_i$ $(i,j=1,2,3)$, where $\overline{a_i,b_j}$ is a tangent of Ω when $a_i=b_j$. Hexagons have three distinct *diagonal points* $\overline{a_i,b_j} \cap \overline{a_j,b_i}$ $(i\neq j)$ and are called *pascalian* when these three points are collinear. The famous theorem of Buekenhout [4] may be stated as:

If each inscribed hexagon in a projective oval Ω is pascalian then Ω is a conic in a projective pappian plane.

Let Ω be a projective oval in a projective plane Π. A line ℓ of Ω

is called Ω-*pascalian* if each hexagon inscribed in Ω which has two diagonal points on ℓ also has the third diagonal point on ℓ. The figure $P(\Omega)$ formed by all of the pascalian lines of Π which are secant or tangent to Ω is called Ω-*pascalian configuration* of Π.

1.1. Oval double loops

It has been observed by many authors ([4], [6], [7]), that a projective oval may be used to define operations of addition $\oplus$ and multiplication ^{0} on its points. We will describe a method for doing this which is a slight modification of Buekenhout's procedure [4, p. 373]. The resulting algebraic system $(Q_T;\oplus,{}^0)$ will called an *oval double loop*.

Let Ω be a projective oval in a projective plane Π. Arbitrarily select three points on Ω and label them y,o,i and then label the point of intersection of the tangents at y and o with x. The points of Ω, other than y, are then arbitrarily assigned symbols with the restriction that o is assigned *o* and i is assigned *1*. If *a* and *b* are the symbols assigned to the points a and b of $\Omega\setminus\{y\}$, we define the sum $a\oplus b$ in the following way:

if $a\neq o$ then the line $\overline{a,x}$ will meet Ω in a point a' other than o; if a=o then let a'=o. If $b\neq o$ then the line $\overline{o,b}$ meets $\overline{x,y}$ in a point z other than x; if b=o then let z=x. Let c be the point of intersection of the line $\overline{a',z}$ with $\Omega\setminus\{a'\}$; if $\{\overline{a',z}\cap\Omega\}\setminus\{a'\}=\emptyset$ then c=a'. Letting *c* be the symbol assigned to this point we define $a\oplus b=c$.

If $a,b\in\Omega\setminus\{o,y\}$, we define the product $a{}^0b$ as follows:

the tangent at i meets the secant $\overline{o,y}$ in a point j; if $b\neq i$ then the line $\overline{b,j}$ will meet Ω in another point $b'\neq i$; if b=i then let b'=i. Let the intersection of the line $\overline{a,i}$ with the line $\overline{o,y}$ be h. Let c be the point of intersection of the line $\overline{h,b'}$ with the set $\Omega\setminus\{b'\}$; if $\{\overline{h,b'}\cap\Omega\}\setminus\{b'\}=\emptyset$ then c=b'. Letting *c* be the symbol assigned to this point we define $a{}^0b=c$.

If we call the set of symbols used Q, then it is easily seen that

Q together with the operations defined above is a double loop which is denoted by $(Q_S;\oplus,{}^0)$ which is also called an *oval double loop*, where $S:=\{o,y,i\}$. This leads us to

LEMMA 1. [6]- The loop $(Q_S,\oplus)$ is an abelian group if, and only if, the line $\overline{x,y}$ is a Ω-pascalian line; the loop $(Q_S^+,{}^0)$ is an abelian group if, and only if, the line $\overline{o,y}$ is a Ω-pascalian line, where $Q_S^+:=Q\setminus\{o\}$.

It is not difficult to verify that every point $p\in\Pi\setminus\Omega$ can be identified with a involutorial permutation $I(p)$ of the points of Ω as follows:

if $p\in\Pi\setminus\Omega$, two points of Ω are a pair in the involutorial permutation $I(p)$ if they lie on the same line through p.

LEMMA 2. [4]- For a line ℓ of a projective plane containing a projective oval Ω, the following are equivalent:

1) ℓ is Ω-pascalian, and
2) for each triple of involutions $I(p)$, $I(q)$, $I(r)$ with centers on ℓ, the composition $I(p)I(q)I(r)$ is also an involution with center on ℓ.

An *automorphism* of a projective oval Ω is a permutation ψ of the points of Ω which preserves the involutorial permutations $I(p)$, where $p\in\Pi\setminus\Omega$, that is to say:

$$\forall\ p\in\Pi\setminus\Omega,\ \exists!\ q\in\Pi\setminus\Omega\ :\ \psi\, I(p)\psi^{-1}=I(q).$$

The automorphisms of Ω form a group. Denote this group by $\mathrm{Aut}\Omega$. The following is easily proven:

each collineation of Π that permutes Ω into itself induces an automorphism of Ω.

We also have the following result.

LEMMA 3. [4]- Let Ω be a projective oval of a projective plane Π and let α be a collineation of Π that permutes Ω into itself. The line

$\overline{x,y}$, where $x,y\in\Omega$, is a Ω-pascalian line if, and only if, the line $\overline{\alpha(x),\alpha(y)}$ is a Ω-pascalian line.

1.2. <u>The near-field of order nine</u> (André [1])

Let $x^2=-1$ an irridducible quadratic over GF(3). Let K be the set of all elements of the form $a+bi$ as a and b vary over GF(3), where we assume $i^2=-1$. We wish to define an addition and a product on K in such a way that, using the field GF(3) addition, K will be a near-field. We define the addition as follows

$$(a+bi)+(c+di):=(a+c)+(b+d)i, \text{ for all } a,b,c,d\in GF(3).$$

We define the product in the following way:

$ai=ia$, for all $a\in GF(3)$

$a(b+c)=ab+ac$, for all $a,b,c\in K$

$ab+ba=0$, for all $a,b\in K\setminus GF(3)$, where $a\neq b$ and $a+b\neq 0$.

It is evident that $(K,+)$ is an abelian group and that $K\setminus\{0\}$ is a group.

Given $a,b,c\in K$ $(a+b\neq 0)$, there is a unique $x\in K$ such that

$ax+bx+c=0$.

Finally, $a^2=-1$ for all $a\in K\setminus GF(3)$.

1.3. <u>The non-desarguesian translation plane of order</u> 9 (André [1])

From the near-field of order 9 K we may now construct a translation affine plane of order nine, denoted by T^o, as follows (see, for example, [1]):

- points are the pairs (x,y) for all $x,y\in K$;
- lines are defined as sets of points (x,y) whose coordinates x,y satisfy an equation of one of the forms

  (1) $x=a$ $(a\in K)$, (2) $y=ax+b$ $(a,b\in K)$.

There is, up to isomorphism, a unique projective plane T such that $T^o=T\setminus\{d\}$ for a line d of T, where d is called the line at infinity of T^o and its points are called points at infinity of the same T^o.

If p is the point at infinity of $y=ax+b$ then it will be denoted by (a). If p is the point at infinity of $x=a$ then it will be denoted by (∞).

It has been shown by Denniston [5] and Nizette [14] that in the translation non-desarguesian plane of order nine T (and in its dual T') the ovals fall into a single transitivity class under the collineation group. The self-duality property make it unnecessary to study T and T' separately; so the following example of projective oval in T will suffices:

$$R=\{(1,0),(-1,0),(0,-1),(i,i),(-i,i),(i,-i),(-i,-i),(0,1),(i),(-i)\}.$$

Rodriguez [16] discovered the oval R and Nizette [14] has studied the group $\mathrm{Aut}R$ of 32 collineations that leaves invariant an oval of T and proved that $\mathrm{Aut}R$ have generators

$$\begin{cases} x'=ix-iy \\ y'=ix+iy \end{cases} \qquad \begin{cases} x'=-x \\ y'=y \end{cases} \qquad \begin{cases} x'=x^\sigma \\ y'=y^\sigma \end{cases}$$

with $i^\sigma=-i$, $\sigma\varepsilon$ Aut K.

1.4. <u>The Hughes plane of order nine</u> (Zappa [22])

From the near-field of order 9 K we may now construct a projective plane, denoted by H as follows:

- the points of H are the triplets (x_1,x_2,x_3), where $x_i\varepsilon K$, other than (0,0,0) with the identification $(x_1,x_2,x_3)=(kx_1,kx_2,kx_3)$ for all non-zero k in K;
- the lines of H will now be the sets of points (x,y,z) which satisfy an equation of the form $x+yt+z=0$, $t\varepsilon K$, such that if σ is any automorphism of K then the mapping

$$\begin{cases} x'=a_1x^\sigma+b_1y^\sigma+c_1z^\sigma \\ y'=a_2x^\sigma+b_2y^\sigma+c_2z^\sigma \\ z'=a_3x^\sigma+b_3y^\sigma+c_3z^\sigma \end{cases} \qquad (\forall a_i,b_i,c_i\varepsilon GF(3),\ i=1,2,3)$$

with $\det(a_i,b_i,c_i)\neq 0$, is a collineation of H.

Denniston [5] and Nizette [14] have discovered that in the Hughes plane H, ovals fall into two transitivity classes under the collineation group of H. An oval, $\mathcal{D}$, in one of these classes is invariant under 48 collineations, as against 16 collineations for the other class. So the following examples of non-isomorphic projective ovals of H will suffice:

$\mathcal{D}$={(-1,1,1),(1,-1,1),(1,1,-1),(1,1,1),(1,i,0),(1,-i,0),(0,i,1),
(C,i,-1),(1,0,i),(1,0,-i)}

$\mathcal{N}$={(1,i,0),(1,-i-1,0),(1,-1,i+1),(1,-1,-i-1),(0,i,1),(0,i,-1),
(1,0,-i-1),(1,0,i+1),(1,1,i),(1,1,-i)}.

In [14], Nizette proved also that Aut$\mathcal{N}$ have generators

$$\begin{cases} x'=x-y \\ y'=-y \\ z'=z \end{cases} \qquad \begin{cases} x'=-x \\ y'=x+y \\ z'=z \end{cases} \qquad \begin{cases} x'=x^{\sigma}+y^{\sigma} \\ y'=x^{\sigma}-y^{\sigma} \\ z'=z^{\sigma} \end{cases}$$

(with $i^{\sigma}=-i-1$).

2. PASCALIAN CONFIGURATIONS IN T AND IN T'

Let R be the Rodriguez oval of the non-desarguesian translation plane of order nine T. First we show that each non-exterior line (to R) through the point (0,0) is a R-pascalian line. Let t denote the tangent y=ix and label the points of R in the following way:

(-i)=0, (i)=∞, (-1,0)=1, (1,0)=2, (0,1)=i, (0,-i)=-i,
(i,-i)=2i+1, (-i,i)=2+i, (i,i)=2+2i, (-i,-i)=1+i .

Letting Q be the symbol assigned to the set $R\setminus\{(i)\}$ (i.e. Q is coinciding with the set of elements of the near-field K), we select the following triplet of points on R :

S={(-i),(i),(-1,0)}.

By 1.1, the algebraic system $(Q_S;\oplus,{}^0)$ is an oval double loop. Now, with a straightforward proof which we omit for shortness, we can to check that

$$a\oplus b=a+b, \text{ for all } a,b\in Q \text{ (i.e. for all } a,b\in K).$$

Since $(K,+)$ is an abelian group, we have that $(Q_S,\oplus)$ is an abelian group. Therefore, by Lemma 1, we have that the tangent at $(i)=\infty$ is a R-pascalian line.

Also, since Aut$R$ is a transitive permutation group on the set $\{(i),(-i)\}\subset R$ (see [4]), by Lemma 3, we have that the tangent $y=-ix$ is R-pascalian.

Now, in order to show that the secant $x=0$ is a R-pascalian line, it is only necessary to prove that, for $S=\{(0,1),(0,-1),(i)\}$, the loop $(Q_S^+,{}^0)$, where $Q_S^+=R\setminus\{(0,1),(0,-1)\}$, is an abelian group. A very long, but straightforward computation, shows it. It is well known that (see [14]) Aut$R$ fixes the point $(0,0)$ and that it is transitive on the points of $R\setminus\{(i),(-i)\}$. Thus, from Lemma 3, we have that all non-exterior lines through $(0,0)$ are R-pascalian.

Now we will prove the non-existence of non-exterior R-pascalian lines not passing through the point $(0,0)$. The points of R may, for shortness, be denoted by digits from 0 to 9 as follows:

$(i)=0, (-i)=1, (-1,0)=2, (1,0)=3, (i,-i)=4, (-i,i)=5, (i,i)=6, (-i,-i)=7,$

$(0,1)=8$ and $(0,-1)=9$.

By [4, p. 383] and [20, table 32/34], it follows that, if we denote by $G(8)$ the group of all elements in Aut$R$ which fix the point 8, then $|G(8)|=4$ and that $G(8)=\{f_1,f_2,f_3,f_4\}$, where

$$f_1=\mathrm{Id},\ f_2=(23)(46)(57),\ f_3=(01)(45)(67),\ f_4=(01)(23)(47)(56).$$

Finally, since Aut$R$ acts transitively on $R\setminus\{(i),(-i)\}$, by Lemma 3, the only thing remaining to be shown is that the lines

$$\overline{8,8}\ ,\ \overline{5,8}\ ,\ \overline{0,8}\ ,\ \overline{2,8}\ ,\ \overline{0,1}$$

are not R-pascalian.

First of all, consider the following points of $\Pi \setminus R$:

$p_1=\overline{1,9}\cap\overline{8,8}$, $p_2=\overline{1,6}\cap\overline{8,8}$, $p_3=\overline{5,8}\cap\overline{0,0}$, $p_4=\overline{5,8}\cap\overline{0,2}$, $p_5=\overline{0,8}\cap\overline{1,1}$,

$p_6=\overline{0,8}\cap\overline{1,2}$, $p_7=\overline{2,8}\cap\overline{0,0}$, $p_8=\overline{2,8}\cap\overline{0,3}$, $p_9=\overline{0,1}\cap\overline{2,2}$, $p_{10}=\overline{0,1}\cap\overline{2,4}$,

$p_{11}=\overline{0,1}\cap\overline{2,6}$.

Now, without giving the proofs (which are straightforward but time-consuming) we remark that $p_1,p_2,p_3\in\overline{8,8}$ but $I(p_1)I(p_2)I(p_3)$ is not a involutory permutation of R with center on $\overline{8,8}$; thus, by Lemma 2, it follows that the line $\overline{8,8}$ is not R-pascalian.

Repeating this process, replacing $I(p_1)I(p_2)I(p_3)$ by $I(p_3)I(p_4)I(p_3)$, gives that the line $\overline{5,8}$ is not R-pascalian.

Similarly: $I(p_5)I(p_6)I(p_5)$, $I(p_7)I(p_8)I(p_7)$, $I(p_9)I(p_{10})I(p_{11})$ are not involutory permutations of R with center in $\overline{0,8}$, $\overline{2,8}$, $\overline{0,1}$, respectively. Hence these lines are not R-pascalian.

The R'-pascalian configuration of the dual T' of T is again of the same type and we omit the analogous proof.

3. PASCALIAN CONFIGURATIONS IN H

In [8], Hughes reproduces the plane H in the useful following way:

- the points are the symbols $A_i, B_i, C_i, D_i, E_i, F_i, G_i$, $i=0,1,\ldots,12$;
- seven of the lines are the following sets of points

1) $\{A_0,A_1,A_3,A_9,B_0,C_0,D_0,E_0,F_0,G_0\}$
2) $\{A_0,B_1,B_8,D_3,D_{11},E_2,E_5,E_6,G_7,G_9\}$
3) $\{A_0,C_1,C_8,E_7,E_9,F_3,F_{11},G_2,G_5,G_6\}$
4) $\{A_0,B_7,B_9,D_1,D_8,F_2,F_5,F_6,G_3,G_{11}\}$
5) $\{A_0,B_2,B_5,B_6,C_3,C_{11},E_1,E_8,F_7,F_9\}$

6) $\{A_0, C_7, C_9, D_2, D_5, D_6, E_3, E_{11}, F_1, F_8\}$

7) $\{A_0, B_3, B_{11}, C_2, C_5, C_6, D_7, D_9, G_1, G_8\}$;

- the remaining lines are found by successively adding one to the sub-scripts, reducing modulo 13.

In this notation, we remark that (see [5] and [14]) the ovals $\mathcal{D}$ *and* N of section 1.4 are the following sets of symbols:

$$\mathcal{D}=\{A_4, A_5, A_{11}, A_{12}, B_0, E_0, C_6, D_6, C_7, D_7\}$$

$$N=\{B_0, C_0, C_4, G_4, C_6, D_6, B_7, F_7, B_{11}, E_{11}\} .$$

We first show that the $\mathcal{D}$-pascalian configuration is the empty set.

The suggestive term *real* is used for the points A_i of H, then in $\mathcal{D}$ there are four real points and six *imaginary* points. It is well known that (see [5], [14]) $|\mathrm{Aut}\mathcal{D}|=48$ and we note further important properties:

1) $\mathrm{Aut}\mathcal{D}$ is generated by: $(A_{11}A_{12})(B_0E_0)(C_6C_7)(D_6D_7)$,

$(A_4A_5A_{11}A_{12})(B_0C_6E_0D_6)$, $(B_0E_0)(C_6D_6)(C_7D_7)$;

2) $\mathrm{Aut}\mathcal{D}$ is transitive on the set of real points of $\mathcal{D}$;

3) $\mathrm{Aut}\mathcal{D}$ is transitive on the set of imaginary points of $\mathcal{D}$;

4) $|(\mathrm{Aut}\mathcal{D})_x|=12$ for all real point $x\in\mathcal{D}$;

5) $(\mathrm{Aut}\mathcal{D})_x$ is transitive on the set of imaginary points of $\mathcal{D}$ for all real point $x\in\mathcal{D}$;

6) if x is a real point of $\mathcal{D}$, then $(\mathrm{Aut}\mathcal{D})_x$ is transitive on the set of real points of $\mathcal{D}\setminus\{x\}$;

7) if y is a imaginary point of $\mathcal{D}$ then $|(\mathrm{Aut}\mathcal{D})_y|=8$;

8) $(\mathrm{Aut}\mathcal{D})_{B_0}=(\mathrm{Aut}\mathcal{D})_{E_0}$;

9) $(\mathrm{Aut}\mathcal{D})_{B_0}$ is transitive on the set $\{C_6, D_6, C_7, D_7\}$.

We omit the proof which is very long, but not difficult.

By the above properties of $\mathrm{Aut}\mathcal{D}$ and Lemma 3, the only thing remaining to be shown is that no one of the lines

$$\overline{A_4,A_4},\ \overline{A_{12},A_{12}},\ \overline{A_4,A_5},\ \overline{B_0,E_0},\ \overline{B_0,C_6},\ \overline{A_4,A_{12}}$$

is $\mathcal{D}$-pascalian. As in the proof of section 2, it is sufficient to exhibit some appropriate involutorial permutation of the points of $\mathcal{D}$. First of all, consider the following points of $\Pi\setminus\mathcal{D}$:

$$A_1,A_2,B_1\varepsilon\overline{A_4,A_4},\ F_1,A_7,E_8\varepsilon\overline{A_{12}A_{12}},\ A_7,B_4,C_4\varepsilon\overline{A_4,A_5}$$

$$B_{11},D_5,D_{12}\varepsilon\overline{A_4,A_{12}},\ A_1,A_9,D_0\varepsilon\overline{B_0,E_0},B_1,F_4\varepsilon\overline{B_0,C_6}.$$

Now we remark that the following permutations

$I(A_1)I(A_2)I(B_1)$, $I(F_1)I(A_7)I(E_8)$, $I(A_7)I(B_4)I(C_4)$,

$I(B_{11})I(D_5)I(D_{12})$, $I(A_1)I(A_9)I(D_0)$, $I(B_1)I(F_4)I(B_1)$

are not involutory permutations of type $I(p)$ with the centers p in $\overline{A_4,A_4}$, $\overline{A_{12},A_{12}}$, $\overline{A_4,A_5}$, $\overline{A_4,A_{12}}$, $\overline{B_0,E_0}$, $\overline{B_0,C_6}$ respectively. Hence, by Lemma 2, these lines are not $\mathcal{D}$-pascalian.

Finally, we must show that the N-pascalian configuration is the empty set too.

Also in this case, it is well known that (see [5] and [14]) $|\mathrm{Aut}N|=16$ and it is not difficult to check that:

1) Aut$N$ is generated by $\lambda=(C_4C_6F_7E_{11}G_4D_6B_7B_{11})$ and $\mu=(B_0C_0)(C_6E_{11})(D_6B_{11})(B_7F_7)$;

2) Aut$N$ fixes the set $\{B_0,C_0\}$;

3) Aut$N$ is transitive on the sets $\{B_0,C_0\}$ and $I=N\setminus\{B_0,C_0\}$ respectively;

4) $(\mathrm{Aut}N)_{B_0}$ is transitive on I;

5) $(\mathrm{Aut}N)_{B_0}=(\mathrm{Aut}N)_{C_0}$;

6) $(\mathrm{Aut}N)_{C_4}=(\mathrm{Aut}N)_{G_4}=\{\mathrm{Id},\mu\}$.

Hence, in order to prove that $P(N)=\emptyset$, it is sufficient to show that no one of the following lines is N-pascalian:

$\overline{B_0,B_0}$, $\overline{C_4,C_4}$, $\overline{B_0,C_0}$, $\overline{B_0,C_4}$, $\overline{C_4,G_4}$, $\overline{C_4,C_6}$, $\overline{C_4,D_6}$, $\overline{C_4,B_7}$.

A repetition of the arguments used in the earlier proof of this section shows that the permutations

$I(A_7)I(B_9)I(B_{12})$, $I(F_1)I(A_3)I(F_1)$, $I(A_0)I(A_1)I(D_0)$,

$I(A_2)I(B_5)I(A_2)$, $I(A_7)I(F_4)I(D_4)$, $I(E_8)I(E_0)I(E_8)$,

$I(C_5)I(G_0)I(C_5)$, $I(E_2)I(A_1)I(E_2)$

are not involutory permutations of the points of N with center in the above mentioned lines, respectively, wile we have that:

$A_7,B_9,B_{12}\varepsilon\overline{B_0,B_0}$, $F_1,A_3\varepsilon\overline{C_4,C_4}$, $A_0,A_1,D_0\varepsilon\overline{B_0,C_0}$, $A_2,B_5\varepsilon\overline{B_0,C_4}$, $A_7,F_4,D_4\varepsilon\overline{C_4,G_4}$, $E_8,E_0\varepsilon\overline{C_4,C_6}$, $C_5,G_0\varepsilon\overline{C_4,D_6}$, $E_2,A_1\varepsilon\overline{C_4,B_7}$.

Hence, by Lemma 2, no one of these lines is a N-pascalian line.

REFERENCES

1. J. ANDRE', *Uber nicht-Desarguessche Ebenen mit transitiver Translationgruppe*, Math. Zeit. 60 (1954), 156-186.
2. A. BARLOTTI, *Un'osservazione intorno ad un teorema di B. Segre sui q-archi*, Le Matematiche 21 (1966), 23-29.
3. U. BARTOCCI, *Considerazioni sulla teoria delle ovali*, Tesi di Laurea, Università di Roma (1967).
4. F. BUEKENHOUT, *Etude intrinsèque des ovales*, Rend. Mat. (5) 25 (1966), 333-393.
5. R.H.F. DENNISTON, *On arcs in projective planes of order 9*, Manuscripta Math. 4 (1971), 61-89.
6. G. FAINA, *Sul doppio cappio associato ad un ovale*, Boll. Un. Mat. Ital. (5) 15-A (1978), 440-443.
7. G. FAINA, *Un raffinamento della classificazione di Buekenhout per gli ovali astratti*, Boll. Un. Mat. Ital. (5) 16-B (1979), 813-825.
8. D.R. HUGHES, *A class of non-Desarguesian projective planes*, Canad. J. Math. 9 (1957), 378-388.
9. Z. JANKO, TRAN VAN TRUNG, *The classification of projective planes of order 9 which posses an involution*, J. Comb. Theory 33 (1982), 65-75.

10. G. KORCHMÁROS, *Sulle ovali di traslazione in un piano di Galois di ordine pari*, Rend. Accad. Naz. XL (5) 3 (1977-78), 55-65.

11. G. KORCHMÁROS, *Una generalizzazione del teorema di Buekenhout sulle ovali pascaliane*, Boll. Un. Mat. Ital. (5) 18-B (1981), 673-687.

12. R. MAGARI, *Le configurazioni parziali chiuse contenute nel piano P sul quasicorpo associativo di ordine 9*, Boll. Un. Mat. Ital. (3) 13 (1958), 128-140.

13. G. MENICHETTI, *Sopra i k-archi completi nel piano grafico di traslazione di ordine 9*, Le Matematiche 21 (1966), 150-156.

14. N. NIZETTE, *Determination des ovales du plan de translation non arguesien et du plan de Hughes d'ordre neuf*, Bull. Soc. Math. Belg. 23 (1971), 436-446.

15. T.G. OSTROM, *Semi-translation planes*, Trans. Amer. Math. Soc. 111 (1964), 1-18.

16. G. RODRIGUEZ, *Un esempio di ovale che non è una quasi-conica*, Boll. Un. Mat. Ital. (3) 14 (1959), 500-503.

17. L.A. ROSATI, *Su una nuova classe di piani grafici*, Ricerche di Mat. 13 (1964), 39-55.

18. B. SEGRE, *Lectures on modern geometry*, Cremonese, Roma 1961.

19. J. TITS, *Ovoïdes à translation*, Rend. Mat. (5) 21 (1962), 37-59.

20. A.D. THOMAS, G.V. WOOD, *Group Tables*, Shiva Pub., Orpington 1980.

21. A. WAGNER, *On perspectivities of finite projective planes*, Math. Zeit. 71 (1959), 113-123.

22. G. ZAPPA, *Sui gruppi di collineazioni dei piani di Hughes*, Boll. Un. Mat. Ital. (3) 12 (1957), 507-516.

Annals of Discrete Mathematics 30 (1986) 217–224

MONOMIAL CODE - ISOMORPHISMS

Pavel Filip and Werner Heise (*)

Mathematisches Institut der Technischen Universität München
P.O.Box 20 24 20, D - 8000 München 2
Germany

Let C and D be two linear subspaces of $GF(q)^n$ and $\varphi : C \to D$ a Hamming weight preserving linear bijection. It will be proved that φ is the restriction of a monomial transformation of $GF(q)^n$, i. e. an $n \times n$-matrix over $GF(q)$, which is the product of a diagonal and a permutation matrix. An application shows that the group of all Hamming weight preserving linear bijections of the q-ary Hamming code $HAM(r,q)$, $r \geq 3$, of length $1+q+q^2+\ldots+q^{r-1}$ is isomorphic to the general linear group $GL_r(q)$.

Let F be a finite set consisting of $q \geq 2$ elements and $n \geq 1$ be an integer. For $i \in \mathbb{Z}_n := \{1,2,\ldots,n\}$ we denote the ith *projection* of the n-fold cartesian product F^n by $\pi_i : F^n \to F$; $(x_1,x_2,\ldots,x_n) \to x_i$. With respect to the *Hamming distance* $\rho : F^n \times F^n \to \mathbb{N}_0$; $(\vec{x},\vec{y}) \to |\{i \in \mathbb{Z}_n ; \pi_i(\vec{x}) \neq \pi_i(\vec{y})\}|$, which associates to every pair $(\vec{x},\vec{y}) \in F^n \times F^n$ the number $\rho(\vec{x},\vec{y})$ of positions in which the words $\vec{x}$ and $\vec{y}$ differ, the cartesian product F^n becomes a metric space. A non empty metric subspace $C \subset F^n$ is called a *(block) code of length* n *over* F. Its *minimal distance* $d(C)$ is defined as $d(C) := \min\{\rho(\vec{x},\vec{y}); \vec{x},\vec{y} \in C, \vec{x} \neq \vec{y}\}$ if $|C| \geq 2$ and as $d(C) := n+1$ or $d(C) := \infty$ if C consists of only one codeword.

Let C and D be two codes of length n over F. In contrast to some more pragmatic definitions in the coding theoretical literature we take a somewhat mathematically puristic attitude. We define a *code - isomorphism* $\varphi : C \to D$ as a bijection φ from C onto D which preserves the Hamming distance, i. e. $\rho(\varphi(\vec{x}),\varphi(\vec{y})) = \rho(\vec{x},\vec{y})$ for all $\vec{x},\vec{y} \in C$. Extending the range of φ from D to F^n we also say that $\varphi : C \to F^n$ is a *code - monomorphism* of C. Clearly, the restriction $\Phi|_C$ of a code - automorphism Φ from the whole space F^n (which forms a trivial code) to a code $C \subset F^n$ is always a code - monomorphism of C.

(*) The authors gratefully thank prof. Luca - Maria Abatangelo for her excellent organisation of the congress and Sergio Povia for his extraordinary care during our sojourn at the hotel Riva del Sole in Giovinazzo.

Every permutation ψ of the symmetric group S_n of $\mathbb{Z}_n$ induces on F^n a code-automorphism $\tilde{\psi}: F^n \to F^n$. We set $\pi_i(\tilde{\psi}(\vec{x})) := \pi_{\psi(i)}(\vec{x})$ for all $i \in \mathbb{Z}_n$ and all $\vec{x} \in F^n$. Two codes $C, D \subset F^n$ are called *equivalent*, if there exists a permutation ψ of $\mathbb{Z}_n$ with $\tilde{\psi}(C) = D$. Sometimes (e.g. in [2] for the binary case $q = 2$) the group $G := \{\psi \in S_n ; \tilde{\psi}(C) = C\}$ is called the automorphism group of the code C. Othertimes (e.g. in [1]) its factor group $\tilde{G} := \{\tilde{\psi} ; \psi \in G\}$ is called the automorphism group of C. From our puristic point of view, we do not agree with these definitions (*), because there are other types of Hamming distance preserving bijections.

Let $\lambda := (\lambda_1, \lambda_2, \ldots, \lambda_n) \in S_F^n$ be a vector of length n, whose components $\lambda_i \in S_F$, $i \in \mathbb{Z}_n$, are permutations of the set F. We define a code-automorphism $\hat{\lambda}$ of F^n by setting $\pi_i(\hat{\lambda}(\vec{x})) := \lambda_i(\pi_i(\vec{x}))$ for all $i \in \mathbb{Z}_n$ and all $\vec{x} \in F^n$.

From now on let q be a power of a prime number and let F be the Galois field of order q, $F = GF(q)$. The Hamming metric ρ on the vector space $F^n = V_n(q)$ is induced by the *Hamming weight*, i.e. the norm $\gamma : F^n \to \mathbb{N}_0 ; \vec{x} \to |\{i \in \mathbb{Z}_n ; \pi_i(\vec{x}) \neq 0\}|$, in the usual way. For $\vec{x}, \vec{y} \in F^n$ we have $\rho(\vec{x}, \vec{y}) = \gamma(\vec{x} - \vec{y})$. In case $n = 1$, the Hamming weight $\gamma : F \to \{0,1\}$ is the ordinary characteristic function of the set $F^* := F \setminus \{0\}$. Using the Kronecker symbol $\delta_{i,j}$ we have $\gamma(x) = 1 - \delta_{0,x}$ for all $x \in F$. For a subset $C \subset F^n$ we abbreviate $\gamma(C) := \sum_{\vec{c} \in C} \gamma(\vec{c})$.

By a *linear* (n,k)-*code* C *over* F we understand a metric subspace $C \subset F^n$, which is also a k-dimensional linear subspace of F^n. Let $C \subset F^n$ be a linear (n,k)-code. For $i \in \mathbb{Z}_n$ the restriction $\pi_i|_C : C \to F$ of the ith projection $\pi_i : F^n \to F$ from F^n to the linear code C is a linear form. This linear form $\pi_i|_C$ is *non-trivial* if and only if the *derivation* $A_i(C) := \ker(\pi_i|_C)$, i.e. $A_i(C) = \{\vec{c} \in C ; \pi_i(\vec{c}) = 0\}$, *of* C *in the position* i is a $(k-1)$-dimensional linear subspace of C. A code-monomorphism $\varphi : C \to F^n$ will be called *linear*, if it is a linear mapping. A linear injective mapping $\varphi : C \to F^n$ is a code-monomorphism, if and only if it preserves the Hamming weight, i.e. iff $\gamma(\varphi(\vec{c})) = \gamma(\vec{c})$ for all $\vec{c} \in C$.

For $i \in \mathbb{Z}_n$ we denote by $\vec{e}_i := (\delta_{i,1}, \delta_{i,2}, \ldots, \delta_{i,n})$ the ith *canonical unit vector* of F^n. Now let $\psi \in S_n$ be again a permutation of $\mathbb{Z}_n$. Then

$$\tilde{\psi}(\vec{x}) = \tilde{\psi}\Big(\sum_{i=1}^{n} \pi_i(\vec{x}) \cdot \vec{e}_i\Big) = \sum_{i=1}^{n} \pi_{\psi(i)}(\vec{x}) \cdot \vec{e}_i = \sum_{i=1}^{n} \pi_i(\vec{x}) \cdot \vec{e}_{\psi^{-1}(i)} \quad \text{for all } \vec{x} \in F^n .$$

Hence $\tilde{\psi}$ is a linear transformation of F^n, $\tilde{\psi} \in GL_n(F)$, which can be represented by the permutation matrix $(\delta_{i,\psi^{-1}(j)})_{1 \le i,j \le n}$ with respect to the canonical basis $\{\vec{e}_i ; i \in \mathbb{Z}_n\}$ of F^n. Now let $\lambda = (\lambda_1, \lambda_2, \ldots, \lambda_n) \in (F^*)^n$ be a vector

(*) Remark by W. Heise: What do I care about my rubbish said yesterday!

where all components are non-zero. For $i \in \mathbb{Z}_n$ we identify the element $\lambda_i \in F$ with the permutation $\lambda_i \in S_F$ defined by $\lambda_i(x) := \lambda_i \cdot x$ for all $x \in F$. Then $\hat{\lambda}(\vec{x}) = \hat{\lambda}(\sum_{i=1}^{n} \pi_i(\vec{x}) \cdot \vec{e}_i) = \sum_{i=1}^{n} \pi_i(\vec{x}) \cdot \lambda_i \cdot \vec{e}_i$ for all $\vec{x} \in F^n$. Thus, $\hat{\lambda}$ is a linear transformation of F^n, $\hat{\lambda} \in GL_n(F)$, which can be represented by the diagonal matrix $(\lambda_i \cdot \delta_{i,j})_{1 \le i,j \le n}$ with respect to the canonical basis of F^n.

A linear transformation $\Phi \in GL_n(F)$ of F^n is called *monomial*, if there exists a vector $\lambda = (\lambda_1, \lambda_2, \ldots, \lambda_n) \in (F^*)^n$ and a permutation $\psi \in S_n$ such that $\Phi = \hat{\lambda} \circ \tilde{\psi}$, i.e. if $\pi_i(\Phi(\vec{x})) = \lambda_i \cdot \pi_{\psi(i)}(\vec{x})$ for all $i \in \mathbb{Z}_n$ and all $\vec{x} \in F^n$. Obviously, monomial linear transformations of F^n preserve the Hamming weight of every vector $\vec{x} \in F^n$. Let C be a linear (n,k)-code over F. A linear code-monomorphism $\varphi : C \to F^n$ will be called *monomial*, if it can be extended to a monomial linear transformation $\Phi : F^n \to F^n$; i.e. if there exists a monomial transformation $\Phi \in GL_n(F)$, whose restriction to C is φ, $\Phi|_C = \varphi$. A code-monomorphism $\varphi : C \to F^n$ is monomial, if and only if there exist n elements $\lambda_1, \lambda_2, \ldots, \lambda_n \in F^*$ and a permutation $\psi \in S_n$ such that $\pi_i(\varphi(\vec{c})) = \lambda_i \cdot \pi_{\psi(i)}(\vec{c})$ for all $i \in \mathbb{Z}_n$ and all $\vec{c} \in C$.

F. J. MacWilliams and N. J. A. Sloane [2;p.238] define the automorphism group of a linear (n,k)-code C over F as the group of all those monomial linear transformations $\Phi \in GL_n(F)$ of the whole vector space F^n, which leave the code C invariant, i.e. for which we have $\Phi(\vec{c}) \in C$ for all $\vec{c} \in C$. In our terminology, the group of linear code-automorphisms of C consists of all linear transformations $\varphi \in GL_k(F)$ of the k-dimensional vector space C, which preserve the Hamming weight, i.e. for which $\gamma(\varphi(\vec{c})) = \gamma(\vec{c})$ for all $\vec{c} \in C$. So the group of all linear code-automorphisms of any linear *equidistant* (n,k)-code (i.e. $\gamma(\vec{c}) = d(C)$ for all $\vec{c} \in C \setminus \{\vec{0}\}$ is trivially the group $GL_k(F)$. Extending problem 33 of [2;p.231f] from the binary, $q = 2$, to the general case we prove in this paper - with our different conception of a code-isomorphism - that every linear code-isomorphism is monomial. This then gives evidence for the fact, that our puristic attitude is not too far from F. J. MacWilliams' and N. J. A. Sloane's pragmatic position. For the proof we make some auxiliary propositions.

Let C be a linear (n,k)-code over $F = GF(q)$, $\varphi : C \to F^n$ a linear code-monomorphism and $D := \varphi(C)$.

The first proposition, as an easy consequence of the rank formula for matrices, whose lines form a basis of the linear code C, does not make use of the metric structure of the code C.

PROPOSITION 1 . Let t be an integer with $0 \leq t \leq k$ and $i(1), i(2), \ldots, i(t) \in \mathbb{Z}_n$ pairwise different indices. The t projections $\pi_{i(h)}|_C$, $h = 1,2,\ldots,t$ are linearly independent if and only if for every choice of t (not necessarily distinct) elements $\alpha_1, \alpha_2, \ldots, \alpha_t \in F$ there are exactly q^{k-t} codewords $\vec{c} \in C$ with $\pi_{i(h)}(\vec{c}) = \alpha_h$, $h = 1,2,\ldots,t$.

Note, that the case $t = k$ provides us with a characterisation of the linear so-called *MDS-codes*, cf. [1;p.164f.] or [2;p.317ff.]. In this paper proposition 1 is only used in the case $t = 1$. In this case proposition 1 is a mere reformulation of the "Satz über die Gleichverteilung der Zeichen in linearen Codes" from [1;p.210].

PROPOSITION 2 . Let s be an integer with $0 \leq s \leq n-k$. If precisely s of the n projections $\pi_i|_C$, $i \in \mathbb{Z}_n$, are trivial, then also precisely s of the n projections $\pi_j|_D$, $j \in \mathbb{Z}_n$, are trivial.

Proof. For each non-trivial projection $\pi_i|_C$, $i \in \mathbb{Z}_n$, by proposition 1 we have $\sum_{\vec{c} \in C} \gamma(\pi_i(\vec{c})) = q^{k-1} \cdot (q-1)$. For each trivial projection $\pi_i|_C$, $i \in \mathbb{Z}_n$, we have $\sum_{\vec{c} \in C} \gamma(\pi_i(\vec{c})) = 0$. Therefore $\gamma(C) = \sum_{i=1}^{n} \sum_{\vec{c} \in C} \gamma(\pi_i(\vec{c})) = (n-s) \cdot q^{k-1} \cdot (q-1)$. Exactly in the same manner we prove $\gamma(D) = (n-\sigma) \cdot q^{k-1} \cdot (q-1)$, where σ is the number of the trivial projections $\pi_j|_D$, $j \in \mathbb{Z}_n$. The bijection $\varphi : C \to D$ preserves the Hamming weight, so we have $\gamma(C) = \gamma(D)$, whence $\sigma = s$.

The proof of proposition 2 is essentially the only place in which we make use of the fact, that $F = GF(q)$ is a finite field. However Hans Kellerer (Math. Inst. d. TU München, not Hans Kellerer, Math. Inst. d. LMU München) showed that proposition 2 holds also for codes over an infinite field F : Denote by $n(h)$, $h = 1,2,\ldots,s$, those s indices from $\mathbb{Z}_n$ for which the projections $\pi_{n(h)}|_C$ are trivial. By σ we denote the number of the trivial projections $\pi_j|_D$, $j \in \mathbb{Z}_n$. Since F is infinite, there is a codeword $\vec{c} \in C$ with $\pi_i(\vec{c}) \neq 0$ for each $i \in \mathbb{Z}_n \setminus \{n(1), n(2), \ldots, n(s)\}$, hence $\gamma(\varphi(\vec{c})) = \gamma(\vec{c}) = n-s$ and thus $n-s \leq n-\sigma$. Since there is also a codeword $\vec{d} \in D$ with $\gamma(\vec{d}) = n-\sigma$ we get $\gamma(\varphi^{-1}(\vec{d})) = \gamma(\vec{d}) = n-\sigma$ and thus $n-\sigma \leq n-s$, whence $\sigma = s$. Independently Ludwig Staiger (ZKI d. Akad. d. Wiss. d. DDR) gave another proof of the same fact which will be included in his forthcoming paper "On covering codewords" in the Atti del Seminario Matematico e Fisico dell' Università di Modena. As a consequence the theorem of this paper does not depend on the finiteness of the underlying field F .

We denote by $N = \{n(1), n(2), \ldots, n(s)\}$ the set of those indices $n(h) \in \mathbb{Z}_n$, $h = 1,2,\ldots,s$, for which the projection $\pi_{n(h)}|_C$ is trivial. Then, by proposition 2 there is a set $M = \{ m(1),m(2),\dots,m(s) \} \subset \mathbb{Z}_n$ of s indices, such that the projections $\pi_{m(h)}|_D$, $h = 1,2,\dots,s$, are trivial. For $i \in \mathbb{Z}_n \setminus M$ the projection $\pi_i|_D$ is always non-trivial. For each index $j \in \mathbb{Z}_n \setminus N$ we indicate by $K(j) := \{ j(1),j(2),\dots,j(r_j) \}$ the set of all those r_j indices $j(h) \in \mathbb{Z}_n \setminus N$, $h = 1,2,\dots,r_j$, such that there exists an element $\alpha_h \in F^*$ with $\pi_{j(h)}|_C = \alpha_h \cdot \pi_j|_C$. Of course, for all $j \in \mathbb{Z}_n \setminus N$ we have $1 \le r_j \le n-k+1$. We denote by R a system of representatives of the sets $K(j)$, $j \in \mathbb{Z}_n \setminus N$. Then $\{N\} \cup \{ K(j) ; j \in R \}$ is a partition of $\mathbb{Z}_n$ and we have $n = \sum_{j \in R} r_j + s$.

<u>PROPOSITION 3</u> . Let $j \in R$ be an index. Then there is a set $L(j) = \{ i(1),i(2),\dots,i(r_j) \} \subset \mathbb{Z}_n \setminus M$ of r_j indices and there are r_j (not necessarily distinct) elements $\lambda_{i(1)},\lambda_{i(2)},\dots,\lambda_{i(r_j)} \in F^*$ such that any two of the r_j projections $\pi_{i(h)}|_D$, $h = 1,2,\dots,r_j$, are linearly dependent and such that $\pi_{i(h)}(\varphi(\vec{c})) = \lambda_{i(h)} \cdot \pi_{j(h)}(\vec{c})$ for $h = 1,2,\dots,r_j$ and all $\vec{c} \in C$. If $j' \in R \setminus \{j\}$ is another index, then $L(j) \cap L(j') = \emptyset$.

<u>Proof</u>. The derivation $A := A_j(C) = \ker(\pi_j|_C)$ of C in the position j is a $(k-1)$-dimensional linear subspace of the k-dimensional vector space C . It coincides with the derivations $A_{j(h)}(C)$ of C in the positions $j(h)$, $h = 1,2,\dots,r_j$. The r_j projections $\pi_{j(h)}|_A$, $h = 1,2,\dots,r_j$, are trivial as well as the s projections $\pi_{n(h)}|_A$, $h = 1,2,\dots,s$. The linear subspace $B := \varphi(A)$ of the k-dimensional vector space $D = \varphi(C)$ has dimension $k-1$. We apply proposition 2 to the codes A and B and to the linear code-isomorphism $\varphi|_A : A \to B$. There are $r_j + s$ trivial projections $\pi_i|_B$, $i \in \mathbb{Z}_n$, among them the projections $\pi_{m(h)}|_B$, $h = 1,2,\dots,s$. The other r_j trivial projections $\pi_{i(1)}|_B,\pi_{i(2)}|_B,\dots,\pi_{i(r_j)}|_B$ are restrictions of the non-trivial projections $\pi_{i(1)}|_D,\pi_{i(2)}|_D,\dots,\pi_{i(r_j)}|_D$ to the subcode $B \subset D$. Thus the derivations $A_{i(h)}(D)$ of D in the positions $i(h)$, $h = 1,2,\dots,r_j$, all coincide with B . Thus the projections $\pi_{i(h)}|_D$, $h = 1,2,\dots,r_j$, are pairwise linearly dependent. Since C is a linear code we can choose a codeword $\vec{b} \in C \setminus A$ with $\pi_j(\vec{b}) = 1$. By supposition we have $\pi_{j(h)}(\vec{b}) = \alpha_h$, $h = 1,2,\dots,r_j$. It is $\varphi(\vec{b}) \in \varphi(C \setminus A) = D \setminus B$, thus for $h = 1,2,\dots,r_j$ the element $\beta_{i(h)} := \pi_{i(h)}(\varphi(\vec{b})) \in F$ is always non-zero. Each codeword $\vec{c} \in C = \langle \vec{b} \rangle \oplus A$ can uniquely be written as $\vec{c} = \beta \cdot \vec{b} + \vec{a}$, with $\beta \in F$ and $\vec{a} \in A$. For $h = 1,2,\dots,r_j$ we set $\lambda_{i(h)} := \beta_{i(h)}/\alpha_h \in F^*$. Then for $h = 1,2,\dots,r_j$ we have $\pi_{i(h)}(\varphi(\vec{c})) = \pi_{i(h)}(\beta \cdot \varphi(\vec{b}) + \varphi(\vec{a})) = \beta \cdot \pi_{i(h)}(\varphi(\vec{b})) =$
$= \beta \cdot \beta_{i(h)} \cdot \alpha_h / \alpha_h = \beta \cdot \lambda_{i(h)} \cdot \alpha_h \cdot \pi_j(\vec{b}) = \beta \cdot \lambda_{i(h)} \cdot \pi_{j(h)}(\vec{b}) = \lambda_{i(h)} \cdot \pi_{j(h)}(\beta \cdot \vec{b}) =$
$= \lambda_{i(h)} \cdot \pi_{j(h)}(\beta \cdot \vec{b} + \vec{a}) = \lambda_{i(h)} \cdot \pi_{j(h)}(\vec{c})$.

The set $L(j)$ consists of the r_j indices $i(1),i(2),\dots,i(r_j)$ .Analoguosly the set $L(j')$ consists of $r_{j'}$ indices $i'(1),i'(2),\dots,i'(r_{j'})$. Suppose there exists an index $i \in L(j) \cap L(j')$. Then, by what we have seen above, there exist two elements $\eta_i, \eta_i' \in F^*$ with $\pi_i(\varphi(\vec{c})) = \eta_i \cdot \pi_j(\vec{c})$ and $\pi_i(\varphi(\vec{c})) = \eta_i' \cdot \pi_{j'}(\vec{c})$ for all $\vec{c} \in C$. Hence $\pi_j|_C$ and $\pi_{j'}|_C$ are linearly dependent, contradictory to our supposition that j and j' are two distinct indices from R.

THEOREM. The linear code-monomorphism φ is monomial.

Proof. The system $\{M\} \cup \{L(j); j \in R\}$ is a partition of $\mathbb{Z}_n$. We define a permutation $\psi \in S_n$ of $\mathbb{Z}_n$ by pasting together the bijections $\psi_N : M \to N; m(h) \to n(h)$ and $\psi_j : L(j) \to K(j); i(h) \to j(h)$, $j \in R$. For $h = 1,2,\ldots,s$ we set $\lambda_{m(h)} := 1$ (we could choose as well any other element of F^*) and get
$\pi_{m(h)}(\varphi(\vec{c})) = 0 = \lambda_{m(h)} \cdot \pi_{\psi(m(h))}(\vec{c})$ for $h = 1,2,\ldots,s$ and all $\vec{c} \in C$, by proposition 2. By proposition 3 we get $\pi_{i(h)}(\varphi(\vec{c})) = \lambda_{i(h)} \cdot \pi_{\psi(i(h))}(\vec{c})$ for all $j \in R$, for $h = 1,2,\ldots,r_j$ and for all $\vec{c} \in C$. Thus $\varphi = \hat{\lambda} \circ \tilde{\psi}$.

Note that the permutation ψ is uniquely determined, if $r_j = 1$ for all $j \in R$ and if $s = 0$ or $s = 1$. In case $r_j = 1$ for all $j \in R$ and $s = 0$; i.e. if all the n projections $\pi_j|_C$, $j \in \mathbb{Z}_n$, are pairwise linearly independent then the vector $\lambda = (\lambda_1, \lambda_2, \ldots, \lambda_n) \in (F^*)^n$ is uniquely determined, also. The same is true if $r_j = 1$ for all $j \in R$, $s = 1$ and $q = 2$.

In F^n we use the usual scalar product $F^n \times F^n \to F$; $(\vec{x},\vec{y}) \to \vec{x}\cdot\vec{y} := \sum_{i=1}^{n} \pi_i(\vec{x}) \cdot \pi_i(\vec{y})$.
The *dual* (a more appropriate but unusual name would be "orthogonal") $C^\perp := \{\vec{x} \in F^n; \vec{x}\cdot\vec{c} = 0\}$ *of* a linear (n,k)-code C over F is a linear $(n,n-k)$-code over F, which is not necessarily a complement to C in F^n.

Now let C be a linear (n,k)-code over F, such that any two of the n projections $\pi_j|_C$, $j \in \mathbb{Z}_n$, are linearly independent. By the "Untere Abschätzung des Minimalabstands linearer Codes" [1;p.227] this condition is equivalent to the fact that the minimal distance of the dual $C^\perp$ of C is at least three, $d(C^\perp) \geq 3$. (Other codes are in many respects fairly uninteresting.) Then there is for every linear code-automorphism φ of C precisely one monomial transformation $\Phi = \hat{\lambda} \circ \tilde{\psi} \in GL_n(F)$ of F^n with $\lambda = (\lambda_1, \lambda_2, \ldots, \lambda_n) \in (F^*)^n$, $\psi \in S_n$ and $\Phi|_C = \varphi$. Therefore in case $d(C^\perp) \geq 3$ there is no difference between MacWilliams' and Sloane's concept of the linear automorphism group of the code C and the author authors' concept. These groups are isomorphic. The transformation $\Phi' = \hat{\lambda}^{-1} \circ \tilde{\psi}$ as a monomial transformation of F^n preserves the Hamming weight of every codeword $\vec{d} \in C^\perp$. From $\Phi(C) = C$ we deduce $\Phi'(C^\perp) = C^\perp$. Thus the map $\varphi \to \Phi'|_{C^\perp}$ is a homomorphism from the group of all linear code-automorphisms of C into the group of all linear code-automorphisms of $C^\perp$. If any two of the n projections $\pi_i|_{C^\perp}$, $i \in \mathbb{Z}_n$, are also linearly independent, i.e. if $d(C) \geq 3$, then the rôles of $C = (C^\perp)^\perp$ can be interchanged in our argumentation and the groups of all linear code-automorphisms of C and $C^\perp$ are isomorphic.

For $r = 3,4,\ldots$ the *simplex code* $HAM^{\perp}(r,q)$ is defined as a linear (n,r)-code over $F = GF(q)$ of length $n = (q^r - 1)/(q-1)$, which as a linear subspace of F^n is generated by the rows of a $r \times n$-matrix over F, whose columns form a system of representatives of the one-dimensional linear subspaces of F^r (cf. e.g. [1;p.232]). Of course, there are many code-isomorphic versions of $HAM^{\perp}(r,q)$. One can change the order of the columns and get equivalent codes. One can also choose other systems of representatives. Note that every linear code C with $d(C^{\perp}) \geq 3$ can be obtained by puncturing (i.e. deleting the components in some fixed positions in all codewords) a suitable simplex code. The dual of $HAM^{\perp}(r,q)$ is the q-ary *Hamming code* $HAM(r,q)$. The simplex code is equidistant with minimal distance $d(HAM^{\perp}(r,q)) = q^{r-1} \geq 4$. Indeed, any line of its generator matrix has $(q^{r-1} - 1)/(q-1)$ zero entries and any non-zero codeword of $HAM^{\perp}(r,q)$ can be interpreted as a line in a generator matrix of $HAM^{\perp}(r,q)$ which is obtainable from the original generator matrix by applying only elementary operations on the lines. The group of all linear code-automorphisms of $HAM^{\perp}(r,q)$ is the general linear group $GL_r(F)$. Since by definition any two of the linear forms $\pi_i|_{HAM^{\perp}(r,q)}$, $i \in \mathbb{Z}_n$, are linearly independent (in fact it is $d(HAM(r,q)) = 3$) the group of all linear code-automorphisms of the Hamming code $HAM(r,q)$ is isomorphic to $GL_r(F)$.

Finally, we make a bow to projective geometry and remark that the theorem of this paper applies mutatis mutandis to "semi-linear code-isomorphisms".

REFERENCES

[1] Heise, W. and Quattrocchi, P., Informations- und Codierungstheorie (Springer, Berlin-Heidelberg-New York-Tokyo, 1983).

[2] MacWilliams, F. J. and Sloane, N. J. A., The theory of error-correcting codes (North-Holland, Amsterdam-New York-Oxford, 1977).

Annals of Discrete Mathematics 30 (1986) 225–242

ON THE CROSSING NUMBER OF GENERALIZED PETERSEN GRAPHS

S. Fiorini

Department of Mathematics, University of Malta

ABSTRACT The Generalized Petersen Graph $P(n,k)$ is defined to be the graph on $2n$ vertices labelled $\{a_1, a_2, \dots, a_n, b_1, b_2, \dots, b_n\}$ and edges $\{a_ib_i, a_ia_{i+1}, b_ib_{i+k}$: $i = 1,2,\dots,n$; subscripts modulo $n\}$. The crossing numbers $\nu(n,k)$ of $P(n,k)$ are determined as follows: $\nu(9,3) = 2$, $\nu(3h,3) = h$, $\nu(3h+2,3) = h+2$, $h+1 \leq \nu(3h+1,3) \leq h+3$, $\nu(4h,4)=2h$; various conjectures are formulated.

PRELIMINARIES All graphs $G = (V(G),E(G))$ considered will be simple, i.e. contain no loops or multiple edges.

The Generalized Petersen Graph $P(n,k)$ is defined to be the graph of order $2n$ with vertices labelled $\{a_1a_2,\dots,a_n,b_1,b_2,\dots,b_n\}$ and edges $\{a_ib_i, a_ia_{i+1}, b_ib_{i+k}: i=1,2,\dots,n$; subscripts modulo n, $1 \leq k \leq n-1\}$. The derived Generalized Petersen Graph denoted $P'(n,k)$ is obtained from $P(n,k)$ by contracting all edges of form a_i,b_i, called spokes; edges of form b_ib_{i+k} in $P(n,k)$ are then called chords of the n-circuit $a_1,a_2,\dots,a_n,a_1$. A drawing of a graph in a surface is a mapping of the graph into the surface in such a way that vertices are mapped to points of the surface and edges vw to arcs in the surface joining the image-points of v and w and the image of no edge contains that of any vertex. In our case, the only surface we consider is the plane and all our drawings will be good in the sense that no two arcs which are images of adjacent edges have a common point other than the image of the common vertex, no two arcs have more than one point in common, and no point other than the image of a vertex is common to more than two arcs. A common point of two arcs other than the image of a common vertex is called crossing. A drawing is said to be optimal if it minimizes the number of crossings; clearly, an optimal drawing is necessarily good. The number of crossings in an optimal drawing of a graph G is denoted by $\nu(G)$; the number of crossings in a drawing $\mathcal{D}$ of G is denoted by $\nu_{\mathcal{D}}(G)$.

TECHNIQUES The technique of proving that the crossing number of some graph G is some positive integer k is quite standard. Some good drawing is exhibited whereby an upper bound for k is established. By some *ad hoc* method it is then shown that this number is also a lower bound. Embodied in the theorems of this section we present some conclusions of a general nature which hopefully could be used also in determining the lower bounds of crossing numbers of other graphs.

THEOREM 1 If two graphs G and H are homeomorphic, then their crossing numbers are identical. //

COROLLARY 1 (The Monotone Theorem) If $\alpha = (k,n)$, the greatest common divisor of k and n, and if $2 \leq \alpha \leq k < \frac{1}{2}n$,

then
$$\nu_{n,k} \geq \nu_{n-k,k}$$

and
$$\nu_{n,k} \geq \nu_{n-n/\alpha,k-k/\alpha},$$

where $\nu_{n,k}$ denotes $\nu(P(n,k))$.

PROOF Let H be obtained from P(n,k) by deleting k successive spokes and let K be obtained from P(n,k) by deleting every k'th spoke in the case $\alpha \leq 2$. Then H is homeomorphic to P(n - k,k) and if $\alpha \leq 2$, then K is homeomorphic to $P(n-n/\alpha, k-k/\alpha)$. The result follows from Theorem 1. //

If G is a graph and $X \subseteq V(G) \cup E(G)$ then the subgraph induced by X is denoted by $\langle X \rangle$.

THEOREM 2 (The Decomposition Theorem) Let $\mathcal{D}$ be an optimal drawing of a graph G and let $E(G) = X \cup Y \cup Z$, $X \cap Y = Y \cap Z = Z \cap X = \emptyset$.
If $\nu_{\mathcal{D}}\langle X \rangle = 0$, then
$$\nu(G) = \nu\langle X \cup Y \rangle + \nu\langle X \cup Z \rangle$$

PROOF $\nu(G) = \nu_{\mathcal{D}}(G) = \nu_{\mathcal{D}}\langle X \cup Y \rangle + \nu_{\mathcal{D}}\langle X \cup Z \rangle - \nu_{\mathcal{D}}\langle X \rangle + k$, where k is the number of crossings of form Y x Z

$$\geq \nu_{\mathcal{D}}\langle X \cup Y \rangle + \nu_{\mathcal{D}}\langle X \cup Z \rangle - \nu_{\mathcal{D}}\langle X \rangle$$
$$= \nu_{\mathcal{D}}\langle X \cup Y \rangle + \nu_{\mathcal{D}}\langle X \cup Z \rangle, \text{ since } \nu_{\mathcal{D}}\langle X \rangle = 0$$
$$\geq \nu\langle X \cup Y \rangle + \nu\langle X \cup Z \rangle \quad //$$

The following corollary readily follows by induction on k:

COROLLARY 2 Let $\mathcal{D}$ be an optimal drawing of a graph G in which some subset X of E(G) makes 0 contribution to $\nu_{\mathcal{D}}(G)$.

If $X \cup \left[\bigcup_{i=1}^{k} Y_i\right]$, $Y_i \cap Y_j = \emptyset$ $(i \neq j)$ is a decomposition of E(G) then $\nu(G) \geq \sum_{i=1}^{k} \nu\langle X \cup Y_i \rangle$. //

THEOREM 3 (The Deletion Theorem) Let α be the least number of edges of a graph G whose deletion from G results in a planar subgraph H of G. Then $\nu(G) \geq \alpha$.

PROOF Assuming on the contrary that $\nu < \alpha$, then deleting the (at most) ν edges being intersected results in a planar subgraph of G, contradicting the minimality of α. //

We often make use of this simple conclusion in conjunction with Euler's polyhedral formula as in the following:

THEOREM 4 $\nu(9,3) = 2$

PROOF The graph of Figure 1 (i) shows that 2 is an upper bound for $\nu(9,3)$. To show that it is also a lower bound we note that $P(9,3)$, contains as subgraph a homeomorph of the graph G of Figure 1 (ii); (the subgraph is obtained by deleting an edge from each of the three triangles of $P(9,3)$.) G has 12 vertices, 18 edges and girth 5, so that if α edges are deleted to obtain a planar subgraph H, Euler's formula for H implies that

$$5(8 - \alpha) \leq 2(18 - \alpha).$$

Thus, $\nu(9,3) \geq \alpha \geq \lceil 4/3 \rceil = 2$. //

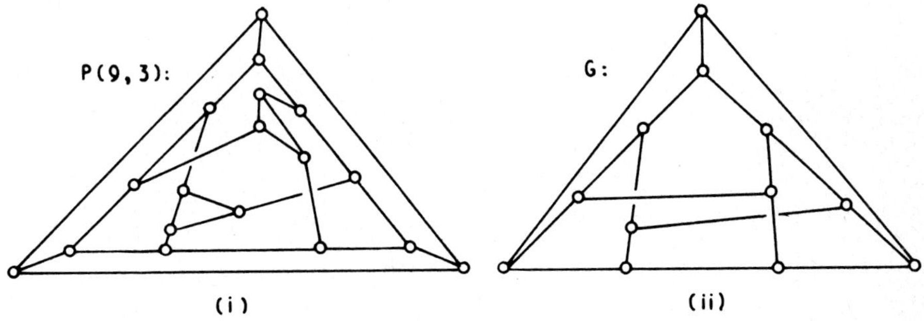

Fig. 1.

THEOREM 5 (The Contraction Theorem) Let $\mathcal{D}$ be a drawing of a graph G and let $e \in E(G)$ make 0 contribution to $\nu_{\mathcal{D}}(G)$. Let G^e be the graph obtained from G by contracting the edge $e = uv$ to a single vertex $u = v$ and let $\mathcal{D}'$ be the drawing of G^e induced by $\mathcal{D}$. Then $\nu_{\mathcal{D}}(G) \geq \nu_{\mathcal{D}'}(G^e)$.

PROOF Let $wv = f \in E(G)$ such that f is adjacent to $e = uv$.

(i) If $f \notin E(G^e)$ (e.g $uw \in E(G)$), then any crossing involving f in $\mathcal{D}$ is missing in $\mathcal{D}'$;

(ii) If $f \in E(G^e)$ and f is crossed by some edge tu in $\mathcal{D}$, then this crossing is also missing in $\mathcal{D}'$.

Since all other crossings are unaffected, in all cases

$$\nu_{\mathcal{D}}(G) \geq \nu_{\mathcal{D}'}(G^e). \quad //$$

COROLLARY 1 If $\mathcal{D}$ is an optimal drawing in which $e \in E(G)$ makes 0 contribution to $\nu_{\mathcal{D}}(G)$, then $\nu(G) \geq \nu(G^e)$.

PROOF By the Contraction Theorem,

$$\nu(G) = \nu_{\mathcal{D}}(G) \geq \nu_{\mathcal{D}'}(G^e) \geq \nu(G^e). \quad //$$

Repeated use of the Contraction Theorem yields the following:

COROLLARY 2 Let $\langle e_1,\ldots,e_t \rangle$ be a sequence of edges of G each of which makes 0 contribution to $\nu_{\mathcal{D}}(G)$ in some drawing $\mathcal{D}$. If we define recursively $G^0 = G$, $\mathcal{D}^0 = \mathcal{D}$, $G^i = (G^{i-1})^{e_i}$, $\mathcal{D}^i$ the drawing of G^i induced by $\mathcal{D}^{i-1}$, then

$$\nu_{\mathcal{D}^0}(G^0) \geq \nu_{\mathcal{D}^1}(G^1) \geq \ldots \geq \nu_{\mathcal{D}^t}(G^t). \quad //$$

COROLLARY 3 Let $\mathcal{D}$ be an optimal drawing of G and let H be a subgraph of G such that for each edge e of H, e makes 0 contribution to $\nu_{\mathcal{D}}(G)$. Then $\nu(G) \geq \nu(G^H)$, where G^H is obtained from G by contracting each edge e of H.

PROOF We order the edges of H and apply Corollaries 1 and 2. //

REMARKS

(i) In the Contraction Theorem and its corollaries, the condition that "e makes 0 contribution to ν" is vital. Consider the graph G obtained from K_7 by "expanding" any vertex into two adjacent vertices u,v, of valency 4. It is not difficult to see that $7 \geq \nu(G) \ngeq \nu(G^{uv}) = \nu(K_7) = 9$.

(ii) The reverse inequality $\nu(G^e) \geq \nu(G)$ cannot in general be proved, even if other conditions (e.g; if G contains no triangles) are imposed.

Consider $$1 = \nu(K_{3,3}) \quad > \quad \nu(K^e_{3,3}) = 0$$

THEOREM 6 If G_k denotes the derived graph $P'(3k,3)$, then $\nu(G_k) = k$ and there exists an optimal drawing in which the $(3k)$-circuit C does not intersect itself.

PROOF That $\nu(G_k) \leq k$ follows from the drawing of Figure 2:

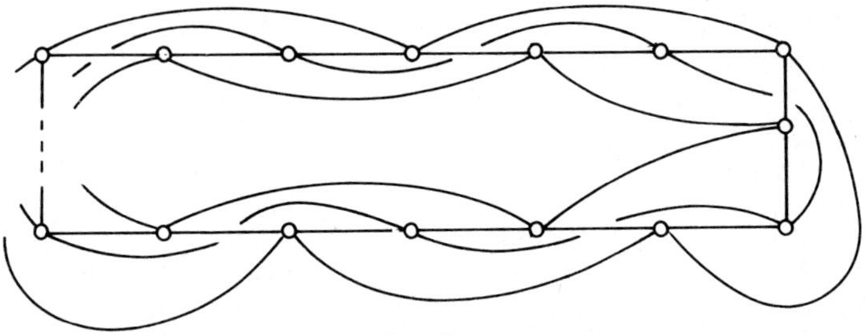

Fig. 2

To establish the reverse inequality we note that the deletion of any three successive edges of C yields a subgraph homeomorphic to G_{k-1}. The statement is now proved by induction on k. To start the induction we apply the Deletion Theorem to G_4, for which $m = 24$, $n = 12$ and $g = 4$, so that

$$f = (24 - \alpha) - 12 + 2 = 14 - \alpha$$

$$\Rightarrow \quad 4(14 - \alpha) \leq 2(24 - \alpha)$$

$$\Rightarrow \quad \nu \geq \alpha \geq 4,$$

and the statement is valid in this case.

We now consider an optimal drawing $\mathcal{D}$ of G_k and assume, for contradiction, that $\nu_{\mathcal{D}}(G_k) = k-1$. If C does not intersect itself in $\mathcal{D}$, then by the Decomposition Theorem with $\langle X \rangle = C$ and Y_i the i'th set of three successive chords $(i = 1,2,\ldots,k)$, we conclude that $\nu(G_k) = k$, since $(X \cup Y_i) = 1$. It follows that in this case $\nu(G_k) = k$ and there exists a drawing in which C does not intersect itself. If, on the other hand, C intersects itself in some edge e, then by deleting e and two successive edges of C, we obtain G_{k-1} for which the inductive hypothesis implies:

$$k - 1 = \nu(G_{k-1}) \leq \nu(G_k) - 1 = k - 1 - 1,$$

a contradiction. //

The same argument, only slightly modified, holds for $P'(3k+h,3)$ $(h = 1,2)$ and determines this crossing number as $k + h$. However, since the inductive argument fails in its initial step for $h = 1$ (the girth of $P'(7,3) = 3$), we start with $k = 3$ for this case.

THEOREM 7 If G_k denotes the derived graph $P'(3k+h,3)$, then for $h = 1$, $k \geq 3$ and for $h = 2$, $k \geq 2$, $\nu(G_k) = k + h$. Further, there exists an optimal drawing in which the $(3k+h)$-circuit C does not intersect itself.

PROOF That $\nu(G_k) \leq k + h$ follows from the drawings of Fig. 3. To establish the reverse inequality we proceed by induction and note that for $(h,k) = (1,3)$ or $(2,2)$ the girth is 4 and $(n,m) = (10,20)$ and $(8,16)$ respectively. The Deletion Theorem, then yields:

$$f = 12 - \alpha \text{ and } f = 10 - \alpha, \text{ respectively, so that}$$

$$4(12 - \alpha) \leq 2(20 - \alpha) \text{ and } 4(10 - \alpha) \leq 2(16 - \alpha),$$

respectively; in either case $\nu \geq \alpha \geq 4 = k + h$.

Now suppose that C makes 0 contribution to $\nu_{\mathcal{D}}$ in some drawing $\mathcal{D}$. Then C is planarly embedded and all chords either lie in Int(C) or in Ext(C).

Case (i) If all adjacent chords lie in different regions, then two distinct sub-cases arise none of which is optimal;

Case (ii) If some pair of adjacent chords $a_{i-3}a_i$, a_ia_{i+3} both lie in the same region, then two further sub-cases, according as $a_{i-2}a_{i+1}$ lies in the same or in different regions as these, arise. In all cases that locate $a_{i-1}a_{i+2}$, a re-drawing is possible which both does not increase ν and in which some chord intersects C.

We conclude that in all cases there exists an optimal drawing in which C is intersected in some edge e. Assuming for contradiction that $\nu(G_k) < k + h$, deleting the edge e and two successive edges, we obtain a homeomorph of G_{k-1} for which the inductive hypothesis implies:

$$k + h - 1 = \nu(G_{k-1}) \leq \nu(G_k) - 1 \leq k + h - 1 - 1,$$

a contradiction.

The drawings of Figure 3 are then seen to be both optimal and in which C does not intersect itself. //

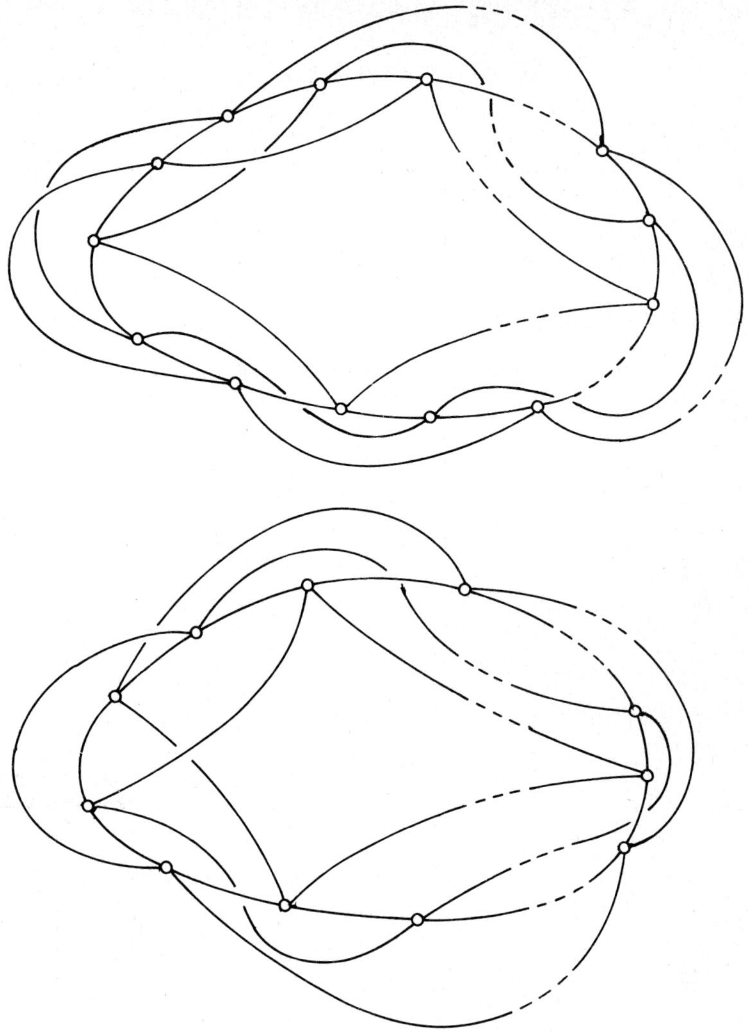

Fig. 3.

THEOREM 8 $k + 3 \geq \nu(3k + 1,3) \geq k + 1$

PROOF That $\nu(3k + 1,3) \leq k + 3$ follows from the drawing of Figure 4.

To show that the lower bound also holds, we consider two cases for a minimal counterexample:

Case (i) If there exists an optimal drawing in which no spoke is intersected, then the Contraction Theorem implies that

$$\nu(3k + 1,3) \geq \nu'(3k + 1,3)$$
$$= k + 1 \text{ (By Theorem 7), for } k \geq 3.$$

That $\nu(7,3) = 3$ follows from the work of Exoo, Harary and Kabell.

Case (ii) If some spoke is intersected, then deleting that spoke and two successive spokes, we obtain a homeormorph of $P(3k - 2,3)$ whose crossing number is k, by the minimality of k.

But then,

$$\nu(3k + 1,3) \geq \nu(3k - 2, 3) + 1 = k + 1,$$

a contradiction. //

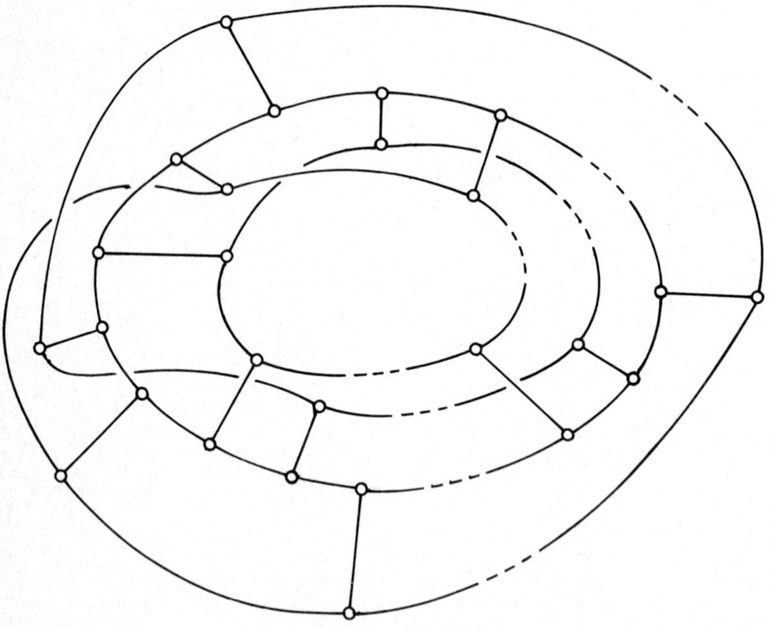

Fig.4.

The remaining two cases: $\nu(3k + h,3) = k + h$ $(h = 0,2)$ are established in exactly the same way once we prove that

$$\nu(8,3) = 4 = \nu(12,3).$$

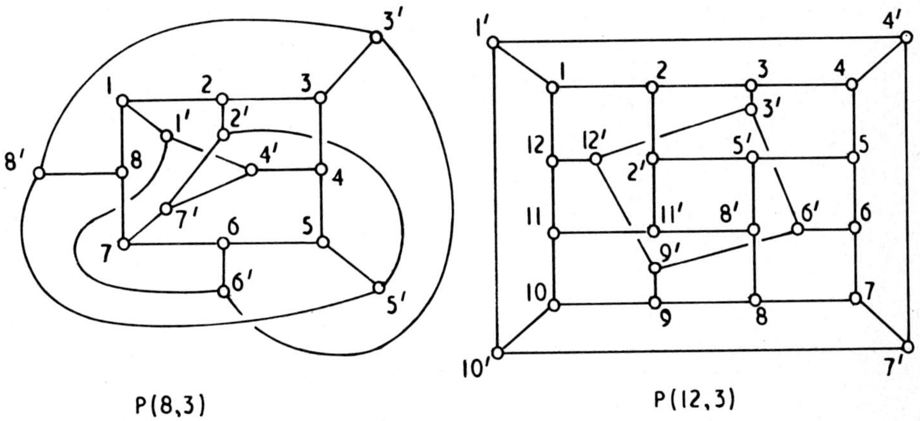

Fig.5

Proofs which are not case-by-case are elusive. To facilitate presentation we sketch the method of procedure. We assume, for contradiction, that the crossing number is at most 3 and consider separately the cases where (i) no crossing is a spoke intersection, (ii) where all three crossings, (iii) two of the crossings, and (iv) exactly one crossing is a spoke intersection. The Contraction Theorem deals with (i) whereas Theorem 1 deals with (ii). Thereafter the argument takes the following sequence: A large (usually Hamiltonian) circuit H is chosen in the graph. If H is planarly embedded in some optimal drawing of the 2-spoke-deleted graph, then a contradiction is obtained by virtue of the Decomposition Theorem with H = X. If not, then H must intersect itself in exactly two of its edges to yield a 2-looped drawing of itself. A contradiction is obtained for each pair of edges. To this end heavy use is made of the following remarks. We define the planarization induced by a drawing of a graph G to be the planar graph obtained by replacing each crossing by a new vertex with 4 incident edges, in the obvious way. We also define a pair of parallel chords of a circuit C to be a pair of edges (a,b),(c,d) in $G \setminus C$ such that $a<b<c<d<a$, where "order" is defined clockwise along C. (In this sense, parallelism is not a transitive relation).

Remark 1: If C intersects itself in edges (p,q) and (r,s), and if (a,b), (c,d) are parallel edges separating (p,q), (r,s), ie; $a \leq p < q \leq b < c \leq r < s \leq d < a$, then one of (a,b),(c,d) must cross some edge.

Remark 2: Each loop of C must contain at least two vertices in any good drawing of G.

Remark 3: If C together with chords (a,b), (c,d),(e,f) is homeomorphic to $K_{3,3}$ then a 2-looped planarization of H together with these edges also contains a homeomorph of $K_{3,3}$ if the crossed edges of H both lie in a segment of H containing at most one of the vertices {a,b,c,d,e,f}

THEOREM 9 $\nu(8,3) = 4$

PROOF: That $\nu(8,3) \leq 4$ follows from the drawing of Figure 5(i). To establish the reverse inequality we note that since n = 16, m = 24 and the girth g = 6, then the Deletion Theorem implies that

$$6(10 - \alpha) = 2(24 - \alpha)$$

so that $$\nu \geq \alpha \geq 3.$$

If there exists an optimal drawing in which no spokes intersect, then the Contraction Theorem together with Theorem 7 imply that

$$\nu\ (8,3) \geq \nu'(8,3) = 4.$$

Thus we can assume there exists an optimal drawing in which exactly three crossings occur one of which is a spoke intersection:

Case (i) If all three intersections are spoke crossings, then deleting the three crossed spokes should result in a planar graph. But deleting these three spokes (in all possible ways) together with an appropriate fourth spoke we can always obtain a homeormorph of the graph obtained from P(5,2) by deleting a spoke, which is non-planar; a contradiction.

Case (ii) If exactly two spokes are intersected, then deleting these two spokes should result in a graph with crossing number 1 and in each optimal drawing of which no spoke is intersected. However, if the two spokes are either successive (at distance 1 on the rim) or alternate (at distance 2), then deleting an appropriate third spoke results in a homeomorph of P(5,2) whose crossing number is 2. If the distance is 3 (resp. 4) then the resulting deleted graph contains a homeomorph of the graph of Figure 6 (i) (resp. (ii)); both these graphs are seen to possess the Hamiltonian circuit H = $\langle 1,2,3,\ldots,10,11,12,1\rangle$. If there exists an optimal drawing in which H is planarly embedded, then by the Decomposition Theorem with X = H, Y = {(1,5),(2,7),(4,9)} and Z = {(6,11),(8,12),(10,3)}, the first graph is seen to have crossing number 2; for the second graph we take Y = {(1,5),(2,10),(3,12)} and Z = {(4,8),(6,9),(7,11)}. In each case $\langle X\cup Y\rangle = K_{3,3} = \langle X\cup Z\rangle$. Thus we may assume that the only optimal drawings are those in which H intersects itself exactly once in non-spoke edges. For the first graph, it S_1 denotes the segment $\langle 6,7,\ldots,12\rangle$ and $S_2 = \langle 1,2,\ldots,7\rangle$, and if we assume that (6,7) = $S_1 \cap S_2$ is not crossed, then some edge in $S_1 \setminus (6,7)$ must cross some edge in $S_2 \setminus (6,7)$, by Remark 3. Now, for all pairs of edges $(e_1,e_2) \in S_1 \times S_2$, there exists a pair of parallel edges separating them except for $e_1 = (5,6)$, which is a spoke, anyway. Thus (6,7) must intersect some edge in $\langle 8,9,\ldots,5\rangle$, by Remark 2.
But (6,7) is separated from each edge in $\langle 10,11,\ldots,3\rangle$ by parallel edges (3,10),(4,9), so it cannot cross any of them, by Remark 1. Of the remaining edges, the only non-spokes are (8,9) and (4,5), both of which cases are dispensed with by Remark 3. As to the second graph, some edge in segment

$S_1 = \langle 6,7,\ldots,11\rangle$ must cross some edge in segment $S_2 = \langle 12,1,\ldots,5\rangle$ by Remark 3, since H together with chords (6,9),(7,11),(8,4) is $K_{3,3}$ and the same holds for chords (2,10),(3,12),(1,5). Now for each pair of edges $(e_1,e_2)\varepsilon(S_1 \times S_2)$, there exists a pair of parallel edges separating them. Thus by Remark 1, H cannot intersect itself.

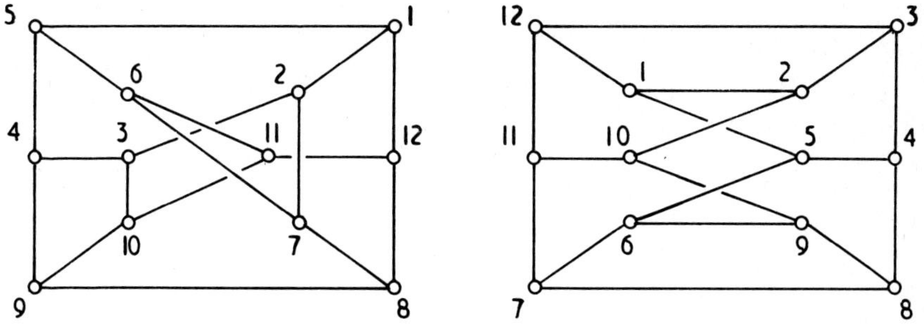

(i) (ii)

Fig. 6.

Case (iii): If exactly one spoke is intersected, then deleting this spoke, (1,1') say, should result in a graph whose crossing number is 2; as in the previous case we show that this leads to a contradiction. The circuit (Fig 7 (i)) $C = \langle 1,2,3,3',6',1',4',4,5,6,7,7',2',5',8',8,1\rangle$ is seen to be a Hamiltonian circuit in this graph and has chords (2,2'),(5,5'),(7,8) which together with C are homeomorphic to $K_{3,3}$. Thus, if C is planarly embedded in some optimal drawing, then one or other of the spokes (2,2'),(5,5') is necessarily crossed, contrary to assumptions. We conclude that C must intersect itself; if exactly once, then this intersection must occur in the segment $\langle 2,1,\ldots,6,5\rangle$; otherwise either (2,2') or (5,5') is crossed. This intersection must also occur in the segment $\langle 5',2',\ldots,4',6'\rangle$, since as before, C together with chords (6,6'),(5,5'),(4',7) is homeomorphic to $K_{3,3}$. We conclude that the crossing must occur in the intersection of these segments, ie; in the segment $\langle 5',2',7',7,6,5\rangle$. Since (7,7') is a spoke and a loop must contain at least two vertices four cases arise according as the crossing occurs in : (a) (a) $(5,6) \times (7',2')$, (b) $(5,6) \times (2',5')$, (c) $(6,7) \times (7',2')$ or (d) $(6,7) \times (2',5')$.

In the unique embedding of the planarization of C spokes (2,2'),(5,5'),(6,6') can be drawn uncrossed in only one way in cases (a),(d), in two ways in case (c) and in no way in (b). In (a) (7,8) and one (3,4) or (3',8') must each contribute 1 to ν ; in (d), (7',4') and one of (3,4) or (3',8') each contribute 1 to ν; and in (c) each of (7,8) and (7',4') contribute 1 to ν in each embedding. Since all cases yield a contradiction we conclude that C must be twisted twice, in three loops, in such a way that all other edges can be drawn in without further crossings.

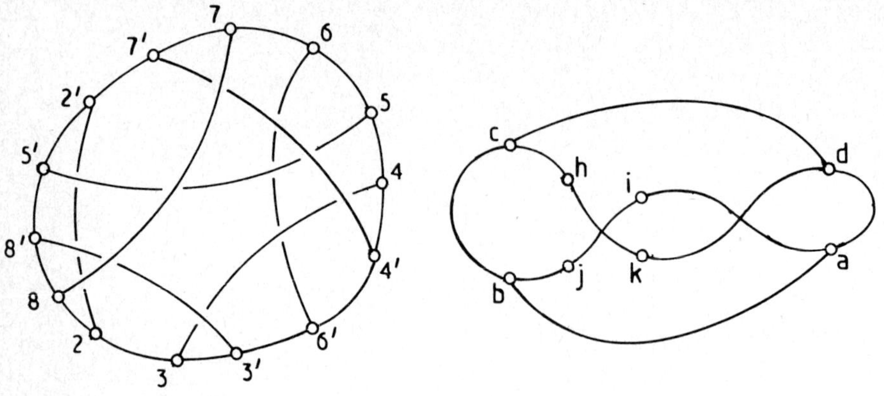

(i) (ii)

Fig. 7.

To identify the two pairs of intersecting edges of C we make use of the following remarks:

Remark 1: If (a,b) and (c,d) (a<b<c<d) are a pair of parallel chords and if (i,j) (a≤i<j≤b) crosses (h,k) (c≤h<k≤d), then unless a second intersection occurs between some edge in segment <a,...,i> with some edge in segment <k,...,d> or between some edge in <c,...,h> with an edge in <j,...,b>, then edges (a,b) and (c,d) cannot be drawn without crossing each other. Furthermore the planarization of C and the parallel edges must be drawn as in Figure 7 (ii) ie; with both end loops in the exterior (or equivalently, the interior) of the middle loop.

Remark 2: In a good drawing with crossing number ν, each of the end-loops must contain at least 2 vertices. If the middle loop has at most two vertices, other than vertices of chords of that loop, then there exists another drawing with at most ν crossings in which C does not intersect itself; this reduces to a previous case.

If both twists occur in the segment <2',7',...6',3'> then the chords (7,8), (2,2'),(3',8') necessarily intersect. so that one twist involves one of the edges (2',5'),(5'8'),(8,2),(2,3) . We investigate each of these separately. The first (in a counter-clockwise sense) candidate to cross (2',5') is (8,2) which is separated from it by the parallel chords (5,5'),(8',3'). Hence the second twist must occur between one of (5',2'),(2',7'),(7,6),(6,5) and one of (8,2),(2,3) , by virtue of Remark 1. Of the first set, (7,6) and (6,5) are excluded since they are also separated by the parallel chords (8',3'),(7',4'). The remaining cases are disposed of by Remark 2. Arguing in this way we conclude that the set of edges that can cross (2',5') is empty. For the other edges, the only crossings that need discussing are (5',8') x (8,2),(5',8') x (2,3), (5',8') x (7',2'), (2,2) x (3',6'). We assume without loss of generality that vertices 8 and 8' lie on the left loop and consider three possible locations of 3': on the right loop, on the lower branch of the middle loop, and on the upper branch. In each case (8',3') can be drawn in uniquely. In the last case, 2 must lie on the right loop, so that (2,2') is necessarily crossed. In the other cases, 7 must lie on the upper branch of the middle loop, so that one of (2,2') or (5,5') is crossed. The other cases are similarly dealt with. //

Theorem 10 $\nu(12,3) = 4$.

Proof: That $\nu(12,3) \leq 4$ follows from the drawing of Figure 5(ii). To establish the reverse inequality we assume for contradiction that $\nu \leq 3$ and note that if none of the spokes intersect in some optimal drawing, then applying the Deletion Theorem to the derived graph for which m = 24, n = 12 and girth is 4, we get

$$4(14 - \alpha) \leq 2(24 - \alpha)$$

so that

$$\nu \geq \alpha \geq 4.$$

Thus some spoke is intersected and we proceed to consider three cases according as the number of spokes involved is exactly 3,2 or 1.

Case (i): If all three intersections are spoke intersections, then deleting three spokes should result in a planar graph. If two of the deleted spokes are either consecutive (at distance 1 on the 12-circuit, C) or alternate (at distance 2 on C), then deleting a fourth appropriate spoke results in a homeomorph of P(9,3) less a spoke, which is non-planar. Thus we assume that the distance on C between deleted spokes is at least 3. If they are equally spaced, then the resulting planar graph contains a homemorph of $K_{3,3}$, a contradiction. There remains therefore two sub-cases according as the successive distances on the rim are (3,3,6) or (3,4,5).

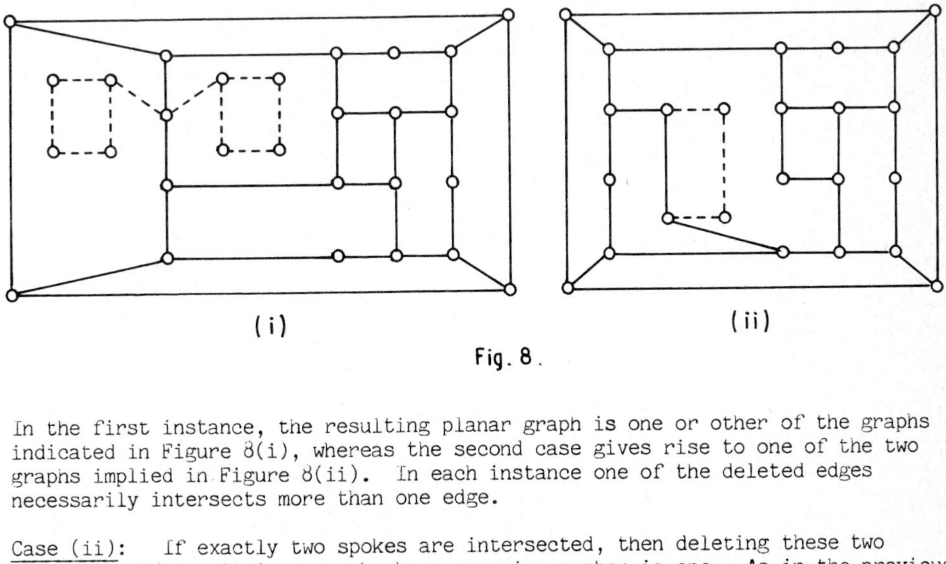

Fig. 8.

In the first instance, the resulting planar graph is one or other of the graphs indicated in Figure 8(i), whereas the second case gives rise to one of the two graphs implied in Figure 8(ii). In each instance one of the deleted edges necessarily intersects more than one edge.

Case (ii): If exactly two spokes are intersected, then deleting these two spokes should result in a graph whose crossing number is one. As in the previous case, if the distance between the deleted spokes is either 1 or 2 along the 12-circuit C, then deleting a further adjacent spoke appropriately yields a homeomorph of P(9,3) whose crossing number is 2. The remaining cases are dealt with separately. In each case we delete the spoke (1,1') and one of (4,4'), (5,5'),(6,6'),(7,7'), respectively. Each of the first three graphs is seen to possess a Hamiltonian circuit H as follows: <1,2,2',5',5,4,3,3',6',6,7,7',4',1', 10',10,11,11',8',8,9,9',12',12,1>, <1,2,2',5',8',11',11,10,10',1',4',7',7,8,9, 9',6',6,5,4,3,3',12',12,1>, <1,2,2',11',11,10,9,8,8',5',5,6,7,7',10',1',4',4,3, 3',6',9',12',12,1>. We assume that H is planarly embedded is some optimal drawing.

Then, in the first case the pair of chord triples ((5',8'),(7,8),(9,10)) and ((3',12'),(11,12),(2',11')) each contribute 1 to the crossing number so that by the Decomposition Theorem the crossing number is at least 2. In the second case, the triple of chords ((4,4'),(8,8'),(9,10)) necessarily yields a crossing involving one or other of the spokes (4,4') or (8,8'), reducing to Case (i); similarly, in the third case, a crossing must arise among the triple of chords ((9,9'),(10,10'),(8',11')), again reducing to Case (i). Since all cases imply a contradiction we conclude that H must intersect itself, giving rise to exactly two loops , in such a way that all remaining edges can be drawn in without further crossings. We shall need the following:

Remark: If H together with chords (a,b),(c,d),(e,f) is homeomorphic to $K_{3,3}$ then the planarization of H obtained from intersecting itself once contains a homeomorph of $K_{3,3}$ if the crossed edges both lie in a segment of H containing at most one vertex in {a,b,c,d,e,f}. Thus in the first example, H together with chords (9,10),(7,8),(5',8') is homeomorphic to $K_{3,3}$ so that one of the crossed edges must lie in the segment <7,7',...,8,9>, non-spoke edges in this segment being: (9,8),(8',11'),(11,10),(10',1'),(1',4'),(4',7'). Taking these in turn, (9,8) is separated from each of the edges in the segment <9',12',...,6') by the parallel chords (9',6'),(9,10) and from those in segment <10',1',4',7'> by (9,10),(10',7') so that the only possible edges crossing it are: (8',11'),(10,11) and (6,7). The only other possible crossings are similarly found to be: (8',11') x (11,10), (8',11') x (6,7),(10,11) x (6,7) and (10',1') x (4'7'). Of these the first and last are disposed of by virtue of Remark (above).

As for the remaining cases, in (9,8) x (10,11) and (9,8) x (6,7) the resulting graphs are indeed planar, but in the unique plane embedding either (1,1') or (4,4') crosses at least two edges, in all the rest, the planarization of H together with (6',9'),(7,8),(8',5') is uniquely embedded in the plane, but then (11',2') necessarily intersects some edge.

In the second instance, H together with (8,8'),(6,7),(9,10) is homeomorphic with $K_{3,3}$ so that some intersecting edge of H must lie in the segment <6,6',...,10>. In the third instance, the intersecting edge must lie in segment <10',7',...,11') since here H and chords (10',10),(9,9'),(8',11') is $K_{3,3}$. Of the six non-spokes in the segment of the former, it is readily verified that none qualify to intersect any other edge and of the seven in the latter case only one, (8,9) x (10,11). In this case, the resulting graph is indeed planar but the deleted spokes (1,1'),(6,6') cross at least twice each in the unique embedding. There remains to consider the fourth graph obtained by deleting (1,1'),(7,7') which is not Hamiltonian. We consider instead its subgraph obtained by deleting 7' and incident edges. This graph K has a circuit H = <12',9',9, 10,11,11',8',8,7,6,6',3',3,4,5,5' ,2',2,12> that includes all vertices except 4',1',10', which lie on a chain joining vertices 4 and 10 on H. If this chain is not intersected in some optimal drawing in which H is planarly embedded, so that it lies in Int(H) without loss of generality, then all chords (9,8),(11,12), (2,3),(5,6) must lie in Ext (H), yielding at least two crossings. If on the other hand, H intersects itself, then one of the crossed edges must lie in the segment <12,12',...,8> and the other in <6,6',...,2>. But each non-spoke in the first is separated from each non-spoke in the second by the parallel edges (2',11'),(5',8') except for (11',8') and (5',2') which are in turn separated by (8,9) and (3',12'). We conclude that the chain <4,4',1',10',10> is intersected in either (1',4') or (1',10'). But then any edge intersecting one of these edges must also intersect one of (7',4'), (7',10') in the corresponding drawing of P(12,3).

Case (iii): If exactly one spoke is intersected, then deleting this spoke, (12,12') say, should result in a graph whose crossing number is 2. Figure 9 shows that this graph possesses a Hamiltonian circuit H.

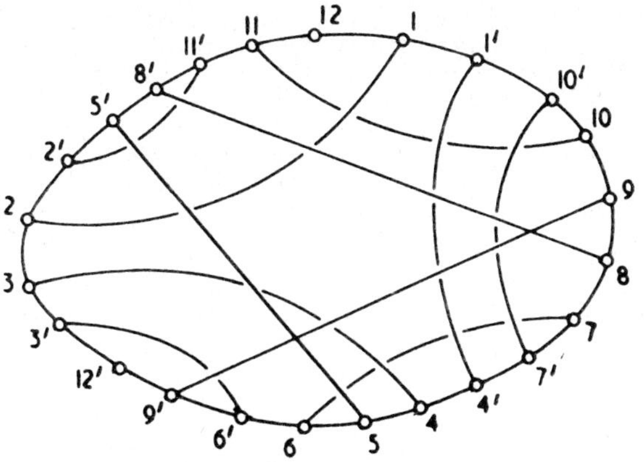

Fig. 9.

Now H together with chords (5,5'),(9,9'),(3,4) is $K_{3,3}$, so that if H is planarly embedded in some optimal drawing, then one of the spokes (5,5'),(9,9') must be crossed. Thus, H crosses itself and we assume that it intersects itself exactly once in two loops. In the induced planarization, we show that two further crossings cannot be avoided. By Remark 3, one of the crossed edges of H, e say, must be in segment S_1 = <8',11',...,7'>; otherwise, one of the spokes (8,8'),(9,9') is intersected. The other, f, say, must be in segment S_2 = <4,5,...,5'>; otherwise, one of (5,5'),(8,8') is crossed. Since parallel spokes (5,5'),(8,8') separate e and f for all (e,f) ε (S_1 \ (8,7)) x (S_2 \ (4,5)), other than spokes, we need only consider (4,5) $\in S_1$ and (8,7) ε S_2. In the first case, (3,4) must cross in each of the three distinct locations of vertices 8 and 9. A further crossing arises from either (6,7) or (1',4'). In the second case, (7',10') must cross and a further crossing arises from either (1',4') or (2',11'). We conclude that H intersects itself exactly twice in such a way that all other edges can be drawn in without further crossings.

As before, some crossed edge, e lies in S_1 and some edge f in S_2. If e crosses f, then by the above reasoning, either e = (4,5) or f = (8,7). If e = (4,5) and f ε S = <8',11',...,8>, then e and f are separated by one or other of the parallel pairs (8,8'),(6,7) and (8,8'),(3,4), so that if g and h are the crossed edges, then g = (4,5) or (5,6), by Remark 1 of Theorem 9. But g = (5,6) and h ε S are also separated by (8,8'),(5,5') which do not separate e and h, so that g = e and f is at distance at least 3 from h along H by Remark 2 of Theorem 9. The set of edge pairs in S satisfying these conditions and separated from (5,6) by precisely the same sets of parallel edges is seen to be empty. Similarly, f = (8,7) cannot cross any e ε <5,6,...,5'> $\cup$ (4,5). Thus, e crosses g $\neq$ f, g $\in$ T = <7',4',...,8'>, and f crosses h $\neq$ e, h ε U = <5',8',...,4>. We note that T $\cap$ U = (5',8'),,(4',7') and the spoke (4,4') which we ignore. If g = (8',5'), then g cannot cross any edge in the sub-segment <4,5,...,3>, since they are separated by parallel chords (2',11'),(3,4) and no edge in T is bounded by (3,4). If g crosses e = (2,3), then h is either (5',8') or (8',11') and f ε $\emptyset$, so that g $\neq$ (8',5'). If h = (8',5') crosses some edge in <7',7,...,11>, these are separated by (2',11') and one of (4',1'),(10,11) which bound no edge in <5',2',...,4>; thus (8',5') crosses no edge.

If g = (4',7'), then g crosses no edge in <4,5,...,5'> since these edges are separated by (1',4') and either (3,4) or (5,5'), which bound no edge in <7',7, ...,8'>. Similarly, h = (4',7') cannot cross any edge in <8,9,...,8'> since they are separated by (6,7),(8,8') which bound no edge in <4,5,...,5'>. If h = (4',7') and f = (7,8) then e and g cannot be separated by parallel edges since h and f are not. Thus (f,h) are either ((2,3),(5,6)) or ((2,3),(4,5)). In the unique planarization of each case (7',10') is necessarily crossed. Thus $\{e,g\} \subseteq$ <4,5,...,5'> and $\{f,h\} \subseteq$ <7',7,...,8'>. If e, say, is (5',2'), then $g \in$ <2,3,...,4> and f = (8',11'), $h \in$ <1,1',...,7'>. But (5',2') is separated from $g \in$ <3,3',...,4> by (2',11'),(3,4) which do not separate f, h, so that g = (2,3). But each h in <1,1',...,7'> is separated from (8',11') by (2',11') and another parallel edge which do not separate e and g, so that neither e nor g is (5',2'). If e = (2,3), then $g \notin$ <3',9',...,4> excluding <3',...,6'>, since these latter edges are separated from (2,3) by (3',6'),(3,4) which do not separate f,h. Thus g is either (6,5) or (5,4) both of which are separated from e by (6,7),(9,9'). But then no edge other than the spoke (7,7') qualifies as either f or h, so that neither e nor g lies in (2,3) $\cup$ <3',...,6'>. Since these edges account for all distinct possibilities in <5',2',...,4>, the result is proved. //

Comments and Problems:

It is known [3] that

(i) if $1 \leq k, m \leq n-1$ and $km \equiv 1 \pmod{n}$, then $P(n,m) \cong P(n,k)$;

(ii) $P(n,k) \cong P(n,n-k)$.

It follows from our conclusions that

$\nu(3h+1,3) = \nu(3h+1,h) = \nu(3h+1,2h+1) = \nu(3h+1,3h-2) = h+1$

$\nu(3h+2,3) = \nu(3h+2,h+1) = \nu(3h+2,2h+1) = \nu(3h+2,3h-1) = h+2$

The following table of known values for $\nu(n,k)$ can be drawn up:

k \ n	1	2	3	4	5	6	7	8	9	10	11	12	13	14
1			0	0	0	0	0	0	0	0	0	0	0	0
2			0	0	2	0	3	0	3	0	3	0	3	0
3				0	2	1	3	4	2	4	5	4	5	6
4					0	0	3	1	2	?	5	*	5	?
5						0	3	4	3	1	2	?	?	6
6							0	0	2	?	2	1	3	?
7								0	3	4	5	?	3	1
8									0	0	5	*	?	?
9										0	3	4	5	6
10											0	0	5	?
11												0	3	6
12													0	0
13														0

Regarding entries marked (*), the following can be said:

P(kt,t)

The drawing of P(kt,t) in which the kt-circuit is planarly drawn and the t"k-helms" are drawn successively alternately in the interior and exterior of the kt-circuit gives the following upper bound for the crossing number ξ_t:

$$\xi_{2n+1} = kn^2 \geq \nu(P(k(2n+1),2n+1))$$

$$\xi_{2n} = kn(n-1) \geq \nu(P(k(2n),2n)).$$

It is readily verified that if this estimate is valid for a particular odd value of t, then it is also valid for t+1. The same cannot be said for even t. (It is of interest to note that a similar situation obtains for the complete bipartite graphs:cf.[2]). We conclude that $\nu(4k,4) = 2k$.

References:

1. G. Exoo, F. Harary, J. Kabell, The Crossing numbers of some Generalized Petersen Graphs, Math. Scand. 48 (1981) 184-188.

2. R. Guy, The decline and fall of Zarankiewicz's Theorem, Proof Techniques in Graph Theory.(F. Harary, ed.)

3. M. Watkins, A Theorem on Tait Colourings..., J.C.T.(B) 6 (1969) 152-164.

Annals of Discrete Mathematics 30 (1986) 243–250

COMPLETE ARCS IN PLANES OF SQUARE ORDER

J.C. Fisher[1], J.W.P. Hirschfeld[2] and J.A. Thas[3]

[1]Department of Mathematics, University of Regina, Regina, Canada, S4S OA2.

[2]Mathematics Division, University of Sussex, Brighton, U.K. BN1 9QH.

[3]Seminar of Geometry and Combinatorics, University of Ghent, 9000 Gent, Belgium.

Large arcs in cyclic planes of square order are constructed as orbits of a subgroup of a group whose generator acts as a single cycle. In the Desarguesian plane of even square order, this gives an example of an arc achieving the upper bound for complete arcs other than ovals.

1. *INTRODUCTION*

Our aim is to demonstrate the existence of complete $(q^2 - q + 1)$-arcs in a cyclic projective plane $\Pi(q^2)$ of order q^2. The only such plane known is $PG(2,q^2)$, the plane over the field $GF(q^2)$. These arcs were found incidentally by Kestenband [5], using different methods, as one of the possible types of intersection of two Hermitian curves in $PG(2,q^2)$. The importance of these arcs, not observed in [5], is Segre's result that for q even, a complete m-arc in $PG(2,q)$ with $m < q + 2$ satisfies $m \le q - \sqrt{q} + 1$. Thus, this example of a complete arc attains the upper bound for q even. As a by-product of the investigation, it is shown that a Hermitian curve in $PG(2,q^2)$ is the disjoint union of $q + 1$ of these arcs.

2. *NOTATION*

Let $\Pi = \Pi(q^2)$ be a cyclic projective plane of order q^2. One can identify its points with the elements i of $\mathbb{Z}_v$, $v = q^4 + q^2 + 1$, so that the cyclic group is generated by the automorphism σ with $\sigma(i) = i + 1$, $i \in \mathbb{Z}_v$, [3], §4.2. The lines are obtained from a perfect difference set $\ell_0 = \{d_0, d_1, \ldots d_{q^2}\}$ as the sets $\sigma^j(\ell_0)$, $j = 0,1,\ldots, v - 1$.

Let $b = q^2 + q + 1$ and $k = q^2 - q + 1$; then $v = bk$. Since b and k are relatively prime, $\mathbb{Z}_v = \mathbb{Z}_b \times \mathbb{Z}_k$. For i in $\mathbb{Z}_v$, we write

$$i = (r,s) \quad \text{where } i \equiv r(\text{mod } b),\ i \equiv s(\text{mod } k).$$

In this notation $\sigma(i) = (r + 1, s + 1)$, where the sum of the first component is taken modulo b and the second modulo k. The notation extends in a natural way to any arithmetical operation in $\mathbb{Z}_v$.

By the multiplier theorem of Hall [2], q^3 is a multiplier of Π; this means that the mapping ψ given by $\psi(i) = q^3 i$ is an automorphism of Π. Since $q^6 \equiv 1 \pmod{v}$, so ψ is an involution. Indeed, ψ is a Baer involution since it fixes all b points of $k\mathbb{Z}_v = \{(r,0) : r \in \mathbb{Z}_b\}$; this is because $q^3 r - r = r(q^3 - 1) \equiv 0 \pmod{b}$. If we define

$$B_s = \{(r,s) : r \in \mathbb{Z}_b\} \text{ for } s = 0,1,\ldots, k-1,$$

then $\sigma(B_s) = B_{s+1}$ and the $q^2 - q + 1$ Baer subplanes B_s partition Π. A *line of a Baer subplane* B_s is a line of Π meeting B_s in $q + 1$ points. Similarly define

$$K_r = \{(r,s) : s \in \mathbb{Z}_k\} \text{ for } r = 0,1,\ldots, b-1,$$

whence $\sigma(K_r) = K_{r+1}$ and the K_r also partition Π. Thus $i = (r,s) = B_s \cap K_r$. It will turn out that K_r is a complete $(q^2 - q + 1)$-arc.

3. *COMPLETE k-ARCS*

LEMMA 3.1: For each $i = (r,s)$ *in* $\mathbb{Z}_v = \mathbb{Z}_b \times \mathbb{Z}_k$, *we have that* $\psi(i) = (r, k - s)$.

Proof: It was noted in §2 that ψ fixes the first component r of i. Now, for each s in $\mathbb{Z}_k$,

$$q^3 s + s = s(q+1)k \equiv 0 \pmod{k},$$

whence

$$q^3 s \equiv -s \equiv k - s \pmod{k}. \ \square$$

LEMMA 3.2: For any line ℓ *of the Baer subplane* B_s, *with* $(r,s) = B_s \cap K_r$,

$$|\ell \cap K_r| \text{ is } \begin{cases} \textit{odd if } (r,s) \in \ell \\ \textit{even if } (r,s) \notin \ell . \end{cases}$$

Proof: By lemma 3.1, the involution ψ fixes exactly one point of K_r, namely the point $(r,0)$ where it meets B_0; the other points of K_r are interchanged in pairs. If ℓ is a line of B_0 it is fixed by ψ, which implies that the number of points of $\ell \cap K_r$ outside B_0 is even. Thus the parity of $|\ell \cap K_r|$ varies as $\ell \cap K_r \cap B_0$ is empty or the point $(r,0)$. For a line ℓ of B_s, apply the same argument to $\sigma^{-s}(\ell)$, which is a line in B_0. $\square$

LEMMA 3.3: Let S *be an automorphism group that acts regularly on the points of some projective plane* $\Pi(n)$ *of order* n, *and suppose that* $V_0, V_1, \ldots, V_t$ *are the orbits of the points under the action of a subgroup* G *of* S. *If* ℓ *is a line of* $\Pi(n)$ *and* $\lambda_j = |\ell \cap V_j|$, *then*

$$\sum_{j=1}^{t} \lambda_j(\lambda_j - 1) = |G| - 1 .$$

Proof: To each of the $n^2 + n$ elements γ of $S\backslash\{1\}$ there corresponds a unique pair of points P, Q of ℓ for which $\gamma(P) = Q$; in fact, $P = \gamma^{-1}(\ell) \cap \ell$ and $Q = \ell \cap \gamma(\ell)$. If there was another such pair on ℓ, then S would not act regularly on the lines of $\Pi(n)$. Now we count the set

$$J = \{(P,Q,\gamma) : P, Q \in \ell ; \ \gamma \in G ; \ \gamma(P) = Q ; \ P \neq Q\}$$

in two ways. First, each γ other than the identity gives a unique pair (P,Q), whence $|J| = |G| - 1$. Second, ℓ is a disjoint union of the sets $\ell \cap V_j$, and to each pair (P,Q), $P \neq Q$, in $\ell \cap V_j$ there is a unique γ in G such that $\gamma(P) = Q$; hence $|J| = \sum_{\lambda_j > 1} \lambda_j (\lambda_j - 1)$ and so $|J| = \sum_{j=1}^{t} \lambda_j (\lambda_j - 1)$. □

We are now ready to prove the main result. In §4, an alternative proof is provided that makes use of the properties of perfect difference sets.

THEOREM 3.4: For $q > 2$, *each orbit* K_r *is a complete* k*-arc with* $k = q^2 - q + 1$ *in* $\Pi(q^2)$. *Furthermore, the lines through* $B_s \cap K_r = (r,s)$ *that lie in* B_s *are the* $q + 1$ *tangents to* K_r *at* (r,s).

Proof: Fix a Baer subplane B_s and let ℓ be one of its lines. For each orbit K_{r_j} $(j = 0,1,\ldots,q)$ that meets $\ell \cap B_s$, set $\alpha_j + 1 = |\ell \cap K_{r_j}|$; for the remaining orbits, set $\beta_j = |\ell \cap K_{r_j}|$, $j = q + 1, q + 2, \ldots, b - 1$. By lemma 3.2 both α_j and β_j are even.

By definition,

$$\sum_{j=q+1}^{b-1} \beta_j = q^2 + 1 - \sum_{j=0}^{q} (\alpha_j + 1) = q^2 - q - \sum_{j=0}^{q} \alpha_j .$$

By lemma 3.3,

$$\sum_{j=q+1}^{b-1} \beta_j (\beta_j - 1) + \sum_{j=1}^{q} (\alpha_j + 1) \alpha_j = q^2 - q ,$$

whence subtraction yields

$$\sum_{j=q+1}^{b-1} \beta_j (\beta_j - 2) = - \sum_{j=0}^{q} \alpha_j^{\,2} .$$

Since β_j is even, so $\sum_{j=q+1}^{b-1} \beta_j(\beta_j - 2) \geq 0$.

Hence $-\sum_{j=0}^{q} \alpha_j^2 \geq 0$ and $\alpha_j = 0$ for $j = 0,1,\ldots, q$.

Consequently $\beta_j \in \{0,2\}$ for $j \geq q + 1$.

Summarily, for any line ℓ of the subplane B_s, either

(i) $(r_j,s) \in \ell \cap K_{r_j}$ so that $\alpha_j = 0$, $|\ell \cap K_{r_j}| = 1$ and ℓ is tangent to K_{r_j} at the point $i = (r_j,s)$

or

(ii) $\ell \cap K_{r_j} \cap B_s = \emptyset$ and ℓ meets K_{r_j} in 0 or 2 points.

Since each line of Π is a line of exactly one of the subplanes B_s, it follows that no line meets K_{r_j} in more than two points; that is, K_{r_j} is a $(q^2 - q + 1)$-arc. From (i) it is clear that, for each point (r_j,s) of K_{r_j}, the $q + 1$ lines of B_s through (r_j,s) are the $q + 1$ tangents of K_{r_j} at this point.

For $q \geq 4$, a simple counting argument suffices to show that the k-arc K_r is complete. Assume the contrary. Then there is a point P through which pass $q^2 - q + 1$ tangents of K_r, one from each of its points. Since $P \in K_{r'}$ for some $r' \neq r$, it follows that through each point of $K_{r'}$ there are $q^2 - q + 1$ tangents of K_r (because $K_{r'}$ and K_r are orbits under the action of the group generated by σ^b). Since $K_{r'}$ is itself a k-arc, none of these tangent lines is counted more than twice, whence K_r has at least $\frac{1}{2}(q^2 - q + 1)^2$ tangents. But as K_r has exactly $(q + 1)(q^2 - q + 1)$ tangents, we have $\frac{1}{2}(q^2 - q + 1)^2 \leq (q + 1)(q^2 - q + 1)$, a contradiction for $q \geq 4$.

When $q = 3$, it must first be observed that PG(2,9) is the unique cyclic plane of order 9, Bruck [1]. Then the only 7-arc of PG(2,9) whose automorphism group contains an element of order 7 is a complete arc, [3], §14.7. The case $q = 2$ is a genuine exception: a 3-arc is never complete. □

Remark

A theorem of Segre [3], §10.3, states that a complete m-arc in PG(2,q), q even, is either an oval, that is a $(q + 2)$-arc, or $m \leq q - \sqrt{q} + 1$. So, for q an even square, theorem 3.4 gives an example of a complete $(q - \sqrt{q} + 1)$-arc and shows that Segre's theorem cannot be improved in this case. For q odd, [3], §10.4, the comparable theorem states that a complete m-arc in $PG(2,q)$, q odd, is either a conic, that is a $(q + 1)$-arc, or $m \leq q - \sqrt{q}/4 + 7/4$. This result has been slightly improved by the third author. However, the existence of a complete 8-arc in $PG(2,9)$ shows that $q - \sqrt{q} + 1$ is not the best bound for all odd q, [3], §14.7.

4. *LINES IN* $\Pi(q^2)$

THEOREM 4.1: Any line of B_0 *consists of*

(i) $q + 1$ *points of the form* $(d,0)$, *where* d *is an element of a perfect difference set* D *for* $\mathbb{Z}_b$;

(ii) $\frac{1}{2}(k - 1) = \binom{q}{2}$ *pairs of points of the form* (r_j,j) *and* $(r_j, k - j)$ *for* $j = 1,2,\ldots, \frac{1}{2}(k - 1)$, *with the* r_j *distinct elements of* $\mathbb{Z}_b \backslash D$.

Proof: B_0 is itself a cyclic plane of order q, since σ^k generates a cyclic group for B_0. Each of its lines ℓ therefore contains $q + 1$ elements $(d,0)$ where d is an element of a perfect difference set D for $\mathbb{Z}_b$.

A line ℓ of B_0 meets any other subplane B_s of the partition in exactly one point. Thus each element j of $\mathbb{Z}_k \backslash \{0\}$ occurs as the second component of exactly one point of ℓ. Lemma 3.1 shows that ψ interchanges (r_j,j) and $(r_j, k - j)$. Since ℓ is fixed by ψ, it follows that both j and $k - j$ are paired with the same element r_j of $\mathbb{Z}_b$.

It remains to show that $r_j \notin D$ and that no r_j in $\mathbb{Z}_b \backslash D$ can appear more than twice among the points of ℓ. This follows from the fact that the points of ℓ constitute a perfect difference set for $\mathbb{Z}_v$: each of the $k - 1$ differences of the form $i = (0,s)$, $s \neq 0$, must occur exactly once, and these are accounted for by the $k - 1$ differences $\pm((r_j,j) - (r_j, k - j))$. □

The description of a line of B_0 given in the theorem is essentially the alternative proof that K_r is a k-arc whose tangents are the lines through (r,s) that lie in B_s. The description of points of type (i) and (ii) shows that any line of B_0 is a tangent to those K_t that meet it in a point of B_0; it meets the other K_t in 0 or 2 points. The proof is completed by noting that $\sigma^s(B_0) = B_s$, which either coincides with B_0 or has no points or lines in common with it.

5. *HERMITIAN CURVES*

The only known cyclic planes are the Desarguesian ones and, in this section, we restrict our attention to $PG(2,q^2)$.

It is convenient to distinguish one line ℓ_0 of B_0 and define $\ell_j = \sigma^{-j}(\ell_0)$. Then i and ℓ_j are incident exactly when $i + j \pmod v$ is an element of ℓ_0. In particular, $\ell_0 \cap B_0 = \{i = (d,0) : d \in D\}$ as in (i) of theorem 4.1; now, D is a distinguished, perfect difference set for $\mathbb{Z}_b$.

THEOREM 5.1: *(i) The set* $H = \{(d/2,\ s) : d \in D,\ s \in \mathbb{Z}_k\}$ *is a Hermitian curve and is the disjoint union of the* $q + 1$ *complete* k*-arcs* $K_{d/2}$.

(ii) $H \cap B_s$ *is a conic or a line of* B_s *according as* q *is odd or even, whence* H *is a disjoint union of* k *subconics or* k *sublines accordingly.*

Proof: Define the correlations $\phi : i \leftrightarrow \ell_i$ and $\rho : (r,s) \leftrightarrow \ell_{(r,-s)}$. Then ϕ is an ordinary polarity for q odd and a pseudo polarity for q even, [3], §8.3. Thus, with ψ as in §2, we have that $\rho = \psi\phi = \phi\psi$. In fact, ρ is a Hermitian polarity since the self-conjugate points of ρ are the $q^3 + 1$ points (r,s) satisfying $(r,s) + (r,-s) = (d,0)$ for d in D. From this (i) follows.

In B_0 the points are $(r,0)$ while the lines are $\ell_{(r,0)}$, both with r in $\mathbb{Z}_b$. So B_0 is self-polar with respect to ρ and meets H in the $q + 1$ self-conjugate points of the polarity ϕ induced on B_0 by ρ. These self-conjugate points form a subconic when q is odd and a subline when q is even. Given s, there exists s' such that $bs' \equiv s \pmod k$ since b and k are coprime. Thus $H \cap B_s = \sigma^{bs'}(H \cap B_0) = \{(d/2,\ s) : d \in D\}$ is a conic or a line of B_s according as q is odd or even, and the last part of (ii) follows. □

THEOREM 5.2: The tangents to any complete $(q^2 - q + 1)$-arc *in* $PG(2,q^2)$, q *even, form a dual Hermitian arc.*

Proof: See Thas [6].

THEOREM 5.3: The tangents to any of the complete $(q^2 - q + 1)$-arcs K_r *in* $PG(2,q^2)$ *form a dual Hermitian curve if and only if* q *is even.*

Proof: Let q be even and consider the arc K_0, where D has been chosen so that $2D = D$ (which is always possible since 2 is a Hall multiplier and each multiplier of B_0 fixes at least one line of B_0). Then the tangents to K_0 at $(0,0)$, namely the lines of B_0 containing $(0,0)$, have the form $\ell_{(d,0)}$ with d in D. Since $\sigma^{bs'}$, which takes $(0,0)$ to $(0,s)$, takes $\ell_{(d,0)}$ to $\ell_{(d,-s)}$, the set of tangents to K_0 is $\{\ell_j : j = (d,s),\ d \in D,\ s \in \mathbb{Z}_k\}$. From the assumption that $D = 2D$ these lines are the self-conjugate lines of the polarity ρ determined by H. Thus the set of tangents to K_0 coincides with the set of tangents to H.

Now let q be odd. Since the number of tangents from a point P not in K_r to K_r has the parity of $q^2 - q + 1$ and so is odd, this number is never $q + 1$. Hence the tangents to K_r do not form a dual Hermitian arc. □

THEOREM 5.4: Each of the $(q^2 - q + 1)$-arcs K_r *is the intersection of two Hermitian curves.*

Proof: First, let q be even. Then as in theorem 5.1, the arc K_r is contained in a Hermitian curve H, which determines a polarity ρ. Let $\rho(H) = \hat{H}$ and let H^* be the dual Hermitian curve of theorem 5.3 that is formed by the tangents to K_r. Then $\rho(K_r) = H^* \cap \hat{H}$, whence $K_r = \rho(H^*) \cap H$.

Now, let q be even or odd. Then, as in theorem 5.1,

$$H = H_0 = \{(d/2,s) : d \in D,\ s \in \mathbb{Z}_k\}$$

is a Hermitian curve. Hence

$$H_r = \{(d/2 + r,s) : d \in D,\ s \in \mathbb{Z}_k\}$$

is also a Hermitian curve. In fact, since there exists r' such that $kr' \equiv r \pmod{b}$, we have that $H_r = \sigma^{kr'}(H_0)$. Since D is a perfect difference set in $\mathbb{Z}_b$, so $|H_{r_1} \cap H_{r_2}| = k$ for any $r_1 \neq r_2$. Also $H_{r_1} \cap H_{r_2} = K_t$, where $t = \tfrac{1}{2}d_1 + r_1 = \tfrac{1}{2}d_2 + r_2$, since there exist unique d_1 and d_2 in D such that $d_1 - d_2 \equiv 2(r_2 - r_1) \pmod{b}$. □

THEOREM 5.5: Let H *be a Hermitian curve in* $PG(2,q^2)$, q *even, and let* K *be an* m*-arc contained in* H.

(i) *If there is no* $(m+1)$*-arc in* H, *then*

 (a) $m = q^2 - q + 1$ *if* $q > 2$;

 (b) $m = 4$ *if* $q = 2$.

(ii) *If* $m = q^2 - q + 1$ *and* $q > 2$, *then* K *is complete.*

Proof: (i) Suppose $m > q^2 - q + 1$. Then by Segre's theorem ([3], theorem 10.3.3, corollary 2), K is contained in an oval O, that is a $(q^2 + 2)$-arc. Now, count the pairs (P,Q) such that $P \in K$, $Q \in O$, $P \neq Q$ and

PQ is tangent to H. There are at most two points P for a given Q, since three would be collinear. So

$$m \leq 2(q^2 + 2 - m) .$$

Hence $3m \leq 2q^2 + 4$, and $3q^2 - 3q + 3 < 2q^2 + 4$ implies that $q = 2$. This gives the result.

(ii) Suppose K is not complete, then the same argument as (i) gives

$$q^2 - q + 1 \leq 2(q + 1) ,$$

whence $q^2 - 3q - 1 \leq 0$; that is, $q = 2$. □

Remark: For q odd, the points of B_0 together with the $q^2 + q + 1$ conics $C_r = \{(\frac{1}{2}d + r, 0) : d \in D\}$, $r \in \mathbb{Z}_b$, form a plane of order q. This plane is isomorphic to PG(2,q) via the isomorphism θ given by $\theta(x,0) = (\frac{1}{2}x, 0)$. For all q, this configuration of conics also appears as the section by a plane π of the $q^2 + q + 1$ quadric surfaces through a twisted cubic T in PG(3,q), where π is skew to T; see [4], theorem 21.4.5.

REFERENCES

[1] Bruck, R.H., Quadratic extensions of cyclic planes, *Proc. Sympos. Appl. Math.* 10 (1960), 15-44.

[2] Hall, M., Cyclic projective planes, *Duke Math. J.* 14 (1947), 1079-1090.

[3] Hirschfeld, J.W.P. Projective Geometries over Finite Fields (Oxford University Press, Oxford, 1979).

[4] Hirschfeld, J.W.P., Finite Projective Spaces of Three Dimensions (Oxford University Press, Oxford, to appear).

[5] Kestenband, B., Unital intersections in finite projective planes, *Geom. Dedicata* 11 (1981), 107-117.

[6] Thas, J.A., Elementary proofs of two fundamental theorems of B. Segre without using the Hasse-Weil theorem, *J. Combin. Theory Ser. A.* 34 (1983), 381-384.

Annals of Discrete Mathematics 30 (1986) 251–262

ON THE MAXIMUM NUMBER OF *SQS(v)* HAVING A PRESCRIBED *PQS* IN COMMON*

Mario Gionfriddo[1], Angelo Lizzio[1], Maria Corinna Marino[2]

Summary. *We determine some results regarding the parameter $D(v,u)$, where $D(v,u)$ is the maximum number of $SQS(v)$s such that any two of them intersect in u quadruples, which occuring in each of the $SQS(v)$s.*

1. Introduction

A *partial quadruple system* (PQS) is a pair (P,s), where P is a finite set having v elements and s is a family of 4-subsets of P such that every 3-subset of P is contained in at most an element of s. If (P,s_1) and (P,s_2) are two PQSs, they are said to be *disjoint* and *mutually balanced* (DMB) if $s_1 \cap s_2 = \emptyset$ and any triple $\{x,y,z\} \subset P$ is contained in an element of s_1 if and only if it is contained in an element of s_2. If (P,s_1) and (P,s_2) are DMB, then $|s_1| = |s_2|$. If (P,s) is a PQS such that every 3-subset of P is contained in exactly one element of s, then (P,s) is said a *Steiner quadruple system* (SQS). The number $|P| = v$ is the *order* and it is well-known that an $SQS(v)$ there exists if and only if $v \equiv 2$ or 4 (mod. 6).

In what follows an $SQS(v)$ will be denoted by (Q,q). We have $|q| = q_v = v(v-1)(v-2)/24$.

On of the most important problem in the theory of SQSs is the determination of the parameter:

$$D(v,u) = Max\ \{h : \exists\ h\ SQS(v)\ (Q,q_1),\ldots,(Q,q_h) / q_i \cap q_j = A\ ,\quad \forall\ i,j\ ,\ i \neq j\ ,\ |A| = u\}\ .$$

---

*Lavoro eseguito nell'ambito del GNSAGA e con contributo del MPI (1983).

[1]Dipartimento di Matematica, Università, Viale A. Doria 6, 95125 Catania, Italy.

[2]Dipartimento di Matematica, Università, Via C. Battisti 90, 98100 Messina, Italy.

In [2] J. Doyen has pointed out this problem for Steiner triple systems.

In this paper we prove some results regarding $D(v,u)$ for $SQSs$.

2. Known results

Let (P,s) be a PQS . We will say that an element $x \in P$ has *degree* $d(x)=r$ if x belongs to exactly r quadruples of s . If $x,y \in P$, $x \neq y$, we will indicate by $(x,y)_r$ a pair $\{x,y\} \subset P$ contained in exactly k quadruples of s . We have $\sum_{x \in P} d(x) = 4|s|$.

The *degree-set* of a PQS (P,s) is the set $DS = [d(x),d(y),\dots]$ where $x,y,\dots$ are the elements of P . If r_i elements of P have degree h_i , for $i=1,2,\dots,p$, we will write $DS = \left[(h_1)_{r_1}, (h_2)_{r_2},\dots,(h_s)_{r_s}\right]$, where $r_1+\dots+r_s = |P|$.

If X is a finite set, let $K_{|X|}$ be the complete graph on X . An *1-factorization* (of $K_{|X|}$) on X is a family $F=\{F_1,\dots,F_h\}$, where F_i is a *factor* [1] of $K_{|X|}$ (on X) and, further, $F_i \cap F_j = \emptyset$ for every $i,j=1,2,\dots$, $i \neq j$. It is $h=|X|-1$. If $1 \leq h < |X|-1$, F is called a *partial* 1-factorization (of $F_{|X|}$) on X .

A partial 1-factorization $F^* = \{F_1^*,F_2^*,\dots,F_h^*\}$ on a set Y is *embedded* in an 1-factorization $F=\{F_1,\dots,F_k\}$ on X , if and only if $Y \subseteq X$, and every $F_i^* \in F^*$ is contained in a $F_j \in F$.

Let X and Y be two finite sets such that $|X|=|Y|=v$ and $X \cap Y = \emptyset$. If $F=\{F_1,\dots,F_{v-1}\}$ is an 1-factorization on X , $G=\{G_1,\dots,G_{v-1}\}$ an 1-factorization on Y , α a permutation on $\{1,2,\dots,v-1\}$, then (F,G,α) indicates the set of the quadruples $\{x_1,x_2,y_1,y_2\} \subseteq X \cap Y$ such that

$$\{x_1,x_2,y_1,y_2\} \in \Gamma(F,G,\alpha) \Leftrightarrow \begin{cases} \{x_1,x_2\} \in F_i \\ \{y_1,y_2\} \in G_j \\ \alpha(i)=j \end{cases} \quad \forall\, i,j=1,2,\dots,v-1$$

It is well-known that, if (X,A) and (Y,B) are two $SQS(v)$, with $X \cap Y = \emptyset$, then $(Q,q) = [X \cup Y](A,B,F,G,\alpha)$, where $Q = X \cup Y$ and $q = A \cup B \cup \Gamma(F,G,\alpha)$, is an $SQS(2v)$.

In [3], [4], [6] M. Gionfriddo has constructed, to within isomorphism, all *DMB PQS* having $m = 8,12,14,15$ (i.e. $m \leq 15$) quadruples.

These results are the following:

$m = 8$

$DS = [(4)_8]$

q_1	q_2
1,2,3,4	1,2,3,5
1,2,5,6	1,2,4,6
1,3,5,7	1,3,4,7
1,4,6,7	1,5,6,7
2,3,5,8	2,3,4,8
2,4,6,8	2,5,6,8
3,4,7,8	3,5,7,8
5,6,7,8	4,6,7,8

$m = 15$

$DS = [(7)_4,(4)_8]$

q_1	q_2
1,2,3,4	1,2,3,5
1,2,5,6	1,2,4,6
1,3,5,7	1,3,4,7
1,4,6,7	1,5,6,7
2,3,5,8	2,3,4,8
2,4,6,8	2,5,6,8
3,4,7,8	3,5,7,8
5,6,7,9	4,6,7,8
5,6,8,0	5,6,9,0
5,7,8,A	5,7,9,A
6,7,8,B	6,7,9,B
5,9,0,A	5,8,0,A
7,9,A,B	6,8,0,B
6,9,0,B	7,8,A,B
8,0,A,B	9,0,A,B

$m = 12$

$DS = [(6)_4,(4)_6]$

q_1	q_2
1,2,3,4	1,2,3,5
1,2,5,6	1,2,4,6
1,3,5,7	1,3,4,7
1,4,6,7	1,5,6,7
2,3,5,8	2,3,4,8
2,4,6,8	2,5,6,8
3,4,7,9	3,5,7,9
5,6,7,9	4,6,7,9
3,4,8,0	3,4,9,0
3,5,9,0	3,5,8,0
4,6,9,0	4,6,8,0
5,6,8,0	5,6,9,0

$DS = [(6)_6,(4)_3]$

q_1	q_2
1,4,5,6	1,4,5,7
1,4,7,8	1,4,6,8
1,5,7,9	1,5,6,9
1,6,8,9	1,7,8,9
2,4,5,7	2,6,8,9
2,6,7,8	2,4,5,9
2,6,5,9	2,4,7,8
2,4,8,9	2,6,5,7
3,4,6,8	3,5,7,9
3,5,6,7	3,4,5,6
3,4,5,9	3,4,8,9
3,7,8,9	3,6,7,8

$DS = [(6)_8]$

q_1	q_2
1,2,3,4	1,2,3,5
1,2,5,6	1,2,4,7
1,2,7,8	1,2,6,8
1,3,5,7	1,3,4,6
1,4,6,7	1,5,6,7
1,3,6,8	1,3,7,8
2,3,5,8	2,3,4,8
2,4,5,7	2,4,5,6
2,4,6,8	2,5,7,8
3,4,5,6	3,4,5,7
3,4,7,8	3,5,6,8
5,6,7,8	4,6,7,8

$m = 14$

$DS = [(7)_2, (6)_3, (4)_6]$

q_1	q_2
1,2,3,4	1,2,3,5
1,2,5,6	1,2,4,6
1,3,5,7	1,3,4,7
1,4,6,7	1,5,6,7
2,3,5,8	2,3,4,8
2,4,6,8	2,5,6,8
3,4,7,8	3,5,7,8
5,6,7,9	4,6,7,9
5,6,8,0	4,7,8,A
5,7,8,A	4,6,8,0
5,9,0,A	5,6,9,0
4,7,9,A	5,7,9,A
4,8,0,A	5,8,0,A
4,6,9,0	4,9,0,A

$DS = [(7)_4, (6)_2, (4)_4]$

q_1	q_2
1,2,3,4	1,2,3,5
1,2,5,6	1,2,4,6
1,3,5,7	1,3,4,7
1,4,6,7	1,5,6,7
2,3,5,8	2,3,4,8
2,4,6,8	2,5,6,8
3,4,7,9	3,4,9,0
3,4,8,0	3,6,8,0
3,6,9,0	3,5,7,8
3,6,7,8	3,6,7,9
5,6,8,0	4,6,7,8
5,6,7,9	5,6,9,0
4,5,9,0	4,5,8,0
4,5,7,8	4,5,7,9

$DS = [(7)_2, (6)_7]$

q_1	q_2
1,3,4,5	1,3,4,6
1,3,6,7	1,3,5,8
1,3,8,9	1,3,7,9
1,4,6,8	1,4,5,9
1,5,7,8	1,5,6,7
1,4,7,9	1,6,8,9
1,5,6,9	1,4,7,8
2,3,4,6	2,3,4,5
2,3,5,8	2,3,7,6
2,3,7,9	2,3,8,9
2,4,5,9	2,4,6,8
2,5,6,7	2,5,7,8
2,6,8,9	2,4,7,9
2,4,7,8	2,5,6,9

$DS = [(7)_8]$

q_1	q_2
1,2,3,4	1,2,3,5
1,2,5,6	1,2,4,7
1,2,7,8	1,2,6,8
1,3,5,7	2,4,5,8
1,4,7,6	2,5,6,7
1,3,6,8	2,3,7,8
1,4,5,8	2,3,4,6
2,3,5,8	1,3,4,8
2,4,5,7	1,4,5,6
2,4,6,8	1,5,7,8
2,3,6,7	1,3,6,7
3,4,5,6	3,4,5,7
3,4,7,8	4,6,7,8
5,6,7,8	3,5,6,8

3. The value of $D(v, q_v - m)$ for some classes of $SQS(v)$

We prove the following theorems.

THEOREM 3.1. *Let* $(P, s_1), \ldots, (P, s_h)$ *be* h *DMB PQS . If there exist an* $i \in \{1,\ldots,h\}$ *and a pair* $\{x,y\} \subset P$ *such that it is* $(x,y)_k$ *in* (P,s_i), *then* $h \leq 2k-1$.

Proof. It follows $(x,y)_k$ in (P,s_j), for every $j \in \{1,2,\ldots,h\}$.

If $\{x,y,a_{11},a_{12}\},\{x,y,a_{21},a_{22}\},\ldots,\{x,y,a_{k1},a_{k2}\} \in s_i$, let F_i be the 1-factors $\{\{a_{11},a_{12}\},\{a_{21},a_{22}\},\ldots,\{a_{k1},a_{k2}\}\}$. It follows that $F = \{F_1,F_2,\ldots,F_h\}$ is a partial 1-factorization of K_{2k} on the set $A = \{a_{11},a_{12},a_{21},a_{22},\ldots,a_{k1},a_{k2}\}$ (in the case $h = 2k-1$ F is exactly an 1-factorization of K_{2k} on A).

Since the set of the 1-factors F_i can (at most) be an 1-factorization on $2k-1$ elements, it follows $h \leq 2k-1$ necessarily.

We have $|F| = h \leq 2k-1$. ■

THEOREM 3.2. *Let* X *and* Y *be two sets such that* $|X| = |Y| = 2k$ *and* $X \cap Y = \emptyset$. *Further, let* F *and* G *be two 1-factorizations of* K_{2k} *on* X *and* Y *respectively, and let* α *be a permutation on* $\{1,2,\ldots,2k-1\}$. *If there exists an* $SQS(v)$ *containing* $\Gamma(F,G,\alpha)$, *then* $D(v,q_v-k^2(2k-1)) \geq 2k-1$.

Proof. Let (Q,q) be an $SQS(v)$ containing the family (F,G,α). It is

$$|\Gamma(F,G,\alpha)| = k^2(2k-1) .$$

If

$$\alpha_0 = \alpha = \begin{pmatrix} 1 & 2 & \ldots & 2k-1 \\ \\ a_1 & a_2 & \ldots & a_{2k-1} \end{pmatrix}, \quad \alpha_i = \begin{pmatrix} 1 & 2 & \ldots & 2k-1 \\ \\ a_1+i & a_2+i & \ldots & a_{2k-1}+i \end{pmatrix} \quad [a_j+i \in Z_{2k-1}]$$

for $i = 1,2,\ldots,2k-2$, then the quadruples of the families $\Gamma(F,G,\alpha_i)$, where $i = 0,1,2,\ldots,2(k-1)$, form $2k-1$ *DMB PQS* $(P,s_0),(P,s_1),\ldots$ $\ldots,(P,s_{2(k-1)})$, all embeddable in an $SQS(v)$. Hence $D(v,q-k^2(2k-1)) \geq 2k-1$. ■

THEOREM 3.3. *If* $k \in N$ *is such that* $2k \equiv 2$ *or* 4 (mod. 6), *then* $D(4k,q_{4k}-k^2(2k-1)) \geq 2k-1$.

Proof. If $k \in N$ is such that $2k \equiv 2$ or 4 (mod. 6), then it is possible to construct an SQS of order $2k$. Let (Q_1, q_1) and (Q_2, q_2) be two $SQS(2k)$ with $Q_1 \cap Q_2 = \emptyset$, and let F and G be two 1-factorizations of K_{2k} on Q_1 and Q_2 respectively. From Theorem 3.2 we can construct exactly $2k-1$ $SQS(4k)$ having $q_{4k} - k^2(2k-1)$ quadruples in common. Hence

$$D(4k, q_{4k} - k^2(2k-1)) \geq 2k-1 . \blacksquare$$

THEOREM 3.4. *For every* $k \in N$, $k \geq 2$, *let* $w = min\{v \in N : v \geq 4k,\ v \equiv 2$ *or* 4 (mod. 6)$\}$. *It follows* $D(2w, q_{2w} - k^2(2k-1)) \geq 2k-1$.

Proof. Let X and Y be two finite sets with $|X| = |Y| = 2k \geq 4$ and $X \cap Y = \emptyset$. Further, let $F = \{F_i\}_{i=1,\dots,2k-1}$ and $G = \{G_i\}_{i=1,\dots,2k-1}$ be two 1-factorizations of K_{2k} on X and Y respectively. From Theorem 8 of [8], there exists an 1-factorizations $F' = \{F'_i\}$ on a set X', such that $X \subset X'$, $|X'| \geq 2|X| = 4k \geq 8$, with F embedded in F'. Let $|X'| = w = min\{v \in N : v \geq 4k \geq 8,\ v \equiv 2$ or 4 (mod. 6)$\}$. If Y' is a set, containing Y, such that $X' \cap Y' = \emptyset$ and $|Y'| = |X'|$, φ a bijection $X' \to Y'$, $G' = \{G'_i\}_{i=1,\dots,w}$ the 1-factorization on Y' such that

$$\{x, y\} \in G'_i \Leftrightarrow \{\varphi^{-1}(x), \varphi^{-1}(y)\} \in F'_i ,$$

then G is embedded in G'. Further, if (X', q_1), (Y', q_2) are two $SQS(w)$ and α is a permutation of $\{1, 2, \dots, w-1\}$, then we can construct an $SQS(2w) = [X' \cup Y'](q_1, q_2, F', G', \alpha)$ containing $\Gamma(F', G', \alpha)$.

From Theorem 3.2 it follows $D(2w, q_{2w} - k^2(2k-1)) \geq 2k-1$. $\blacksquare$

COROLLARY. *From the same hypotheses of Theorem* 3.4 *it follows* $D(2v, q_w - k^2(2k-1)) \geq 2k-1$, *for every* $v \geq w$, $v \equiv 2$ *or* 4 (mod. 6).

Proof. The statement follows from proof of Theorem 3.4 and from Theorem 8 of [8]. $\blacksquare$

From previous theorems we have the following scheme:

k	$\omega \geq 4k$	$q_{2\omega}-k^2(2k-1)$	$D(2\omega,q_{2\omega}-k^2(2k-1))$	$v \geq \omega$, $v \equiv 2$ or 4 (mod. 6)	
2	8	$q_{16}-12$	≥ 3	$D(v,q_v-12) \geq 3$	, $\forall\, v \geq 16$
3	14	$q_{28}-45$	≥ 5	$D(v,q_v-45) \geq 5$	, $\forall\, v \geq 28$
4	16	$q_{32}-112$	≥ 7	$D(v,q_v-112) \geq 7$	, $\forall\, v \geq 32$
5	20	$q_{40}-225$	≥ 9	$D(v,q_v-225) \geq 9$	, $\forall\, v \geq 40$
6	26	$q_{52}-396$	≥ 11	$D(v,q_v-396) \geq 11$	, $\forall\, v \geq 52$
7	28	$q_{56}-637$	≥ 13	$D(v,q_v-637) \geq 13$	, $\forall\, v \geq 56$
8	32	$q_{64}-960$	≥ 15	$D(v,q_v-960) \geq 15$	, $\forall\, v \geq 64$
.	..	....	...	....	....
.	..	....	...	....	....

It is easy to see that $\lim_{v \to +\infty} D(v,q_v-k^2(2k-1)) = +\infty$.

4. The value of $D(v,q_v-m)$ for $m = 8,14,15$ and the value of $D(8,q_8-12)$

In this section we determine $D(v,q_v-m)$ for $m = 8,14,15$ and $D(8,q_8-12)$.

THEOREM 4.1. *Let* (P,s_i) *be* h *DMB PQS (for* $i = 1,2,\dots,h$*). If in a* s_i $(i \in \{1,\dots,h\})$ *there exist three elements* $x,y,z \in P$ *and a quadruple* b *such that* $(x,y)_2,(x,z)_2,\{x,y,z\} = b \in s_i$, *then* $h \leq 2$.

Proof. From Theorem 2.1 it is $h \leq 3$. If

$$\begin{matrix} \{x,y,z,a\} & & \{x,y,z,b\} & \\ \{x,y,b,c\} \in s_1 & , & \{x,y,a,c\} \in s_2 & , \\ \{x,y,b,d\} & & \{x,z,a,d\} & \end{matrix}$$

and $h = 3$, then $\{x,y,z,c\},\{x,y,a,b\} \in s_3$. It follows $\{x,z,c,e\} \in s_1$, hence $(x,z)_{\geq 3}$.

THEOREM 4.2. *It is not possible to construct three DMB PQS with* $m = 8,14,15$ *quadruples.*

Proof. It is easy to see that in the unique pairs of *DMB PQS* with $m = 8$ and $m = 15$, and in the pairs of *DMB PQS* with $m = 14$ and $DS = [(7)_2, (6)_7]$, $DS = [(7)_4, (6)_2, (4)_4]$, $DS = [(7)_2, (6)_3, (4)_6]$, there exist (in every case) at least three elements $x, y, z \in P$ such that $(x,y)_2$, $(x,z)_2$ and a quadruple $b = \{x,y,z\}$ (see § 2). Therefore, from Theorem 4.1, in these cases it is $h = 2$. Consider the case $m = 14$ and $DS = [(7)_8]$. If it is $h = 3$, since

s_1	s_2
1,2,3,4	1,2,3,5
1,2,5,6	1,2,4,7
1,2,7,8	1,2,6,8
1,3,5,7	1,3,4,8
1,4,6,7	1,4,5,6
1,3,6,8	1,3,6,7
1,4,5,8	1,5,7,8 ,

then $\{1,2,3,x\} \in s_3$ with $x \in \{6,7,8\}$. But, $x = 6$ [resp. $x = 7$] implies $\{1,2,4,8\}, \{1,2,5,7\} \in s_3$ [resp. $\{1,2,4,6\}, \{1,2,5,8\} \in s_3$], with $\{1,4,6,y\} \in s_3$ [$\{1,3,8,y\} \in s_3$] and $y \notin \{1,2,\dots,8\}$. From $x = 8$ it follows $\{1,3,5,6\}, \{1,3,4,7\} \in s_3$, with $\{1,5,8,y\} \in s_3$ and $y \notin \{1,2,\dots,8\}$. Therefore, it is $h = 2$. ∎

THEOREM 4.3. *There exist three DMB PQS with* $m = 12$ *quadruples. Their degree-set is* $DS = [(6)_8]$.

Proof. In the pairs of *DMB PQS* with $m = 12$ having $DS = [(6)_6, (4)_3]$ or $DS = [(6)_4, (4)_6]$ there exist three elements x, y, z such that: $(x,y)_2$, $(x,z)_2$ and a quadruple $b = \{x,y,z\}$ (see § 2). Therefore, for them it is $h = 2$ (Theor. 4.1). Consider the last pair of *DMB PQS* with $m = 12$. It has degree-set $DS = [(6)_8]$. Let $F = \{F_1, F_2, F_3\}$, $G = \{G_1, G_2, G_3\}$ be the following *1*-factorizations of K_4 on $X = \{1,4,5,8\}$ and $Y = \{2,3,6,7\}$ respectively:

$F =$

F_1	F_2	F_3
1,4	1,5	1,8
5,8	4,8	4,5

$G =$

G_1	G_2	G_3
2,3	2,6	2,7
6,7	3,7	3,6

Further, let α_i be the following permutations on $\{1,2,3\}$

$$\alpha_1 = \begin{pmatrix} 1 & 2 & 3 \\ 1 & 2 & 3 \end{pmatrix}, \quad \alpha_2 = \begin{pmatrix} 1 & 2 & 3 \\ 3 & 1 & 2 \end{pmatrix}, \quad \alpha_3 = \begin{pmatrix} 1 & 2 & 3 \\ 2 & 3 & 1 \end{pmatrix}.$$

If $P=\{1,2,\dots,8\}$, we can verify that $(P,\Gamma(F,G,\alpha_1))$, $(P,\Gamma(F,G,\alpha_2))$, $(P,\Gamma(F,G,\alpha_3))$ are three DMB PQS with $m=12$ and $DS=[(6)_8]$.

Further, $\Gamma(F,G,\alpha_1)$ and $\Gamma(F,G,\alpha_2)$ are the two families indicated in § 2. Since it is $(1,4)_2$, it follows that it is not possible to construct $h \geq 3$ DMB PQS with $m=12$ and $DS=[(6)_8]$. Hence, it follows the statement.■

THEOREM 4.4. $D(8,q_8-12)=3$.

Proof. Let (X,A) and (Y,B) be two $SQS(4)$, where $X=\{1,4,5,8\}$ and $Y=\{2,3,6,7\}$. Further, let $F=\{F_1,F_2,F_3\}$, $G=\{G_1,G_2,G_3\}$ be the two 1-factorizations of K_4 on X and Y, and let α_i $(i=1,2,3)$ be the permutations, defined in Theorem 4.3. It is known [9] that the pair $(Q,q_i)=[X\cup Y](A,B,F,G,\alpha_i)$, for every $i=1,2,3$, in an $SQS(8)$.

We have:

q_1	q_2	q_3
1,4,5,8 2,3,6,7	1,4,5,8 2,3,6,7	1,4,5,8 2,3,6,7
1,4,2,3 1,4,6,7 5,8,2,3 5,8,6,7	1,4,2,7 1,4,3,6 5,8,2,7 5,8,3,6	1,4,2,6 1,4,3,7 5,8,2,6 5,8,3,7
1,5,2,6 1,5,3,7 4,8,2,6 4,8,3,7	1,5,2,3 1,5,6,7 4,8,2,3 4,8,6,7	1,5,2,7 1,5,3,6 4,8,2,7 4,8,3,6
1,8,2,7 1,8,3,6 4,5,2,7 4,5,3,6	1,8,2,6 1,8,3,7 4,5,2,6 4,5,3,7	1,8,2,3 1,8,6,7 4,5,2,3 4,5,6,7

We can see immediately that:

$$q_1 \cap q_2 = q_1 \cap q_3 = q_2 \cap q_3 \quad , \quad |q_i \cap q_j| = 2$$

for every $i,j \in \{1,2,3\}$, $i \neq j$.

Hence $D(8,q_8-12) \geq 3$. From Theorem 3.1, in the case $k=2$, we have $D(8,q_8-12)=3$.

THEOREM 4.5. *We have* $D(v,q_v-8)=D(v,q_v-14)=D(v,q_v-15)=2$, *for every* $v=2^{n+2}$, $v=5\cdot 2^n$, $v=7\cdot 2^n$, *and* $n \geq 2$.

Proof. Since for $v=2^{n+2}$, $v=5\cdot 2^n$, $v=7\cdot 2^n$ it is possible to construct at least two $SQS(v)$ with q_v-8 or q_v-14 or q_v-15 quadruples in common (see [6], [7], [13]), the statement follows from Theorem 4.2, directly.∎

REFERENCES

[1] C. Berge, *Graphes et hypergraphes*, Dunod, Paris, 1970.

[2] J. Doyen, *Constructions of disjoint Steiner triple systems*, Proc. Amer. Math. Soc., 32 (1972), 409-416.

[3] M. Gionfriddo, *On some particular disjoint and mutually balanced partial quadruple systems*, Ars Combinatoria, 12 (1981), 123-134.

[4] M. Gionfriddo, *Some results on partial Steiner quadruple systems*, Combinatorics 81, Annals of Discrete Mathematics, 18 (1983), 401-408.

[5] M. Gionfriddo, *On the block intersection problem for Steiner quadruple systems*, Ars Combinatoria, 15 (1983), 301-314.

[6] M. Gionfriddo, *Construction of all disjoint and mutually balanced partial quadruple systems with 12 , 14 or 15 blocks*, Rendiconti del Seminario Matematico di Brescia, 7 (1984), 343-354.

[7] M. Gionfriddo and C.C. Lindner, *Construction of Steiner quadruple systems having a prescribed number of blocks in common*, Discrete Mathematics, 34 (1981), 31-42.

[8] C.C. Lindner, E. Mendelsohn, and A. Rosa, *On the number of 1-factorizations of the complete graph*, J. of Combinatorial Theory, 20 (B) (1976), 265-282.

[9] C.C. Lindner and A. Rosa, *Steiner quadruple systems - A survey*, Discrete Mathematics, 22 (1978), 147-181.

[10] A. Lizzio, M.C. Marino, F. Milazzo, *Existence of* $S(3,4,v)$, $v=5\cdot 2^n$ *and* $n\geq 3$, *with* q_v-21 *and* q_v-25 *blocks in common*, Le Matematiche

[11] A. Lizzio, S. Milici, *Constructions of disjoint and mutually balanced partial Steiner triple systems*, Boll. Un. Mat. Ital. (6) 2-A (1983), 183-191.

[12] A. Lizzio, S. Milici, *On some pairs of partial triple systems*, Rendiconti Ist. Mat. Un. Trieste, (to appear).

[13] G. Lo Faro, *On the set* $J(v)$ *for Steiner quadruple systems of order* $v=7\cdot 2^n$ *with* $n\geq 2$, Ars Combinatoria, 17 (1984), 39-47.

[14] A. Rosa, *Intersection properties of Steiner quadruple systems*, Annals of Discrete Mathematics, 7 (1980), 115-128.

Annals of Discrete Mathematics 30 (1986) 263–268

ON FINITE TRANSLATION STRUCTURES WITH PROPER DILATATIONS

Armin Herzer

Fachbereich Mathematik
Johannes Gutenberg-Universität
Mainz, Germany

Recently, Biliotti and the author obtained a certain number of results on translation structures with proper dilatations including structure- and characterisation-theorems, which here will be reformulated in a different manner, throwing a new light on some of the regarded questions.

1. GROUPS OF EXPONENT p AND CLASS ≤ 2.

Let K be a (commutative) field of characteristic $p > 0$ with automorphism γ and V a vector space over K. For a subspace W of V we consider mappings $f: V \times V \to W$ with property (*); namely f is alternating, vanishing on $V \times W$ and bisemilinear with automorphism γ, i.e. f satisfies the following conditions:

(*) (i) $f(u,v) = -f(v,u)$

(ii) $f(u_1+u_2,v) = f(u_1,v)+f(u_2,v)$

(iii) $f(uk,v) = f(u,v)k^\gamma$

(iv) $f(u,u) = 0 = f(u,w)$

for all $u,u_1,u_2,v \in V$, $w \in W$, $k \in K$.

Clearly f is bilinear iff $\gamma=1$.

A group $G = (G,\cdot)$ is called of exponent p, if $x^p = 1$ for all x G holds, and G is called of (nilpotency) class ≤ 2, if the commutator subgroup of G is contained in the center of G:

$$G' \leq Z(G).$$

We define a multiplication $\circ$ on V by

$$x \circ y := x+y+f(x,y) \quad \text{for all} \quad x,y \in V.$$

We write (V,f) for the structure consisting of the set V and the multiplication $\circ$ on it, where f has property (*).

PROPOSITION 1. $G = (V,f)$ is a group of exponent p and class ≤ 2.

Proof: The neutral element is o, the inverse of x is -x, and an easy computation shows

$$(x \circ y) \circ z = x+y+z+f(x,y)+f(x,z)+f(y,z) = x \circ (y \circ z).$$

Moreover $x^n = x+\dots+x$ (n times) holds and so $x^p = o$, since K has characteristic p. At last for the commutator of x and y we have

$$[x,y] = x^{-1} \circ y^{-1} \circ x \circ y = 2f(x,y),$$

and so $G' \leq W \leq Z(G)$ is valid.

Conversely the following is true:

PROPOSITION 2. Every group of prime exponent p and class ≤ 2 is isomorphic to a group (V,f) as defined before.

Proof: Let G be such a group. We define an abelian group $\overline{G} = (G,+)$ in the following manner. For p=2 let be $x+y=xy$; for $p\neq 2$ we define

$$x+y := xy[x,y]^{\frac{p-1}{2}} \qquad \text{for all } x,y\in G.$$

Then $\overline{G}$ is a K-vector space for some field K of characteristic p (at least K=GF(p)). Defining

$$f(x,y) := [x,y]^{\frac{p+1}{2}} \qquad \text{for p odd,}$$

and f(x,y)=o for p=2, the mapping $f: G\times G \to G'$ has property (*) with $\gamma=1$, and $G=(\overline{G},f)$ holds.

It is easy to construct such mappings f with property (*). Let V,W,K, γ be as before and $v_1,\ldots v_h$ a base of a complement of W in V. We choose elements $w_{ij}\in W$ for $1\leq i<j\leq h$. Given $u,v\in V$ we have $x_i,y_j\in K$ and $w,w'\in W$, such that

$$u = w + \sum_{i=1}^{h} v_i x_i, \qquad v = w' + \sum_{j=1}^{h} v_j y_j$$

holds. Then we define

$$f(u,v) = \sum_{1\leq i<j\leq h} w_{ij}(x_i y_j - x_j y_i)^{\gamma},$$

and obtain f with property (*).

For p an odd prime there is also a connection between such groups and the kinematic algebras: We look for A a local K-algebra with $A= K \oplus M$, where M is the maximal ideal of A satisfying $x^2=0$ for all $x\in M$. Then defining f on M by f(a,b)=ab for all $a,b\in M$ one can show that the bilinear map f has property (*) with $\gamma=1$, since char $K \neq 2$. On the other hand for A* the group of units of A the factor group $A^*/K^*=\{(1+a)K^* \mid a\in M\}$ is a two-sided affine linearly fibered incidence group[1], and is isomorphic to (M,f). For, since (1+a)(1+b)=1+a+b+f(a,b) holds, the mapping $M \to A^*/K^*;\ a \sim (1+a)K^*$ is an isomorphism of groups, ([6]). Conversely from the preceeding results every group of prime exponent p and class ≤ 2 has (in at least one way) the structure of such an affine incidence group as cited before.

2. GROUPS WITH A PARTITION π AND A NON TRIVIAL π-AUTOMORPHISM.

A (non-trivial) partition π of a group G is a system of subgroups of G called <u>components</u> with the properties:

(1) G and {1} are no components

(2) every element of G different from 1 is contained in exactly one component.

An endomorphism α of G is called a π-endomorphism, if $U^{\alpha} \leq U$ does hold for every component U of π.

For an abelian group G with partition π we have the following RESULT 1 (André,[1]). The π-endomorphisms form a ring K without zero-divisors. So in the finite case K is a field (called the kernel of π) and G is a K-vector space, whereas the components are K-subspaces.

More general we have

PROPOSITION 3 (Biliotti/Scarselli [4], Herzer [7]). Let G be a finite group with a partition π and a non-trivial π-automorphism α. Then G is of prime exponent p and class ≤ 2.

Now suppose we have a group G = (V,f) for some K-vector space V, and there is some $a \in K$ with $0 \neq a \neq 1$, and α defined by

$$v\alpha = va \qquad \text{for every } v \in V$$

is π-automorphism of G for a partition π of G. This forces $(x \circ y)a = (xa) \circ (ya)$ and so

$$f(x,y)a = f(xa,ya) = f(x,y)a^{2\gamma}.$$

Thus, if G is non-abelian and therefore $f(x,y) \neq o$ for appropriate x,y, we have $a = a^{2\gamma}$ which is only possible for $\gamma \neq 1$.
(For every odd number n one can construct such an α of period n for an infinite series of finite fields K and suitable automorphism γ of K, see [7], pg.387.)

In a certain sense also the converse of the above statement is true:
PROPOSITION 4 (Herzer [7]). Let G be a finite non-abelian group with partition π and non-trivial π-automorphism α. Then there is a field K, an automorphism γ of K, a K-vector space V and a mapping f with property (*), such that G is isomorphic to (V,f). Moreover there is some $a \in K$ with $v\alpha = va$ for all $v \in V$, and every component is a K-subspace of V.

3. THE GEOMETRIES BELONGING TO THOSE GROUPS.

(Most of the following ideas can be found in Biliotti/Herzer [3].)
Let P be a set the elements of which are called points, and $\mathcal{B}$ a set of subsets of P called blocks with the property, that any two points x and y are contained in exactly one block denoted by (x,y) and moreover every block contains at least two elements and is different from P. We call $\mathfrak{S} = (P,\mathcal{B},\parallel)$ a <u>parallel structure</u>, if $\parallel$ is a parallelism of $\mathfrak{S}$, i.e. an equivalence relation on $\mathcal{B}$ the classes of which are partitions of P. A permutation σ of P is called a <u>dilatation</u>, if for every block B also $B\sigma$ and $B\sigma^{-1}$ are blocks and moreover $B \parallel B\sigma$ holds. A dilatation is called a <u>translation</u>, if it has no fixed points or is the identity map, and is called a <u>proper</u> dilatation, if it has exactly one fixed point. A parallel structure $\mathfrak{S} = (P,\mathcal{B},\parallel)$ is called a <u>translation structure</u>, if there is a group of translations of $\mathfrak{S}$ acting transitively on P.

Let G be a group with a partition π.Then $\mathfrak{S}(G,\pi) = (G,\mathcal{B},\parallel)$ is a translation structure, where $\mathcal{B}$ is defined to be the set of all right cosets of the components and any two blocks are parallel, if they are right cosets of the same component:

$$\mathcal{B} = \{Ux \mid U \in \pi,\ x \in G\}; \qquad Uy \parallel Wz \Leftrightarrow U = W.$$

A transitive translation group of $\mathfrak{S}(G,\pi)$ is G acting on the pointset G by right multiplication.

PROPOSITION 5. $\mathfrak{S}(G,\pi)$ possesses proper dilatations if and only if G possesses at least one non-trivial π-automorphism.
(So far we have repeated basic concepts, which can be found in André's paper [1] and as a short survey also e.g. at the beginning of Biliotti [2].)

Now, by the preceding results, if G is finite and possesses a non-trivial π-automorphism, then G is of prime exponent p and class ≤ 2; moreover G has the structure of a K-vector space such that the components are subspaces. Then the concept of parallel structure leads to the following definitions:

The parallel structure $\mathcal{S} = (P,\mathcal{B},\|)$ is called affinely embedded, if there is an affine space $\mathcal{A}$ such that P is the set of points of $\mathcal{A}$ and $\mathcal{B}$ is a set of affine subspaces of $\mathcal{A}$(, whereas $\|$ in general is not the natural parallelism of $\mathcal{A}$). For different points x,y of $\mathcal{A}$ the line of $\mathcal{A}$ joining x and y is expressed by $x \vee y$.

Let $\mathcal{R}$ be the set of all lines of $\mathcal{A}$ and $\|r$ a parallelism on $\mathcal{R}$. For $B \in \mathcal{B}$ and $x \in P$ define $C = \Pi(x,B)$ by $C \in \mathcal{B}$ and $B \| C$ and $x \in C$. Similarly for $R \in \mathcal{R}$ and $y \in P$ define $S = \Pi r(y,R)$ by $S \in \mathcal{R}$ and $R \|r S$ and $y \in S$. Let the parallel strucure $\mathcal{A} = (P,\mathcal{B},\|)$ be affinely embedded in the affine space $\mathcal{A}$ as defined before. A parallelism $\|r$ on $\mathcal{R}$ is called compatible with $\mathcal{S}$, if for every $B \in \mathcal{B}$ and $R \in \mathcal{R}$ with $|B \cap R| = 1$ the following hold:

(i) $|\Pi(x,B) \cap \Pi r(y,R)| = 1$ for every $x \in R$ and every $y \in B$

(ii) Let $x \in R$ and $y_i \in B$ and define z_i by $\{z_i\} = \Pi(x,B) \cap \Pi r(y_i,R)$ for $i=1,2,3$. Then y_1,y_2,y_3 are collinear in $\mathcal{A}$ if and only if z_1, z_2,z_3 are collinear in $\mathcal{A}$.

(If an affinely embedded parallel structure $\mathcal{S} = (P,\mathcal{B},\|)$ possesses a compatible parallelism $\|r$ on $\mathcal{R}$ for $B \in \mathcal{B}$ and $R \in \mathcal{R}$ with $|B \cap R| = 1$, the set

$$\bigcup_{x \in R} \Pi(x,B) = \bigcup_{y \in B} \Pi r(y,R)$$

is a generalized affine Segre variety, and the set $\{X \in \mathcal{B} \mid X \| B,\ X \cap R \neq \emptyset\}$ is the projection of an affine d-regulus, where d is the dimension of B in $\mathcal{A}$, as defined in Herzer [8].)

PROPOSITION 6. An affinely embedded parallel structure $\mathcal{S}$ possesses at most one parallelism $\|r$ on $\mathcal{R}$ beeing compatible with $\mathcal{S}$.

The parallel structure $\mathcal{S} = (P,\mathcal{B},\|)$ is called an affine microcentral translation structure (of odd order), if $\mathcal{S}$ in an embedding in an affine space $\mathcal{A}$ (of odd order) is a translation structure with translation group T, such that the elements of T also are affinities of $\mathcal{A}$ and there is a parallelism $\|r$ on $\mathcal{R}$ which satisfies for every $\tau \in T$ with $\tau \neq 1$ the condition

$$x \vee x\tau \;\|r\; y \vee y\tau \qquad \text{for all } x,y \in P.$$

PROPOSITION 7. For an affine microcentral translation structure $\mathcal{S}$ the parallelism $\|r$ on $\mathcal{R}$ as given by the definition is compatible with $\mathcal{S}$.

Let $\mathcal{S}$ be a finite affinely embedded parallel structure possessing a parallelism $\|r$ on $\mathcal{R}$ compatible with $\mathcal{S}$. We define the following incidence proposition (D) (- a kind of Desargues' theorem -):

(D) For i=1,2,3 let x_i,y_i be points no three of which are collinear.
If then $x_1 \vee y_1 \;\|r\; x_j \vee y_j$ and
$(x_1,x_j) \| (y_1,y_j)$ hold for j=2,3,
so also $(x_2,x_3) \| (y_2,y_3)$.

PROPOSITION 8. Under the conditions for $\mathcal{S}$ given in the preceding section $\mathcal{S}$ is an affine microcentral translation structure, if and only if (D) is valid for $\mathcal{S}$.

By omitting the more complicated case p=2 (which can be found in [3]) all this together gives the following characterisation:

THEOREM. For the finite parallel structure $\mathfrak{S}$ the following statements are equivalent:

a) There is an affine embedding of $\mathfrak{S}$ in an affine space of odd order, such that $\mathfrak{S}$ possesses a compatible parallelism $\|r$ on R, for which (D) is valid

b) $\mathfrak{S}$ is an affine microcentral translation structure of odd order

c) There is a group G of prime exponent $p > 2$ and class ≤ 2 and a partition π of G, such that $\mathfrak{S} = \mathfrak{S}(G,\pi)$ holds.

Concerning the proof the implications a)$\Rightarrow$b) and c)$\Rightarrow$a) follow from the preceding propositions. The implication b)$\Rightarrow$c) follows from the fact, that $\mathfrak{S}$ fulfilling b) can be considered as an affine two-sided linearly fibered incidence group[1], which then is represented by an affine kinematic algebra $A= K \oplus M$ as defined at the end of chapter 1.

EXAMPLE. Let V be a K-vector space of dimension at least 2, and for a subspace W of V and automorphism γ of K let $f: V\times V \to W$ have property (*). Define $G = (V,f)$ and $\pi = \{vK | o\neq v\in V\}$. Then the translation structure $\mathfrak{S} = \mathfrak{S}(G,\pi)$ is even a microcentral affine translation Sperner space (i.e. all blocks have the same cardinality): For any subfield F of the fixed field of γ we can choose $\mathfrak{A} = AG(V/F)$ as an embedding affine space of $\mathfrak{S}$. (In the last chapter of [3] these special geometries of most simple shape in their class are characterized.) The properties of f imply $f(v,u)=o$ for some fixed $u\neq o$ and all $v\in V$. So in $\mathfrak{S}$ there is at least one parallel class of blocks which is also a parallel class of subspaces of $\mathfrak{A}$ w.r.to the natural parallelism in $\mathfrak{A}$. (The analogic thing holds for the lines and their parallelism $\|r$.)

REMARK. Since the parallel structures $\mathfrak{S}$ of the theorem are the only candidates for finite translation structures with proper dilatations, one could ask how to characterize this additional property. But for a parallel structure $\mathfrak{S}$ which is affinely embedded and possesses a parallelism $\|r$ on R compatible with $\mathfrak{S}$, for different points x,y,z of the same block it is easy to give an incidence proposition (Dxyz), which guaranties the existence of a (proper) dilatation δ with fixed point x and $y\delta = z$ (see [3], 3.8).

FOOTNOTE

1. G is an affine two-sided linearely fibered incidence group, if G is the set of all points of an affine space $\mathfrak{A}$, such that for every element a of G by left multiplication and right multiplication a on $\mathfrak{A}$ are induced affinities and moreover all lines through 1 are subgroups of G. For $\mathfrak{A} = AG(V/K)$ every such group can be represented by means of a local K-algebra $A = K \oplus M$ with $x^2=0\ \forall x\in M$, see [5] or [9].

REFERENCES

[1] André, J., Über Parallelstrukturen I.II., Math.Z.76 (1961) 85-102 and 155-163.

[2] Biliotti, M., Sulle strutture di traslazione, Boll.U.M.I.(5) 14 - A (1978) 667-677.

[3] Biliotti, M. and Herzer, A., Zur Geometrie der Translationsstrukturen mit eigentlichen Dilatationen, Abh.Math.Sem.Hamburg (1984) 1-27.

[4] Biliotti, M. and Scarselli, A., Sulle strutture di traslazione dotatedi dilatazioni proprie, Atti Acc.Naz.Lincei, Rend. Cl.Sci.Fis.Mat.Nat. (8) 67 (1979) 75-80.

[5] Bröcker, L., Kinematische Räume, Geom.Ded. 1 (1973) 241-268.

[6] Herzer, A., Endliche translationstransitive postaffine Räume, Abh. Math.Sem.Hamburg 48 (1979) 25-33.

[7] Herzer, A., Endliche nichtkommutative Gruppen mit Partition Π und fixpunktfreiem Π-Automorphismus, Arch.Math.34 (1980) 385-392.

[8] Herzer, A., Varietà di Segre generalizzate, Rend.Mat.(Roma) to appear 1986.

[9] Karzel, H., Kinematic Spaces, Symposia Mathematica XI (1973) 413-439 (Istituto Nazionale di Alta Matematica).

Annals of Discrete Mathematics 30 (1986) 269–274

Sharply 3-transitive groups generated by involutions

Monika Hille and Heinrich Wefelscheid

The set $J := \{\gamma \in G \mid \gamma^2 = id, \gamma \neq id\}$ of involutions of a group G which operates sharply 3-transitively on a set M, induces, to a large extent the structure of this group G. For example the characteristic of G and the planarity of G are expressible as properties of J^2.

In this paper we are looking for sharply 3-transitive permutation-groups G which are generated by their involutions. It is shown that: $G = J^2 \Leftrightarrow G \cong PGL(2,K) \wedge K \neq \mathbb{Z}_2$ and results on groups G with $G = J^3$ are given. Also a class of examples of groups with $G = J^3$ are presented The question, for what $n \in \mathbb{N}$ there exist groups G such that $G = J^n$ but $G \neq J^{n-1}$ is still open. On the other side there do exist sharply 3-transitive groups G which are not generated by their involutions; e.g. all finite such groups G which are not isomorphic to a PGL(2,K) for some K have this property.

For understanding the following we need the notion of a KT-field and the basic representation theorem:

<u>Definition 1:</u> $F(+,\cdot,\sigma)$ is called a KT-field if the following axioms are valid:

<u>FB 1</u> $(F,+)$ is a loop (with identity 0) which has the properties: $a + x = 0 \Rightarrow x + a = 0$. (we put $x := -a$)
For each pair of elements $a,b \in F$ there exists an element $d_{a,b} \in F$ such that $a + (b + x) = (a + b) + d_{a,b} \cdot x$ for each $x \in F$.

<u>FB 2</u> $(F^*,\cdot)$ is a group (with identity 1; $F^* = F \setminus \{0\}$)

<u>FB 3</u> $a\cdot(b + c) = a\cdot b + a\cdot c$ and $0 \cdot a = 0$ for all $a,b,c \in F$

<u>KT</u> σ is an involutary automorphism of the multiplicative group $(F^*,\cdot)$ which satisfies the functional equation:

$$\sigma(1+\sigma(x)) = 1-\sigma(1+x) \quad \text{for all } x\in F\setminus\{0,1\}.$$

<u>Representation-Theorem 2:</u> Let F be a KT-field, ∞ an element not in F and denote by $\bar{F} := F\cup\{\infty\}$.
The transformations of type α and β of $\bar{F}$ onto $\bar{F}$:

$$\alpha:\ x\to a+m\cdot x \qquad \beta:\ x\to a+\sigma(b+m\cdot x)$$

with $a,b,m\in F$, $m\neq 0$ and $\alpha(\infty)=\infty$, $\sigma(\infty)=0$, $\sigma(0)=\infty$, form a group $T_3(\bar{F})$ which operates sharply 3-transitively on $\bar{F}$. Conversely each sharply 3-transitive group is isomorphic as a permutation group to the groups $T_3(F)$ of a uniquely determined KT-field.

In the following let G be a sharply 3-transitive group as being represented in the form $T_3(\bar{F})$ The involutions of G are:

$$\alpha_o:\ x\ \to\ a-x \qquad a\in F \text{ (if char } F=2, \text{ then } a\neq 0)$$

$$\beta_o:\ x\ \to\ -b+n\sigma(b+x) \qquad b\in F \text{ and } n\in S := \{s\in F^*\,|\,\sigma(s)\cdot s=1\}$$

In case of char $F=2$ then α_o has the only fixed point ∞; in case of char $F\neq 2$ then α_o has the two fixed points ∞ and $a\cdot 2^{-1}$, whereas β_o possesses fixed points x_1,x_2 if and only if $n\in R := \{z\cdot\sigma(z^{-1})\,|\,z\in F^*\}$. If that is the case $x_1,x_2 = -b\pm z$ with $n=z\cdot\sigma(z^{-1})$.

<u>Definition 3:</u> A transformation $\delta\in G$ with $\delta^2\neq \mathrm{id}$ is called a pseudo-involution if there exists two different points $a,b\in\bar{F}$ with $\delta(a)=b$ and $\delta(b)=a$

<u>Lemma 4:</u> If G contains a pseudo-involution δ then $\delta\notin J^2$ and $S\neq F^*$.

Proof: Let be $\delta\in G$ a pseudo-involution with $\delta(0)=\infty$ and $\delta(\infty)=0$. Then δ has the form $\delta(x)=\sigma(m\cdot x)=\sigma(m)\cdot\sigma(x)$ with $m\in F^*\setminus S$. Suppose there exist two involutions $\beta_1,\beta_2\in J$ such that $\delta=\beta_1\beta_2$. Then we have:

$$\infty=\delta(0)=\beta_1\beta_2(0) \quad\Rightarrow\quad \beta_1(\infty)=\beta_2(0)$$

$$0=\delta(\infty)=\beta_1\beta_2(\infty) \quad\Rightarrow\quad \beta_1(0)=\beta_2(\infty)$$

1. Case: Let be $\beta_1(\infty)=\beta(0)=z\neq 0,\infty$.

Then $\delta\delta(z)=\delta\delta\beta_2(0)=\delta\beta_1\beta_2\beta_2(0)=\delta\beta_1(0)=\delta\beta_2(\infty)=\beta_1\beta_2\beta_2(\infty)=$

$= \beta_1(\infty) = z. \Rightarrow \sigma(m) \cdot m \cdot z = \delta\delta(z) = z \Rightarrow \sigma(m) \cdot m = 1 \Rightarrow m \in S$ contrary to our assumption $m \in F^* \setminus S$.

2. Case: $\beta_1(\infty) = \infty$

Then we get $\beta_2(0) = \beta_1(\infty) = \infty. \Rightarrow 0 = \beta_2(\infty) = \beta_1(0) \Rightarrow \beta_1$ has the fixed points 0 and $\infty \Rightarrow \beta_1(x) = -x$. β_2 interchanges 0 and ∞. Therefore β_2 has the form $\beta_2(x) = n \cdot \sigma(x)$ with $n \in S$. $\Rightarrow \sigma(m) \cdot \sigma(x) = \delta(x) = \beta_1\beta_2(x) = -n \cdot \sigma(x). \Rightarrow \sigma(m) = -n \Rightarrow m \in S$ since $n \in S$ and -1 lies in the center of $(F^*, \cdot)$. Contradiction to $m \in F^* \setminus S$.

3. Case: $\beta_1(\infty) = 0$

$\Rightarrow \beta_2(0) = \beta_1(\infty) = 0$. This leads to the same proof as in case 2, if we interchange β_1 and β_2. □

<u>Theorem 5:</u> Let G be a sharply 3-transitive group. Then $G = J^2$ if and only if $G \cong PGL(2,K)$ and $|K| > 2$.

Proof: "$\Leftarrow$" $PGL(2,K) \cong J^2$ if $|K| > 2$ is commonly known and can be verified by calculation.

"$\Rightarrow$" Let be $G = J^2$. Because of lemma 4 the group G cannot contain any pseudo-involution. Therefore $F^* \setminus S = \emptyset$. A KT-field with $F^* = S$ is a commutative field (cf. [3], Satz 1.6). □

Now we turn to groups G with $G = J^3$.

<u>Definition 6:</u> A KT-field $(F,+,\cdot\sigma)$ is called planar, if the equation $ax + bx = c$ with $a \neq -b$ always has a solution $x \in F$.

Planar KT-fields already are nearfields; i.e. $(F,+)$ is an abelian group. (cf. [1], 5.6).

<u>Theorem 7:</u> Let be $(F,+,\cdot,\sigma)$ a KT-field with $|F| > 2$ and let be G the induced sharply 3-transitive group. Then

$$G_a := \{\gamma \in G \mid \gamma(a) = a\} \subset J^2 \text{ for an } a \in \bar{F}$$

is valid if and only if F is planar and $S \cdot S = F^*$ with $S := \{s \in F^* \mid s \cdot \sigma(s) = 1\}$.

Proof: We conduct the demonstration in several steps:

A: Let be F a planar KT-field and $S\cdot S = F^*$.

<u>1.</u> $G_{\infty,o} \subset J^2$

Proof: The elements of the group $G_{\infty,o} := \{\gamma \in G \mid \gamma(0) = 0,\ \gamma(\infty) = \infty\}$ have the form:

$$G_{\infty,o} = \{\mu \in G \mid \mu := \rightarrow kx \text{ with } k \in F^*\}.$$

Because of $S\cdot S = F^*$ there exist $n_1, n_2 \in S$ such that $n_1 \cdot n_2 = k$. Then $\mu(x) = n_1 n_2(x) = n_1 n_2(x) = n_1 \cdot \sigma\sigma(n_2 x) = \beta_1\beta_2(x)$ with $\beta_1(x) = n_1 \cdot \sigma(x)$, $\beta_2(x) = \sigma(n_2 x)$ and $\beta_1, \beta_2 \in J$.

<u>2.</u> $G_{\infty,a} \subset J_2$ for each $a \in F$.

Proof: Let be $\alpha: x \rightarrow a+x$. Then $G_{\infty,a} = \alpha\, G_{\infty,o}\, \alpha^{-1} \subset \alpha J^2 \alpha^{-1} = J^2$ because of $\alpha J \alpha^{-1} = J$.

<u>3.</u> $G_\infty = \{\gamma \in G \mid \gamma(x) = c + m\cdot x,\ c, m \in F,\ m \neq 0\} \subset J^2$

Proof: Let be $T := \{\tau \in G \mid \tau(x) = x \Rightarrow x = \infty\}$.
Then we have

$$G_\infty = \bigcup_{a \in F} G_{\infty,a} \cup T$$

Since F is planar, the set T consists of transformations of the form:

$$T = \{\tau \in G_\infty \mid \tau(x) = b + x \text{ with } b \in F^*\}$$

But for $|F| > 2$ we have $\tau = \alpha_2\alpha_1 \in J^2$ with $\alpha_i(x) = a_i - x$ and $a_2 - a_1 = b$. Hence $T \subset J^2$.

B. Let be $G_a \subset J^2$ for some $a \in \bar{F}$.

Then $G_\infty = \gamma G_a \gamma^{-1} \subset \gamma J^2 \gamma^{-1} = J^2$ for some $\gamma \in G$ with $\gamma(a) = \infty$.
Let now be $\alpha \in G_\infty$ with $\alpha(x) = c + mx$, $m \neq 1$ and $\alpha = \beta_2\beta_1 \in J^2$ with $\beta_i(x) = -b_i + n_i \cdot \sigma(b_i + x)$, $i = 1,2$. Because of $\beta_i \in J$ we have $n_i \in S$.

We show that α possesses another fixed point different from ∞. We have

$$\infty = \alpha(\infty) = \beta_2\beta_1(\infty) = -b_2 + n_2\sigma(b_2 + (-b_1 + n_1\sigma(b_1 + \infty)))$$

$$= -b_2 + n_2\sigma(b_2 - b_1)$$

$$\Rightarrow b_2 = b_1 =: b. \Rightarrow \alpha(x) = -b + (n_2 n_1^{-1} b + n_2 n_1^{-2} \cdot x)$$

$\Rightarrow$ The second fixed point of α is $-b$.

Since each $\alpha \in G_\infty$ has two fixed points, F is planar whereas follows that (F,+) is an abelian group.

$$\Rightarrow \alpha(x) = (-b+n_2n_1^{-1}b)+n_2n_1^{-1}\cdot x \quad \Rightarrow m = n_2n_1^{-1} \in S\cdot S$$

Thus $F^* = S\cdot S$ since $m \in F^*$ was arbitrarily chosen. □

Remark: If $G_a \subset J^2$ then $G \subset J^3$. Wether the converse is true, we do not know.

In conclusion we give examples of sharply 3-transitive groups G with $G = J^3$ but $G \neq J^2$. We use the following general method for constructing sharply 3-transitive groups due to Kerby (cf.[2], p.60):

Let $(F,+,\cdot)$ be a commutative field and let $\sigma(x) = x^{-1}$. Suppose $A < (F^*,\cdot)$ such that:

(1) $Q := \{a^2 \in F^* \mid a \in F^*\} \subset A$

(2) There exists a monomorphism $\pi: F^*/A \to \mathrm{Aut}(F,+,\cdot)$

(3) $\tau(x) \in xA$ for all $\tau \in \pi(F^*/A)$ and all $x \in F^*$.

Define $0\circ b = 0$, and for $a \neq 0$, $a\circ b = a\cdot a_\varphi(b)$ where $a_\varphi = \pi(a\cdot A)$. Then $(F,+,\circ,\sigma)$ is a KT-field. (To be exact: $(F,+,\circ)$ is a strongly coupled Dickson-nearfield, which in addition is planar (cf.[5]))

(1) induces $[F^*:A] = 2^{|I|}$ where I denotes an appropriate index set. One can construct for arbitrary index sets such KT-field for which the induced group G satisfies $G = J^3$ but $G \neq J^2$. We illustrate this method of construction for $I = \{1,2\}$:

Let be K a commutative field and $L := K(x_1,x_2)$ a transcendental extension of K. Consider the following automorphisms $\alpha_1,\alpha_2 \in \mathrm{Aut}_K L$:

$$\alpha_1: \begin{cases} x_1 & \to & 1-x_1 \\ x_2 & \to & x_2 \\ k\in K & \to & k \end{cases} \qquad \alpha_2: \begin{cases} x_2 & \to & 1-x_2 \\ x_1 & \to & x_1 \\ k\in K & \to & k \end{cases}$$

α_1 and α_2 are involutary and satisfy $\alpha_1\alpha_2 = \alpha_2\alpha_1$.

Take $F := L(t_1,t_2)$ where the t_i are transcendental over L. Let be $\tau_i \in \mathrm{Aut}\, F$ the following continuation of α_1 for $h \in L[t_1,t_2]$:

$$\tau_i: h = \Sigma\, a_{\mu\nu}t_1^\mu t_2^\nu \longrightarrow \Sigma\, \alpha_i(a_{\mu\nu})t_1^\mu t_2^\nu$$

and define $\tau_i(\frac{h_1}{h_2}) := \frac{\tau_i(h_1)}{\tau_i(h_2)}$ for $f = \frac{h_1}{h_2}$, $h_1, h_2 \in L[t_1, t_2]$

Also let be $\text{grad}_i f = \text{grad}_i \frac{h_1}{h_2} := \text{grad}_i h_1 - \text{grad}_i h_2$ the degree function of the polynomials $h_1, h_2 \in L[t_1, t_2]$ with respect to t_i.

Now we define for $f, g \in F$:

$$f \circ g := f \cdot f_\varphi(g) \quad \text{with } f_\varphi := \tau_1^{\text{grad}_1 f} \tau_2^{\text{grad}_2 f}$$

Then $F, +, \circ, \sigma)$ is a planar KT-field (to be exact: $(F, +, \circ, \sigma)$ is a planar KT-nearfield; $\sigma(a) = a^{-1}$ the inverse with respect to $(\cdot)$).

We note that $A = \text{Kern } \varphi = \{f \in F^* \mid \text{grad}_i f \equiv 0 \text{ mod } 2 \text{ for } i=1,2\} \subset S$ and $[F^* : A] = 4 = 2^{|I|}$. The 3 other cosets are: $t_1 \cdot A$, $t_2 \cdot A$, $t_1 \cdot t_2 \cdot A$

The Dickson-group $\Gamma := \{f_\varphi \in \text{Aut } F \mid f \in F^*\}$ consists of the 4 automorphisms: $\Gamma = \{id, \tau_1, \tau_2, \tau_1\tau_2\}$. For any $\tau \in \Gamma$ and any $z \in \{t_1, t_2, t_1 \cdot t_2\}$ we have $\tau(z) = z$ and therefore $z \circ \sigma(z) = 1$. $\Rightarrow$ $z \in S$ and $z \cdot A = z \circ A \subset S \circ S$ for $z \in \{t_1, t_2, t_1 \cdot t_2\}$. $\Rightarrow$ $F^* = S \circ S$.

This shows that this example satisfies the conditions of Theorem 7, whence the induced group G fulfills $G_\infty \subset J^2$ and therefore $G = J^3$.

References

[1] H. KARZEL: Zusammenhänge zwischen Fastbereichen, scharf 2-fach transitiven Permutationsgruppen und 2-Strukturen mit Rechtecksaxiom, Abh. Math. Sem. Hamburg, 31(1967) 191-208

[2] W. KERBY: Infinite sharply multiply transitive groups. Hamburger Mathematische Einzelschriften. Neue Folge Heft 6, Göttingen 1974, Vandenhoek und Ruprecht

[3] W. KERBY und H. WEFELSCHEID: Über eine scharf 3-fach transitiven Gruppen zugeordnete algebraische Struktur. Abh. Math. Seminar Hamburg 37(1972) 225-235

[4] H. WEFELSCHEID: ZT-subgroups of sharply 3-transitive groups, Proceedings of the Edinburgh Mathematical Society 1980, Vol. 23, 9-14

[5] H. WEFELSCHEID: Zur Planarität von KT-Fastkörpern, Archiv der Mathematik, Vol. 36(1981) 302-304

Fachbereich Mathematik
der Universität Duisburg
D - 4100 Duisburg 1

Annals of Discrete Mathematics 30 (1986) 275–284

ON THE GENERALIZED CHROMATIC NUMBER

Florica Kramer and Horst Kramer
Computer Technique Research Institute
str.Republicii 109
3400 Cluj-Napoca
ROMANIA

The chromatic number relative to distance p, denoted by $\gamma(p,G)$ is the minimum number of colors sufficing for coloring the vertices of G in such a way that any two vertices of distance not greater than p have distinct colors. We give upper bounds for the chromatic number $\gamma(3,G)$ for bipartite (planar) graphs and generalize a result of S.Antonucci giving a lower bound for $\gamma(2,G)$.

1. INTRODUCTION

In the following we shall use some concepts and notions introduced in the book [5] of C.Berge. We restrict ourselves to simple graphs, i.e. connected undirected finite graphs which have no loops or multiple edges. Let $G=(V,E)$ be a simple graph, where V is the vertex set and E is the edge set of G. If $x,y \in V$, we shall denote by $g(x)$ the degree of the vertex x, by $\Delta(G)=\max\{g(x); x \in V\}$ the maximum degree of the vertices of G, by $d_G(x,y)$ the distance between the vertices x and y in the graph G, i.e. the length of the shortest path connecting vertices x and y in the graph G (the length of a path is considered to be the number of edges which form the path). K_n denotes the complete graph with n vertices and $K_{m,n}$ the complete bipartite graph.

In 1969 we have considered in the papers [11] and [12] the following coloring problem of a graph:

DEFINITION 1. Let p be a given integer, $p \geqq 1$. An admissible k-coloring relative to distance p of the graph $G=(V,E)$ is a function $f: V \rightarrow \{1,2,\dots,k\}$ such that

$$\forall x,y \in V \text{ with } 1 \leqq d_G(x,y) \leqq p \implies f(x) \neq f(y).$$

The smallest integer k for which there exists an admissible k-coloring relative to distance p is denoted by $\gamma(p,G)$ and is called the chromatic number relative to distance p of the graph G.

DEFINITION 2. Let $G=(V,E)$ be a given graph and p an integer, $p \geq 1$. We shall denote by $G^p=(V,E_p)$ the graph, which has the same vertex set V as the graph G whereas the edge set E_p is defined by:

$$(x,y) \in E_p \text{ if and only if } 1 \leqq d_G(x,y) \leqq p.$$

There exists then the following relation between the chromatic number $\gamma(p,C)$ and the ordinary chromatic number $\gamma(G)$:

$$\gamma(p,G) = \gamma(1,G^p) = \gamma(G^p).$$

In the papers [11], [12] and [13] we obtained some results relative to the chromatic number $\gamma(p,G)$, especially one characterizing those graphs G for which we have $\gamma(p,G)=p+1$.

The problem of coloring a graph relative to a distance s was reconsidered in 1975 by F.Speranza [16] under the name of L_s-coloring of a graph; the same problem was also studied by M.Gionfriddo [6],[7] and [8], S.Antonucci [1] and by G.Wegner [17].

2. UPPER BOUNDS FOR THE CHROMATIC NUMBER $\gamma(3,G)$

G.Wegner [17] proved the following theorem:

THEOREM 1. Let $G=(V,E)$ be a simple planar graph such that $\Delta(G) \leqq 3$. We have then: $\gamma(2,G) \leqq 8$.

A still open problem is that of finding a planar graph G with maximal degree 3 such that we have $\gamma(2,G)=8$ or else to prove $\gamma(2,G) \leqq 7$.

First we shall give some bounds for the generalized chromatic number $\gamma(3,G)$ of bipartite graphs. Let $G=(A,B;E)$ be a given bipartite graph. We shall denote by $G_A=(A,E_A)$ the graph, which has the vertex set A and the edge set E_A is defined by:

$$(x,y) \in E_A \text{ if and only if } d_G(x,y)=2.$$

The graph G_A can be obtained from the graph G by the elimination of the vertices of B and by the application of the following three operations:

a) If we have a "3-star" with the center in a vertex y of B (it follows that the vertices x_1,x_2,x_3 adjacent to y are in A) then we construct in G_A the circuit $x_1x_2x_3x_1$.

b) If a vertex y of B has degree 2 in G and x_1,x_2 are the two vertices of A adjacent to y in G, then we replace the path x_1yx_2 of length 2 in G by an edge (x_1,x_2) in G_A.

c) If by the application of the preceding two operations appear multiple edges between two vertices x and y, we shall retain only one

of these edges and delete the others.
Analogously we introduce the graph $G_B=(B,E_B)$ with the vertex set B and the edge set E_B defined by:

$(x,y)\in E_B$ if and only if $d_G(x,y)=2$.

An upper bound for the chromatic number $\gamma(3,G)$ in function of the ordinary chromatic numbers $\gamma(G_A)$ and $\gamma(G_B)$ is given in the following:

THEOREM 2. Let $G=(A,B;E)$ be a bipartite graph. Then

$$\gamma(3,G) \leqq \gamma(G_A) + \gamma(G_B). \tag{1}$$

Proof. $\gamma(G_A)$ being the ordinary chromatic number of the graph G_A there is a coloring $f_1: A \to \{1,2,\ldots,\gamma(G_A)\}$ of its vertices such that $(x,y)\in E_A$ implies $f_1(x) \neq f_1(y)$. We consider also a coloring with $\gamma(G_B)$ colors of the vertices of the graph G_B $f_2: B\to\{\gamma(G_A)+1, \gamma(G_A)+2,\ldots,\gamma(G_A)+\gamma(G_B)\}$ such that $(x,y)\in E_B$ implies $f_2(x) \neq f_2(y)$. The function $f: A\cup B \to \{1,2,\ldots,\gamma(G_A)+\gamma(G_B)\}$ defined by

$$f(x) = \begin{cases} f_1(x), & \text{if } x\in A \\ f_2(x), & \text{if } x\in B \end{cases} \tag{2}$$

is then a coloring of the vertices of the bipartite graph $G=(A,B;E)$ with $\gamma(G_A) + \gamma(G_B)$ colors. We shall now verify that the so obtained coloring is an admissible $(\gamma(G_A)+\gamma(G_B))$-coloring relative to distance 3 of the graph G, i.e. that $x,y \in A\cup B$ and $1\leqq d_G(x,y) \leqq 3$ implies $f(x) \neq f(y)$. Really, if $d_G(x,y)=1$ or $d_G(x,y)=3$ it follows that one vertex, let it be x, belongs to A and the other, y, belongs to B. Then obviously $f(x) \neq f(y)$. On the other hand if $d_G(x,y)=2$ it follows from the bipartite character of G that both vertices x and y are in A, or both are in B. Consider now the cases:

a) if $x,y\in A$ and $d_G(x,y)=2$, it results that $(x,y)\in E_A$ and therefore $f(x) = f_1(x) \neq f_1(y) = f(y)$.

b) if $x,y\in B$ and $d_G(x,y)=2$, it results that $(x,y)\in E_B$ and therefore $f(x) = f_2(x) \neq f_2(y) = f(y)$.

Thus we have proved inequality (1).

THEOREM 3. Let $G=(A,B;E)$ be a simple planar bipartite graph with $\Delta(G)\leqq 3$. Then we have

$$\gamma(3,G) \leqq 8. \tag{3}$$

This bound is sharp in the sense that there are cubic planar bipartite graphs for which $\gamma(3,G)=8$.

Proof. From the planarity of the initial graph G and the construction way of the graphs G_A and G_B results the planarity of the graphs G_A and G_B. By the four-color theorem proved by K.Appel and W.Haken([2],

[3] and [4]) we have $\gamma(G_A) \leqq 4$ and $\gamma(G_B) \leqq 4$. By Theorem 2 follows then immediately inequality (3).

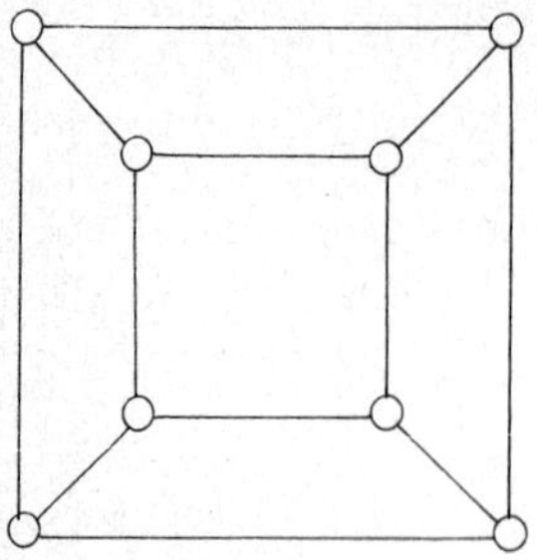

Fig. 1

The fact that the so obtained bound for $\gamma(3,G)$ can not be lowered, can be seen from the following example: Let $G=(A,B;E)$ be the graph of the hexahedron, which is a bipartite planar cubic graph with 8 vertices and diameter 3. It results that we have $\gamma(3,G) = |A \cup B| = 8$.

If we renounce to the planarity of G in Theorem 3 we obtain:

THEOREM 4. Let $G=(A,B;E)$ be a bipartite graph such that $\Delta(G) \leqq 3$. Then we have:

$$\gamma(3,G) \leqq 14. \tag{4}$$

The obtained bound is sharp in the sense that there exist bipartite cubic graphs for which $\gamma(3,G) = 14$.

Proof. Consider a bipartite graph $G=(A,B;E)$ with $\Delta(G) \leqq 3$ and the corresponding graphs G_A and G_B defined above. From $\Delta(G) \leqq 3$ results immediately $\Delta(G_A) \leqq 6$ and $\Delta(G_B) \leqq 6$. A well-known theorem (see for instance [15] vol.I, Satz IV.2.1) asserts that for any graph G, the ordinary chromatic number is at most one greater than the maximum degree $\Delta(G)$. We have then $\gamma(G_A) \leqq 7$ and $\gamma(G_B) \leqq 7$. Applying again Theorem 2 we have $\gamma(3,G) \leqq 14$.

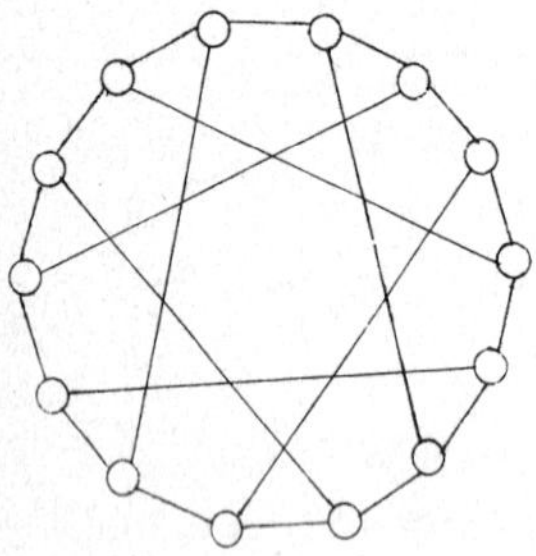

Fig. 2

The fact that this bound is sharp, can be seen from the example of the well-known Heawood graph (see Fig.2), which is a cubic bipartite graph with 14 vertices and of diameter 3. We have then: $\gamma(3,G) = |A \cup B| = 14$.

A direct generalization of Theorem 4 is the following theorem.

THEOREM 5. Let G be a simple bipartite graph with maximum degree $\Delta = \Delta(G)$. Then we have:

$$\gamma(3,G) \leqq 2(1 + \Delta(\Delta - 1)). \tag{5}$$

Proof. From $\Delta = \max\{g(x);\ x \in A \cup B\}$ we obtain immediately that the number of vertices y which are at distance 2 from a given vertex x in the graph G is at most $\Delta(\Delta-1)$. We have then $\Delta(G_A) \leqq \Delta(\Delta-1)$ and $\Delta(G_B) \leqq \Delta(\Delta-1)$ and thereby $\gamma(G_A) \leqq \Delta(\Delta-1)+1$ and respectively $\gamma(G_B) \leqq \Delta(\Delta-1)+1$. Applying now Theorem 2 we obtain inequality (5).

COROLLARY 1. Let $G=(A,B;E)$ be a simple Δ-regular bipartite graph of diameter 3. Then we have:

$$|A| + |B| \leqq 2(1 + \Delta(\Delta - 1). \tag{6}$$

Examples of graphs for which we have the equality sign in (5) and also in (6) are: 1) for $\Delta = 2$ the circuit C_6 of length 6; 2) for $\Delta = 3$ the above discussed Heawood graph and 3) for $\Delta = 4$ the 4-regular bipartite graph of diameter 3 with 26 vertices from Fig. 3. $D(G) = 3$ implies again $\gamma(3,G) = |A \cup B| = 26$. An improvement of Theorem 5 and of Theorem 4 can be obtained if we apply Brooks theorem (see for instance [9] Theorem 12.3): If $\Delta(G) = \Delta$ then G is Δ-colorable unless:
(i) $\Delta = 2$ and G has a component which is an odd cycle, or (ii) $\Delta > 2$ and $K_{\Delta+1}$ is a component of G.

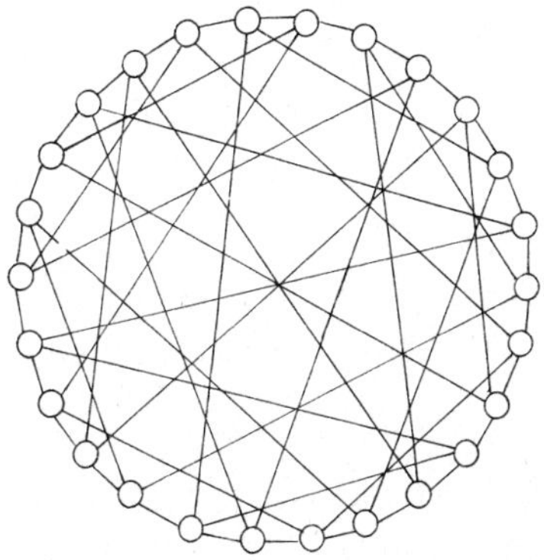

Fig. 3

The case $\Delta = 2$ was discussed in [11], [12] and [13]. Let us assume that $\Delta > 2$. The graphs G_A and G_B are connected since G is connected. If now at least one of G_A, G_B - say, G_A - is not a $K_{\Delta(\Delta-1)+1}$ then by Brooks theorem, $\gamma(G_A) \leqq \Delta(\Delta-1)$ and by Theorem 2 we have the inequality $\gamma(3,G) \leqq 2\Delta(\Delta-1)+1$. Thus if equality holds in Theorem 5 we have $|A| = |B| = \Delta(\Delta-1)+1$ and G is a bipartite Δ-regular graph of diameter 3. If none of G_A, G_B is a $K_{\Delta(\Delta-1)+1}$ then $\gamma(G_A) \leqq \Delta(\Delta-1)$, $\gamma(G_B) \leqq \Delta(\Delta-1)$ and by Theorem 2 $\gamma(3,G) \leqq 2\Delta(\Delta-1)$. We can now enounce:

COROLLARY 2. Let $G=(A,B;E)$ be a simple bipartite graph such that $\Delta(G) = \Delta > 2$ and $\min(|A|,|B|) > \Delta(\Delta-1)+1$. Then we have:

$$\gamma(3,G) \leqq 2\Delta(\Delta-1).$$

In particular, if $\Delta = 3$ and $|A| > 7, |B| > 7$ then we have:

$$\gamma(3,G) \leqq 12.$$

3. LOWER BOUNDS FOR THE CHROMATIC NUMBER $\gamma(2,G)$

First we shall prove a result needed in the sequel.

THEOREM 6. Let $G=(V,E)$ be a simple graph with the properties:
(i) the degree of each vertex is at least 2,
(ii) the diameter of the graph $D(G)=2$,
(iii) G does not contain circuits of length 3 and 4.
Then G is a Moore graph.

Proof. Property (i) implies the existence of at least one circuit in the graph G. By (ii) and (iii) results that the girth of G is 5. G is then a Moore graph by a result of R.Singleton [14] which asserts that a simple graph with diameter $k \geq 1$ and girth $2k+1$ is also regular and hence a Moore graph.

In 1978 S.Antonucci [1] obtained the following lower bound for the chromatic number $\gamma(2,G)$ as a function of the number of vertices and the number of edges of the graph G:

THEOREM 7. Let $G=(V,E)$ be a simple graph with n vertices and m edges and without circuits of length 3 and 4. Then we have:

$$\gamma(2,G) \geq \frac{n^3}{n^3-4m^2} . \tag{7}$$

But S.Antonucci didn't give any example of graphs for which this bound is attained. We shall prove the following theorem:

THEOREM 8. The only graphs of diameter $D(G) \leq 2$, without circuits of length 3 and 4, with n vertices and m edges for which we have

$$\gamma(2,G) = \frac{n^3}{n^3-4m^2} \tag{8}$$

are the graphs K_1, K_2 and the Moore graphs of diameter 2.

Proof. The only graph of diameter $D(G)=0$ is the graph K_1, for which we have $n=1$, $m=0$, $\gamma(2,G)=1$. K_1 verifies then evidently (8).
If $D(G)=1$, G is a complete graph K_n with $n \geq 2$. The only complete graph K_n, $n \geq 2$, without circuits of length 3 is the graph K_2, for which we have $n=2$, $m=1$, $\gamma(2,K_2)=2$ and therefore (8) is verified.
If $D(G)=2$, we have to distinguish two cases:
a) $\delta = \min\{g(x);\ x \in V\} = 1$. Then there is a vertex a_1 of degree 1 and the vertex b adjacent to a_1 has to be adjacent to all the other vertices of V because $D(G)=2$. But as G doesn't contain circuits of

length 3, G is a (n-1)-star with $n \geqq 3$, i.e. $G=(V,E)$ with $V=\{a_1,a_2,\dots,a_{n-1},b\}$ and $E=\{(a_i,b),\ i=1,2,\dots,n-1\}$. We have then $m=n-1$ and $\gamma(2,G)=n$. Relation (8) becomes $n(n^3-4(n-1)^2)=n^3$. The only solutions of this equation are $n_1=0$, $n_2=1$ and $n_3=n_4=2$, none of which corresponds because as we have seen above we have $n \geqq 3$. The conclusion is that we can not have $\delta = 1$.

b) $\delta = \min\{g(x);\ x\in V\} \geqq 2$. G is then a Moore graph of diameter 2 by Theorem 6. A δ-regular Moore graph of diameter 2 has $n = 1+\delta^2$ vertices and $m = n\cdot\delta/2 = \delta(1+\delta^2)/2$ edges. Because $D(G)=2$ we have $\gamma(2,G) = n = 1+\delta^2$. It follows that

$$\frac{n^3}{n^3-4m^2} = \frac{(1+\delta^2)^3}{(1+\delta^2)^3-\delta^2(1+\delta^2)^2} = 1+\delta^2 = \gamma(2,G).$$

With that Theorem 7 is proved.

REMARK. By a well-known result of A.J.Hoffman and R.R.Singleton [10] a Moore graph of diameter 2 has one of the degrees 2, 3, 7 or 57; for each of the degrees 2, 3, 7 there is exactly one Moore graph of diameter 2 (it is not known whether or not there is a Moore graph of diameter 2 and degree 57).

A lower bound for $\gamma(2,G)$ similar to that obtained by S.Antonucci can also be deduced for graphs which have circuits of length 3 or 4.

THEOREM 9. Let $C=(V,E)$ be a simple connected graph with n vertices and m edges in which we denote by:

(i) c_3 the number of circuits of length 3 in G;

(ii) c_4^o the number of circuits of length 4, for which no pair of opposite vertices in the circuit are adjacent in G;

(iii) c_4^1 the number of circuits of length 4, for which one pair of opposite vertices in the circuit are adjacent in G and the other pair of opposite vertices are not adjacent in G.

If G doesn't contain any subgraph of the type $K_{2,3}$ then the chromatic number $\gamma(2,G)$ verifies the inequality

$$\gamma(2,G) \geqq \left[\frac{n^3}{n^3+n(6c_3+4c_4^o+2c_4^1)-4m^2}\right]^* . \tag{9}$$

This bound is sharp in the sense that there exists graphs verifying the hypotheses of the theorem and for which we have the equality sign in (9).

Proof. The proof of this theorem can be obtained by a modification of the proof given by S.Antonucci for Theorem 7. As we have observed above we have $\gamma(2,G) = \gamma(1,G^2) = \gamma(G^2)$, where $G^2=(V,E_2)$ is the square of the graph G. If we denote by m_2 the cardinality of the edge set E_2, then we have by a Theorem of C.Berge ([5], p.321)

(10) $$\gamma(2,G) = \gamma(G^2) \geqq \frac{n^2}{n^2-2m_2}.$$

The number of all possible paths of length 2 in the graph G is given by the sum $\sum_{i=1}^{n} \binom{g(x_i)}{2}$. If we introduce corresponding to each path xyz of length 2 in G an edge (x,z) we shall obtain a graph G"=(V,E"). Obviously $E_2 \subset E"$, but there may be edges in E_2 which are multiple edges in E". Let $a,b \in V$ be a pair of vertices, which is connected in G by at least one path of length 2. We have to distinguish the cases:
1) a and b are adjacent vertices in G. Then the edge (a,b) is contained in at least one circuit of length 3 in G and the order of multiplicity of the edge (a,b) in E" will be equal with the number of circuits of length 3 which contain the edge (a,b). As each circuit of length 3 contributes to the increase of the multiplicity of each edge of this circuit by one unity, we have to delete $3c_3$ edges from E" in order to make all edges apartaining to circuits of length 3 simple edges.
2) a and b are not adjacent in G. Because G does not contain any subgraph of the type $K_{2,3}$, the vertices a and b can be connected in G by at most two paths of length 2. We distinguish then thesubcases:
2a) a and b are connected in G by exactly one path of length 2. Then (a,b) is obviously a simple edge of the graph G".
2b) a and b are connected in G by two paths of length 2. The edge (a,b) will be a double edge in G". But in this case a and b form a pair of opposite vertices in a circuit of length 4 in G. Because a circuit of length 4 of the type (ii) leads to the duplicating of both diagonals of the circuit, and a circuit of length 4 of the type (iii) leads to the duplicating of only one diagonal, in order to obtain the graph G^2 we have to delete $2c_4^o+c_4^1$ edges from E" beside the $3c_3$ edges already deleted. We have thus

$$m_2 = m + \sum_{i=1}^{n} \binom{g(x_i)}{2} - (3c_3+2c_4^o+c_4^1).$$

It results then:

$$m_2 = m + \frac{1}{2}\sum_{i=1}^{n} g^2(x_i) - \frac{1}{2}\sum_{i=1}^{n} g(x_i) - (3c_3+2c_4^o+c_4^1) =$$

$$= \frac{1}{2}\sum_{i=1}^{n} g^2(x_i) - (3c_3+2c_4^o+c_4^1) \geqq \frac{1}{2n}\left(\sum_{i=1}^{n} g(x_i)\right)^2 - (3c_3+2c_4^o+c_4^1) =$$

$$= \frac{2m^2}{n} - (3c_3+2c_4^o+c_4^1).$$

This inequality and (10) yields

$$\gamma(2,G) \geqq \frac{n^3}{n^3+2n(3c_3+2c_4^o+c_4^1)-4m^2}$$

As $\gamma(2,G)$ is an integer we obtain immediately the inequality (9) in which $[r]^*$ denotes the smallest integer $\geqq r$.

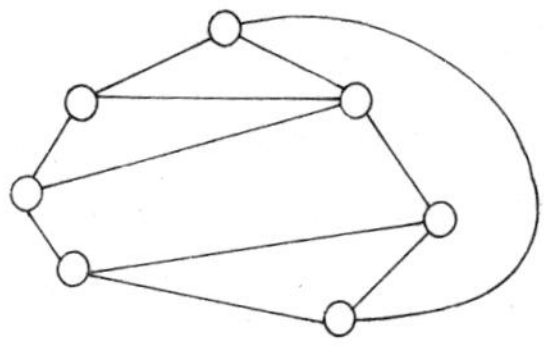

Fig. 4

An example of a graph for which we have the equality sign in (9) is the graph from Fig. 4 for which we have $n=7$, $m=11$, $c_3=3$, $c_4^1=1$, $c_4^o=2$, $D(G)=2$ and $\gamma(2,G)=n=7$. It is easy to verify that for this graph we have the relations:

$$\left[\frac{n^3}{n^3+n(6c_3+4c_4^o+2c_4^1)-4m^2}\right]^* = [343/55]^* = 7 = \gamma(2,G).$$

ACKNOWLEDGMENT. The authors wish to thank the referee for the helpful comments. The second author would like to thank also to the Alexander von Humboldt-Stiftung for the financial support during the years 1981-1982.

REFERENCES

[1] Antonucci,S., Generalizzazioni del concetto di cromatismo d'un grafo, Boll.Un.Mat.Ital. (5)15-B (1978) 20-31.

[2] Appel,K.,Haken,W., Every planar map is four colorable, Bull. Amer.Math.Soc. 82(1976) 711-712.

[3] Appel,K.,Haken,W., Every planar map is four colorable,Part I. Discharging, Illinois J.Math. 21 (1977) 429-490.

[4] Appel,K.,Haken,W.,Koch,J., Every planar map is four colorable, Part II. Reducibility, Illinois J.Math. 21 (1977) 491-567.

[5] Berge,C., Graphes et hypergraphes (Dunod, Paris, 1970).

[6] Gionfriddo,M., Sulle colorazioni L_s d'un grafo finito, Boll.Un. Mat.Ital. (5)15-A (1978) 444-454.

[7] Gionfriddo,M., Alcuni risultati relativi alle colorazioni L_s d'un grafo, Riv.Mat.Univ.Parma (4) 6 (1980) 125-133.

[8] Gionfriddo,M., Su un problema relativo alle colorazioni L_2 d'un grafo planare e colorazioni L_s, Riv.Mat.Univ.Parma (4) 6 (1980)

151-160.

[9] Harary,F., Graph Theory (Addison-Wesley Publ.Comp.,Mass. 1969).

[10] Hoffman,A.J.,Singleton,R.R., On Moore graphs with diameters 2 and 3, IBM J.Res.Develop. 4 (1960) 497-504.

[11] Kramer,F.,Kramer,H., Un probleme de coloration des sommets d'un graphe, C.R.Acad.Sci.Paris 268 A (1969) 46-48.

[12] Kramer,F.,Kramer,H., Ein Färbungsproblem der Knotenpunkte eines Graphen bezüglich der Distanz p, Revue Roum.Math.Pures et Appl. 14 (1969) 7, 1031-1038.

[13] Kramer,F., Sur le nombre chromatique K(p,G) des graphes, R.A.I. R.O., R-1 (1972) 67-70.

[14] Singleton,R.R., There is no irregular Moore graph, American Math.Monthly 75 (1968) 42-43.

[15] Sachs,H., Einführung in die Theorie der endlichen Graphen, vol.I. (Teubner Verlagsg., Leipzig, 1970).

[16] Speranza,F., Colorazioni di specie superiore d'un grafo, Boll. Un.Mat.Ital. (4) 12 Suppl.fasc. 3 (1975) 53-62.

[17] Wegner,G., Graphs with given diameter and a coloring problem, (Preprint, Dortmund, 1976).

Annals of Discrete Mathematics 30 (1986) 285–290

A CONSTRUCTION OF SETS OF PAIRWISE ORTHOGONAL F-SQUARES OF COMPOSITE ORDER

Paola Lancellotti and Consolato Pellegrino

Dipartimento di Matematica
Via Campi, 213/B
41100 MODENA (ITALY)

In this note complete systems of orthogonal F-squares are constructed in which the number of symbols is variable and the order n is a prime power. Further, we give an extension of the MacNeish theorem by constructing systems of orthogonal F-squares of composite order $n = p_1^{e_1} p_2^{e_2} \dots p_m^{e_m}$ having a variable number of symbols: such construction improves the results that have been reached so far.

1 - DEFINITIONS AND PRELIMINARY RESULTS

A square matrix $F = [a_{ij}]$ of order n, defined on a set A, is said to be an F-square of type (n,λ) (and we shall write shortly $F(n,\lambda)$ if each element of A appears λ times in each row and in each column of F. Given the F-squares $F_1(n,\lambda_1)$ and $F_2(n,\lambda_2)$ defined respectively on the sets A_1 and A_2, we shall say that they are orthogonal if after superimposing them each pair $(x,y) \in A_1 x A_2$ appears $\lambda_1\lambda_2$ times. In such case we shall write $F_1(n,\lambda_1) \perp F_2(n,\lambda_2)$.

Given the F-squares $F_1(n,\lambda_1)$, $F_2(n,\lambda_2)$, ..., $F_t(n,\lambda_t)$, we shall say that they form an orthogonal system if $F_i(n,\lambda_i) \perp F_j(n,\lambda_j)$ for all $i,j=1,2,\dots,t$, $i \neq j$. If $|\{\lambda_1,\lambda_2,\dots,\lambda_t\}| > 1$ we shall say that the orthogonal system has a variable number of symbols. In [1] the following theorem has been proved:

THEOREM 1.1. The maximal number t of orthogonal F-squares $F_1(n,\lambda_1)$, $F_2(n,\lambda_2)$, ..., $F_t(n,\lambda_t)$ satisfies the following inequality (we set $n = \lambda_i m_i$ for $i=1,2,\dots,t$) :

$$\sum_{i=1}^{t} m_i - t \leqslant (n-1)^2 .$$

An orthogonal system of F-squares $F_1(n,\lambda_1)$, $F_2(n,\lambda_2)$, ..., $F_t(n,\lambda_t)$ is called complete if

$$\sum_{i=1}^{t} m_i - t = (n-1)^2 .$$

Complete systems of F-squares are constructed in [2] , [3] .

2 - CONSTRUCTION OF COMPLETE SYSTEMS OF F-SQUARES

We shall say that a partition $\mathscr{A} = \{X_1, X_2, \ldots, X_r\}$ of an m-set $A = \{a_1, a_2, \ldots, a_m\}$ is h-regular if $|X_i| = h$ $(i=1,2,\ldots,r)$. If $I_r = \{1,2,\ldots,r\}$, we denote by $\Phi = \Phi(\mathscr{A})$ the mapping from A to I_r defined by

$$\Phi(a_i) = j \text{ if and only if } a_i \in X_j .$$

The mapping Φ will be called the canonical mapping associated with the partition $\mathscr{A}$. Given an F-square F of type (n,λ) defined on a set A and the canonical mapping Φ associated with an h-regular partition $\mathscr{A}$ of A, we replace each element x of F by $\Phi(x)$ and obtain an F-square of type $(n,\lambda h)$. We shall call such a square the descendant of F by Φ and denote it by $\Phi(F)$.

OBSERVATION

If F is an F-square over the set A and Φ is the canonical mapping associated with an 1-regular partition of A, we can identify $\Phi(F)$ with F.

PROPOSITION 2.1. Let F_1, F_2 be F-squares of type (n,λ_1) and (n,λ_2) respectively, defined on the sets A_1 and A_2; let Φ_1 be the canonical mapping associated with an h_1-regular partition $\mathscr{A}_1 = \{X_1, X_2, \ldots, X_r\}$ of A_1 and let Φ_2 be the canonical mapping associated with an h_2-regular partition $\mathscr{A}_2 = \{Y_1, Y_2, \ldots, Y_s\}$ of A_2; if $F_1 \perp F_2$ then $\Phi_1(F_1) \perp \Phi_2(F_2)$.

PROOF. Let us observe first of all that the sets $X_i \times Y_j$ $(i=1,2,\ldots,r;\ j=1,2,\ldots,s)$ constitute a partition h_1h_2-regular of $A_1 \times A_2$; further, since each element $(a,b) \in X_i \times Y_j$ appears $\lambda_1\lambda_2$ times in the superimposition of F_1 with F_2, the pair $(i,j) \in I_r \times I_s$ appears $\lambda_1\lambda_2 h_1h_2$ times in the superimposition of $\Phi_1(F_1)$ with $\Phi_2(F_2)$ and that proves the assertion.

PROPOSITION 2.2. Given an orthogonal system $\mathscr{F}$ of F-squares, if one replaces one particular square of $\mathscr{F}$ by a set of descendants which are pairwise orthogonal, one obtains an orthogonal system again.

PROOF. It follows immediately from Proposition 2.1. and from the observation preceding it.

PROPOSITION 2.3. Let F be an F-square of type (n,λ) defined on a set A; let Φ be the canonical mapping associated with an h-regular partition $\mathscr{A} = \{X_1, X_2, \ldots, X_r\}$ of A and let Ψ be the canonical mapping associated with a k-regular partition $\mathscr{B} = \{Y_1, Y_2, \ldots, Y_s\}$ of A; then if the condition

$$\text{(I)} \qquad |X_i \cap Y_j| = \frac{h}{s} = \frac{k}{r} \qquad (i=1,2,\ldots,r;\ j=1,2,\ldots,s)$$

holds, then $\Phi(F) \perp \Psi(F)$ and conversely.

PROOF. Superimposing $\Phi(F)$ and $\Psi(F)$ each pair $(i,j) \in I_r \times I_s$ appears $\lambda|X_i \cap Y_j|$ times in each row and hence $n\lambda|X_i \cap Y_j|$ times altogether; it follows that

$\Phi(F) \perp \Psi(F)$ if and only if the cardinality of the sets $X_i \cap Y_j$ is the same for each $i=1,2,\dots,r$; $j=1,2,\dots,s$ and that proves the assertion.

Two mappings Φ and Ψ satisfying the hypotheses of Proposition 2.3. together with the condition (I) will be called orthogonal and denoted briefly by $\Phi \perp \Psi$.

PROPOSITION 2.4. Let $(G,+)$ be an abelian group; let H, K be sobgroups of G and Φ , Ψ the canonical mappings of G onto the quotient groups G/H ang G/K respectively. If

$$\text{(II)} \qquad H + K = \{ x+y : x\in H , y\in K = G \}$$

then $\Phi \perp \Psi$.

PROOF. Since $H\cap K$ is a subgroup of G , it sufficies to prove that for each $X\in G/H$ and for each $Y\in G/K$ the condition $X\cap Y\in G/H\cap K$ holds. In fact if $X = a + H$ and $Y = b + K$, then it follows from (II) that $X = y + H$ for some element $y\in K$ and $Y = x + K$ for some element $x\in H$. If $z\in X\cap Y$ then we have

$$z = y + x' = x + y'$$

with

$$x' - x = y' - y\in H\cap K \ .$$

Hence

$$z = (x + y' - y + y) \in (x + y) + H\cap K$$

and that means

$$X\cap Y \subseteq (x + y) + H\cap K \ .$$

Similarly one shows that $(x + y) + H\cap K \subseteq X\cap Y$ and that proves the assertion.

PROPOSITION 2.5. For each prime power q and for each F-square F of type (n,λ) defined on a set A of cardinality $m=q^k$, there exists $r = (q^k - 1)/(q - 1)$ pairwise orthogonal descendants of F of type $(n,\lambda q^{k-1})$ defined on the set I_q.

PROOF. Let us consider a k-dimensional vector space V over $GF(q)$; for any two distinct (k-1)-dimensional vector subspaces W_1 , W_2 of V we have $W_1 \oplus W_2 = V$. Since there are $r = (q^k - 1)/(q - 1)$ distinct (k-1)-dimensional vector subspaces of V , it follows from the previous proposition that F admits r pairwise orthogonal descendants of type $(n,\lambda q^{k-1})$.

PROPOSITION 2.6. Let $\mathscr{F}$ be a complete system of t F-squares of order n ; if one of the squares of the system is defined over a cardinality $m = q^k$ for a prime power q , then there exists a complete system of $s = t - 1 + (q^k-1)/(q-1)$ F-squares of order n .

PROOF. An immediate consequence of Proposition 2.2. and 2.5.

The Proposition 2.6. permits, in a simple way, to construct complete systems either of known type or with a variable number of symbols. For istance we can obtain complete systems of both types by replacing one or more latin squares, written in the first column of the following table, with the respective descendant written beside each of them.

$$\begin{array}{cccc}0&1&2&3\\1&0&3&2\\2&3&0&1\\3&2&1&0\end{array}\qquad\begin{array}{cccc}0&0&1&1\\0&0&1&1\\1&1&0&0\\1&1&0&0\end{array}\qquad\begin{array}{cccc}0&1&0&1\\1&0&1&0\\0&1&0&1\\1&0&1&0\end{array}\qquad\begin{array}{cccc}0&1&1&0\\1&0&0&1\\1&0&0&1\\0&1&1&0\end{array}$$

$$\begin{array}{cccc}0&2&3&1\\1&3&2&0\\2&0&1&3\\3&1&0&2\end{array}\qquad\begin{array}{cccc}0&1&1&0\\0&1&1&0\\1&0&0&1\\1&0&0&1\end{array}\qquad\begin{array}{cccc}0&0&1&1\\1&1&0&0\\0&0&1&1\\1&1&0&0\end{array}\qquad\begin{array}{cccc}0&1&0&1\\1&0&1&0\\1&0&1&0\\0&1&0&1\end{array}$$

$$\begin{array}{cccc}0&3&1&2\\1&2&0&3\\2&1&3&0\\3&0&2&1\end{array}\qquad\begin{array}{cccc}0&1&0&1\\0&1&0&1\\1&0&1&0\\1&0&1&0\end{array}\qquad\begin{array}{cccc}0&1&1&0\\1&0&0&1\\0&1&1&0\\1&0&0&1\end{array}\qquad\begin{array}{cccc}0&0&1&1\\1&1&0&0\\1&1&0&0\\0&0&1&1\end{array}$$

3 - EXTENSION OF THE MACNEISH THEOREM

MacNeish in 1922 proved the following theorem (cf. [4]):

THEOREM 3.1. If we let the prime decomposition of a number n be $n = p_1^{e_1} p_2^{e_2} \dots p_m^{e_m}$ then there exists a set of $\min(p_1^{e_1}, p_2^{e_2}, \dots, p_m^{e_m})$ pairwise orthogonal latin squares of order n .

J.P.Mandeli and W.T.Federer in 1983 (cf. [5]) gave a method for constructing orthogonal systems of F-squares of composite order n with a variable number of symbols, using a technique which is an extension of the MacNeish theorem for latin squares. If $n = p_1^{e_1} p_2^{e_2} \dots p_m^{e_m}$ then the system consists of the F-squares defined as follows:

$$L^j_{n_1} \oplus L^j_{n_2} \oplus \dots \oplus L^j_{n_m} \qquad j = 1, 2, \dots , n_1-1$$

$$0_{n_1} \oplus L^j_{n_2} \oplus \dots \oplus L^j_{n_m} \qquad j = n_1, n_1+1, \dots, n_2-1$$

$$\vdots$$

$$0_{n_1 n_2 \dots n_{m-1}} \oplus L^j_{n_i+1} \oplus \dots \oplus L^j_{n_m} \qquad j = n_i, n_i+1, \dots, n_{i+1}-1$$

$$\vdots$$

$$0_{n_1 n_2 \dots n_{m-1}} \oplus L^j_{n_m} \qquad j = n_{m-1}, n_{m-1}+1, \dots, n_m-1 \;;$$

we have denoted by $\mathscr{L}_i = L^1_{n_i}, L^2_{n_i}, \dots, L^{n_i-1}_{n_i}$ a complete system of latin squares of order $n_i := p_i^{e_i}$, by 0_n the square matrix of order n with all entries equal to 0 and by $\oplus$ the direct sum of matrices. By Proposition 2.6. it is possible to replace each square of order $p_i^{e_i}$ with a system of $t=(p_i^{e_i}-1)/(p_i-1)$ F-squares. If s denotes the greatest of the number $s_i=(p_i^{e_i}-p_1^{e_1})((p_i^{e_i}-1)/(p_i-1))$ $i=2,3,\ldots,m$ we can then state the following Theorem

THEOREM 3.2. If the prime decomposition of a number n is given by $n = p_1^{e_1} p_2^{e_2} \cdots p_m^{e_m}$ $(p_1^{e_1} < p_2^{e_2} < \ldots < p_m^{e_m})$ then there exists a system of $t = s + p_1^{e_1} - 1$ F-squares of order n with a variable number of symbols; such system contains the system of pairwise orthogonal latin squares constructed by MacNeish.

EXAMPLE. If $n = 2^3 \cdot 3^3 \cdot 31$ we construct a system of 76 F-squares which contains:

a) the system of 7 pairwise orthogonal latin squares constructed by MacNeish;

b) 23 F-squares of type $(n, 2^3 \cdot 3^2)$;

c) 46 F-squares of type $(n, 2^3 \cdot 3^2 \cdot 31)$.

ACKNOWLEDGEMENTS. Work done within the sphere of GNSAGA of CNR, partially supported by MPI.

REFERENCES

[1] J.P.Mandeli, F.C.H.Lee, W.T.Federer, On the construction of orthogonal F-squares of order n from an orthogonal array (n,k,s,2) and an OL(s,t) set ; J.Statist. Plan.Inference 5 (1981) 267-272.

[2] A.Hedayat, D.Raghavarao, E.Seiden, Further contributions to the theory of F-squares design ; Ann. Statist. 3 (1975) 712-716 .

[3] W.T.Federer, On the existence and construction of a complete set of orthogonal F(4t;2t,2t)-squares design ; Ann.Statist. 3 (1977) 561-564 .

[4] H.F.MacNeish, Euler's squares ; Ann. Math. 23 (1922) 221-227 .

[5] J.P.Mandeli, W.T.Federer, An extension of MacNeish's Theorem to the construction of sets of pairwise orthogonal F-squares of composite order; Util. Math. 24 (1983) 87-96 .

Annals of Discrete Mathematics 30 (1986) 291–296
© Elsevier Science Publishers B.V. (North-Holland)

RIGHT S-n-PARTITIONS OF A GROUP AND REPRESENTATION OF GEOMETRICAL SPACES OF TYPE "n-STEINER"

Domenico Lenzi
Dipartimento di Matematica
Università degli Studi
LECCE (Italy)

SUMMARY. In [9] we gave a generalization of the notions of subgroup and of S-partition of a group, in order to obtain a description of all the linear spaces having a transitive group of collineations, by means of a method that generalizes some methods used in particular cases like transitive projective planes, transitive linear spaces with a parallelism (in the sense of André) and others (see bibliography).

Here, after some considerations on geometrical spaces, we extend our method, by means of the concept of right S-n-partition of a group (see def.6), in order to obtain a description of all geometrical spaces $(E,\mathcal{B})$, where $\mathcal{B} \subseteq \mathcal{P}(E)$, of type "n-Steiner" (i.e.: if $P_1,\dots,P_{n+1}$ are n arbitrary and pairwise different elements of E, then there is a unique $B \in \mathcal{B}$ such that $\{P_1,\dots,P_{n+1}\} \subseteq B$) having a transitive group of automorphisms.

N.1.PRELIMINARIES AND RECALLS. Let G be a group and S a subgroup of G. We shall say that a subset $\mathcal{Q}$ of $\mathcal{P}(G)$ is a partial right S-covering of G if the following properties hold:

(1) $S \subseteq \bigcap_{A \in \mathcal{Q}} A$;

(2) $\forall A \in \mathcal{Q}$, $\forall a \in A$: $Aa^{-1} \in \mathcal{Q}$.

REMARK 1. One can immediately verify that:

(a) $Sa \subseteq A$, for every $A \in \mathcal{Q}$. Indeed $S \subseteq Aa^{-1}$, by (2) and (1).

(b) Let $I := \bigcap_{A \in \mathcal{Q}} A$. Then, by (2), $Ai^{-1} \supseteq I$ for every $A \in \mathcal{Q}$ and $i \in I$; thus $A \supseteq Ii$ and hence I is a submonoid of G. This ensures that the set of the subgroups of G included in I has a maximum element $H_1=\{i\in G: Ii=I\}=\{i\in G: iI=I\}=\{i\in G; \forall A\in\mathcal{Q}, iA=A\}$.

(c) If we set $\mathcal{Q}G:=\{Ag: A\in\mathcal{Q}, g\in G\}$, then the group of the right translations determined by the elements of G is a group of automorphisms of the geometrical space $(G, \mathcal{Q}G)$. Moreover $\mathcal{Q}$ coincides with the set of the elements $B\in\mathcal{Q}G$ such that $1\in B$; as a consequence, for $x\in G$ and $B\in\mathcal{Q}G$, $x\in B$ if and only if $Bx^{-1}\in\mathcal{Q}$.

(d) For every $A, A'\in\mathcal{Q}$, let us set $A\equiv A'$ when $A'=Aa^{-1}$, where a is a suitable element of A. In such a manner we define an equivalence relation on $\mathcal{Q}$.

Now, for an arbitrary right S-covering $\mathcal{C}$ of G, let: $E:=\{Sx: x\in G\}$ (the set of the right cosets of S in G); $\langle B\rangle:=\{Sx: Sx\subseteq B\}$ (for $B\in\mathcal{Q}G$) and $\mathcal{B}:=\{\langle B\rangle : B\in\mathcal{Q}G\}$.

If we set $(Sx)*y=S(xy)$ (for every $x,y\in G$), then G becomes a transitive generalized automorphism group of the geometrical space $(E,\mathcal{B})$ (i.e.: the function that associates to every $g\in G$ the permutation $\hat{g}$ acting on E like g is an homomorphism from G into the automorphism group of $(E,\mathcal{B})$).

Conversely, every geometrical space $(E'\ \mathcal{B}')$ having a transitive generalized automorphism group G can be obtained (but for an isomorphism) in the previous manner. Let indeed P be a fixed element of E' and $S:=\{g\in G: Pg=P\}$ (the stabilizer of P in G); if we associate to every $Q=Pg\in E'$ the right coset Sg, then we obtain a bijective function f from E' onto the set E* of the right cosets of S in G; as a consequence we can construct a geometrical space $(E^*, \mathcal{B}^*)$ (by setting $\mathcal{B}^*:=\mathcal{B}'f$); furthermore G becomes in a natural way (by setting $(Sx)*y:=(Sx)f^{-1}yf=S(xy)$) a transitive generalized automorphism group of $(E^*, \mathcal{B}^*)$. Now let us $B^*_{\cup}:=\bigcup_{Sx\in B^*}Sx$ (for every $B^*\in\mathcal{B}^*$) and $\mathcal{Q}:=\{B^*_{\cup}: 1\in B^*_{\cup}\}$. It is easy to verify that $\mathcal{Q}$ is a partial right S-covering of G and $(E^*,\mathcal{B}^*)$ can be obtained from $\mathcal{Q}$ in the previous manner.

If $(E,\mathcal{B})$ is a geometrical space let us set, for $P\in E$, $\mathcal{B}_P=\{B\in\mathcal{B}: P\in B\}$ and $\bar{P}=\bigcap_{B\in\mathcal{B}_P}B$ (the closure of P in $\mathcal{B}$). Then $(E,\mathcal{B})$ is said to be a T_0 space when, for every $X,Y\in E$, $\bar{X}=\bar{Y}$ implies that $X=Y$; moreover, $(E,\mathcal{B})$ is said to be a T_1 space when, for every $X\in E$, $|\bar{X}|=1$.

REMARK 2. Obviously, if $(E,\mathcal{B})$ has a transitive generalized automorphism group and P is a fixed element of E, then $(E,\mathcal{B})$

is a T_0 space if and only if, for every $X \in E$, $\bar{X}=\bar{P}$ implies that $X=P$; moreover, $(E,\mathcal{B})$ is a T_1 space if and only if $\bar{P} = \{P\}$.

Now let us fix a subgroup S and a right S-covering $\mathcal{A}$ of G. The following propositions hold.

PROPOSITION 3. For every $g \in G$, $\bigcap_{A \in \mathcal{A}} (Ag) = \bigcap_{A \in \mathcal{A}} A$ if and only if $g \in H_1$ (see (b) in remark 1).

PROOF. The assert is obvious, since $\bigcap_{A \in \mathcal{A}}(Ag) = (\bigcap_{A \in \mathcal{A}} A)g = Ig$.

Q.E.D.

PROPOSITION 4. Let $(E,\mathcal{B})$ be the geometrical space associated to $\mathcal{A}$ with respect to S; then:

i) $(E,\mathcal{B})$ is a T_1 space if and only if $S = \bigcap_{A \in \mathcal{A}} A$.

ii) $(E,\mathcal{B})$ is a T_0 space if and only if $S=H_1$ (see (b) in remark 1).

PROOF. i) It is obvious. ii) For every $\langle B \rangle \in \mathcal{B}$, $Sg \in \langle B \rangle$ if and only if $g \in B$; then the closure $\overline{Sg}$ of Sg in $\mathcal{B}$ is the set of the right cosets of S included in $\bigcap_{g \in B \in \mathcal{B}} B = \bigcap_{A \in \mathcal{A}} (Ag)$. As a consequence, $\overline{Sg}$ is equal to the closure of S in $\mathcal{B}$ if and only if $\bigcap_{A \in \mathcal{A}} (Ag) = \bigcap_{A \in \mathcal{A}} A$; whence the thesis by proposition 3 and remark 2.

Q.E.D.

In [9] (see theor. 9) we proved that a group G is a transitive generalized group of automorphism (collineation) of a linear space if and only if it admits a right generalized S-partition. We can say that a subset $\mathcal{A}$ of $\mathcal{P}(G)$ is a right generalized S-partition of G if it is a partial right S-covering of G and the following properties hold:

(3) $\bigcup_{A \in \mathcal{A}} A = G$;

(4) $\forall A_1, A_2 \in \mathcal{A} : A_1 \neq A_2 \Rightarrow A_1 \cap A_2 = S$;

(5) $S \notin \mathcal{A}$.

By (1),(2) and (4) it is easy to verify that, for every element A of a right generalized S-partition $\mathcal{A}$ of G, the following property holds (cfr. [9], def.2):

(6) $S \subseteq A$ and $(\forall x,y \in A: Ax^{-1} \neq Ay^{-1} \Rightarrow Ax^{-1} \cap Ay^{-1} = S)$.

We observe that the foregoing conditions can be reformuled, by virtue of the following

PROPOSITION 5. Let $\mathcal{A} \subseteq \mathcal{P}(G)$ and $I := \bigcap_{A \in \mathcal{A}} A$; furthermore let property (2) hold, and:

$$(4')\ \forall A_1, A_2 \in \mathcal{A}: A_1 \neq A_2 \Rightarrow A_1 \cap A_2 = I.$$

Then I is a subgroup of G.

PROOF. We can suppose $|\mathcal{A}| \neq 1$. Now let A_1, A_2 different elements of $\mathcal{A}$. Then, for every $i,j \in I$, $ij^{-1} \in A_1 j^{-1} \cap A_2 j^{-1}$. But $A_1 j^{-1} \neq A_2 j^{-1}$, hence (by (2) and (4')) $A_1 j^{-1} \cap A_2 j^{-1} = I$, therefore $ij^{-1} \in I$.

Q.E.D.

N.2. THE CASE OF THE GEOMETRICAL SPACES OF TYPE "n-STEINER".

Let us give the following

DEFINITION 6. We shall say that a subset $\mathcal{A}$ of the power set $\mathcal{P}(G)$ of a group G is a right S-n-partition of G if it is a partial right S-covering such that the following properties hold:

(7) $\forall g_1, \ldots, g_n \in G\ \exists A \in \mathcal{A} : \{g_1, \ldots, g_n\} \subseteq A$.

(8) $\forall A_1, A_2 \in \mathcal{A} : A_1 \neq A_2 \Rightarrow A_1 \cap A_2$ is the union of $m \leq n$ right cosets of S in G (where m is depending on A_1 and A_2).

It is easy to verify that if $(E', \mathcal{B}')$ is a geometrical space of type "n-Steiner" (see summary) having a transitive generalized automorphism group G, then the set $\mathcal{A} \subseteq \mathcal{P}(G)$ from which $(E', \mathcal{B}')$ can be obtained (*) is a right S-n-partition of G.

Conversely, let $\mathcal{A}$ be a right S-n-partition of G; then the associated geometrical space $(E, \mathcal{B})$ (see N.1) is of type "n-Steiner". This is an immediate consequence of the following

PROPOSITION 7. If $Sh_1, \ldots, Sh_{n+1}$ are n arbitrary and pairwise different right cosets of S in G, then there is a unique subset B of G of type Ag, with $A \in \mathcal{A}$ and $g \in G$, such that $\bigcup_{i=1}^{n} Sh_i \subseteq B$.

PROOF. By (7) and (1), there is $A \in \mathcal{A}$ such that $\{h_1 h_{n+1}^{-1}, \ldots, h_n h_{n+1}^{-1}, 1\} \subseteq A$, hence $\{h_1, \ldots, h_{n+1}\} \subseteq Ah_{n+1}$, thus $\bigcup_{i=1}^{n+1} Sh_i \subseteq Ah_{n+1}$ (since $SA=A$).

On the other hands, if $A_1 g_1 \cap A_2 g_2 \supseteq \bigcup_{i=1}^{n+1} Sh_i$ (with $A_1, A_2 \in \mathcal{A}$ and $g_1, g_2 \in G$) then $S \subseteq \bigcup_{i=1}^{n+1} Sh_i h_{n+1}^{-1} \subseteq A_1 g_1 h_{n+1}^{-1} \cap A_2 g_2 h_{n+1}^{-1}$. As a consequence (see the second part of (c) in remark 1) $A_1 g_1 h_{n+1}^{-1} = A_2 g_2 h_{n+1}^{-1}$ by virtue of (8), and hence $A_1 g_1 = A_2 g_2$.

Q.E.D.

(*) But for an isomorphism; see N.1.

REMARK 8. We observe that if $\mathcal{A}$ is a right S-n-partition of G, then (by (8) and (2)) for every $A \in \mathcal{A}$ the following property holds:

(9) $\forall x,y \in A: Ax^{-1} \neq Ay^{-1} \Longrightarrow Ax^{-1} \cap Ay^{-1}$ is the union of $m \leq n$ right cosets of S in G (where m is depending on A).

We shall say that a subset A of G is a right S-n-block of G if $S \subseteq A$ and property (9) holds (**); moreover if A is a right S-n-block of G but not a right S-(n-1)-block, we shall say that A is a proper right S-n-block of G.

Obviously, every right S-o-block is a subgroup of G.

PROPOSITION 9. A necessary (and trivially sufficient) condition for a subset A of G to be a right S-n-block of G is that:

(10) $S \subseteq A$,

and

(11) $\forall g_1, g_2 \in G : Ag_1 \neq Ag_2 \Longrightarrow Ag_1 \cap Ag_2$ is the union of $m \leq n$ right cosets of S in G.

PROOF. It is enough to prove that the condition (11) is necessary. Hence let A be a right S-n-block of G, let $Ag_1 \neq Ag_2$ (where $g_1, g_2 \in G$) and let $t \in Ag_1 g_2^{-1} \cap A$. Then $Ag_1 g_2^{-1} t^{-1} \neq At^{-1}$ and $tg_2 g_1^{-1} \in A \cap Ag_2 g_1^{-1}$. Consequently, by (9), $Ag_1 g_2^{-1} t^{-1} \cap At^{-1}$ (and hence also $Ag_1 \cap Ag_2$) is the union of $m \leq n$ right cosets of S in G.

Q.E.D.

We conclude by observing that every right S-n-block A of G determines in a natural manner a right S-n-partition of G. In fact one can consider the set $\mathcal{A} := \mathcal{A}_1 \cup \mathcal{A}_2$, where $\mathcal{A}_1 = \{A_1 \in \mathcal{P}(G) : A_1 = Aa^{-1}$, with $a \in A\}$ and $\mathcal{A}_2$ is the class of the subset A_2 of G such that:

j) $S \subseteq A_2$;

jj) A_2 is the union of n+1 pairwise different right cosets of S in G;

jjj) $\forall A_1 \in \mathcal{A}_1 : A_2 \not\subseteq A_1$.

Consequently, for every group G and for every subgroup S of G,

(**) In |9| and |10| we called right S-blocks the right S-1-blocks (cfr. the previous property (6)) and proved several algebraical and geometrical properties.

we can have several (trivial) geometrical spaces of type n-Steiner, since if $m \leq n+1$ then the union A of m pairwise different right cosets of S in G such that $S \subseteq A$ is a right S-n-block of G.

BIBLIOGRAPHY

[1] Andre' J., Über Parallelstrukturen , Teil I, Teil II, Teil III, Teil IV, Math. Z.76(1961), 85-102,155-163,240-256, 311-333.
[2] Biliotti,M., S-spazi ed O-partizioni, Boll. U.M.I.(5) 14-A(1977), 333-342.
[3] Biliotti,M., Sulle strutture di traslazione, Boll.U.M.I. (5) 15-A(1978), 667,677.
[4] Biliotti, M., Strutture di André ed S-spazi con traslazioni, Geom. Dedicata, 10 (1981), 113,128.
[5] Biliotti, M., Strutture di André ed S-spazi con traslazioni II, Ann. di Mat. pura ed appl., (IV), Vol. CXXXV.(1983), 151-172.
[6] Biliotti, M., Herzer A., Zur Geometrie der Translationsstrukturen mit eigentlichen Dilatationen, Abh. Math.Sem.Hamburg 53 (1983), 1-27.
[7] Karzel, H., Bericht über projektive Inzidenzgruppen, Jber. Deutsch.Math. Verein. 67(1964) 58-92.
[8] Lingenberg,R., Ueber Gruppen projectiver Kollineationen, welche eine perspektive Dualität invariant lassen, Archiv. der Math. 13(1962) 385-400.
[9] Lenzi, D., Representation of a linear space with a transitive group of collineations by a generalized S-partition of a group, Atti Convegno "Geom.combin." La Mendola-Italy (1982), 471-482.
[10] Lenzi,D., On a characterization of finite projective planes having a transitive group of collineations, to appear.
[11] Scarselli, A., Sulle S-partizioni regolari di un gruppo finito, Rend. Acc.Naz.Lincei, LXII (1977), 300-304.
[12] Zappa,G., Sui piani grafici finiti transitivi e quasi transitivi, Ricerche di Matematica,II (1953), 274-287.
[13] Zappa,G., Sugli spazi generali quasi di traslazione, Le Matematiche, 19 (1964), 127-143.
[14] Zappa, G., Sulle S-partizioni di un gruppo finito, Ann.di Mat.74, (1966), 1-24.
[15] Zappa, G., Partizioni ed S-partizioni dei gruppi finiti, Symposia Math. I(1969), 85-94.

Annals of Discrete Mathematics 30 (1986) 297–302

ON BLOCK SHARING STEINER QUADRUPLE SYSTEMS

Giovanni LO FARO (*)

Dipartimento di Matematica dell'Università,
Via C. Battisti 90, 98100 MESSINA, Italy

Abstract.

We determine, for all $v \equiv 4$ *or* 8 (mod. 12), *the set of all those numbers* k *for which there exist two Steiner quadruple systems of order* v *on the same set that share exactly* k *blocks.*

Introduction.

A Steiner quadruple system *(SQS)* is a pair (Q,q) where Q is a finite set and q is a collection of four element subsets of Q (called blocks) such that every three element subset of Q belongs to exactly one block of q.

The number $|Q| = v$ is called the order of the $SQS(Q,q)$ and $v \equiv 2$ or 4 (mod. 6) is an obvious necessary existence condition for an SQS of order v ($SQS(v)$). On the other hand, in 1960 Hanani proved |8| that this condition is also sufficient. Therefore in saying that a certain property concerning $SQS(v)$ is true for all v it is understood that $v \equiv 2$ or 4 (mod. 6).

If (Q,q) is an $SQS(v)$ then $|q| = q_v = v(v-1)(v-2)/24$.

A quadruple system (Q,q) has a proper subsystem if there exist sets $R \subset Q$ and $r \subset q$ such that (R,r) is an SQS with $|r| < |q|$.

A. Hartman |9| proved that for every $v \geq 16$ there exists an $SQS(v)$ with a proper subsystem of order 8.

Earlier results on subsystems of $SQSs$ can be found in |3|.

In their excellent survey of Steiner quadruple systems, C.C. Lindner and A. Rosa |11| rise a series of questions. One of them is: "For a given v, for which $k \leq q_v$ is it possible to construct a pair of $SQS(v)s$ having exactly k blocks in common?".

Denote by $J(v) = \{k : \exists\ SQS(v)\ (Q,q_1),(Q,q_2)$ such that $|q_1 \cap q_2| = k\}$; $I(v) = \{0,1,2,\dots,q_v-14,q_v-12,q_v-8,q_v\}$, $v \geq 8$.

So the author's best knowledge, the only results concerning this problem are:

(i) $J(4)=1$; $J(8)=\{0,2,6,14\}=I(8)$; $J(10)=\{0,2,4,6,8,12,14,30\}$;
$J(14)\supseteq I(14)-\{48,50,52,57,58,60,62,63,\ldots,77,79,83\}$;
$\{q_{14}-8=83 ; q_{14}-12=79 ; q_{14}-14=77 ; q_{14}-15=76\}\cap J(14)=\emptyset$

(ii) $J(v)\subseteq I(v)$, for all $v\geq 8$

(iii) $J(v)=I(v)$, for all $v=2^{n+1} ; 5\cdot 2^n ; 7\cdot 2^n$, $n\geq 2$.

The aim of this paper is the investigation of the block intersection problem for $SQSs$ of order $2v$ obtained by doubling suitable systems of order v . In particular it is obtained that $J(2v)=I(2v)$.

Although he was not able to prove it yet, the author feels that the following conjecture makes sense.

Conjecture: $J(v)=I(v)$, for all $v\geq 16$.

2. Preliminaries.

In this section we describe two constructions for quadruple systems or order $2v$ which are the main tool used in what follows.

Construction A (well known e.g. see [11]).

Let (X,a) and (Y,b) be any two $SQS(v)$ with $X\cap Y=\emptyset$. Let $F=\{F_1,F_2,\ldots,F_{v-1}\}$ and $G=\{G_1,G_2,\ldots,G_{v-1}\}$ be any two 1-factorizations of K_v (the complete graph on v vertices) on X and Y respectively, and let α be any permutation on the set $\{1,2,\ldots,v-1\}$.

Define a collection s of blocks of $S=X\cup Y$, as follows:

(1) Any block belonging to a or b belongs to s ;

(2) If $x_1,x_2\in X$ $(x_1\neq x_2)$ and $y_1,y_2\in Y$ $(y_1\neq y_2)$ then $(x_1,x_2,y_1,y_2)\in s$ if and only if $\{x_1,x_2\}\in F_i$, $\{y_1,y_2\}\in G_j$ and $\alpha(i)=j$.

It is a routine matter to check that (S,s) is an $SQS(2v)$. We will denote (S,s) by $[X\cup Y][a,b,F,G,\alpha]$.

Construction B (compare [3]).

Let (Q,q) be an $SQS(v)$, Q' a finite set such that $|Q|=|Q'|$, $Q\cap Q'=\emptyset$ and let φ be a bijection from Q onto Q' with $x_i'=\varphi(x_i)$, for every $x_i\in Q$.

Obviously, (Q',q') is an $SQS(v)$ (the $SQS(v)$ obtained from (Q,q)) where $q'=\varphi(q)=\{\varphi(c) : c\in q\}$.

If $q_1\subset q$ we define a collection $p(q_1)$ of blocks of $P=Q\cup Q'$ as follows:

(1) Any block belonging to q_1 or q_1' $(=\varphi(q_1))$ belongs to $p(q_1)$;

(2) $\{(x_1,x_2,x_3',x_4');(x_1,x_3,x_2',x_4');(x_1,x_4,x_2',x_3');(x_2,x_3,x_1',x_4');(x_2,x_4,x_1',x_3');$ $(x_3,x_4,x_1',x_2')\}\subset p(q_1)$ if and only if $(x_1,x_2,x_3,x_4)\in q_1$;

(3) $\{(x_1,x_2,x_3,x_4');(x_1,x_2,x_4,x_3');(x_1,x_3,x_4,x_2');(x_2,x_3,x_4,x_1');(x_1',x_2',x_3',x_4);$ $(x_1',x_2',x_4',x_3);(x_1',x_3',x_4',x_2);(x_2',x_3',x_4',x_1)\}\subset p(q_1)$ if and only if $(x_1,x_2,x_3,x_4)\in q-q_1$;

(4) $(x_1,x_2,x_1',x_2')\in p(q_1)$, for every $x_1,x_2\in Q$, $x_1\neq x_2$.

It is a routine matter to check that $(P,p(q_1))$ is an $SQS(2v)$. We will denote $(P,p(q_1))$ by $((Q\cup Q'),(q,q_1))$. If $q_1=\emptyset$ then we denote $(P,p(\emptyset))$ by (P,p).

3. $SQS(2v)$s with blocks in common.

In this section we will determine $J(2v)$.

Take $X=\{1,2,\ldots,v\}$, v an even positive integer, and let F and G be two 1-factorizations of X where $F=\{F_1,F_2,\ldots,F_{v-1}\}$ and $G=\{G_1,G_2,\ldots,G_{v-1}\}$. We will say that F and G have k edges in common if and only if $k=\sum_{i=1}^{v-1}|F_i\cap G_i|$.

In |12| C.C. Lindner and W.D. Wallis gave a complete solution to the intersection problem for 1-factorizations. In particular, they showed that for any $v\geq 8$ there exist two 1-factorizations F and G with k edges in common for every $k\in\{0,1,2,\ldots,\frac{v(v-1)}{2}=n\}-\{n-1,n-2,n-3,n-5\}$.

Take $Q=\{1,2,\ldots,v\}$ and $Q'=\{1',2',\ldots,v'\}$ with $Q\cap Q'=\emptyset$ and let $F=\{F_1,F_2,\ldots,F_{v-1}\}$ be 1-factorization of Q, then $F'=\{F_1',F_2',\ldots,F_{v-1}'\}$ is a 1-factorization of Q' (called the 1-factorization derived from F) where $\{x_i',x_h'\}\in F_j'$ if and only if $\{x_i,x_h\}\in F_j$.

THEOREM 3.1. *Assume* $v\geq 8$. *If* $H=\{0,1,2,\ldots,\frac{v(v-1)}{2}=n\}-\{n-1,n-2,n-3,n-5\}$ *then* $H\subset J(2v)$.

Proof. Let (Q,q) be an $SQS(v)$. If $k\in H$ then there exist two 1-factorizations $F=\{F_1,F_2,\ldots,F_{v-1}\}$ and $G=\{G_1,G_2,\ldots,G_{v-1}\}$ such that $\sum_{i=1}^{v-1}|F_i\cap G_i|=k$.

If e is the identity permutation, it is a routine matter to see that $[Q\cup Q'][q,q',F,G',e]$ and $((Q\cup Q'),(q,\emptyset))$ are two $SQS(2v)$s with exactly k blocks in common. The statement follows.

It is well known [1] that if x and y are even positive integers and $x\geq 2y$, then there exists a 1-factorization of order x containing a sub-1-factorization of order y (a 1-factorization of order n is a 1-factorization of K_n).

THEOREM 3.2. *Let* $n=\frac{v(v-1)}{2}$, $v\geq 16$. *If* $k\in\{n-1,n-2,n-3,n-5,n+1,n+2,\ldots$ $\ldots,n+98,n+100,n+104,n+112)\}$, *then* $k\in J(2v)$.

Proof. Let (Q,q) be an $SQS(v)$ containing an $SQS(8)$ (R,r) as a subsystem.

It is straightforward to see that the $SQS(2v)$ $((Q\cup Q'),(q,\emptyset))=(P,p)$ contains the $SQS(16)$ $((R\cup R'),(r,\emptyset))=(T,z)$.

Let $F=\{F_1,F_2,\ldots,F_{v-1}\}$ and $F^*=\{F_1^*,F_2^*,\ldots,F_7^*\}$ be two 1-factorizations of Q and R respectively with $F_i^*\subseteq F_i$, for every $i=1,2,\ldots,7$. It is equally easy to see that the $SQS(2v)$ $[Q\cup Q']$ $[q,q',F,F',i]=(P,s)$ contains the $SQS(16)$ $[R\cup R']$ $[r,r',F^*,F^{*\prime},i]=(T,t)$.

Obviously, we have:

$$(1)\qquad |p\cap s|=\frac{v(v-1)}{2}\quad;\qquad (2)\qquad |z\cap t|=28$$

Let $k\in J(16)=I(16)$. It is possible to construct two $SQS(16)$ (T,a) and (T,b) such that $|a\cap b|=k$.

If $(p-z)\cup a=p'$ and $(s-t)\cup b=s'$, then (P,p') and (P,s') are two $SQS(2v)$s with exactly $\frac{v(v-1)}{2}-28+k$ blocks in common.

This completes the proof of the theorem.■

LEMMA 3.3. $0\leq h\leq k\leq\frac{v(v-1)(v-2)}{24}=q_v$. If $l=q_{2v}-8(k-h)$ then $l\in J(2v)$.

Proof. Let (Q,q) be an $SQS(v)$ and take q_1,q_2 with $q_1\subseteq q_2\subseteq q$, $|q_1|=h$, $|q_2|=k$. Clearly, the number of distinct blocks of $(P,p(q_1))$ and $(P,p(q_2))$ is: $2(k-h)+6(k-h)=8(k-h)$ and then $|p(q_1)\cap p(q_2)|=q_{2v}-8(k-h)$.

REMARK. Let $q_1=\emptyset$ and $q_2\subseteq q$, $|q_2|=t\leq q_v$; from Lemma 3.3, it follows that $q_{2v}-8t \in J(2v)$, for every $t \in \{0,1,2,\ldots,q_v\}$.

THEOREM 3.4. *Assume* $v \geq 16$. *If* $\frac{v(v-1)}{2} + 98 < l < q_{2v}-140$, *then* $l \in J(2v)$.

Proof. We start noticing that if $q_{2v}-l = 8t+h$, $h = 0,1,2,\ldots,7$, then

$$17 + \frac{4-h}{8} < t < q_v - 12 - \frac{h+2}{8}$$

Let (Q,q) be an $SQS(v)$ containing an $SQS(8)$ (R,r) and let $r \subset q_1 \subset q$ with $|q_1| = t+12(<q_v)$. By Lemma 3.3 $(P,p(r))$ and $(P,p(q_1))$ are two $SQS(2v)$s such that $|p(r) \cap p(q_1)| = q_{2v}-8(t-2)$.

It is straightforward that both $(P,p(r))$ and $(P,p(q_1))$ contain the $SQS(16)$ $((R \cup R'),(r,r))$; therefore $q_{2v}-8(t-2)-(140-k) \in J(2v)$, for all $k \in J(16) = I(16)$.

The statement follows by choosing $k \in \{117,118,\ldots,124\} \subset J(16)$.■

THEOREM 3.5. *Assume* $v \geq 16$. *If* $l \in I(2v)$ *and* $l \geq q_{2v}-140$ *then* $l \in J(2v)$.

Proof. It is well known [1] that if there exists an $SQS(v)$ with a sub-$SQS(u)$, then there exists an $SQS(2v)$ with a sub-$SQS(2u)$. Thus, there exists an $SQS(2v)$ with a sub-$SQS(16)$.

It is easy to check that $q_{2v}-(140-h) \in J(2v)$, for all $h \in J(16) = I(16)$.

This completes the proof.■

Combining together (iii) and Theorems 3.1, 3.2, 3.4, 3.5 we get our main result:

THEOREM 3.6. $J(v) = I(v)$ *for all* $v \equiv 4$ *or* $8 \pmod{12}$.

REFERENCES

[1] A. CRUSE, *On embedding incomplete symmetric latin squares*, J. Comb. Theory Ser. A (1974), 19-22.

[2] J. DOYEN and A. ROSA, *An updated bibliography and survey of Steiner systems*, Annals of Discrete Math. 7 (1980), 317-349.

[3] J. DOYEN and M. VANDENSAVEL, *Non isomorphic Steiner quadruple systems*, Bull. Soc. Math. Belgique 33 (1971), 393-410.

[4] M. GIONFRIDDO, *On the block intersection problem for Steiner quadruple systems*, Ars Combinatoria 15 (1983), 301-314.

|5| M. GIONFRIDDO and C.C. LINDNER, *Construction of Steiner quadruple systems having a prescribed number of blocks in common*, Discrete Math. 34 (1981), 31-42.

|6| M. GIONFRIDDO and G. LO FARO, *On Steiner systems* $S(3,4,14)$, to appear.

|7| M. GIONFRIDDO and M.C. MARINO, *On Steiner Systems* $S(3,4,20)$ *and* $S(3,4,32)$, Utilitas Math., 25 (1984), 331-338.

|8| H. HANANI, *On quadruple systems*, Canad. J. Math. 12 (1960), 145-157.

|9| A. HARTMAN, *Quadruple systems containing* $AG(3,2)$, Discrete Math., 39 (1982) (3), 293-299.

|10| C.C. LINDNER, *On the construction of non-isomorphic Steiner quadruple systems*, Colloq. Math. 29 (1974), 303-306.

|11| C.C. LINDNER and A. ROSA, *Steiner quadruple systems - a survey*, Discrete Math. 21 (1978), 147-181.

|12| C.C. LINDNER and W.D. WALLIS, *A note on one-factorizations having a prescribed number of edges in common*, Annals of Discrete Math. 12 (1982), 203-209.

|13| G. LO FARO, *On the set* $J(v)$ *for Steiner quadruple systems of order* $v = 7\cdot 2^n$ *with* $n \geq 2$, Ars Combinatoria, 17 (1984), 39-47.

|14| G. LO FARO, *Steiner quadruple systems having a prescribed number of quadruples in common*, to appear.

|15| G. LO FARO and L. PUCCIO, *Sull'insieme* $J(14)$ *dei sistemi di quaterne di Steiner*, to appear.

(*) Lavoro eseguito nell'ambito del GNSAGA e con finanziamento MPI (1984, 40%).

Annals of Discrete Mathematics 30 (1986) 303–310

ROOTS OF AFFINE POLYNOMIALS

Giampaolo Menichetti (^)
Dipartimento di Matematica, Università di Bologna, Italy

INTRODUCTION. Let $F = GF(q)$ be a Galois field of order $q = p^h$, where p is a prime, and let $K = GF(q^n)$ be an algebraic extension of a given degree $n \geqslant 1$. An *affine polynomial* (of $K[x]$ over F) is a polynomial of type

(1) $$P(x) = L(x) - b \ , \ b \in K,$$

with

(2) $$L(x) = \sum_{i=0}^{n-1} l_i x^{q^i} \ , \ l_i \in K.$$

If a basis $\{u_0, u_1, \dots, u_{n-1}\}$ of K over F is fixed then we can put $x = \sum_{i=0}^{n-1} x_i u_i$. Hence, the determination of (eventual) roots of the polynomial (1) in K can be reduced to the determination of solutions of a linear system of equations in indeterminates x_i and with coefficients in F (cf.[1], Chap.11). This procedure is, however, tedious also in the most simply cases and does not decise "a priori" how many roots in K exist.

In this paper, we prove that the equation (1) has roots in K if and only if the following system of linear equations

(3) $$\begin{aligned} & l_0 y_0 + l_1 y_1 + \dots + l_{n-1} y_{n-1} = b \\ & l_{n-1}^q y_0 + l_0^q y_1 + \dots + l_{n-2}^q y_{n-1} = b^q \\ & \cdots\cdots\cdots\cdots\cdots\cdots \\ & l^{q^{n-1}} y_0 + l_2^{q^{n-1}} y_1 + \dots + l_0^{q^{n-1}} y_{n-1} = b^{q^{n-1}} \end{aligned}$$

has solutions. Moreover, if (3) is solvable and if r is the rank of the matrices belonging to (3), then q^{n-r} gives us the number of solutions of (1) in K. Besides this, we show that the roots of (1) are expressible as functions of certain solutions of the linear system (3). In particular, the obtained results are useful also in the case that the coefficients of the polynomial (1) are not constant (cf.f.e.[2]).

(^) This research was supported in part by a grant from the M.P.I.(40% funds).

1. An $n\times n$ matrix of the type

$$\underline{A}(a_0,a_1,\dots,a_{n-1})=\begin{pmatrix} a_0 & a_1 & \dots & a_{n-1}\\ a_{n-1}^q & a_0^q & \dots & a_{n-2}^q\\ \dots\dots & & & \\ a_1^{q^{n-1}} & a_2^{q^{n-1}} & \dots & a_0^{q^{n-1}}\end{pmatrix},\quad a_i\in K,$$

is called *autocirculant*. Each row of $\underline{A}(a_0,a_1,\dots,a_{n-1})$ is obtained by the previous one if we permute the elements under the cyclic permutation $(0\ 1\dots n-1)$ and then apply the field automorphism $\alpha : K\to K$, $a\mapsto a^q$.

The sum and the product of autocirculant matrices are autocirculant matrices. In addition, if $\underline{A}$ is autocirculant, the transpose $\underline{A}^t$ is also autocirculant.

Let

$$\underline{T}=\underline{A}(0,1,0,\dots,0)$$

and $\underline{A}^q(a_0,a_1,\dots,a_{n-1})=\underline{A}(a_0^q,a_1^q,\dots,a_{n-1}^q)$. It is easy to verify

(4) $\quad \underline{A}^q=\underline{T}\,\underline{A}\,\underline{T}^{-1}$.

LEMMA 1. *An autocirculant matrix* $\underline{A}$ *has rank* r, $1\leq r\leq n-1$, *if and only if its first* r *rows (columns) are linearly* K*-independent and its* $(r+1)$*-th row (column) is a* K*-linear combination of the preceding rows (columns).*

Proof. The transpose of an autocirculant matrix is itself autocirculant. Thus it is sufficient to prove the statement for the columns of $\underline{A}$.

The assertion is an obvious consequence of the following observation.

Let $\underline{A}=(\underline{A}_0,\underline{A}_1,\dots,\underline{A}_{n-1})$. If

(5) $\quad \underline{A}_s=\sum_{i=0}^{s-1}k_i\underline{A}_i\ ,\ k_i\in K,$

$0<s<n-1$, then $\underline{A}_s^q=\sum_{i=0}^{s-1}k_i^q\underline{A}_i^q$ and therefore

$$\underline{A}_{s+1}=\sum_{i=0}^{s-1}k_i^q\underline{A}_{i+1}=\sum_{i=0}^{s-2}k_i^q\underline{A}_{i+1}+k_{s-1}^q\sum_{i=0}^{s-1}k_i\underline{A}_i=\sum_{i=0}^{s-1}k_i'\underline{A}_i\ ,k_i'\in K.$$

If we raise both sides of (5) to the powers $q^2,q^3,\dots,q^{n-s-1}$ and use previous arguments, we see that the columns $\underline{A}_{s+2}$, $\underline{A}_{s+3},\dots,\underline{A}_{n-1}$ can be expressed as a K-linear combination of $\underline{A}_0$, $\underline{A}_1,\dots,\underline{A}_s$. □

PROPOSITION 2. *If* $\underline{A}$ *is an autocirculant matrix of rank* r *then the homogeneous linear system*

(6) $\quad \underline{A}\,\underline{y}=\underline{0}\ ,\ \underline{y}=(y_0\ y_1\ \dots\ y_{n-1})^t$

has solutions of the type $\underline{z}=(z_0\ z_1\dots\ z_r\ 0\ \dots\ 0)^t$ *with* $z_r\neq 0$. *Furthermore, given* $\underline{A}'=\underline{A}(z_0,z_1,\dots,z_r,0,\dots,0)$, *we have* $\operatorname{rank}(\underline{A}')=n-r$ *and* $\underline{A}\,\underline{A}'^t=\underline{0}$.

Proof. If one has $\underline{A}=(\underline{A}_0,\underline{A}_1,\dots,\underline{A}_{n-1})$ then Lemma 1 guaranties the existence of the elements $a_i'\in K$ such that $\underline{A}_r=\sum_{i=0}^{n-1}a_i'\underline{A}_i$ holds. And, this proves the first part of the assertion.

Let $\underline{A}'^{t} = (\underline{A}'_0, \underline{A}'_1, \ldots, \underline{A}'_{n-1})$. By definition, one has $\underline{A}'_0 = \underline{z}$ and therefore $\underline{A}'_1 = \underline{T}^{-1}\underline{z}^q, \ldots, \underline{A}'_{n-1} = \underline{T}^{-(n-1)}\underline{z}^{q^{n-1}}$.

If we consider the q-th power of $\underline{A}\,\underline{z} = \underline{0}$ and take in account also (4), we deduce $\underline{0} = \underline{A}^q\underline{z}^q = \underline{T}\,\underline{A}\,\underline{A}'_1$. Thus $\underline{A}\,\underline{A}'_1 = \underline{0}$. Analogously, we prove $\underline{A}\,\underline{A}'_2 = \underline{0}, \ldots, \underline{A}\,\underline{A}'_{n-1} = \underline{0}$, and have finally

$$\underline{A}(\underline{A}'_0, \underline{A}'_1, \ldots, \underline{A}'_{n-1}) = \underline{A}\,\underline{A}'^{t} = \underline{0}.$$

We deduce, in particular, that every element $\underline{A}'_0, \underline{A}'_1, \ldots, \underline{A}'_{n-1} \in K^n$ is a solution of the linear system (6) and hence, it follows $\mathrm{rank}(\underline{A}'^t) \leqslant n-r$.

To see the inequality $\mathrm{rank}(\underline{A}') \geqslant n-r$, we remember that the matrix which agrees in the first n-r rows and last n-r columns with $\underline{A}'$ is non-singular. ▯

LEMMA 3. *The elements* $w_i \in K$, $i = 0,1,\ldots,s$, *are linearly* F-*independent if and only if the vectors* $\underline{w}_i = (w_i\ w_i^q\ \ldots\ w_i^{q^{n-1}})^t \in K^n$, $i = 0,1,\ldots,s$, *are linearly* K-*independent.*

Proof. Let us examine the condition

$$\sum_{i=0}^{s} k_i\underline{w}_i = \underline{0},\ k_i \in K, \tag{7}$$

under the hypothesis that the elements w_i are F-independent.

If we suppose that at least one coefficient k_i, for example k_0, is not zero, then we can determine $k \in K$ such that $h_0 = kk_0$, $\mathrm{tr}(h_0) \neq 0$ [1] holds. From (7), we obtain

$$\sum_{i=0}^{s} h_i\underline{w}_i = \underline{0},\ h_i = kk_i.$$

Raising the left side of this equation to the powers q^j we obtain $\sum_{i=0}^{s} h_i^{q^j}\underline{w}_i = \underline{0}$, $j = 0,1,\ldots,n-1$. If we add these n expressions, we find $\sum_{i=0}^{s}(\mathrm{tr}(h_i))\underline{w}_i = \underline{0}$; in particular

$$\sum_{i=0}^{s}(\mathrm{tr}(h_i))w_i = 0,\ \mathrm{tr}(h_0) \neq 0,$$

in contrast with the hypothesis.

It is evident how the second part of the thesis may be proved. ▯

COROLLARY 4. *The* $n \times n$ *matrix* $\underline{u} = (\underline{u}_0, \underline{u}_1, \ldots, \underline{u}_{n-1})$, $\underline{u}_i = (u_i\ u_i^q\ \ldots\ u_i^{q^{n-1}})^t$, *is non-singular if and only if* $\{u_0, u_1, \ldots, u_{n-1}\}$ *is a basis of the vector* F-*space* K. ▯

For any polynomial (2), the set

$$Z(L) = \{x \in K:\ L(x) = 0\}$$

is obviously a vector subspace of K. Moreover, if $Z(P) = \{x \in K:\ P(x) = 0\} \neq \emptyset$ then

$$Z(P) = x_0 + Z(L),\ x_0 \in Z(P). \tag{8}$$

---

[1] $\mathrm{tr}(x) = \mathrm{tr}_F(x) = x + x^q + \ldots + x^{q^{n-1}}$, $\forall\, x \in K$.

Given an affine polynomial (1), let

$$\underline{A}(P) = \underline{A}(L) := \underline{A}(l_0, l_1, \dots, l_{n-1}).$$

PROPOSITION 5. *If* $\mathrm{rank}(\underline{A}(L)) = r$ *then* $\dim_F Z(L) = n-r$.

Proof. Let $\{w_0, w_1, \dots, w_s\}$ be a basis of $Z(L)$ and let $V \subseteq K^n$ be the solution space of the homogeneous linear system

$$\text{(9)} \qquad \underline{A}(L)\underline{y} = \underline{0},\ \underline{y} = (y_0\ y_1\ \dots\ y_{n-1})^t.$$

From Lemma 3, it follows that the vectors $\underline{w}_i = (w_i\ w_i^q\ \dots\ w_i^{q^{n-1}})^t, i=0,1,\dots,s$, are linearly K-independent and it is easily verified that each of them is a solution of (9).

Thus,

$$\text{(10)} \qquad \dim_F Z(L) \leqslant \dim_K V = n-r.$$

Let $\underline{A}'$ be an autocirculant matrix which satisfies the conditions

$$\text{(11)} \qquad \underline{A}(L)\underline{A}'^t = \underline{0},\ \mathrm{rank}(\underline{A}') = n-r$$

(cf. Prop.2) and let $\underline{U} = (\underline{u}_0, \underline{u}_1, \dots, \underline{u}_{n-1})$ be an $n \times n$ non-singular matrix (cf. Coroll.4).

From (11), we deduce

$$\underline{A}(L)(\underline{A}'^t\underline{U}) = \underline{0},\ \mathrm{rank}(\underline{A}'^t\underline{U}) = n-r.$$

With the observation that $\underline{A}'^t\underline{U} = (\underline{u}'_0, \underline{u}'_1, \dots, \underline{u}'_{n-1})$, $\underline{u}'_i = (u'_i\ u'^q_i \dots u'^{q^{n-1}}_i)^t$, we can conclude that $u'_i \in Z(L)$, $i = 0,1,\dots,n-1$ and $\dim_F < u'_0, u'_1, \dots, u'_{n-1} > = n-r$ (cf. also Lemma 3). From this and (10), the Proposition 5 follow now immediately.□

COROLLARY 6. *Suppose* $\mathrm{rank}(\underline{A}(L)) = r$. *If* $\underline{z} = (z_0\ z_1 \dots z_r\ 0\ 0 \dots 0)^t$ *is a solution of the linear system* (9) *for any choose of the basis* $\{u_0, u_1, \dots, u_{n-1}\}$ *of the vector* F*-space* K, *the elements*

$$\text{(12)} \qquad x_i = z_0 u_i + z_r^{q^{n-r}} u_i^{q^{n-r}} + z_{r-1}^{q^{n-r+1}} u_i^{q^{n-r+1}} + \dots + z_1^{q^{n-1}} u_i^{q^{n-1}},\ i=0,1,\dots,n-1,$$

form a set of generators of $Z(L)$. *Hence, one has*

$$\text{(12)}' \qquad Z(L) = \{x = z_0 k + z_r^{q^{n-r}} k^{q^{n-r}} + \dots + z_1^{q^{n-1}} k^{q^{n-1}} : k \in K\}.$$

Proof. The Corollary follows from the proof of the previous Proposition if one observes that $\underline{A}'^t = \underline{A}(z_0, 0, \dots, 0, z_r^{q^{n-r}}, z_{r-1}^{q^{n-r+1}}, \dots, z_1^{q^{n-1}})$.□

PROPOSITION 7. *If* $\underline{z} = (z_0\ z_1 \dots z_{n-1})^t \in K^n$ *is a solution of the linear system*

$$\text{(13)} \qquad \underline{A}(L)\underline{y} = \underline{b},\ \underline{y} = (y_0\ y_1\ \dots\ y_{n-1})^t,\ \underline{b} = (b\ b^q\ \dots\ b^{q^{n-1}})^t,$$

then, for every $v \in K$ *with* $\mathrm{tr}(v) \neq 0$,

$$\text{(14)} \qquad x_0 = (z_0 v + z_{n-1}^q v^q + z_{n-2}^{q^2} v^{q^2} + \dots + z_1^{q^{n-1}} v^{q^{n-1}})/\mathrm{tr}(v)$$

is a root of the polynomial (1).

Proof. Let $\underline{A}(\underline{z}) = \underline{A}(z_0, z_1, \ldots, z_{n-1})$. Then

$$\underline{A}^t(\underline{z}) = (\underline{z}, \underline{T}^{-1}\underline{z}^q, \underline{T}^{-2}\underline{z}^{q^2}, \ldots, \underline{T}^{-(n-1)}\underline{z}^{q^{n-1}}).$$

Raising $\underline{A}(L)\underline{z} = \underline{b}$ to the power q, we obtain $\underline{A}^q(L)\underline{z}^q = b^q = (b^q\ b^{q^2} \ldots b^{q^{n-1}}\ b)^t$. Using (4), we have, therefore, $\underline{T}\,\underline{A}(L)\underline{T}^{-1}\underline{z}^q = \underline{b}^q$ or $\underline{A}(L)(\underline{T}^{-1}\underline{z}^q) = \underline{b}$. Iterating this, we find

$$\underline{A}(L)(\underline{T}^{-2}\underline{z}^{q^2}), \ldots, \underline{A}(L)(\underline{T}^{-(n-1)}\underline{z}^{q^{n-1}}) = \underline{b}.$$

Thus, it follows,

$$\underline{A}(L)\,\underline{A}^t(z) = (\underline{b}\ \underline{b}\ \ldots\ \underline{b}).$$

Now, the right multiplication of this equation by $\underline{v} = (v\ v^q \ldots\ v^{q^{n-1}})^t$ gives

$$\underline{A}(L)\,\underline{v}' = (tr(v))\underline{b}\ ,\underline{v}' = (v'\ v'^q \ldots v'^{q^{n-1}})^t$$

where $v' = z_0 v + z_{n-1}^q v^q + \ldots + z_1^{q^{n-1}} v^{q^{n-1}}$. ▯

COROLLARY 8. *The polynomial* (1) *has roots in* K *if and only if* $\mathrm{rank}(\underline{A}(L)) = \mathrm{rank}(\underline{A}(L) \,\vdots\, \underline{b}) = r$. *If this condition holds, one has* $|Z(P)| = q^{n-r}$.

Proof. If (1) has a root $x_0 \in K$ then raising both sides of the equality

$$l_0 x_0 + l_1 x_0^q + \ldots + l_{n-1} x_0^{q^{n-1}} = b$$

to the powers $q, q^2, \ldots, q^{n-1}$, we find

$$l_{n-1}^q x_0 + l_0^q x_0^q + \ldots + l_{n-2}^q x_0^{q^{n-1}} = b^q,$$

$$\ldots\ldots\ldots\ldots\ldots\ldots\ldots$$

$$l_1^{q^{n-1}} x_0 + l_2^{q^{n-1}} x_0^q + \ldots + l_0^{q^{n-1}} x_0^{q^{n-1}} = b^{q^{n-1}}.$$

Thus, $\underline{x}_0 = (x_0\ x_0^q \ldots x_0^{q^{n-1}})^t$ is a solution of (13). From here and from Prop.7, it follow that (1) has roots in K if and only if (13) has solutions.

Taking in account (8), the last part of the assertion follows from Prop.5. ▯

In particular, we find the following

RESULT (Dickson [3]). *The map* L: K → K, x ↦ L(x) *is a permutation on* K *if and only if* $\det(\underline{A}(L)) \neq 0$.

Moreover, we observe that if $\det(\underline{A}(L)) \neq 0$, the only root $x_0 \in K$ of the polynomial (1) can be determined using Cramer's rule, that is

$$x_0 = \det(\underline{b}, \underline{A}_1, \ldots, \underline{A}_{n-1}) / \det(\underline{A}_0, \underline{A}_1, \ldots, \underline{A}_{n-1}),$$

$(\underline{A}_0, \underline{A}_1, \ldots, \underline{A}_{n-1}) = \underline{A}(L)$.

In general, the affine subvariety of K consisting of the solutions of polynomial (1) is given by (8) with x_0 and Z(L) expressed by (14) and (12)' respectively.

From Corollary 8, we deduce the following useful

OBSERVATION. *A polynomial* (1) *with* $\deg(L(x)) = q^d$, $0 \leq d \leq n-1$, *is completely reducible in* K *if and only if* $\operatorname{rank}(\underline{A}(L) \,\vdots\, b) = \operatorname{rank}(\underline{A}(L)) = n-d$.

Another consequence is the following

PROPOSITION 9. *Two affine polynomials,* (1) *and* $P'(x) = L'(x) - b'$, *have common roots in* K *if and only if the equations of the linear sistems* (13) *and* $\underline{A}(L')\,\underline{y} = \underline{b}'$ *are compatible.*

Proof. If $x_0 \in K$ is a common root of both $P(x)$ and $P'(x)$ then $\underline{x}_0 = (x_0\ x_0^q \ldots x_0^{q^{n-1}})^t$ is a solution for both linear systems in the assertion. Conversely, if the equations of both systems are compatible, we find, by (14), a common root for the given polynomials.□

It is easy to prove that, when the condition of the previous proposition is satisfied, the set of common roots for $P(x)$ and $P'(x)$ is an affine subvariety of K whose dimension is $n-r'$, where

$$r' = \operatorname{rank}\begin{pmatrix} \underline{A}(L) \\ \hline \underline{A}(L') \end{pmatrix} = \operatorname{rank}\left(\begin{array}{c:c} \underline{A}(L) & \underline{b} \\ \hdashline \underline{A}(L') & \underline{b}' \end{array}\right).$$

Now we want to use the previous results to discuss the equation

$$(15) \qquad x^{q^m} - x = b,\ b \in K,\ 1 \leq m \leq n-1\ .$$

First we observe that, given $d = (n,m)$ and $k = n/d$, the integers $im + j$, $i = 0,1,\ldots,k-1$, $j = 0,1,\ldots,d-1$, are pairwise incongruent modulo n.

In this case, the linear system (13) becomes

$$(16) \qquad \begin{array}{lll} y_{m+j} & - y_j & = b^{q^j} \\ y_{2m+j} & - y_{m+j} & = b^{q^{j+m}} \\ \multicolumn{3}{c}{\cdots\cdots\cdots\cdots} \\ y_{(k-1)m+j} & - y_{(k-2)m+j} & = b^{q^{j+(k-2)m}} \\ y_j & - y_{(k-1)m+j} & = b^{q^{j+(k-1)m}} \end{array} \quad , \ j = 0,1,\ldots,d-1,$$

and thus its equations are compatible if and only if $\sum_{i=0}^{k-1} b^{q^{j+im}} = (\sum_{i=0}^{k-1} b^{q^{im}})^{q^j} = 0$.

From this, we deduce that (15) has some roots in K if and only if

$$(17) \qquad \sum_{i=0}^{k-1} b^{q^{im}} = tr_{F'}(b) = 0,$$

where $F' = GF(q^d) \subseteq GF(q^n)$ [2].

---

[2] The integers hd, $h = 0,1,\ldots,k-1$, and im, $i = 0,1,\ldots,k-1$, are congruent modulo m and therefore $\sum_{i=0}^{k-1} b^{q^{im}} = \sum_{h=0}^{k-1} b^{q^{hd}}$.

$L(x) = x^{q^m} - x$ implies obviously

(18) $Z(L) = GF(q^d)$, $d = (n,m)$.

Therefore, we can determine a root $x_0 \in K$ of (15) using Prop.7 and supposing that (17) is satisfied.

From (16), by successive substitutions, we find

$$y_{im+1} = y_j + \Big(\sum_{h=0}^{i-1} b^{q^{hm}} \Big)^{q^j}, \quad i = 1,2,\dots,k-1, \; j = 0,1,\dots,d-1,$$

and by (17)

$$y_{im+j} = \lambda_j - \Big(\sum_{h=i}^{k-1} b^{q^{hm}} \Big)^{q^j}, \quad \lambda_j \in K, \; i = 0,1,\dots,k-1, \; j = 0,1,\dots,d-1.$$

Let us consider the particular solution

$$z_{im+j} = - \Big(\sum_{h=i}^{k-1} b^{q^{hm}} \Big)^{q^j}, \quad i = 0,1,\dots,k-1, \; j = 0,1,\dots,d-1,$$

obtained for $\lambda_j = 0$, $j = 0,1,\dots,d-1$.

From (14) we obtain

$$x_0 tr(v) = \sum_{h=0}^{n-1} z_{n-h}^{q^h} v^{q^h} = \sum_{i=0}^{k-1} \sum_{j=0}^{d-1} z_{im+j}^{q^{n-(im+j)}} v^{q^{n-(im+j)}} .$$

Hence, setting $v = w^{q^{d-1}}$, we have

$$x_0 tr(w) = \sum_{i=0}^{k-1} \sum_{j=0}^{d-1} z_{im+j}^{q^{n-(im+j)}} w^{q^{n-(im+j)+d-1}}$$

where $tr(w) = tr(v) \neq 0$.

If we observe that

$$z_{im+j}^{q^{n-(im+j)}} = - \sum_{h=i}^{k-1} b^{q^{n+(h-i)m}} = - \sum_{r=0}^{k-i-1} b^{q^{rm}},$$

then, substituting into the previous equality, one has

$$\begin{aligned} x_0 tr(w) &= - \sum_{i=0}^{k-1} \sum_{r=0}^{k-i-1} b^{q^{rm}} \sum_{j=0}^{d-1} w^{q^{n-im+(d-1-j)}} \\ &= - \sum_{i=0}^{k-1} \sum_{r=0}^{k-i-1} b^{q^{rm}} \sum_{s=0}^{d-1} w^{q^{n-im+s}} \\ &= - \sum_{i=0}^{k-1} \sum_{r=0}^{k-i-1} b^{q^{rm}} \Big(\sum_{s=0}^{d-1} w^{q^s} \Big)^{q^{n-im}} . \end{aligned}$$

From here, putting

$$a = \sum_{s=0}^{d-1} w^{q^s}$$

and observing

$$tr(w) = \sum_{i=0}^{k-1} \sum_{j=0}^{d-1} w^{q^{im+j}} = \sum_{i=0}^{k-1} a^{q^{im}} = tr_{F'}(a),$$

we deduce

$$x_0 tr_{F'}(a) = -\sum_{i=0}^{k-1}\sum_{r=0}^{k-i-1} b^{q^{rm}} a^{q^{n-im}}$$

$$= -\sum_{h=1}^{k}\sum_{r=0}^{h-1} b^{q^{rm}} a^{q^{hm}}$$

$$= -\sum_{h=1}^{k-1}\sum_{r=0}^{h-1} b^{q^{rm}} a^{q^{hm}} .$$

Therefore: *The equation* (15) *has roots in* $K = GF(q^n)$ *if and only if* b *satisfies the condition* (17). *If such condition is satisfied, the set of roots is the affine subvariety* (8) *in which* Z(L) *is given by* (18) *and*

$$x_0 = -\frac{1}{tr_{F'}(a)}\sum_{h=1}^{k-1}\sum_{r=0}^{h-1} b^{q^{rm}} a^{q^{hm}} , \quad tr_{F'}(a) \neq 0.$$

If $(k,p) = 1$ (p = char K) then $tr_{F'}(1) = k \neq 0$ and therefore, we can set $a=1$.

The previous result allows us to determine the roots of a second degree equation in a field K of char 2. In fact, for $q = 2$, $m = 1$, we find the well-known condition $tr(b) = 0$ in order that the equation

$$x^2 + x + b = 0$$

has a root in $K = GF(2^n)$. Moreover, from (18) and (19), we deduce that the roots of the above equation are

$$x_0 = -\frac{1}{tr(a)}\sum_{h=1}^{n-1}\sum_{r=0}^{h-1} b^{2^r} a^{2^h} \quad \text{and} \quad x_0 + 1,$$

where $a \in K$ is a fixed element with $tr(a) \neq 0$.

REFERENCES

[1] Berlekamp, E.R., *Algebraic coding theory* (Mc Graw Book Company,New York,1968).

[2] Biliotti M. and Menichetti G., *On a generalization of Kantor's likeable planes*, Geom. Dedicata, 17 (1985) 253-277.

[3] Dickson, L.E., *Linear Groups with an exposition of the Galois field theory* (Teubner, Leipzig. Reprint Dover, New York, 1958).

Annals of Discrete Mathematics 30 (1986) 311–330

On the parameter $D(v,t_v-13)$ for Steiner triple systems (*)

Salvatore Milici (**)

Abstract. *Let* $D(v,k)$ ([1],[8]) *be the maximum number of Steiner Triple Systems of order* v *that can be constructed in such a way that any two of them have exactly* k *blocks in common, these* k *blocks being moreover in each of the* $STS(v)_s$. *Let* $t_v = v(v-1)/6$. *In this paper we prove that* $D(v,t_v-13) = 3$ *for every (admissible)* $v \geq 15$.

1. Introduction and definitions.

A *Partial Triple System* *(PTS)* is a pair $(P,\mathcal{P})$ where P is a finite non-empty set and $\mathcal{P}$ is a collection of 3-subset of P, called blocks, such that any 2-subset of P is contained in at most one block of $\mathcal{P}$.

Using graph theoretic terminology, we will say that an element x of P has *degree* $d(x)=h$ if x belongs to exactly h blocks of $\mathcal{P}$. Clearly $\sum_{x\in P} d(x) = 3|\mathcal{P}|$. We will call the *degree-set* *(DS)* of a *PTS* $(P,\mathcal{P})$ the n-uple $DS=[d(x),d(y),\ldots]$, where $x,y,\ldots$ are the elements of P. If there are r_i elements of P having degree h_i , for $i=1,\ldots,s$, we will write $DS=[(h_1)_{r_1},(h_2)_{r_2},\ldots$ $\ldots,(h_s)_{r_s}]$, where $r_1+\ldots+r_s=|P|$. If $r_i=1$, for some i , then we will write $(h_i)_1 = h_i$.

Two *PTSs* $(P,\mathcal{P}_1)$ and $(P,\mathcal{P}_2)$ are said *disjoint* and *mutually balanced* *(DMB)* if $\mathcal{P}_1\cap\mathcal{P}_2=\emptyset$ and a 2-subset of P is contained in a block of $\mathcal{P}_1$ if and only if it is contained in a block of $\mathcal{P}_2$. A set of s *PTSs* $(P,\mathcal{P}_1),(P,\mathcal{P}_2),\ldots,(P,\mathcal{P}_s)$ is said to be a set of

---

(*) Lavoro eseguito nell'ambito del GNSAGA (CNR) e con contributo finanziario MPI (1983).

(**) Dipartimento di Matematica dell'Università, Viale A. Doria, 6 95125 - Catania.

s *DMB PTSs* if (P,P_i) and (P,P_j) are *DMB* for every $i,j \in \{1,2,3,\dots,s\}$ and $i \neq j$. We denote by $(P;P_1,P_2,\dots,P_s)$ a set of s *DMB PTSs*.

Two $(s+1)$-tuples $(P;P_1,P_2,\dots,P_s)$ and $(P;P'_1,P'_2,\dots,P'_s)$ are *isomorphic* if (P,P_i) is isomorphic to (P,P'_1) by the same isomorphism, for every $i=1,2,\dots,s$.

A degree set DS is associated with every $(P;P_1,P_2,\dots,P_s)$. However if $(P;P_1,P_2,\dots,P_s)$ and $(P;P'_1,P'_2,\dots,P'_s)$ are not isomorphic, it is possible that they have the same DS.

A *Steiner triple system of order* v (or more briefly a $STS(v)$) is a PTS $(S,\mathcal{B})$ such that $|S|=v$ and every 2-subset of S is contained in exactly one block of $\mathcal{B}$. It is well-known that a necessary and sufficient condition for the existence of an $STS(v)$ is $v \equiv 1$ or 3 (mod 6) (v admissible) and that the number of the blocks is $|\mathcal{B}| = t_v = v(v-1)/6$.

We will put

$$M_r = \{x \in P : d(x)=r\},$$

$$A(a) = \left\{ x \in P : x \neq a \text{ and exist } R \in \bigcup_{i=1}^{s} P_i \text{ with } x \in R \right\},$$

$$A(a,\{a,b,c\}) = A(a) - \{b,c\} \text{ with } \{a,b,c\} \in \bigcup_{i=1}^{s} P_i .$$

The PTS (P,P) is said to be *embedded* in the triple system $(S,\mathcal{B})$ provided that $P \subseteq S$ and $P \subseteq \mathcal{B}$.

Given an integer k such that $0 \leq k \leq t_v$, let us denote by $D(v,k)$ the maximum number of $STS(v)$s that can be constructed on a set of cardinality v in such a way that any two of them have exactly k blocks in common, these k blocks being moreover in each of the $D(v,k)$ systems. In [1], Doyen asked to determine $D(v,k)$, (see also [8]).

Much results are already known in the case $k=0$ [8]. For $k \neq 0$, same results are contained in [3], [4], [5], [6], [7]. In particular, in these papers $D(v,k)$ has been determined for every v admissible and $k = t_v - m$ with $m \leq 12$.

In this paper we prove that $D(v,t_v-13)=3$ for every $v \geq 15$.

2. Preliminar results.

We now give some properties which will be used in the remaining sections.

Let $\lfloor x \rfloor$ be the greatest integer lesser than or equal to x. In any $(P;P_1,P_2,\dots,P_s)$ we have [3]:

(2.1) $|P_1| = |P_2| = \dots = |P_s| \geq 4$, $|P| \geq 6$; we will put $n = |P|$ and $m = |P_i|$ $(i = 1,2,\dots,s)$;

(2.2) If $h = max\,\{d(x) : x \in P\}$, then $m \geq 2h$ and $n \geq 2h+1$;

(2.3) $d(x) \geq 2$, $\mu = min\,\{d(x) : x \in P\} \leq \left\lfloor \frac{3m}{n} \right\rfloor$ and $s \leq 2\mu-1$;

(2.4) $s \leq 2d(u) - \eta - 1$, where $\eta = |A(u,\{u,v,w\}) - A(v,\{u,v,w\})|$;

(2.5) If R is a block such that $|R \cap M_r| \geq 2$ with $r = 2,3$, then $s \leq 2r-2$.

Lemma 2.1. *Let* $R = \{1,2,3\} \subseteq M_3$, $X = \bigcup_{j=1}^{3} A(j,\{1,2,3\})$ *and* $B = \bigcap_{j=1}^{3} A(j,\{1,2,3\})$. *In a* $(P;P_1,P_2,P_3,P_4)$ *if* $R \in P_i$, *for some* $i = 1,2,3,4$, *then:*

i) $|X| = |B| = 4$ *or* ii) $|X| = 5$ and $|B| = 2$.

Proof. Let $R \in P_1$, without loss of generality. If $A(1) = \{1,2,3,4,5,6,a\}$ and $A(2) = \{1,3,4,5,6,b\}$, we have necessarily

i)

P_1	P_2	P_3	P_4
1 2 3	1 2 4	1 2 5	1 2 6
1 4 a	1 5 a	1 4 3	1 4 5
1 5 6	1 6 3	1 6 a	1 3 a
2 4 6	2 3 5	2 4 b	2 4 5
2 5 b	2 6 b	2 3 6	2 3 b
...	...	...	...
...	...	...	...

or

ii)

P_1	P_2	P_3	P_4
1 2 3	1 2 4	1 2 5	1 2 6
1 4 a	1 3 5	1 4 6	1 4 3
1 5 6	1 6 a	1 3 a	1 5 a
2 4 6	2 5 6	2 3 4	2 3 5
2 5 b	2 3 b	2 6 b	2 4 b
...	...	...	...
...	...	...	...

In case i) we obtain $3 \in M_4$. This is impossible. In case ii) we obtain $A(3) = \{1,2,4,5,a,b\}$ and this complete the proof of the lemma.■

Lemma 2.2. *Let* $R = \{1,2,a\}$ *, with* $1,2 \in M_3$ *and* $a \in M_4$ *. In a* $(P;P_1,P_2,P_3,P_4)$ *, if* $R \in P_i$ *, for some* $i = 1,2,3,4$ *, then*

i) $A(1,\{1,2,a\}) = A(2,\{1,2,a\})$ *and* $|A(a,\{1,2,a\}) \cap A(j,\{1,2,a\})| = 3,4$ *, for every* $j = 1,2$ *.*

or

ii) *if* $x_1, x_2 \in P$ *are such that* $x_j \in A(j,\{1,2,a\})$ *and* $x_j \notin A(3-j,\{1,2,a\})$ *, for every* $j = 1,2$ *, then necessarily* $x_1, x_2 \in A(a)$ *.*

Proof. Let $R = \{1,2,a\} \in P_1$, without loss of generality. If $A(1,\{1,2,a\}) = A(2,\{1,2,a\}) = \{4,5,6,7\}$, we have necessarily

P_1	P_2	P_3	P_4
1 2 a	1 6 a	1 6 4	1 4 7
1 4 5	1 7 5	1 7 a	1 5 a
1 6 7	1 2 4	1 2 5	1 2 6
2 4 6	2 5 a	2 4 7	2 5 4
2 5 7	2 6 7	2 6 a	2 7 a
...	...	...	...

or

P_1	P_2	P_3	P_4
1 2 a	1 2 4	1 6 4	1 2 7
1 4 5	1 6 a	1 7 a	1 4 a
1 6 7	1 7 5	1 2 5	1 5 6
2 4 6	2 5 6	2 4 a	2 4 5
2 5 7	2 7 a	2 7 6	2 6 a
...	...	...	...

or

P_1	P_2	P_3	P_4
1 2 a	1 2 4	1 2 6	1 2 7
1 4 5	1 6 5	1 4 a	1 4 6
1 6 7	1 7 a	1 5 7	1 5 a
2 4 6	2 5 a	2 5 4	2 4 a
2 5 7	2 6 7	2 7 a	2 6 5
...	...	...	...

or

P_1	P_2	P_3	P_4
1 2 a	1 2 5	1 2 6	1 2 7
1 4 5	1 6 a	1 4 a	1 4 6
1 6 7	1 4 7	1 5 7	1 5 a
2 4 6	2 4 a	2 5 a	2 4 5
2 5 7	2 6 7	2 7 4	2 6 a
...	...	...	...

It is a routine matter to see that

$$|A(a,\{1,2,a\})\cap A(j,\{1,2,a\})| = 3,4 \qquad \text{with} \quad j = 3,4 \ .$$

If $A(1,\{1,2,a\}) = \{4,5,6,x_1\}$ and $A(2,\{1,2,a\}) = \{4,5,6,x_2\}$, with $x_1 \neq x_2$, we have necessarily

P_1	P_2	P_3	P_4
1 2 a	1 2 4	1 2 5	1 2 6
1 4 x_1	1 5 a	1 4 6	1 4 a
1 5 6	1 6 x_1	1 a x_1	1 5 x_1
2 5 x_2	2 5 6	2 4 a	2 5 a
2 4 6	2 a x_2	2 6 x_2	2 4 x_2
...	...	...	...

or

P_1	P_2	P_3	P_4
1 2 a	1 2 4	1 2 5	1 2 6
1 4 x_1	1 5 x_1	1 4 a	1 4 5
1 5 6	1 6 a	1 6 x_1	1 a x_1
2 5 x_2	2 5 a	2 4 x_2	2 4 5
2 4 6	2 6 x_2	2 6 a	2 a x_2
...	...	...	...

Hence $x_1, x_2 \in A(a)$. At this point the proof of the lemma is complete. ∎

3. *STS* with blocks in common

In this section we will prove that $D(v,t_v-13) \geq 3$ for every $v \geq 15$.

Let $P = \{1,2,\ldots,8,a,b,c\}$ and let P_1, P_2 and P_3 be the following three sets of 13 triples each:

P_1 =

1 2 3
1 a b
1 4 c
2 a c
2 4 b
3 b c
3 4 a
a 5 6
a 7 8
b 5 7
b 6 8
c 5 8
c 6 7

P_2 =

1 2 4
1 a c
1 3 b
2 3 a
2 c b
3 4 c
4 a b
a 5 7
a 6 8
b 5 8
b 6 7
c 5 6
c 7 8

P_3 =

1 2 c
1 3 a
1 4 b
2 3 4
2 b a
3 b c
4 c a
a 5 8
a 6 7
b 5 6
b 7 8
c 5 7
c 6 8

Lemma 3.1. *$D(v,t_v-13) \geq 3$, for $v = 15$ and for every $v \geq 31$.*

Let $S = \{1,2,...,8,9,0,a,b,c,d,e\}$ and

T =

a 9 0	1 5 9	2 7 0	4 5 0
a d e	1 6 0	2 8 d	4 6 d
b 9 e	1 7 d	3 5 d	4 7 e
b d 0	1 8 e	3 6 e	4 8 9
c 9 d	2 5 e	3 7 9	
c e 0	2 6 9	3 8 0	

Clearly, $(S,P_i \cup T)$, $i = 1,2,3$, are three $STS(15)$s such that any two of them intersect in the same block-set T with $|T| = t_{15}-13$.

It follow that $D(15,t_{15}-13) \geq 3$. In [2], J. Doyen and R.M. Wilson have shown that any $STS(v)$ can be embedded into an $STS(u)$ for every $u \geq 2v+1$. Then $D(v,t_v-13) \geq 3$ for every $v \geq 31$. ∎

Lemma 3.2. *$D(19,t_{19}-13) \geq 3$.*

Let $S = \{1,2,...,8,9,0,a,b,c,d,e,x,y,z,t\}$ and

F =

a 9 0	c 9 d	1 d x	3 5 x	4 7 e	7 9 x
a d e	c e 0	1 e y	3 6 y	4 8 d	7 0 y
a x y	c x t	2 5 0	3 7 d	4 9 t	8 9 y
a z t	c y z	2 6 9	3 8 e	4 0 z	8 0 x
b 9 e	1 5 9	2 7 t	3 9 z	5 d z	
d b 0	1 6 0	2 8 z	3 0 t	5 e t	
b x z	1 7 z	2 d y	4 5 y	6 d t	
b y t	1 8 t	2 e x	4 6 x	6 e z	

Clearly, $(S, P_i \cup F)$, $i = 1,2,3$, are three $STS(19)s$ such that any two of them intersect in the same block-set F with $|F| = t_{19}-13$. Then $D(19, t_{13}-13) \geq 3$. ■

Lemma 3.3. $D(21, t_{21}-13) \geq 3$.

Let $S = \{1,2,\ldots,9,0,a,b,c,x,y,z,d,e,f,g,h\}$ and

$L =$

x y z	y d c	z d b	d 1 8	e 3 5	b h 0
x d a	y a e	z e c	d 5 h	f 8 g	c g 0
x e b	y b f	z a f	d 2 9	f 7 h	c 9 h
x f c	y 3 g	z 4 g	d 3 0	f 1 9	g 2 5
x 1 g	y 4 h	z 3 h	d 7 4	f 2 0	h 1 6
x 2 h	y 6 9	z 8 9	e 7 g	f 3 6	9 3 7
x 5 9	y 5 0	z 6 0	e 8 h	f 4 5	4 8 0
x 7 0	y 2 8	z 2 7	e 4 9	a 9 0	
x 3 8	y 1 7	z 1 5	e 1 0	a g h	
x 4 6	d e f	d 6 g	e 2 6	b 9 g	

Clearly, $(S, P_i \cup L)$, $i = 1,2,3$, are three $STS(21)s$ such that any two of them intersect in the same block-set L with $|L| = t_{21}-13$. Then $D(21, t_{21}-13) \geq 3$. ■

Lemma 3.4. $D(27, t_v-13) \geq 3$.

Let $S = \{1,2,\ldots,8,9,0,x,y,z,a,b,c,d,e,f,g,i,p,q,r,s,t,u\}$ and

$M =$

a 0 9	b x z	c u q	9 z s	1 5 0	2 5 9	3 5 x	4 5 y	5 d u	6 r p
a d e	b y t	c i p	9 t u	1 6 9	2 6 0	3 6 y	4 6 x	5 e i	6 s q
a x y	b u p	c r f	d x f	1 7 d	2 7 e	3 7 z	4 7 t	5 z f	7 0 f
a z t	b q i	c s g	d y g	1 8 e	2 8 d	3 8 t	4 8 z	5 t g	7 9 g
a u s	b r g	0 x r	d z p	1 x u	2 x s	3 0 p	4 0 q	5 r q	7 x p
a i r	b s f	0 y s	d t q	1 y i	2 y u	3 9 q	4 9 p	5 s p	7 y q
a f p	c 0 e	0 z u	e x g	1 z r	2 z i	3 d r	4 d s	6 d i	7 u i
a g q	c d 9	0 t i	e y f	1 t s	2 t r	3 e s	4 e r	6 e u	7 r s
b 0 d	c x t	9 x i	e z 7	1 f g	2 f q	3 u g	4 i g	6 z g	8 0 g
b 9 e	c y z	9 y r	e t 5	1 p q	2 p g	3 i f	4 u f	6 t f	8 9 f
						8 x q	8 y p	8 u r	8 i s

Clearly, $(S, P_i \cup M)$, $i = 1,2,3$, are three $STS(27)s$ such that any two of them intersect in the same block-set M with $|M| = t_{27}-13$. Then $D(27, t_{27}-13) \geq 3$. ■

Lemma 3.5. $D(25, t_{25}-13) \geq 3$.

Let $\bar{P} = \{1,2,\ldots,9,0\}$ and let $\bar{P}_1, \bar{P}_2, \bar{P}_3$ be the following three sets of 13 triples each:

$\bar{P}_1 =$	$\bar{P}_2 =$	$\bar{P}_3 =$
0 1 4	0 1 2	0 1 3
0 2 5	0 3 4	0 2 6
0 3 6	0 5 6	0 4 5
7 8 1	7 8 3	7 8 9
7 9 2	7 9 6	7 1 5
8 9 3	8 9 2	7 2 3
7 3 4	7 1 4	7 4 6
7 5 6	7 2 5	8 1 4
8 2 6	8 1 5	8 2 5
8 4 5	8 4 6	8 3 6
9 1 5	9 1 3	9 1 2
9 4 6	9 4 5	9 3 4
1 2 3	2 3 6	9 5 6

Let $S = \{1,2,\ldots,9,0,a,b,c,d,e,f,g,h,i,l,n,p,r,s,t\}$ and

$Q =$

1 6 a	2 4 b	3 5 c	7 0 a	8 0 b	9 0 c	a g d	c i l	4 i g	5 l r	6 s t	
1 b d	2 a l	3 a n	7 b e	8 a i	9 a t	a f r	c t f	4 h r	5 i t	0 d i
1 c g	2 c p	3 b r	7 c h	8 c s	9 b f	a e h	c n r	4 n p	6 b c	0 e p
1 e r	2 t d	3 d f	7 d s	8 n d	9 d h	a s p	c d e	5 a b	6 d r	0 f g
1 f n	2 e f	3 e s	7 f i	8 e g	9 e l	b g h	4 a c	5 p d	6 e i	0 h t
1 h i	2 g r	3 g t	7 g n	8 f l	9 g p	b l p	4 d l	5 e n	6 f p	0 l n
1 l s	2 h s	3 h l	7 l t	8 h p	9 s n	b i s	4 t e	5 f h	6 g l	0 s r
1 p t	2 i n	3 p i	7 p r	8 r t	9 r i	b n t	4 f s	5 g s	6 h n	

Clearly, $(S, \bar{P}_i \cup Q)$, $i = 1,2,3$, are three $STS(25)s$ such that any two of them intersect in the same block-set Q with $|Q| = t_{25}-13$. Then $D(25, t_{25}-13) \geq 3$.■

In conclusion, by Lemmas 3.1, 3.2, 3.3, 3.4 and 3.5 we obtain the following theorem.

Theorem 3.1. $D(v, t_v-13) \geq 3$ *for every* $v \geq 15$.

4. $D(v, t_v-13)$ for every $v \geq 15$.

In this section we will prove that there does not exist a

$(P;P_1,P_2,P_3,P_4)$ with $m=13$. Further we will determine $D(v,t_v-13)$.

From Property 2.3, the existence of a $(P;P_1,P_2,P_3,P_4)$ implies $M_2=\emptyset$. It is easy to see that a $(P;P_1,P_2,P_3,P_4)$, with $M_2=\emptyset$ and $m=13$, can have the following parameters:

1) $n=10$ and $DS=[(4)_9,3]$;

2) $n=11$ and $DS=[(4)_6,(3)_5]$, or $DS=[5,(4)_4,(3)_6]$, or $DS=[(5)_2,(4)_2,(3)_7]$, or $DS=[(5)_3,(3)_8]$;

3) $n=12$ and $DS=[(4)_3,(3)_9]$ or $DS=[5,4,(3)_{10}]$;

4) $n=13$ and $DS=[(3)_{13}]$.

Lemma 4.1. *There is no* $(P;P_1,P_2,P_3,P_4)$ *with* $DS=[(4)_9,3]$.

Proof. Suppose that there exists a $(P;P_1,P_2,P_3,P_4)$ with $DS=[(4)_9,3]$. Let $M_3=\{x\}$, $M_4=\{1,2,...,9\}$, $P=M_3\cup M_4$ and $A=A(x)=\{1,2,3,4,5,6\}$.

If $\{7,8,9\}\in P_i$, for some $i=1,2,3,4$, then necessarily

$$P_i = \begin{array}{|l|l|l|l|l|} \hline \left.\begin{matrix} x\\ x\\ x\end{matrix}\right\} F_1(A) & \left.\begin{matrix} 7\\ 7\\ 7\end{matrix}\right\} F_2(A) & \left.\begin{matrix} 8\\ 8\\ 8\end{matrix}\right\} F_3(A) & \left.\begin{matrix} 9\\ 9\\ 8\end{matrix}\right\} F_4(A) & 7\ 8\ 9 \\ \hline \end{array}$$

where $F_i(A)$, $i=1,2,3,4$, there are four distinct 1-factors on A.

If $\{7,8,9\}\notin P_i$, then necessarily there exists a block $R\in P_i$ such that $R\subseteq A$. Let $R=\{1,2,3\}$, we have

$$P_i = \begin{array}{|ccc|ccc|ccc|ccc|ccc|} \hline 7&8&1 & 7&3&4 & 8&4&5 & x&1&\alpha & 1&2&3 \\ 7&9&2 & 7&5&6 & 9&1&5 & x&2&\beta & && \\ 8&9&3 & 8&2&6 & 9&4&6 & x&3&\gamma & && \\ \hline \end{array}$$

with $(\alpha=4\ ,\ \beta=5\ ,\ \gamma=6)$ or $(\alpha=6\ ,\ \beta=4\ ,\ \gamma=5)$.

Particularly, if $\{7,8,9\}\notin P_1$ we obtain

$$\begin{array}{|ccc|}\hline x & 1 & 4\\ x & 2 & 5\\ x & 3 & 6\\ \hline\end{array} \in P_1 \qquad \text{or} \qquad \begin{array}{|ccc|}\hline x & 1 & 6\\ x & 2 & 4\\ x & 3 & 5\\ \hline\end{array} \in P_1$$

Let $F = \{F_i \mid i = 1,2,\ldots,5\}$ be the 1-factorization on A given by:

F_1	F_2	F_3	F_4	F_5
1,4	1,2	1,3	1,5	1,6
2,5	3,4	2,6	2,3	2,4
3,6	5,6	4,5	4,6	3,5

It follows that $A(x) = F_1$ or $A(x) = F_5$.

i) Let $A(x) = F_1$. Then, up to isomorphism, we obtain

P_1	P_2				
x 1 4		x 1 2			
x 2 5		x 3 4			
x 3 6		x 5 6			
7 8 1	7 8 5	7 8 3	7 8 9		
7 9 2	7 9 4	7 9 6	7		
8 9 3	8 9 1	8 9 2	7	F_{j_1}	
7 3 4	7 1 3	7 1 4	7		
7 5 6	7 2 6	7 2 5	8		
8 2 6	8 2 3	8 1 5	8	F_{j_2}	
8 4 5	8 4 6	8 4 6	8		
9 1 5	9 2 5	9 1 3	9		
9 4 6	9 3 6	9 4 5	9	F_{j_3}	
1 2 3	1 4 5	2 3 6	9		

P_3			P_4		
	x 1 3			x 1 5	
	x 2 6			x 2 3	
	x 4 5			x 4 6	
7 8 5	7 8 6	7 8 9	7 8 3	7 8 2	7 8 9
7 9 1	7 9 3	7	7 9 1	7 9 6	7
8 9 2	8 9 4	7 $\}F_{r_1}$	8 9 4	8 9 5	7 $\}F_{s_1}$
7 2 3	7 1 4	7	7 2 6	7 1 3	7
7 4 6	7 2 5	8	7 4 5	7 4 5	8
8 1 4	8 1 5	8 $\}F_{r_2}$	8 1 2	8 1 4	8 $\}F_{s_2}$
8 3 6	8 2 3	8	8 5 6	8 3 6	8
9 3 4	9 1 2	9	9 3 6	9 1 2	9
9 5 6	9 5 6	9 $\}F_{r_3}$	9 2 5	9 3 4	9 $\}F_{s_3}$
1 2 5	3 4 6	9	1 3 4	2 5 6	9

where $j_i \in \{1,3,4\}$, $r_i \in \{1,2,4\}$, $s_i \in \{1,2,3\}$.

ii) Let $A(x) = F_5$. Then, up to isomorphism, we obtain

P_1	P_2		
x 1 6		x 1 2	
x 2 4		x 3 4	
x 3 5		x 5 6	
7 8 1	7 8 5	7 8 4	7 8 9
7 9 2	7 9 3	7 9 6	7
8 9 3	8 9 1	8 9 2	7 $\}\bar{F}_{j_1}$
7 3 4	7 1 6	7 1 5	7
7 5 6	7 2 4	7 2 3	8
8 2 6	8 2 3	8 1 6	8 $\}\bar{F}_{j_2}$
8 4 5	8 4 6	8 3 5	8
9 1 5	9 2 6	9 1 3	9
9 4 6	9 4 5	9 4 5	9 $\}\bar{F}_{j_3}$
1 2 3	1 3 5	2 4 6	9

P_3			P_4		
	x 1 3			x 1 5	
	x 2 6			x 2 3	
	x 4 5			x 4 6	
7 8 5	7 8 2	7 8 9	7 8 2	7 8 3	7 8 9
7 9 1	7 9 3	7	7 9 6	7 9 5	7
8 9 6	8 9 4	7 $\}\bar{F}_{r_1}$	8 9 1	8 9 4	7 $\}\bar{F}_{s_1}$
7 2 3	7 1 5	7	7 1 3	7 1 6	7
7 4 6	7 4 6	8	7 4 5	7 2 4	8
8 1 2	8 1 6	8 $\}\bar{F}_{r_2}$	8 3 4	8 1 2	8 $\}\bar{F}_{s_2}$
8 3 4	8 3 5	8	8 5 6	8 5 6	8
9 2 4	9 1 2	9	9 2 4	9 1 3	9
9 3 5	9 5 6	9 $\}\bar{F}_{r_3}$	9 3 5	9 2 6	9 $\}\bar{F}_{s_3}$
1 5 6	2 3 4	9	1 2 6	3 4 5	9

where $j_1 \in \{1,3,4\}$, $r_i \in \{1,2,4\}$, $s_i \in \{1,2,3\}$.

It is a routine matter to see that, in i) and ii), there is no a $(P;P_1,P_2,P_3,P_4)$ with $DS = [3,(4)_9]$.

Lemma 4.2. *There is no* $(P;P_1,P_2,P_3,P_4)$ *with* $DS = [(4)_6,(3)_5]$.

Proof. Suppose that there exists a $(P;P_1,P_2,P_3,P_4)$ with $DS = [(4)_6,(3)_5]$. Let $M_3 = \{1,2,\dots,5\}$, $M_4 = \{a,b,c,d,e,t\}$ and $P = M_3 \cup M_4$.

At first, suppose that there exists a block $R \subseteq M_3$. Let $R = \{1,2,3\} \in P_1$, without loss of generality. Applying Lemma 2.1, we obtain $\left|\bigcup_{i=1}^{3} A(i,\{1,2,3\})\right| = 4,5$. Let $Y = P - \bigcup_{i=1}^{3} A(i)$, we obtain $|Y| = 3,4$, with $|Y \cap M_4| \geq 1$. Then it must be $|P_1| \geq 14$. This is impossible.

Now, suppose that there exists a block R such that $|R \cap M_3| = 2$. Let $\{1,2,a\} \in P_1$, $X = A(1) \cup A(2) \cup A(a) - \{1,2,a\}$ and $Y = P - \{A(1) \cup A(2) \cup A(a)\}$. It follows that $|X| = 6,7,8$.

At first, suppose $|X| = 8$. Let $X - A(a) = \{x_1,x_2\}$, from Lemma

2.2 we obtain $x_1, x_2 \in A(i,\{1,2,a\})$, for some $i=1,2$ and hence $|A(i,\{1,2,a\}) \cap A(a,\{1,2,a\})| \le 2$. This is impossible.

Now, suppose $|X|=6$. Since $Y \cap M_4 \ne \emptyset$ implies $|P_1| \ge 14$,then we have $Y=\{3,4\}$. Hence $\{3,4,t\} \in P_1$ with $t \in A(1) \cap A(2) \cap A(a)$, since otherwise $m > 13$. Further, from Property 2.4 it follows that $|A(3,\{3,4,t\}) \cap A(4,\{3,4,t\}) \cap (M_4 - \{a,t\})| \ge 3$ and $|A(1,\{1,2,a\}) \cap$ $\cap A(2,\{1,2,a\}) \cap (M_4 - \{a,t\})| \ge 2$. Then $|M_4| \ge 7$. This is impossible.

Now, suppose $|X|=7$. Since $n=11$, it follows that $Y \cap M_4 = \emptyset$. Let $Y=\{3\}$, then, since $n=11$, $(3,4) \subseteq P_i$ for every $i=1,2,3,4$. Let $\{3,4,t\} \in P_1$. Since $4 \notin Y$, we have $4 \in A(1) \cup A(2) \cup A(a)$. It follows that $5 \notin A(4)$, otherwise, for some $j=2,3,4$, $R_1 = \{3,4,5\} \in P_j$ with $R_1 \subseteq M_3$. Since $1,2,a \notin A(3)$, it follows that $a \in A(4)$ and $1,2 \notin A(4)$. In fact, let $x \in \{1,2\}$, if $x, a \in A(4)$ we must have $|A(3,\{3,4,t\}) \cap A(4,\{3,4,t\})| \le 2$. If $x \in A(4)$ and $a \notin A(4)$, we must have $|A(4,\{4,x,\gamma\}) \cap A(x,\{4,x,\gamma\})| \le 2$ with $\gamma \in M_4 - \{a,t\}$. Let $\{4,a,b\},\{4,c,d\} \in P_1$. Since $3 \notin A(a)$, we have $\{b,c,d\} \subseteq A(3)$ and $\{c,d,t\} \subseteq A(a)$. Further, since $c,d \notin A(1) \cap A(2)$ (otherwise $c,d \in M_5$), it follows that $\{b,t,e\} \subseteq A(1) \cap A(2)$. Let $R_2 = \{u,v,r\} \in P_1$ with $\{u,v,r\} \subseteq \{5,b,c,d,e\}$. At this point we have neither $c,d \notin A(1) \cup A(2)$ nor $c,d \in A(1) \cup A(2)$, otherwise $(c,d) \subseteq R_2$ or $u=v=5$ or $u=v=e$.

It follows that $5,\alpha \in A(1) \cup A(2)$ and $\beta \notin A(1) \cup A(2)$ with $(\alpha,\beta)=(c,d)$. If $\alpha = c$ and $\beta = d$ [risp. $\alpha = d$ and $\beta = c$] , then $R_2 = \{5,e,d\}$. From Property 2.4 it is not possible that $|A(x,\{x,5,\gamma\}) \cap A(5,\{x,5,\gamma\})| < 3$ for $x=1,2$ and $\gamma \in M_4$. Then $\{5,a,t\} \in P_1$ or $\{5,a,b\} \in P_1$.

This is impossible and the proof of the lemma is complete. ■

Lemma 4.3. *There is no* $(P;P_1,P_2,P_3,P_4)$ *with* $DS = [5,(4)_4,(3)_6]$.

Proof. Suppose that there exists a $(P;P_1,P_2,P_3,P_4)$ with $DS=[5,(4)_4,(3)_6]$. Let $M_3=\{1,2,\dots,6\}$, $M_4=\{a,b,c,d\}$, $M_5=\{x\}$ and $P = \bigcup_{i=3}^{5} M_i$.

At first, suppose that there exists a block $R \subseteq M_3$. Let,

$R=\{1,2,3\}\in P_1$ without loss of generality. Applying Lemma 2.1 we obtain $\left|\bigcup_{i=1}^{3} A(i,\{1,2,3\})\right| = 4,5$. Let $Y = P-(A(1)\cup A(2)\cup A(3))$, then (necessarily) $|Y|=3$ with $Y\subseteq M_3$, since otherwise $|P_1|\geq 14$. It follows that

$P_1 =$

1 2 3	2 x b	x 4 5	4 . .	3 . .
1 x a	2 . .	x 6 d	5 6 b	
1 . .	3 x c	4 6 a	5 . .	

with $d\in A(4)\cup A(5)$, otherwise $|A(6,\{6,d,x\})\cap A(d,\{6,d,x\})|\leq 2$.

Suppose, first, $d\in A(4)$. Necessarily $(5,c,a)\in P_1$. Since $(4,b,d)\notin P_1$ (otherwise $|A(5,\{5,c,a\})\cap A(c,\{5,c,a\})|\leq 2$), it follows that $(4,d,c)\in P_1$ and $b\in\bigcap_{i=1}^{3} A(i)$.

Since $|A(5,\{5,6,b\})\cap A(b,\{5,6,b\})|\geq 3$, it follows that $\{b,1,c\},\{b,3,a\}\in P_1$ and so $d\in M_3$. This is impossible.

Now, suppose $d\in A(4)\cap A(5)$. It follows that $a\notin A(5)$, otherwise $\{4,d,c\}\in P_1$, $c\in\bigcap_{i=1}^{3} A(i)$ and so $\{1,c,b\},\{2,c,a\},\{3,d,b\}\in P_1$ and $|A(1,\{1,a,x\})\cap A(a,\{1,a,x\})|\leq 2$. If $a\notin A(5)$, for Property 2.4 it is not possible that $|A(4,\{4,6,a\})\cap A(a,\{4,6,a\})|<3$. Then, $\{4,b,d\},\{a,2,d\},\{a,3,b\}\in P_1$ and hence $\{1,b,c\}\in P_1$ with $c\in M_3$. This is impossible.

Now, suppose that there exists a block R such that $|R\cap M_3|=2$. Let $\{1,2,a\}\in P_1$, $X=A(1)\cup A(2)\cup A(a)-\{1,2,a\}$ and $Y=P-(A(1)\cup A(2)\cup A(a))$. It follows that $|X|=6,7,8$. From Lemma 4.2 we obtain $|X|\neq 8$.

Now, suppose $|X|=6$. Since $Y\cap M_4=\emptyset$ implies $|P_1|\geq 14$, then we have $Y=\{3,4\}$. Hence $\{3,4,b\}\in P_1$ with $b\in A(1)\cap A(2)\cap \cap A(a)$ since otherwise $m>13$. Further, since $|A(1)\cap A(2)\cap \cap(M_4-\{a\})|\geq 2$ and $|A(3)\cap A(4)\cap(M_4-\{a,b\})|\geq 2$, it follows $|M_4|>4$. This is impossible.

Now, suppose $|X|=7$. Since $n=11$, it follows that $Y\cap M_4=\emptyset$.

Let $Y = \{3\}$, then, since $n = 11$, $(3,4) \subseteq P_i$ for every $i = 1,2,3,4$. It follows that $|A(3) \cap M_3| = 2$, otherwise, for some $j = 2,3,4$, $R \in P_j$ with $R \subseteq M_3$. Let $A(3) = \{4,5,x,b,c,d\}$. Necessarily $\{3,4,x\} \notin P_1$, otherwise $A(4) = \{3,x,a,b,c,d\}$ and so $|A(1,\{1,2,a\}) \cap A(2,\{1,2,a\}) \cap (M_4 \cup M_5)| \le 2$.

Let $\{3,4,b\} \in P_1$. Since $1,2,a \notin A(3)$, it follows that $a \in A(y)$ and $1,2 \notin A(y)$ for $y = 4,5$.

In fact, let $x \in \{1,2\}$, if $a, x \in A(y)$ we must have $|A(3,\{3,y,\delta\}) \cap A(y,\{3,y,\delta\})| \le 2$ with $\delta = b$ for $y = 4$ and $\delta \in \{x,d\}$ for $y = 5$. If $a \notin A(y)$ and $x \in A(y)$ we must have $|A(y,\{y,x,\alpha\}) \cap A(x,\{y,x,\alpha\})| \le 2$ with $\alpha \in (M_4 - \{a\}) \cup M_5$. Let $a \in A(4)$.

Suppose, first, $\{3,5,x\},\{3,c,d\} \in P_1$. It follows that $x,c,d \in A(4)$ and $c,b,d \in A(5)$. Then $\{4,x,c\},\{4,a,d\},\{5,c,a\}$, $\{5,d,b\} \in P_1$, $\{a,x,6\},\{1,x,b\},\{1,6,c\},\{2,x,d\},\{2,6,b\} \in P_1$ and hence $|A(1,\{1,2,a\}) \cap A(2,\{1,2,a\}) \cap (M_4 \cup M_5)| \le 2$. This is impossible.

Now, suppose $\{3,x,c\},\{3,5,d\} \in P_1$. It follows that $\{x,c,d,a\} \subseteq A(4)$ and $\{x,c,b,a\} \subseteq A(5)$ with $\{4,5,c,x\} \subseteq A(a)$. Further it follows that $6 \in A(1) \cap A(2) \cap A(a)$ and hence $|A(i,\{1,2,a\}) \cap A(a,\{1,2,a\}) \cap (M_4 - \{a\})| \le 2$, for some $i = 1,2$. This is impossible and the proof is complete.■

Lemma 4.4. *There is no $(P;P_1,P_2,P_3,P_4)$ with $DS = [(5)_2,(4)_2,(3)_7]$.*

Proof. Suppose that there exists a $(P;P_1,P_2,P_3,P_4)$ with $DS = [(5)_2,(4)_2,(3)_7]$. Let $M_3 = \{1,2,\dots,7\}$, $M_4 = \{a,b\}$, $M_5 = \{x,y\}$ and $P = \bigcup_{i=3}^{5} M_i$. Since $n = 11$, it follows that $(x,y) \subseteq P_1$.

At first, suppose that exists a block $R \subseteq M_3$. Let $R = \{1,2,3\} \in P_1$, without loss of generality. Applying Lemma 2.1, we obtain $\left|\bigcup_{i=1}^{3} A(i,\{1,2,3\})\right| = |X| = 4,5$. Let $Y = P - \bigcup_{i=1}^{3} A(i)$. Then

$$|Y| = \begin{cases} 3 & \text{if } |X| = 5 \\ 4 & \text{if } |X| = 4 \end{cases}$$

. Further, it follows $Y \subseteq M_3$ otherwise $|P_1| \ge 14$.

If $|X| = 4$, we obtain $X = \{a,b,x,y\}$, $Y = \{4,5,6,7\}$,

$P_1 =$

1 2 3	1 a b	2 x b	3 y b	4 6 y	5 6 b	6 7 x
1 y x	2 a y	3 x a	4 5 x	4 7 a	5 7 y	

and hence $|A(7,\{4,7,a\}) \cap A(a,\{4,7,a\})| = 2$. This is impossible.

If $|X| = 5$, we obtain $Y = \{4,5,6\}$ and $X = \{7,a,b,x,y\}$. It follows $7 \in \bigcap_{i=1}^{3} A(i)$, otherwise $(i,j,k) \subseteq A(7)$ with $i,j \in \{1,2,3\}$ and $k \in \{4,5,6\}$ or $i,j \in \{4,5,6\}$ and $k \in \{1,2,3\}$ and hence $|A(7,\{7,k,\gamma\}) \cap A(k,\{7,k,\gamma\})| \leq 2$ with $\gamma \in \{a,b,x,y\}$. Then necessarily we have $z \in A(i)$ and $z \notin A(j)$ with $i \in \{1,2,3\}$, $j \in \{\{1,2,3\} - \{i\}\}$ and $z \in \{a,b\}$.

If $z = b$, we obtain

$P_1 =$

4 5 x	4 b a	5 b y	...
4 6 y	5 6 a	6 b y	...

and hence $|A(i,\{i,b,\gamma\}) \cap A(b,\{i,b,\gamma\})| = 2$ with $\gamma \in \{7,a,x,y\}$.

Now, suppose that there exists a block R such that $|R \cap M_3| = 2$. Let $R = \{1,2,a\} \in P_1$, $X = A(1) \cup A(2) \cup A(a) - \{1,2,a\}$, $Y = P - (A(1) \cup \cup A(2) \cup A(a))$. If follows $|X| = 6,7,8$. From Lemma 4.2 we obtain $|X| \neq 8$. It follows $|X| \neq 6$, since otherwise $Y \subseteq M_3$, with $Y = \{3,4\}$, $\{3,4,b\} \in P_1$ with $b \in A(1) \cap A(2) \cap A(a)$ and hence $|A(3,\{3,4,b\}) \cap \cap A(b,\{3,4,b\}) \cap M_i| \leq 2$ for $i \geq 4$.

Further, it follows $|X| \neq 7$, since otherwise $\{3,4,\delta\} \in P_1$, with $\delta \in M_4 \cup M_5$ and hence $|A(3,\{3,4,\delta\}) \cap A(4,\{3,4,\delta\}) \cap M_i| \leq 2$ for $i \geq 4$.

This completes the proof of the lemma. ■

Lemma 4.5. *There is no* $(P;P_1,P_2,P_3,P_4)$ *with* $DS = [(5)_3,(3)_8]$.

Proof. Suppose that there exists a $(P;P_1,P_2,P_3,P_4)$ with $DS = [(5)_3,(3)_8]$. Let $M_3 = \{1,2,\ldots,8\}$, $M_5 = \{a,b,c\}$ and $P = M_3 \cup M_5$.

Since $n = 11$, it follows that $(x,y) \subseteq P_1$, for $x,y \in \{a,b,c\}$.

Let $\{a,b,c\} \notin P_1$, without loss of generality. Then

$P_1 =$

$a\ b\ 1$	$a\ 3\ \alpha$	$b\ 2\ \beta$	$c\ 1\ \gamma$	$1\ 2\ 3$
$a\ c\ 2$	$a\ .\ .$	$b\ .\ .$	$c\ .\ .$	
$b\ c\ 3$	$a\ .\ .$	$b\ .\ .$	$c\ .\ .$	

with $\alpha, \beta, \gamma \in \{4,5,6,7,8\}$.

Since $\left|\bigcap_{i=1}^{3} (A(i,\{1,2,3\})\right| \geq 3$, from Lemma 2.1 we obtain $\alpha = \beta = \gamma = 4$;

At first, suppose $\{1,2,4\} \in P_2$. Necessarily

i)

P_1	P_2	$\bar{P}_3$	$\bar{P}_4$	$\bar{P}_5$
$1\ 2\ 3$	$1\ 2\ 4$	$1\ 2\ a$	$1\ 2\ c$	$1\ 2\ b$
$1\ a\ b$	$1\ a\ 3$	$1\ 4\ b$	$1\ 4\ a$	$1\ 4\ 3$
$1\ 4\ c$	$1\ b\ c$	$1\ 3\ c$	$1\ 3\ b$	$1\ c\ a$
$2\ a\ c$	$2\ a\ b$	$2\ 4\ c$	$2\ 4\ 3$	$2\ 3\ a$
$2\ 4\ b$	$2\ 3\ c$	$2\ 3\ b$	$2\ b\ a$	$2\ 4\ c$
$3\ 4\ a$	$3\ 4\ b$	$3\ 4\ a$	$3\ a\ c$	$3\ b\ c$
$3\ b\ c$	$4\ a\ c$	...	...	...
...	...	...	...	...

or

ii)

P_1	P_2	$\bar{P}_3$	$\bar{P}_4$	$\bar{P}_5$
$1\ 2\ 3$	$1\ 2\ 4$	$1\ 2\ a$	$1\ 2\ b$	$1\ 2\ c$
$1\ a\ b$	$1\ a\ c$	$1\ 4\ 3$	$1\ 4\ a$	$1\ 3\ a$
$1\ 4\ c$	$1\ 3\ b$	$1\ b\ c$	$1\ 3\ c$	$1\ 4\ b$
$2\ a\ c$	$2\ 3\ a$	$2\ 4\ c$	$2\ 3\ a$	$2\ 4\ 3$
$2\ 4\ b$	$2\ c\ b$	$2\ 3\ b$	$2\ 4\ c$	$2\ b\ a$
$3\ 4\ a$	$3\ 4\ c$	$3\ a\ c$	$3\ 4\ b$	$3\ b\ c$
$3\ b\ c$	...	...	...	...
...	...	...	...	...

In case i) we have $P_1 \cap \bar{P}_5 \neq \emptyset$ and $P_1 \cap \bar{P}_3 \neq \emptyset$. In case ii) we have $P_2 \cap \bar{P}_4 \neq \emptyset$ and $P_1 \cap \bar{P}_5 \neq \emptyset$. Then a $(P;P_1,P_2,P_3,P_4)$ cannot exist.

Now suppose $\{1,2,4\} \notin P_j$ for every $j = 2,3,4$. Necessarily

i)

P_1	P_2	P_3	P_4
1 2 3	1 2 a	1 2 b	1 2 c
1 a b	1 4 3	1 4 a	1 3 a
1 4 c	1 b c	1 3 c	1 4 b
2 a c	2 4 c	2 a 3	2 3 4
2 4 b	2 3 b	2 4 c	2 b a
3 4 a	3 a c	3 4 b	3 b c
3 b c	...	...	...
...	...	...	...

or

ii)

P_1	P_2	P_3	P_4
1 2 3	1 2 a	1 2 b	1 2 c
1 a b	1 4 b	1 3 4	1 4 a
1 4 c	1 3 c	1 c a	1 3 b
2 a c	2 4 c	2 4 a	2 3 4
2 4 b	2 3 b	2 3 c	2 b a
3 4 a	3 4 a	3 a b	3 a c
3 b c	...	...	...
...	...	...	...

It follows that $P_1 \cap P_4 \neq \emptyset$ in case i) and $P_1 \cap P_2 \neq \emptyset$ in case ii). Then a $(P;P_1,P_2,P_3,P_4)$ cannot exists and the proof is complete.■

Lemma 4.6. *There is no* $(P;P_1,P_2,P_3,P_4)$ *with* $DS = [5,4,(3)_{10}]$ *or* $DS = [(4)_3,(3)_9]$.

Proof. Suppose that there exists a $(P;P_1,P_2,P_3,P_4)$ with $DS = [5,4,(3)_{10}]$ or $DS = [(4)_3,(3)_9]$. In every case exists a block $R \subseteq M_3$. Let $\{1,2,3\} \in P_1$ without loss of generality. Applying Lemma 2.1 we obtain $\left|\bigcup_{i=1}^{3} A(i,\{1,2,3\})\right| = |X| = 4,5$. Let $Y = P - \bigcup_{i=1}^{3} A(i)$, clearly $|X| = 4+k$ and $|Y| = 5-k$ for $k = 0,1$. It follows that $|X| \neq 4$, otherwise $|Y| = 5$ and $|P_1| \geq 14$. Then we have $|X| = 5$ and $|Y| = 4$ with $Y \cap M_i = \emptyset$ for $i = 4,5$.

Let $Y = \{4,5,6,7\} \subseteq M_3$. Since $m = 13$, we obtain that $(x,y) \subseteq P_1$

with $x, y \in Y$.

Let $Z = \{z \in P - \{1,2,3,\ldots,7\} : \exists \{z,y,x\} \in P_1$ with $x,y \in Y\}$. It follows that $|Z| = 3$, otherwise $\{4,5,z_1\},\{6,7,z_2\} \in P_1$, with $z_1 \neq z_2$ and hence $|A(4,\{4,5,z_1\}) \cap A(z_1,\{4,5,z_1\})| \leq 2$. Observe that $Z \cap M_3 = \emptyset$. Otherwise we obtain $|A(1,\{1,z_1,\gamma\}) \cap A(z_1,\{1,z_1,\gamma\})| = 0$ with $z_1 \in Z \cap M_3$. At this point we have $M_5 = \emptyset$, $\bigcap_{i=1}^{3} A(i,\{1,2,3\}) = \{8,9\} \subseteq M_3$ and hence $DS = [(4)_3,(3)_9]$.

Let $M_3 = \{1,2,\ldots,9\}$, $M_4 = \{a,b,c\}$ and $P = \bigcup_{i=3}^{4} M_i$.

Then

i) $P_1 =$

1 2 3	2 8 a	3 9 b	4 7 c	6 7 a
1 8 9	2 9 c	4 5 a	5 6 c	
1 a b	3 8 c	4 6 b	5 7 b	

or

ii) $P_1 =$

1 2 3	2 8 b	3 9 a	4 7 c	6 7 a
1 8 a	2 9 c	4 5 a	5 6 c	
1 9 b	3 8 c	4 6 b	5 7 b	

In case i) we have $|A(1,\{1,a,b\}) \cap A(a,\{1,a,b\})| \leq 2$. In case ii), $|A(1,\{1,8,a\}) \cap A(a,\{1,8,a\})| \leq 2$. Then a $(P;P_1,P_2,P_3,P_4)$ cannot exists and this completes the proof.∎

Lemma 4.7. *There is no* $(P;P_1,P_2,P_3,P_4)$ *with* $DS = [(3)_{13}]$.

Proof. The statement follows immediately from Theorem 2.1 of [5].

Theorem 4.1. $D(v,t_v-13) = 3$ *for every* $v \geq 15$.

Proof. Applying Lemmas 4.1, 4.2, 4.3, 4.4, 4.5, 4.6 and 4.7, we obtain that a $(P;P_1,P_2,\ldots,P_s)$ with $m = 13$ and $s > 3$ cannot exist. Then, since the existence of s $STS(v)s$, such that any two of them intersect in t_v-13 blocks (these t_v-13 blocks occurring, moreover, in each of the $STS(v)$s) implies the existence of a $(P;P_1,P_2,\ldots,P_s)$, from Theorem 3.1 we obtain $D(v,t_v-13)=3$ for every $v \geq 15$. ∎

REFERENCES

[1] J. Doyen, *Construction of disjoint Steiner triple systems*, Proc. Amer. Math. Soc., 32 (1972), 409-416.

[2] J. Doyen and R.M. Wilson, *Embeddings of Steiner triple systems*, Discrete Math., 5 (1972), 229-239.

[3] S. Milici and G. Quattrocchi, *Some results on the maximum number of STSs such that any two of them intersect in the same block-set*, preprint.

[4] S. Milici and G. Quattrocchi, *Alcune condizioni necessarie per l'esistenza di tre DMB PTS con elementi di grado 2*, Le Matematiche (to appear).

[5] G. Quattrocchi, *Sul massimo numero di DMB PTS aventi 12 blocchi e immergibili in un STS*, Riv. Mat. Univ. Parma (to appear).

[6] G. Quattrocchi, *Sul parametro $D(13, 14)$ per Sistemi di Terne di Steiner*, Le Matematiche (to appear).

[7] G. Quattrocchi, *Sul parametro $D(v,t_v-10)$, $19 \leq v \leq 33$ per Sistemi di Terne di Steiner*, Quaderni del Dipartimento di Matematica di Catania, Rapporto interno.

[8] A. Rosa, *Intersection properties of Steiner systems*, Annals Discrete Math., 7 (1980), 115-128.

Annals of Discrete Mathematics 30 (1986) 331–334

A NEW CONSTRUCTION OF DOUBLY DIAGONAL ORTHOGONAL LATIN SQUARES

Consolato Pellegrino and Paola Lancellotti

Dipartimento di Matematica
Via Campi, 213/B
41100 MODENA (ITALY).

We give a new simple construction of pairs of doubly diagonal orthogonal Latin squares of order n, DDOLS(n), for some n=3k including the case n=12.

A pair of doubly diagonal orthogonal Latin squares of order n, DDOLS(n), is a pair of orthogonal Latin squares of order n with the property that each square has a transversal both on the front diagonal D_1 and on the back diagonal D_2. The reader is referred to the monograph [1] by J.Dénes and A.D.Keedwell for the definitions which are not given here. W.D.Wallis and L.Zhu proved the existence of 4 DDOLS(12) in [2]. The problem was posed by K.Heinrich and A.J.W.Hilton in [3].

Let Q be a Latin square of order n based on the set $I_n=\{0,1,\ldots,n-1\}$ and let S, T be transversals of Q. We form a permutation $\sigma_{S,T}$ on I_n as follows: to the element of I_n occupying the cell (h,i) of S we associate the element of I_n occupying the cell (k,i) of T (i.e. the cell of T that lies in the same column). We denote by Q(S,T) the Latin square obtained by replacing each entry s of Q with the element $\sigma_{S,T}(s)$. Obviously we have:

(a) if U is a transversal of Q then U is also a transversal of Q(S,T);

(b) if R is a Latin square which is orthogonal to Q then R is also orthogonal to Q(S,T).

Let Q be a Latin square and let h be a symbol; we denote by Q_h the copy of Q obtained by replacing each entry s of Q with the ordered pair (h,s).

THEOREM. For an even positive integer k let A, B be a pair of DDOLS(k) and let T_1, T_2 be two common transversals of A and B. If T_1 and T_2 have no common cell with each other and with each diagonal D_1 and D_2, then there exists a pair of DDOLS(3k).

Proof. Consider the two orthogonal Latin squares of order 3k

$$\bar{A} = \begin{bmatrix} A_0 & A_1 & A_2 \\ A_1 & A_2 & A_0 \\ A_2 & A_0 & A_1 \end{bmatrix} \qquad \bar{B} = \begin{bmatrix} B_0 & B_2 & B_1 \\ B_1 & B_0 & B_2 \\ B_2 & B_1 & B_0 \end{bmatrix}$$

Of course $\bar{A}$ possesses a transversal on the front diagonal, while the back diagonal is a transversal of $\bar{B}$. Starting from $\bar{A}$ and $\bar{B}$ we form the following Latin squares of order 3k

$\tilde{A}$ =

$A_{11}=A_0(D_1,D_2)$	$A_{12}=A_1$	$A_{13}=A_2(D_2,T_2)$
$A_{21}=A_1(T_1,D_1)$	$A_{22}=A_2$	$A_{23}=A_0(T_2,D_1)$
$A_{31}=A_1(D_2,T_1)$	$A_{32}=A_0$	$A_{33}=A_1(D_1,D_2)$

$\tilde{B}$ =

$B_{11}=B_0(D_1,T_1)$	$B_{12}=B_2$	$B_{13}=B_1(D_2,D_1)$
$B_{21}=B_1(T_1,D_2)$	$B_{22}=B_0$	$B_{23}=B_2(T_2,D_2)$
$B_{31}=B_2(D_2,D_1)$	$B_{32}=B_1$	$B_{33}=B_0(D_1,T_2)$

From (a) and (b) it follows immediately that the square $\tilde{A}$ still has a transversal on the front diagonal while $\tilde{B}$ has a transversal on the back diagonal. In addition we have:

(c) each subsquare A_{ij} of $\tilde{A}$ and B_{ij} of $\tilde{B}$ is a doubly diagonal Latin square having T_1,T_2,D_1,D_2 as pairwise disjoint transversals;

(d) $\tilde{A}$ and $\tilde{B}$ are orthogonal.

Since the square $\tilde{A}$ is obtained from $\bar{A}$ by suitably renaming symbols in the subsquares, we have for $j=1,2,\dots,k$:

(e) the set H_j of the entries of the j-th column of $\tilde{A}$ which lie on the transversals D_1 of A_{11}, T_1 of A_{21} and D_2 of A_{31} coincides whit the set H'_j of the entries of the (k+j)-th column of $\tilde{A}$ lying on the transversals D_1 of A_{21}, T_1 of A_{22} and D_2 of A_{32};

(f) the set K_j of the entries of the (k+j)-th column of $\tilde{A}$ which lie on the transversals D_2 of A_{12}, T_2 of A_{22} and D_1 of A_{32} coincides whit the set K'_j of the entries of the (2k+j)-th column of $\tilde{A}$ lying on the transversals D_2 of A_{13}, T_2 of A_{23} and D_1 of A_{33}.

For each $j=1,2,\dots,k$ exchange in $\tilde{A}$ the elements of H_j and H'_j appearing on the same row; similarly we exchange the elements of K_j and K'_j appearing on the same row of $\tilde{A}$ (property (c) implies that the elements of H'_j and K'_j occupy distinct cells): from (e) and (f) it follows immediately that the resulting matrix $\hat{A}$ is a Latin square. Further $\hat{A}$ is doubly diagonal as the construction easily shows.

Observing that $\tilde{B}$ has properties which are analogous to (e) and (f) we can exchange elements in $\tilde{B}$ as we did in deriving $\hat{A}$ from $\tilde{A}$ and thus obtain a doubly diagonal Latin square $\hat{B}$ which is orthogonal to $\hat{A}$ because of property (d). Hence $\hat{A}$ and $\hat{B}$ are pair of DDOLS(3k).

EXAMPLE. Since for each $r \geq 2$ there exists a pair of DDOLS(2^r) satisfying the hypothesis of the previous Theorem, we have that for each $r \geq 2$ we can construct a pair of DDOLS($3 \cdot 2^r$).

ACKNOWLEDGEMENTS. Work done within the sphere og GNSAGA of CNR, partially supported by MPI.

REFERENCES

[1] J.Dénes and A.D.Keedwell, Latin squares and their Applications (Academic Press New York, 1974).

[2] W.D.Wallis and L.Zhu, Four pairwise orthogonal diagonal Latin squares of side 12, Util. Math. 21 (1982) 205-207.

[3] K.Heinrich and A.J.Hilton, Doubly diagonal orthogonal Latin squares, Discr. Math. 46 (1983) 173-182.

Annals of Discrete Mathematics 30 (1986) 335–338

ON THE MAXIMAL NUMBER OF MUTUALLY ORTHOGONAL F-SQUARES

Consolato PELLEGRINO and Nicolina A. MALARA

Dipartimento di Matematica
Via Campi, 213/B
41100 MODENA (ITALY)

In this paper we prove that the upper bound, given by Mandeli and Lee and Federer [3] for the number t of orthogonal $F_1(n;\lambda_1)$, $F_2(n;\lambda_2)$, ... , $F_t(n;\lambda_t)$ squares, also holds for the number t of orthogonal $F_1(n;\lambda_{1,1},\lambda_{1,2},\ldots,\lambda_{1,m_1})$, $F_2(n;\lambda_{2,1},\lambda_{2,2},\ldots,\lambda_{2,m_2})$, ... , $F_t(n;\lambda_{t,1},\lambda_{t,2},\ldots,\lambda_{t,m_t})$ squares.

1 - DEFINITIONS AND PRELIMINARY RESULTS

Hedayat and Seiden, in connection with some results by other authors, give in [1] a generalization of the concept of latin square: the condition that each element appear exactly once in each row and in each column is substituted by the condition that each element appear one and the same fixed number of times in each row and in each column. They call such squares frequency-square or shortly F-squares. More precisely they give the following definition:

DEFINITION 1. Let $F = [a_{i,j}]$ be a $n\times n$ matrix defined on a m-set $A = \{a_1,a_2,\ldots,a_m\}$. We say that F is an F-square of type $(n;\lambda_1,\lambda_2,\ldots,\lambda_m)$, and we write briefly $F(n;\lambda_1,\lambda_2,\ldots,\lambda_m)$, if for each $k=1,2,\ldots,m$ the element a_k of A appears precisely λ_k times $(\lambda_k \geq 1)$ in each row and in each column of F.

In particular if $\lambda_1=\lambda_2=\ldots=\lambda_m=\lambda$ then m is determined uniquely by n and λ, hence we simply write $F(n;\lambda)$. Note that an $F(n;1)$ square is simply a latin square of order n. It is easy to prove that an $F(n;\lambda_1,\lambda_2,\ldots,\lambda_m)$ square exists if and only if $\sum_{i=1}^{m} \lambda_i = n$.

In [1] Hedayat and Seiden also extend to F-squares the concept of orthogonality of latin squares through the following definitions:

DEFINITION 2. Given an $F(n;\lambda_1,\lambda_2,\ldots,\lambda_r)$ square on a r-set $U = \{u_1,u_2,\ldots,u_r\}$ and $F_2(n;\mu_1,\mu_2,\ldots,\mu_s)$ square on a s-set $V = \{v_1,v_2,\ldots,v_s\}$, we say that F_1 is orthogonal to F_2, and write $F_1 \perp F_2$, if upon superimposition of F_1 on F_2 the pair (u_i,v_j) of $U\times V$ appears $\lambda_i\mu_j$ times, for each $i=1,2,\ldots,r$ and for each $j=1,2,\ldots,s$.

DEFINITION 3. Let A_i be a m_i-set, $i=1,2,\ldots,t$. For each i, let F_i be on F-square of type $(n;\lambda_{i,1},\lambda_{i,2},\ldots,\lambda_{i,m_i})$ on the set A_i. We say that $F_1,F_2,\ldots,F_t$ is a set of t mutually (pairwise) orthogonal F-squares if $F_i \perp F_j$, $i\neq j$, $i,j=1,2,\ldots,t$.

In [2] Hedayat, Raghavarao and Seiden proved that the maximal number of mutually orthogonal $F(n;\lambda)$ squares is $(n-1)^2/(m-1)$, where $m=n/\lambda$. In [3] Mandeli, Lee and Federer proved that the maximal number t of mutually orthogonal $F_1(n;\lambda_1), F_2(n;\lambda_2), \ldots, F_t(n;\lambda_t)$ squares (where for each $i=1,2,\ldots,t$ F_i is defined on a m_i-set and $n=\lambda_i m_i$) satisfies the inequality

$$\sum_{i=1}^{t} m_i - t \le (n-1)^2.$$

2. ON THE MAXIMAL NUMBER OF MUTUALLY ORTHOGONAL F-SQUARES

In analogy to [3] we prove the following

THEOREM. Let $F_1(n;\lambda_{1,1},\lambda_{1,2},\ldots,\lambda_{1,m_1})$, $F_2(n;\lambda_{2,1},\lambda_{2,2},\ldots,\lambda_{2,m_2})$, ..., $F_t(n;\lambda_{t,1},\lambda_{t,2},\ldots,\lambda_{t,m_t})$ be t mutually orthogonal F-squares, where for each $i=1,2,\ldots,t$ F_i is defined on the m_i-set A_i and $n=\sum_{j=1}^{m_i}\lambda_{i,j}$. Then the number t satisfies the inequality

$$\sum_{i=1}^{t} m_i - t \le (n-1)^2 .$$

Proof. From $F_h(n;\lambda_{h,1},\lambda_{h,2},\ldots,\lambda_{h,m_h})$ we define a $n^2 x m_h$ matrix $M_h = \left[a^h_{ij,k}\right]$, where $a^h_{ij,k}=1$ if the k-th symbol of A_h occurs in the cell (i,j) $(i,j=1,2,\ldots,n)$ of F_h and 0 otherwise. Let $M = \left[M_1|M_2|\ldots|M_t\right]$. By the property of the F-squares, the number of linearly independent rows in M is at most $(n-1)^2+1$ and so we obtain

$$\operatorname{rank}(M) \le \min\{(n-1)^2+1, \sum_{h=1}^{t} m_h\} .$$

Now, we can write the product of the transpose of M with M in this manner:

$$M'M = \begin{bmatrix} L_1N_1 & L_1J_{m_1xm_2}L_2 & \cdots & L_1J_{m_1xm_t}L_t \\ L_2J_{m_2xm_1}L_1 & L_2N_2 & \cdots & L_2J_{m_2xm_t}L_t \\ \vdots & \vdots & \ddots & \vdots \\ L_tJ_{m_txm_1}L_1 & L_tJ_{m_txm_2}L_2 & \cdots & L_tN_t \end{bmatrix}$$

where $L_i = \left[u^i_{r,s}\right]$ $(i=1,2,\ldots,t)$ is a diagonal matrix of order m_i with $u^i_{r,r}=\lambda_{i,r}$ for each $r=1,2,\ldots,m_i$, $N_i = \left[n^i_{r,s}\right]$ is a diagonal matrix of order m_i with $n^i_{r,r}=n$ for each $r=1,2,\ldots,m_i$ and $J_{m_ixm_j}$ is a matrix of size m_ixm_j $(i,j=1,2,\ldots,t)$ with the element 1 everywhere.

Let

$$\Lambda = \begin{bmatrix} L_1 & 0_{m_1 x m_2} & \cdots & 0_{m_1 x m_t} \\ 0_{m_2 x m_1} & L_2 & \cdots & 0_{m_2 x m_t} \\ \vdots & \vdots & \ddots & \vdots \\ 0_{m_t x m_1} & 0_{m_t x m_2} & \cdots & L_t \end{bmatrix}$$

where $0_{m_i x m_j}$ is the matrix of size $m_i x m_j$ $(i,j=1,2,\ldots,t)$ with the element 0 everywhere. As $\lambda_{i,j} \neq 0$, Λ is invertible and the matrix $M'M$ has the same rank as the matrix

$$\tilde{M} = \Lambda^{-1} M'M = \begin{bmatrix} N_1 & J_{m_1 x m_2} L_2 & \cdots & J_{m_1 x m_t} L_t \\ J_{m_2 x m_1} L_1 & N_2 & \cdots & J_{m_2 x m_t} L_t \\ \vdots & \vdots & \ddots & \vdots \\ J_{m_t x m_1} L_1 & J_{m_t x m_2} L_2 & \cdots & N_t \end{bmatrix}.$$

The eigenvalues of $\tilde{M}$ are tn, n and 0 with respective multiplicities 1, $\sum_{i=1}^{t} (m_i - 1)$ and $t-1$. Then

$$1 + \sum_{i=1}^{t} (m_i - 1) = \operatorname{rank}(\tilde{M}) = \operatorname{rank}(M'M) =$$

$$\operatorname{rank}(M) \leq \min\{(n-1)^2 + 1, \sum_{i=1}^{t} m_i\}.$$

Hence

$$\sum_{i=1}^{t} m_i - t \leq (n-1)^2 .$$

When $\lambda_{i,1} = \lambda_{i,2} = \ldots = \lambda_{i,m_i} = \lambda_i$ $(i=1,2,\ldots,t)$ we have the result by Mandeli, Lee and Federer [3]. Furthermore, the previous theorem suggests that we call a set of mutually orthogonal F-squares $F_1(n;\lambda_{1,1},\lambda_{1,2},\ldots,\lambda_{1,m_1})$, $F_2(n;\lambda_{2,1},\lambda_{2,2},\ldots,\lambda_{2,m_2})$, ... , $F_t(n;\lambda_{t,1},\lambda_{t,2},\ldots,\lambda_{t,m_t})$ a complete set if

$$\sum_{i=1}^{t} m_i - t = (n-1)^2 ,$$

where $n=\lambda_{i,1}+\lambda_{i,2}+\ldots+\lambda_{i,m_i}$ $(i=1,2,\ldots,t)$.

ACKNOWLEDGEMENTS. Work done within the sphere of GNSAGA of CNR, partially supported by MPI.

REFERENCES

[1] A.Hedayat, E.Seiden, F-squares and orthogonal F-squares design: a generalization of latin square and orthogonal latin squares design; Ann. Math. Statist. 41 (1970) 2035-2044.

[2] A.Hedayat, D.Raghavarao, E.Seiden, Further contributions to the theory of F-squares design, Ann. Statist. 3 (1975) 712-716.

[3] J.P.Mandeli,F.C.H.Lee, W.T.Federer, On the construction of orthogonal F-squares of order n from an orthogonal array (n,k,s,2) and an OL(s,t) set, J. Statist. Plann. Inference 5 (1981) 267-272.

Annals of Discrete Mathematics 30 (1986) 339–346

CARTESIAN PRODUCTS OF GRAPHS AND THEIR CROSSING NUMBERS

Giustina Pica +

Dipartimento di Matematica e Applicazioni
Università di Napoli, Naples, Italy

Tomaž Pisanski ++

Oddelek za Matematiko, Univerza v Ljubljani
Ljubljana, Yugoslavia

Aldo G.S.Ventre +

Istituto di Matematica, Facoltà di Architettura
Università di Napoli, Naples, Italy

Kainen and White have determined exact crossing numbers of some infinite families of graphs. Their process uses repeated cartesian products of regular graphs. It is shown how this process can be substantially generalized yielding exact crossing numbers and bounds for various families of graphs.

INTRODUCTION

In this paper graph embeddings and immersions are studied. In order to keep it short we adopt standard definitions of topological graph theory that can be found, say in [2,3,4,5,13]. Usually only normal immersions of graphs into surfaces are considered, i.e. immersions in which no two edges cross more than once and no edge crosses itself. In particular, this means that two edges that are adiacent do not cross. We require in addition the immersion to be a 2-cell immersion which means that the complement of the immersed graph is a disjoint union of open disks (2-cells) and that there exists a set of edges that can be removed from the immersed graph in order to obtain a 2-cell embedding of its spanning subgraph into the same surface. The connected components of the complement of the immersion are called faces. In a 2-cell immersion or embedding all faces are open disks. A face is said to be partial if it has at least one crossing point on its boundary otherwise it is said to be total. We will make use of the definition of an (s,k)-embedding of [11] that we repeat here for convenience (see also [10] and [12]).

A 2-cell embedding of a graph G into a surface S is said to be an (s,k)-embedding if we can partition the set of faces of the embedding into s+1 sets $F_1, F_2, \ldots, F_s, R$ in such a way that the boundary of each set F_i, $1 \leqslant i \leqslant s$, i.e. the union of boundaries of faces belonging to F_i, is an even 2-factor of G, i.e. a spanning subgraph of G consisting of cycles of even lengths; furthermore, k out of the s 2-factors consist of quadrilaterals only and all faces of R (if there are any) are quadrilaterals. R is called the set of residual faces and may be empty. If k=s we are dealing with quadrilateral embedding. If G has no triangles the embedding is also minimal, yielding the genus or nonorientable genus of G (depending on the orientability type of S), see [11] . Let G_1 have an (s,k)-embedding into S' and let G_2 have an (s,k)-embedding into S". We say that the two (s,k)-embeddings agree if there exists a bijection between the vertex sets of G_1 and G_2 which induces a bijection of all s sets of nonresidual faces.

Example 1. Part (a) of Figure 1 shows an (1,0)-embedding of $K_{3,3}$-$2K_2$ into the sphere. The outer face is hexagonal and there are two residual faces. Part (b) of Figure 1 shows an (1,0)-embedding of $K_{3,3}$ into the projective plane. There is one hexagonal face and three residual faces. The two (1,0)-embeddings agree, which is shown by an appropriate numbering of vertices in both graphs.

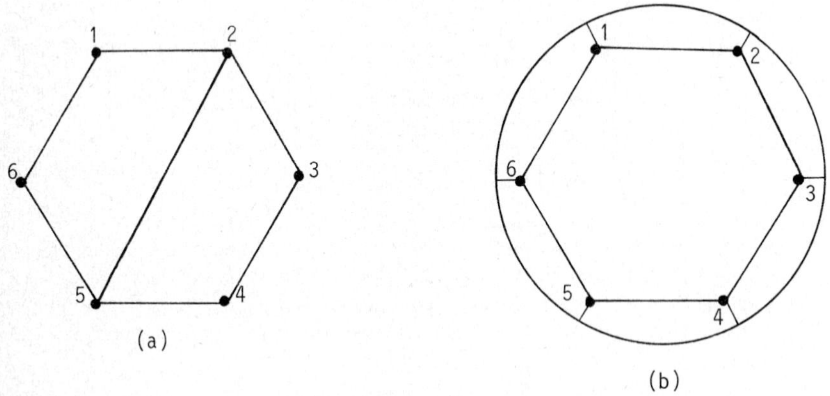

Figure 1

An immersion of a graph G into a surface S is said to be an (s,k,c,e)-immersion if it is a 2-cell immersion with c crossing points and it is possible to obtain an (s,k)-embedding of a spanning subgraph H of G into S by removal of e edges, and by removal of any e-1 edges there remain some crossing points (e is minimal). H is said to be a reduced graph of the (s,k,c,e)-immersion of G. Let G have an (s,k,c,e) -immersion into S and an (s,k,c',e')-immersion into S'. We say that the two immersions agree, if the corresponding (s,k)-embeddings of reduced graphs agree. The following examples help explain the above definitions.

Example 2. Figure 2 represents planar (1,0,3,2)-immersion of $K_{3,3}$. The reduced graph and its (1,0)-embedding is depicted on Figure 1(a). Note that the immersions of $K_{3,3}$ on Figures 2 and 1(b) agree.

The following two examples were first used by Kainen [6,7] and Kainen and White [9].

Example 3. Part (a) of Figure 3 shows an (1,1,4,4)-immersion of $K_{4,4}$ into the sphere. If the edges 1-6, 2-7, 3-8, and 4-5 are removed a (3,3)-embedding which is of course also an (1,1)-embedding of the 3-cube graph Q_3 into the sphere results; see Figure 3(b). Part (c) of Figure 3 represents the well-known genus embedding of $K_{4,4}$ into the torus which is a (4,4)-embedding.

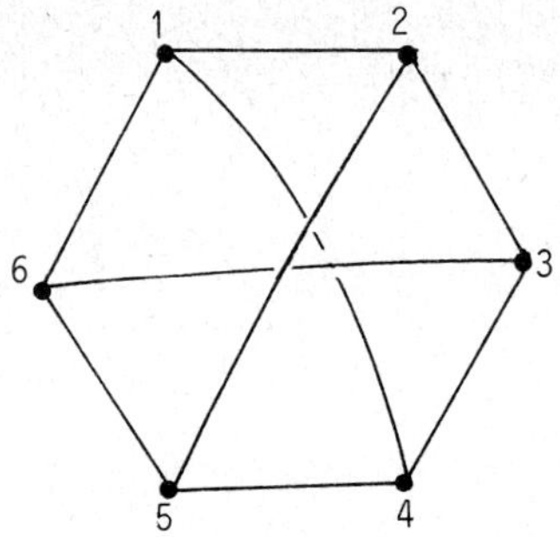

Figure 2

Note that immersions on Figures 3(a) and 3(c) agree as (1,1,4,4)- and (1,1,0,0)-immersions, as they have faces 1-2-3-4 and 5-6-7-8 in common.

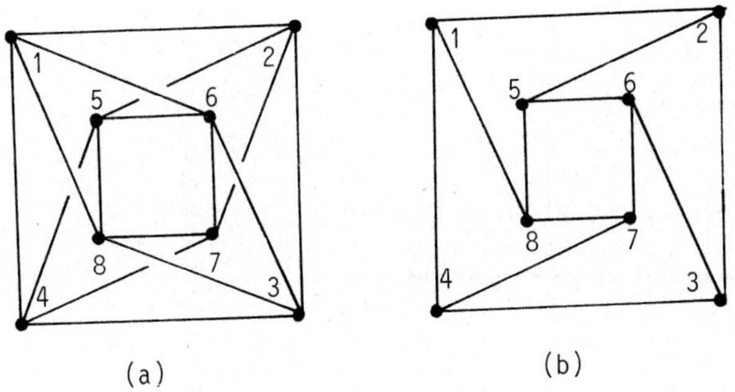

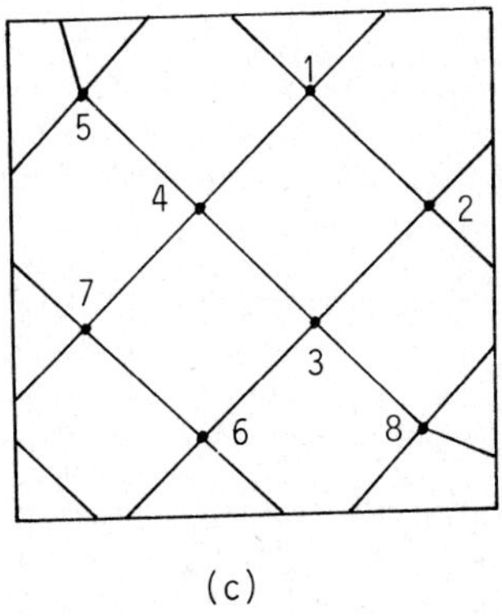

(c)

Figure 3

Example 4. Figure 4(a) represents a (2,2,8m - 8,4)-immersion of the Cartesian product $C_{2m} \times C_4$ into the sphere for the case m = 3. Parts (b) and (c) of Figure 4 are analogous to parts (b) and (c) of Figure 3. Namely, by removing appropriate four edges A-B, C-D, E-F, and G-H we obtain a planar, residual (3,3)-embedding of $P_{2m} \times C_4$ (which is of course also a (2,2)- and even (1,1)-embedding) as depicted by Figure 4(b). Finally, Figure 4(c) represents the familiar toroidal (4,4)-embedding of $C_{2m} \times C_4$. Note that (a) and (c) agree as (1,1,8m - 8,4) - and (1,1,0,0)- immersions (and not as (2,2,p,q)- immersions, for any p and q).

Recently Beineke and Ringeisen have shown [1] that $cr(C_m \times C_4) = 2m$. This means that the immersion of Figure 4(a) is far from optimal.

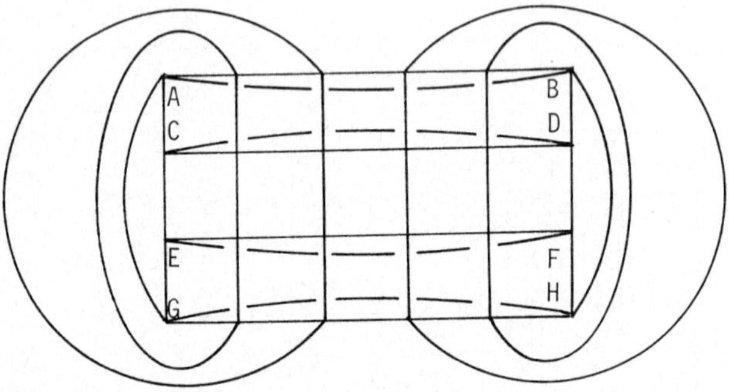

(a)

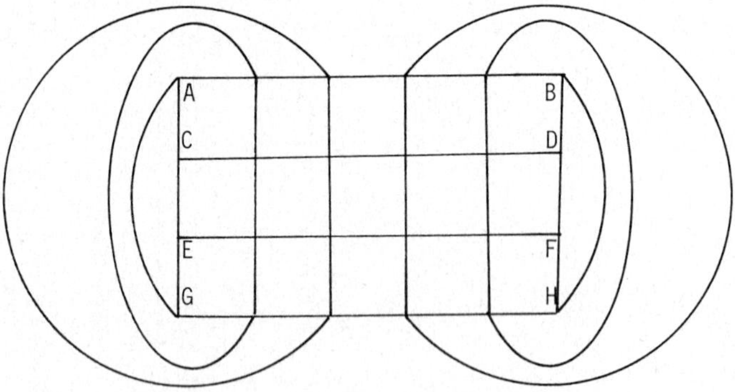

(b)

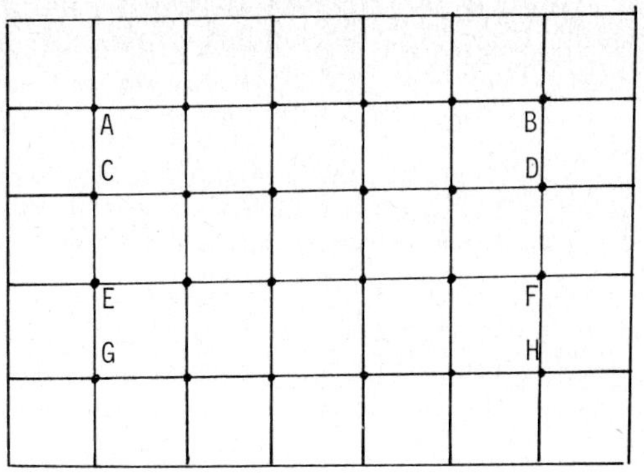

(c)

Figure 4

CONSTRUCTION OF IMMERSIONS

When dealing with crossing numbers on surfaces the following two combinatorial invariants are handy.

$$d_k(G) = q - g(p - 2(1 - k))/(g - 2)$$
$$\bar{d}_k(G) = q - g(p - 2 + k)/(g - 2)$$

Here p denotes the number of vertices, q denotes the number of edges, and g denotes the girth of G. In both cases k is a nonnegative integer representing in the first case the (orientable) genus and in the second case the nonorientable genus of some surface. They are sometimes called Euler deficiencies as they refer to the graph and the surface, and only the Euler characteristic of the surface is involved. They were introduced by Kainen. The orientable version was introduced in [6] while the nonorientable one was defined in [8] and later used by Kainen and White [9]. Euler deficiency tells us the number of superfluous edges which obstruct the embedding of a graph into the surface. The following lemma shows how Euler deficiencies serve as lower bounds for crossing numbers $cr_k(G)$ and $\overline{cr}_k(G)$.

Lemma 5.

$$cr_k(G) \geqslant d_k(G) \quad \text{and} \quad \overline{cr}_k(G) \geqslant \bar{d}_k(G) .$$

For proof of the orientable case see [6]. The nonorientable case is essentially the same; see also [8,9]. To obtain an upper bound for the crossing number we need the following lemma.

Lemma 6. Let G be a connected graph with p vertices and q edges. If G admits an orientable (s,s,c,e)-immersion, then $cr_k(G) \leqslant c$, where $k = 1 - p/2 + (q-e)/4$.

If G admits a nonorientable (s,s,c,e)-immersion, then $\overline{cr}_h(G) \leqslant c$, where $h=2-p+(q-e)/2$.

Proof. By removal of e edges we obtain in both cases a quadrilateral embedding of the reduced subgraph H. Therefore the subscript k(h) corresponds to the genus (nonorientable genus) of the surface in which the immersion of G and the embedding of H takes place. Since there are c crossing points the statement of Lemma follows

The following lemma will help us to find graphs with (s,s,c,e)-immersions, namely the Cartesian products of other graphs. These will be the graphs for which Lemmas 5 and 6 will apply yielding lower and upper bounds for the crossing numbers.

Lemma 7. Let G and H be two connected graphs and $s \geq 1$, $s \geq k \geq 0$ two integers. Let there exist a proper edge coloring of H using at most s colors such that d of the colors used determine d 1-factors of H. Let there be for each vertex v of H an (s,k,c_v,e_v)-immersion of G into some surface S(v). Furthermore, let all the immersions agree. Define $C = \sum c_v$ and $E = \sum e_v$. Then the Cartesian product G x H admits an (s+d, min(k+2d ,s+d), C,E)-immersion into some surface T. (Here T is orientable if and only if all surfaces S(v)are orientable and H is bipartite.)

The proof is essentially that of Lemma 2.3 of [11] and is omitted. Intuitively, the crossing points do not interfere with the argument used in the proof, as they occur only in the residual faces and all embeddings of G agree on all nonresidual faces.

Starting now with graphs such as those of Examples 1,2,3 and 4 and using repeatedly Lemma 7 it is possible to obtain composite graphs that satisfy Lemma 6. In some cases the upper bound of Lemma 6 and the lower bound of Lemma 5 coincide to give exact crossing number; see [9] for examples!

CONSEQUENCES

In this section we apply our lemmas to obtain bound for crossing numbers and exact crossing numbers in some special cases.

Theorem 8. If a connected graph G admits an (s,s,c,c)-immersion into a surface of genus k then $cr_k(G) = c$ if the surface is orientable or $\overline{cr}_h(G) = c$ if it is nonorientable, where $k=1-p/2+(q-e)/4$ and $h=2-p+(q-e)/2$.

Proof. Combining Lemmas 5 and 6 we obtain the inequalities $e \leq cr_k(G) \leq c$ in the orientable case, and the inequalities $e \leq \overline{cr}_h(G) \leq c$ in the nonorientable case. As we have c=e the result follows.

Using our lemmas and the Theorem above it is then possible to prove all theorems of [9] and both theorems of [7]. In addition we obtain the following results.

Theorem 9. Let G be an arbitrary connected bipartite graph. For each $n \geq \Delta(G)-1$ and for each m, such that $0 \leq m \leq 2^n$ there is $cr_{g-m}(K_{4,4} \times Q_n \times G) = 4m$, where g is the genus of $K_{4,4} \times Q_n \times G$.

Proof. We first prove that $K_{4,4} \times Q_n$ admits an orientable (n+1,n+1,4m,4m)-immersion, for each m, $0 \leq m \leq 2^n$. The proof is by induction on n. The statement is certainly true for n = 0, as shown by Figure 3(a,c). The induction step follows by Lemma 7 if we take $K_{4,4} \times Q_n$ for G and K_2 for H. As each bipartite graph is edge-colorable with $\Delta(G)$ colors we apply Lemma 7 to the Cartesian product of $K_{4,4} \times Q_n$ and G to obtain an orientable (n+1,n+1,4m,4m)-immersion of $K_{4,4} \times Q_n \times G$. Using Theorem 8 the result now follows readily.

Note that the result of Theorem 6 in case of $G = K_1$ was obtained already by Kainen and White [9].

Theorem 10. Let G be an arbitrary connected bipartite graph and let $k \geq 2$ be an integer. For each integer $n \geq \Delta(G) - 1$ and for each m such that $0 \leq m \leq 2^n$ there is

$$4_m \leq cr_{g-m}(C_{2k} \times Q_{n+2} \times G) \leq 8(k-1)m ,$$

where g is the genus of $C_{2k} \times Q_{n+2} \times G$.

Proof. Starting with immersions as in Figure 4(a,c) we could easily prove by induction on n that $C_{2k} \times Q_{n+2}$ admits an orientable $(n+1,n+1,8(k-1)m,4m)$-immersion and then as in the proof of Theorem 9 we observe that also $C_{2k} \times Q_{n+2} \times G$ admits an immersion of the same type. By the orientable part of Lemmas 5 and 6 the result follows.

Note that by taking k = 2 in Theorem 10 we obtain the inequality: $4_m \leq cr_{g-m}(Q_{n+4} \times G) \leq 8_m$ which further reduces for $G = K_1$ to the inequality of Kainen [7]. Note one can take a nonbipartite graph with girth at least four in Theorems 9 and 10 in order to obtain results of the same type for $\overline{cr}_{g-m}$.

Theorem 11. Let G be an arbitrary connected graph without triangles. For each $n \geq \Delta(G)$ and for each m such that $0 \leq m < 2^n$ there is $2m \leq \overline{cr}_{\overline{g}-m}(K_{3,3} \times Q_n \times G) \leq 3m$, where g denotes the nonorientable genus of $K_{3,3} \times Q_n \times G$.

Proof. The proof is essentially the same as that of Theorem 9 that is by induction on n. The basis of induction is provided by Figure 1(b) and Figure 2. As $m<2^n$ there has to be at least one nonorientable immersion of $K_{3,3}$ involved and the final surface is nonorientable. Note that if G is not bipartite, then the Theorem can be extended also to the case $m = 2^n$. If G is 1-factorable the bound for n can be lowered by 1.

REFERENCES

[1] L.W.Beineke and R.D.Ringeisen, On the crossing numbers of products of cycles and graphs of order four, J.Graph Theory 4 (1980) 145-155.

[2] R.K.Guy, Crossing numbers of graphs, in "Graph Theory and Applications", Lect. Notes in Mathematics 303 (ed. Y.Alavi, et al.), Springer-Verlag, Berlin, Heidelberg, New York, 1972, 553-569.

[3] R.K.Guy and T.A.Jenkyns, The toroidal crossing number of $K_{n,m}$, J.Combinatorial Theory 6 (1969) 235-250.

[4] M.Jungerman, The non-orientable genus of the n-cube, Pacific J.Math. 76 (1978) 443-451.

[5] P.C.Kainen, Embeddings and orientations of graphs, in "Combinatorial Structures and their Applications", Gordon and Breach, New York, 1970,193-196.

[6] P.C.Kainen, A lower bound for crossing numbers of graphs with applications to K_n,$K_{p,q}$ and Q(d), J.Combinatorial Theory B 12 (1972) 287-298.

[7] P.C.Kainen, On the stable crossing numbers of cubes, Proc. Amer.Math.Soc. 36 (1972) 55-62.

[8] P.C.Kainen, Some recent results in topological graph theory, in "Graphs and Combinatorics", Lect.Notes in Mathematics 406 (ed. R.A.Bari and F.Harary),

Springer-Verlag, Berlin, Heidelberg, New York, 1973, 76-108.
[9] P.C. Kainen and A.T. White, On stable crossing numbers, J. Graph Theory 2 (1978) 181-187.
[10] T. Pisanski, Genus of Cartesian products of regular bipartite graphs, J. Graph Theory 4 (1980) 31-42.
[11] T. Pisanski, Nonorientable genus of Cartesian products of regular graphs, J. Graph Theory 6 (1982) 391-402.
[12] G. Pica, T. Pisanski and A.G.S. Ventre, The genera of amalgamations of cube graphs, Glasnik Mat. 19 (39) (1984) 21-26.
[13] A.T. White, Graphs, groups and surfaces, North-Holland, Amsterdam, London, 1984.

+ Work performed under the auspices of CNR-GNSAGA.

++ The author was partially supported by B. Kidrič Fund, Slovenia, Yugoslavia.

Annals of Discrete Mathematics 30 (1986) 347–354

OVOIDS AND CAPS IN PLANAR SPACES

Giuseppe Tallini (Roma)

1. - INTRODUCTION

Let $(S,\mathscr{L})$ be a linear space, that is a set S whose elements we call points, $\mathscr{L}$ a family of parts in S, whose elements we call lines, such that any line has at least two points and two distinct points are contained in just one line. A subspace in $(S,\mathscr{L})$ is a subset S' in S such that for any $x,y \in S'$, $x \neq y$, the line joining them belongs to S'. Obviously the set theoretical intersection of subspaces in $(S,\mathscr{L})$ is a subspace, so that the family of subspaces is a closure system.

Suppose a family $\mathscr{P}$ of subspaces in $(S,\mathscr{L})$ exists such that $|\mathscr{P}| \geq 2$, every $\pi \in \mathscr{P}$ contains three independent points and through three independent points there is only one element of $\mathscr{P}$. The triple $(S,\mathscr{L},\mathscr{P})$ is called <u>planar space</u>, the elements of $\mathscr{P}$ are called <u>planes</u>. Straightforward examples of planar spaces are: the affine or projective spaces of dimension $n \geq 3$, with respect to their lines and their planes; any subset H in $PG(n,q)$, $n \geq 3$, $[H] = PG(n,q)$, with respect to the intersections of H with its secant lines and planes having three independent points of H. A further example is the following: consider a whatever family $\mathscr{P}'$ of d-dimensional subspaces in $PG(n,q)$, $n \geq 4$, two by two meeting in at most one line, let $\mathscr{P}''$ be the family of planes in $PG(n,q)$ each of them not belonging to any $S_d \in \mathscr{P}'$. Set $S = PG(n,q)$ and $\mathscr{L}$ the family of lines in $PG(n,q)$. The triple $(S,\mathscr{L},\mathscr{P}=\mathscr{P}'\cup\mathscr{P}'')$ is a planar space.

Let $(S,\mathscr{L},\mathscr{P})$ be a planar space. We denote <u>cap</u> H in $(S,\mathscr{L},\mathscr{P})$ a set of points three by three not collinear. We call <u>ovoid</u> in $(S,\mathscr{L},\mathscr{P})$ a cap Ω such that:

(1.1) <u>Given any point</u> $P \in \Omega$ <u>the set theoretical union of the tangent lines to</u> Ω <u>through</u> P <u>is a subspace</u> τ_p <u>in</u> $(S,\mathscr{L})$ <u>such that</u>

<u>every plane through</u> P <u>not in</u> τ_P <u>meets</u> τ_P <u>in a line. The space</u> τ_P <u>is called tangent space to</u> Ω <u>through</u> P .

In the following we suppose that the lines of the planar space $(S,\mathcal{L},\mathcal{P})$ have the same size $k\geq 3$, and the planes have the same size v , that is:

$$(1.2)\qquad \begin{cases} \forall \ell\in\mathcal{L}, & |\ell| = k\geq 3 \\ \forall \pi\in\mathcal{P}, & |\pi| = v \end{cases}.$$

Then $(S,\mathcal{L})$ is a Steiner system $S(2,k,V)$, $|S| = V$ and for any $\pi\in\mathcal{P}$, if $\mathcal{L}_\pi$ denotes the set of lines in π , $(\pi,\mathcal{L}_\pi)$ is a Steiner system $S(2,k,v)$. If R denotes the number of lines through a point of $(S,\mathcal{L})$ and r is the number of lines through a point of $(\pi,\mathcal{L}_\pi)$, it is:

$$(1.3)\qquad R = (V-1)/(k-1) \quad , \quad r = (v-1)/(k-1) \, ,$$

moreover:

$$(1.4)\qquad |\mathcal{L}| = V(V-1)/k(k-1) \quad , \quad |\mathcal{L}_\pi| = v(v-1)/k(k-1) \, .$$

If s is the number of planes through a line in $(S,\mathcal{L},\mathcal{P})$, we easily obtain

$$(1.5)\qquad s = (V-k)/(v-k) = (R-1)/(r-1) \, .$$

Counting in two different ways the number of pairs consisting of a plane π and a line through it, we have: $|\mathcal{P}|\,|\mathcal{L}_\pi| = |\mathcal{L}|\,s$, that is (see (1.4),(1.3), (1.5)):

$$(1.6)\qquad |\mathcal{P}| = V(V-1)s/v(v-1) = VRs/vr = VR(R-1)/vr(r-1) \, .$$

If $|\mathcal{P}_P|$ is the number of planes through a point P , counting in two different ways the number of pairs each of them consisting of a plane through P and a line on it not through P , we obtain: $|\mathcal{P}_P|(|\mathcal{L}_\pi|-r) = |\mathcal{L}|-R$, that is (see (1.4),(1.3),(1.5)):

$$(1.7)\qquad |\mathcal{P}_P| = (V-1)(V-k)/(v-1)(v-k) = Rs/r = R(R-1)/r(r-1) \, .$$

If $(S,\mathcal{L},\mathcal{P})$ is a planar space satisfying (1.2), we prove:

<u>Theorem</u> 1. - <u>If in a</u> $(S,\mathcal{L},\mathcal{P})$ <u>an ovoid</u> Ω <u>exists, it is</u> $(S,\mathcal{L},\mathcal{P}) = PG(3,q)$ <u>and</u> Ω <u>is an elliptic quadric if</u> $q = k-1$ <u>is odd, an ovoid in</u> $PG(3,q)$, <u>if</u> q <u>is even</u>.

Theorem 2. - *Let H be a h-cap of a planar space* $(S,\mathcal{L},\mathcal{P})$ *. Then:*

(1.8) $$h \leq R+1 \quad , \quad r \text{ odd.}$$

If r is even, or if r is odd and $|\pi \cap H| \neq r+1$ *for any* $\pi \in \mathcal{P}$, *we have:*

(1.9) $$h \leq R-s+1 \quad ,$$

the equality holds iff H is an ovoid, that is (see Theorem 1) iff $(S,\mathcal{L},\mathcal{P}) = PG(3,q)$.

Theorem 3. - *If in a planar space* $(S,\mathcal{L},\mathcal{P})$ *a set H of type* $(0,n)$, $n>1$, *with respect to lines exists, then* $m = (r-1)/n$ *must be an integer and* $r-m$ *divides* $s-1$. *Moreover* $n \leq r-m \leq r-1$, *the equalities holding iff* $(S,\mathcal{L},\mathcal{P}) = PG(d,q)$, *H is the complement of a prime in* $PG(d,q)$ *and* $n = q$.

We remark that, if r is odd, a $(R+1)$-cap in $(S,\mathcal{L},\mathcal{P})$ is a set of type $(0,2)$, so by Theorem 3, we have:

Theorem 4. - *If r is odd, in (1.8) the equality never holds if* $(r+1)/2$ *does not divide* $s-1$. *If* $MCD(s-1,(r+1)/2) = 2$ *and in (1.8) equality holds, then* $(S,\mathcal{L},\mathcal{P}) = PG(d,2)$ *and the cap H is the complement of a prime in* $PG(d,2)$.

2. - OVOIDS IN $(S,\mathcal{L},\mathcal{P})$

Let Ω be an ovoid in a planar space $(S,\mathcal{L},\mathcal{P})$ satisfying (1.2). If $P,Q \in \Omega$, $P \neq Q$, any plane π through PQ does not belong to the tangent space τ_P to Ω at P (because $Q \in \pi$ and $P \neq Q$), so it meets τ_P in a line t_P . It follows that the $r-1$ lines through P in π , different from t_P , are secant of $\pi \cap \Omega$ and so $|\pi \cap \Omega| = r$. It follows that every plane of $(S,\mathcal{L},\mathcal{P})$ is either exterior, or tangent, or r-secant to Ω and (being s the number of planes through PQ): $|\Omega| = s(r-2)+2$, i.e. by (1.5):

(2.1) $$|\Omega| = R-s+1 \ .$$

Let t_0 , t_1 , t_r be the numbers of exterior, tangent and r-secant planes to

Ω respectively. We have (see (1.6),(2.1),(1.7)):

$$(2.2)\qquad \begin{cases} t_0 + t_1 + t_r = |\mathscr{P}| = (V/v)(R/r)s , \\ t_1 + r\, t_r = |\Omega|\;|\mathscr{P}_P| = (R-s+1)R\, s/r , \\ r(r-1)t_r = |\Omega|(|\Omega|-1)s = (R-s+1)(R-s)s . \end{cases}$$

$(2.2)_1$ is obvious. $(2.2)_2$ and $(2.2)_3$ follow by computing in two different ways the pairs consisting of a point of Ω and a plane through it and two points of Ω and a plane through them. By (2.2) we have (see (1.5)):

$$(2.3)\qquad \begin{cases} t_r = (R-s+1)(R-s)s/r(r-1) , \\ t_1 = (R-s+1)s(s-1)/r(r-1) , \\ t_0 = (s/r)\left[(V/v)R + s^2 - (R+1)s\right] . \end{cases}$$

Being $t_0 \geq 0$, by $(2.3)_3$ we have:

$$(2.4)\qquad v\left[s^2 - s(R+1)\right] + VR \geq 0 .$$

By (1.3) it is $V = (k-1)R+1$; $v = (k-1)r+1$, whence we obtain:

$$(2.5)\qquad (k-1)\left[R(R-r) - r(s-1)(R-s)\right] - (s-1)(R-s) \geq 0 .$$

By (1.5) we have $s-1 = (R-r)/(r-1)$, $R-s = (rR-2R+1)/(r-1)$, whence we have (see also (2.5) and dividing it by $R-s$):

$$(2.6)\qquad k-1 \geq \left[R(r-2)+1\right]/(R-r) = r-2 + (r-1)^2/(R-r) .$$

Being $(r-1)^2/(R-r) > 0$, by (2.6) it is $k \geqslant r$, so $k = r$ and every plane π in $\mathscr{P}$ is a projective plane of order $q = k-1$. It follows that $(S,\mathscr{L},\mathscr{P})$ is a Galois space $PG(n,q)$, $n \geqslant 3$ and Ω is a $(q^{n-1}+1)$-cap in $PG(n,q)$ (being $|\Omega| = R-s+1 = \sum_{i=0}^{n-1} q^i - \sum_{i=0}^{n-2} q^i + 1 = q^{n-1} + 1$). It is known (see [5]) that in $PG(n,q)$, $n \geqslant 4$, $(q^{n-1}+1)$-caps don't exist, whence $n = 3$ and Ω is a (q^2+1)-cap in $PG(3,q)$. So Theorem 1 is proved.

3. - CAPS IN $(S,\mathscr{L},\mathscr{P})$

Let H be a h-cap in a planar space $(S,\mathscr{L},\mathscr{P})$ satisfying (1.2). Given $\pi\in\mathscr{P}$, $n = |\pi\cap H| \geq 2$, $\pi \cap H$ is a n-arc in π . We easily prove that:

(3.1) $$n \leq r + 1 ,$$

(3.2) $$n = r+1 \Leftrightarrow H \text{ has no tangent lines} \Rightarrow r \text{ odd} .$$

Let $P,Q \in H$, $P \neq Q$. By (3.1) each of the s planes through PQ meet H in at most $r-1$ points different from P and Q , whence $h = |H| \leq (r-1)s+2$, that is (see (1.5)):

(3.3) $$h \leq R + 1 ,$$

the equality holding iff:

$$\pi \in \mathscr{P}, \ |\pi \cap H| \geq 2 \Rightarrow |\pi \cap H| = r+1 ,$$

so that, by (3.2):

(3.4) $$h = R+1 \Leftrightarrow H \text{ has no tangent lines} \Leftrightarrow$$
$$\Leftrightarrow H \text{ is of type } (0,r+1) \text{ with respect to planes.}$$

Let us now suppose:

(3.5) $$\forall \pi \in \mathscr{P} \Rightarrow |\pi \cap H| \neq r+1 ;$$

we remark that, by (3.2), <u>(3.5) is fulfilled, if</u> r <u>is even</u>. For any $P,Q \in H$, $P \neq Q$, by (3.5) and (3.1), each of the s planes through PQ meets H in at most $r-2$ points different from P and Q , whence $h = |H| \leq (r-2)s+2$, that is (see (1.5)):

(3.6) $$h \leq R-s+1 ,$$

the equality holds iff

$$\forall \pi \in \mathscr{P}, \ |\pi \cap H| \geq 2 \Rightarrow |\pi \cap H| = r ,$$

that is:

(3.7) $$h = R-s+1 \Leftrightarrow H \text{ is of class } [0,1,r] \text{ with respect to planes.}$$

If (3.5) holds and $h = R-s+1$, for any plane π through PQ , $\pi \cap H$ is a r-arc having only one tangent at P . As the number of planes through PQ is s , there are just s tangents to H at P . The set theoretical union of such s tangents by (3.7) is a subspace τ_P of $(S,\mathscr{L})$ meeting every plane through P , not in τ_P , in a line. So H is an ovoid. Conversely, if $H = \Omega$ is an ovoid, by (2.1), $|H| = R-s+1$ and (3.5) holds (see sect. 2), i.e.:

$$h = R-s+1 \Leftrightarrow H \text{ is an ovoid} .$$

So Theorem 2 is proved.

4. - SETS OF TYPE (0,n) WITH RESPECT TO LINES IN $(S,\mathcal{L},\mathcal{P})$

Let H be a set of type $(0,n)$ with respect to lines in a planar space $(S,\mathcal{L},\mathcal{P})$ satisfying (1.2). Let S' be a subspace of $(S,\mathcal{L})$ such that $S'\cap H \neq \emptyset$ and $r' = (|S'|-1)/(k-1)$ be the number of lines through a point in S'. If $P\in S'\cap H$, every line in S' through P meets $S'\cap H$ in $n-1$ points different from P, whence:

(4.1) $$|S'\cap H| = r'(n-1) + 1 \quad .$$

If we set $S' = S$, or $S' = \pi$ respectively in (4.1), we have:

(4.2) $$|H| = R(n-1) + 1 \quad ,$$

(4.3) $$\forall\, \pi\in\mathcal{P},\ |\pi\cap H| = \begin{cases} 0 \\ r(n-1)+1 \end{cases} \quad ,$$

whence H is of type $(0,r(n-1)+1)$ with respect to planes.

If $n = 1$, H reduces to a point, so it is not $n = k$. Therefore:

(4.4) $$2\leq n\leq k-1 \quad .$$

We remark that if $n = k-1$, H is the complement of a subspace S' in $(S,\mathcal{L})$, meeting any line not belonging to S' in a point. We denote such a subspace <u>prime</u> of $(S,\mathcal{L})$. A planar space $(S,\mathcal{L},\mathcal{P})$ is called <u>proper</u>, if every $\pi\in\mathcal{P}$ is the linear closure of whatever triplet of its independent points.

We prove:

<u>Theorem</u> 5. - <u>If in a proper planar space</u> $(S,\mathcal{L},\mathcal{P})$ <u>a set</u> H <u>of type</u> $(0,k-1)_1$ <u>exists, then</u> $(S,\mathcal{L},\mathcal{P}) = PG(d,q)$ <u>and</u> H <u>is the complement of a prime in</u> $PG(d,q)$. <u>It follows that a proper planar space is a Galois space iff it contains a prime</u>.

<u>Proof</u> - $S' = S-H$ is a subspace of $(S,\mathcal{L})$ meeting any line not in S' in a point. Let $P,Q\in S'$ and $T\in H$. The plane π joining P,Q,T meets S' in the line PQ, being $(S,\mathcal{L},\mathcal{P})$ proper. Each of the r lines through T in π meets $PQ = S'\cap\pi$ in a point, so $r = k$ and every plane in $(S,\mathcal{L},\mathcal{P})$ is projective, i.e. $(S,\mathcal{L},\mathcal{P}) = PG(d,q)$, with $q = k-1$. Thus the theorem is proved.

Let n be an integer satisfying (4.4) and $\pi\in\mathcal{P}$ such that $\pi\cap H \neq \emptyset$, i.e. (see (4.3)) $|\pi\cap H| = r(n-1)+1$. If $P\in\pi - H$, the number of lines through P

in π , n-secant of $\pi \cap H$, is $u = |\pi \cap H|/n$, so that:

$$u = (r(n-1)+1)/n = r - (r-1)/n \quad ,$$

hence $m = (r-1)/n$ must be an integer. If ℓ is an exterior line of H , the number of planes through ℓ meeting H is $v = |H|/(r(n-1)+1)$, that is (by (4.2),(1.5) and being $r-1 = mn$) we have:

$$v = s - (s-1)/(r-m) \quad , \tag{4.5}$$

therefore $r-m$ divides $s-1$.

It is $r-m \geq n$. In fact, if $r-m < n$ (being $m = (r-1)/n$ and $n \geq 2$) we should have $n+1 > r$; by (4.4), $k \geq n+1$, whence the contradiction $k > r$. So it is $n \leq r-m \leq r-1$, the equalities hold iff $n = r-1$, but by (4.4), $r \geq k \geq n+1$, whence $k = r$, so that every plane is projective. Then $(S,\mathcal{L},\mathcal{P})$ is a Galois space PG(d,q) and H is a set of type $(0,k-1)_1$, that is the complement of a prime. Thus Theorem 3 is completely proved.

REFERENCES

[1] F. Buekenhout, Une caractérisation des espaces affins basée sur la notion de droite, Math. Z. 111 (1969), 367-371.

[2] F. Buekenhout et R. Deherder, Espaces linéaires finis à plans isomorphes, Bull. Soc. Math. Belg. 23 (1971), 348-359.

[3] M.Hall, Jr., Automorphisms of Steiner triple systems, IBM J. Res. Develop. 4 (1960), 460-472 (= Amer. Math. Soc. Proc. Symp. Pure Math. 6 (1962),47-66).

[4] H. Hanani, On the number of lines and planes determined by d points, Technion. Israel Inst. Tech. Sci. Publ. 6 (1954/5), 58-63.

[5] G. Tallini, Sulle k-calotte di uno spazio lineare finito, Ann. Mat. (4) 42 (1956), 119-164.

[6] G. Tallini, La categoria degli spazi di rette, Ist. Mat. Univ. L'Aquila, (1979/80), 1-25.

[7] L. Teirlinck, On linear spaces in which every plane is either projective or affine, Geometriae Dedicata, 4 (1975), 39-44.

Annals of Discrete Mathematics 30 (1986) 355–362

(k,n;f)-ARCS AND CAPS IN FINITE PROJECTIVE SPACES

B. J. Wilson

Chelsea College
University of London
552 King's Road
London SW10 0UA

In §1 some theorems which generalise results of E. d'Agostini are given. Some particular examples of (k,n;f)-arcs are discussed in §2 and two infinite classes are described in §3. In §4 the extension of the results of §1 to higher dimensions is noted.

1. In the papers of Barnabei [2] and d'Agostini [4][5] an account has been given of some results concerning weighted (k,n)-arcs in finite and particularly Galois planes. These objects are also called (k,n;f)-arcs in [4] and [5]. In this note we are concerned with extending the work in [5]. However we first mention that the idea of a weighted (k,n)-arc was originally proposed by M. Tallini-Scafati [16] and that the theme of that paper, namely the embedding of the arc in an algebraic curve was continued by Keedwell in [9] and [10].

We shall work with the definition of a weighted arc given in [5] acknowledging that the definition given in [16] is equivalent and has priority. Thus we are concerned with a set K of $k > 0$ points in PG(2,q) to each point P of which is assigned a natural number f(P) called its weight and such that the total weight of the points on any line does not exceed a given natural number n, i.e. for each line ℓ of PG(2,q) we have

$$\sum_{P \in \ell} f(P) \leqq n$$

A line having total weight i is called an i-secant of K. Points not included in K are assigned the weight zero. Following the notation of [5] we let

$$w = \max_{P \in K} f(P)$$

and use ℓ_j to denote number of points of weight j for $j = 0,1,\ldots,w$.

Following the notation and terminology originating in [13] let t_i denote the number of i-secants of K for which $i = 0,1,\ldots,n$. We call the t_i the characters of K and, if exactly u of them are non-zero we say that the arc K has u characters. If the values of i for which t_i is non-zero are $m_1 < \ldots < n$ then K is said to be of type $(m_1, m_2, \ldots, n)$. In [13], [14], [7] and subsequently in later papers, one of which is [15] considerable attention has been given to (k,n)-arcs having exactly two characters. In particular the connection between such arcs and Hermitian curves has been explored and generalisations into higher dimensions discussed in [15]. A good bibliography is included in [16].

It is proved in [5] that for a $(k,n;f)$-arc K of type (m,n), when $0 < m < n$, it is necessary that

$$w \leqq n - m \tag{i}$$

and

$$q \equiv 0 \bmod (n-m). \tag{ii}$$

Let $W = \sum_{P \in K} f(P) = \sum_{P \in PG(2,q)} f(P)$, and then it may easily be seen [5] that

$$m(q+1) \leqq W \leqq (n-w)q + n. \tag{iii}$$

Arcs for which equality holds on the left are called minimal and arcs for which equality holds on the right are called maximal. The case of $m = n - 2$ was discussed at length in [5]. By (ii) we must have $q = 2^h$ and then (i) requires that $w \leqq 2$. In order to have an arc which is not simply a (k,n)-arc we thus must have $w = 2$ so that (iii) gives

$$(n-2)(q+1) \leqq W \leqq (n-2)(q+1) + 2. \tag{iv}$$

It may easily be shown that $W \neq (n-2)(q+1) + 1$ and the other two possible values of W are discussed in [5]. Such arcs have points having possible weights 0,1 and 2; with this in mind we can state the following results which are generalisations of theorems in [5] and which can be proved by similar methods.

THEOREM 1.

Let K be a $(k,n;f)$-arc of type (m,n) with $n > m > 0$ of minimal weight $W = m(q+1)$ having some points of weight $w = n - m$, some points of weight α for exactly one value of α satisfying both $1 \leqq \alpha \leqq w - 1$ and $(w,\alpha) = 1$ and at least one point of weight 0.

(a) Suppose there is exactly one point of weight 0. Then $\alpha = w - 1$ and the points of weight w form a $(wq + w - q - 1, w)$-arc of which the $(w-1)$-secants are concurrent in the single points of weight 0.

(b) Suppose that $\ell_0 > 1$. Then K consists of $\alpha q/w$ collinear points each of weight w and the q^2 points not collinear with them, each of weight α. Further, $n = \alpha q + w$.

THEOREM 2.

Let K be a $(k,n;f)$-arc of type (m,n) with $n > m > 0$ of maximal weight $W = (n-w)(q+1) + w$ having some points of weight 0, some points of weight α for exactly one value of α satisfying both $1 \leqq \alpha \leqq w - 1$ and $(w,\alpha) = 1$ and at least one point of weight w.

(a) Suppose there is exactly one point of weight w. Then $\alpha = 1$ and the points of weight 0 form a $(wq + w - q - 1, w)$-arc of which the $(w-1)$-secants are concurrent in the single point of weight w.

(b) Suppose $\ell_w > 1$. Then K consists of $\alpha q/w + 1$ collinear points each of weight w and the q^2 points not collinear with them, each of weight α. Further $n = \alpha q + w$.

It has been shown [1] that the existence of a $(n_0 q + n_0 - q - 1, n_0)$-arc with $n_0 > 2$ requires that $q \equiv 0 \pmod{n_0}$ and that in that case the arc possesses exactly $q + 1$ $(n_0 - 1)$-secants which are concurrent in a point N called the nucleus of the arc. The addition of the point N to the arc then gives a $(n_0 q + n_0 - q, n_0)$-arc, known as a maximal

(k,n_0)-arc. For even values of q such arcs have been constructed in Galois planes by Denniston [6]. It has also been shown that no such arcs can be constructed in Galois planes for $n_0 = 3$ by Cossu for $q = 9$ [3] and Thas [19]. Hence the theorems 1 and 2 are of interest only if $n - m \geq 3$.

2. It follows from the results of §1 that for the case $n - m = 3$ we need to consider further in a discussion of maximal and minimal (k,n;f)-arcs the cases

$$\ell_0 > 0;\ \ell_1 > 0;\ \ell_2 > 0;\ \ell_3 = 0 \tag{v}$$

and

$$\ell_0 > 0;\ \ell_1 > 0;\ \ell_2 > 0;\ \ell_3 > 0. \tag{vi}$$

In particular we discuss (v). In that case

$$(n - 3)(q + 1) \leqq W \leqq (n - 2)q + n. \tag{vii}$$

Following the notation of [5] we use the symbol v_i^j to denote the number of i-secants which pass through a point of weight j. Using the arguments in [5] the values of v_n^j and v_{n-3}^j, $j = 0,1,2$ are fixed independently of the point under consideration.

For W minimal, i.e. $W = (n - 3)(q + 1)$ we have in particular

$$\begin{array}{llll} v_{n-3}^0 = q + 1 & & v_n^0 = 0 & \\ v_{n-3}^1 = \dfrac{2q+1}{3} & & v_n^1 = \dfrac{q}{3} & \qquad\text{(viii)} \\ v_{n-3}^2 = \dfrac{q+3}{3} & & v_n^2 = \dfrac{2q}{3} & \end{array}$$

It follows immediately that no point of weight 0 lies on an n-secant. We now attempt to construct examples of (k,n;f)-arcs which satisfy these criteria, i.e. (v), $n - m = 3$, and $W = (n - 3)(q + 1)$. Easy counting arguments give

$$\begin{aligned} t_{n-3} + t_n &= q^2 + q + 1 \\ (n - 3)t_{n-3} + nt_n &= W(q + 1) = (n - 3)(q + 1)^2 \end{aligned} \tag{ix}$$

Solving (ix) gives

$$\begin{aligned} t_n &= (n - 3)q/3 \\ t_{n-3} &= (3q^2 + 6q - nq + 3)/3 \end{aligned} \tag{x}$$

Now let a be a n-secant on which there are no points of weight 0 so we suppose that on a are α points of weight 1 and β points of weight 2. Counting points of a and weights of points on a gives

$$\begin{aligned} \alpha + \beta &= q + 1 \\ \alpha + \beta &= n \end{aligned} \tag{xi}$$

Solving (xi) gives

$$\begin{aligned} \alpha &= 2(q + 1) - n \\ \beta &= n - (q + 1) \end{aligned} \tag{xii}$$

Counting incidences between points of weight 2 and n-secants gives

$$\ell_2 v_n^2 = t_n \beta$$

Hence, using (viii), (x) and (xii) we have

$$\ell_2 = (n-3)(n-q-1)/2 \qquad \text{(xiii)}$$

Similarly counting incidences between points of weight 1 and n-secants gives

$$\ell_1 v_n^1 = t_n \alpha$$

whence, using (viii), (x) and (xii) we have

$$\ell_1 = (n-3)(2q+2-n) \qquad \text{(xiv)}$$

From (xiii) and (xiv), counting the points in the plane we obtain

$$2q^2 + (11-3n)q + n^2 - 6n + 11 - 2\ell_0 = 0 \qquad \text{(xv)}$$

It is thus necessary that

$$(n-q)^2 - (48 - 16\ell_0)$$

should be a square. Whilst it is not the only case for investigation an obvious value to try is $\ell_0 = 3$. The resulting solutions for (xv) are then $n = 2q+1$ and $n = q+5$.

By counting the points of an $(n-3)$-secant containing three collinear points of weight 0 it may easily be seen that for $n = 2q+1$ it is impossible for the three points of weight 0 to be collinear. A simple example may be found in PG(2,3). Assign the weight 0 to the points (1,0,0),(0,1,0) and (0,0,1), the weight 1 to the points (2,1,1),(1,2,1),(1,1,2) and (1,1,1) and the weight 2 to all other points. This yields a (10,7;f)-arc of type (4,7).

For this case of $\ell_0 = 3$ with $n = 2q+1$ we may obtain from (x)

$$t_n = \tfrac{2}{3}q(q-1)$$

Thus by (xiii) the n-secants form the dual of a

$$\left(\tfrac{2}{3}q(q-1),\ \tfrac{2}{3}q\right)\text{-arc}$$

The existence of such an arc would, if $q > 3$, violate the Lunelli-Sce conjecture [11] that for a (k,n_0)-arc with $q \equiv 0 \pmod n$ it is necessary that $k \leq (n_0 - 1)q + 1$. However it was shown by Hill and Mason [8] that counterexamples to this conjecture can be found for an infinite number of values of q.

We now consider the case in which (v) holds, with $\ell_0 = 3$, W minimal and $n = q+5$. If the three points of weight 0 are collinear then it may be shown that the points of weight 2 form a complete $(2q+3,4)$-arc of the type (1,2,4). Examples of such a configuration have been found in $PG(2,3^2)$. If we define $GF(3^2)$ by the relation $\alpha^2 = 2\alpha + 1$ over GF(3) then one such example is constructed as follows:

The three points $(0,1,\alpha^3),(0,1,\alpha^6),(0,1,\alpha^7)$ have weight 0 and the twenty two points

$Q_0(0,0,1)$	$(1,\alpha^2,\alpha^3)$	$(1,\alpha^5,\alpha^4)$
$Q_1(0,1,0)$	$(1,\alpha^2,\alpha^4)$	$(1,\alpha^5,\alpha^5)$
$Q_2(0,1,\alpha^4)$	$(1,\alpha^2,\alpha^5)$	$(1,\alpha^6,1)$
$Q_3(0,1,\alpha^5)$	$(1,\alpha^3,\alpha^3)$	$(1,\alpha^6,\alpha)$
$(1,\alpha,1)$	$(1,\alpha^3,\alpha^4)$	$(1,\alpha^6,\alpha^7)$
$(1,\alpha,\alpha)$	$(1,\alpha^3,\alpha^5)$	$(1,\alpha^7,1)$
$(1,\alpha,\alpha^7)$	$(1,\alpha^5,\alpha^3)$	$(1,\alpha^7,\alpha)$
		$(1,\alpha^7,\alpha^7)$

have weight 2. The remaining points of PG(2,3) are assigned weight 1 giving an (88,14;f)-arc K_0 of type (11,14) with the points of weight 2 forming a complete (22,4)-arc of type (1,2,4). Let v be the line, $x_0 = 0$, of collinearity of the three points of weight 0. On v are four points R_0,R_1,R_2,R_3 of weight 1 and four points Q_0,Q_1,Q_2,Q_3, with coordinates as indicated above, of weight 2. Further, there are precisely three points P_1,P_2,P_3 of weight 1 which do not lie on v and which are joined to the points R_i only by 14-secants. Their line of collinearity passes through Q_0 and the nine lines P_iQ_j are 11-secants meeting in threes at the six points P_1,P_2,P_3,Q_1,Q_2,Q_3 and otherwise only in pairs. This configuration is, in $PG(2,3^2)$, determined completely by five of the points $P_1, \ldots ,Q_3$.

3. The (10,7;f)-arc of type (4,7) in $PG(2,3^2)$ which was constructed in §2 is a particular case of two otherwise distinct infinite classes. Firstly we note that the points of weight 1 form a 4-arc, this being the irreducible conic

$$x_0^2 + x_1^2 + x_2^2 = 0.$$

In PG(2,q), with q odd, let C be an irreducible conic. There are exactly $q+1$ points on C at each of which there is a unique tangent line to C. The points of the plane which are not on C may then be partitioned into the disjoint classes of $q(q+1)/2$ exterior points, through each of which pass exactly two tangents to C and $q(q-2)/2$ interior points, through each of which there are not tangents to C. Assigning weight 0 to each interior point, weight 1 to each point of C and weight 2 to each exterior point gives a minimal $((q^2+3q+2)/2, 2q;f)$-arc of type $(q+1,2q+1)$. By assigning different weights to the points of this configuration other (k,n;f)-arcs may be obtained. For example assigning the weight 0 to each interior point of C, the weight 1 to each exterior point of C and the weight $(q+1)/2$ to each point of C a maximal $((q^2+3q+2)/2,(3q+1)/2;f)$-arc of type $(q+1,2q+1)$ is obtained.

A second infinite class of (k,n;f)-arcs is suggested by regarding the triangle of points of weight 0 in the (10,7;f)-arc of §2 as a subplane. Generally let π_0 be a subplane of order q_0 of a (not necessarily Galois) finite projective plane π of order q with $q \geq q_0^2 + q_0$. In this case there are some lines of π which do not contain any point of π_0. Assign weight 0 to points of π_0, weight u to points of π which are not on lines of π_0 and weight v to the remaining points where $u/v = (q-q_0^2)/(q-q_0-q_0^2)$ in its lowest terms. Then there is formed a minimal

$$((q-q_0)(q+q_0+1),(q_0^2+q_0+1)u+(q-q_0^2-q_0)v;f)\text{-arc}$$

of type $((q+q_0)u,(q_0^2+q_0+1)u+(q-q_0^2-q_0)v)$.

As in the case of the previous example reassignment of other weights to the sets of points involved leads to further (k,n;f)-arcs.

4. The definition of a (k,n;f)-arc given in §1 may be extended to that of a (k,n;f)-cap [5] by substituting PG(r,q) for PG(2,q) with $r > 2$. In [5] it was shown that (k,n;f)-caps of type $(n-2,n)$, with $r \geqq 3$ do not exist. This proof required results listed by Segre [12] p 166 concerning the non-existence of certain k-caps in PG(r,q) with $r \geqq 3$.

If we use the notation Q_r to denote the number of points in PG(r,q) then the results in [12] showed that the number of points on a k-cap cannot be Q_{r-1}. For a (k,n;f)-cap of type $(m-n)$ with $0 < m < n$ the minimal weight is mQ_{r-1}. However it may be shown using analogous arguments to those indicated above that a (k,n;f)-cap of minimal weight mQ_{r-1} and otherwise satisfying the conditions of theorem 1 cannot exist.

A similar result can be obtained for maximal arcs.

REFERENCES

[1] Barlotti, A., Su {k;n}-archi di un piano lineare finito, Boll. Un. Mat. Ital. 11 (1956) 553-556.

[2] Barnabei, M., On arcs with weighted points, Journal of Statistical Planning and Inference, 3 (1979), 279-286.

[3] Cossu, A., Su alcune proprietà dei {k;n}-archi di un piano proiettivo sopra un corpo finito, Rend. Mat. e Appl. 20 (1961), 271-277.

[4] d'Agostini, E., Alcune osservazioni sui (k,n;f)-archi di un piano finito, Atti dell' Accademia della Scienze di Bologna, Rendiconti, Serie XIII, 6 (1979), 211-218.

[5] d'Agostini, E., Sulla caratterizzazione delle (k,n;f)-calotte di tipo (n-2,n), Atti Sem. Mat. Fis. Univ. Modena, XXIX, (1980), 263-275.

[6] Denniston, R.H.F., Some maximal arcs in finite projective planes, J. Combinatorial Theory 6 (1969), 317-319.

[7] Halder, H.R., Über Kurven vom Typ (m;n) und Beispiele total m-regulärer (k,n)-Kurven, J. Geometry 8, (1976), 163-170.

[8] Hill, R. and Mason, J., On (k,n)-arcs and the falsity of the Lunelli-Sce Conjecture, London Math. Soc. Lecture Note Series 49 (1981), 153-169.

[9] Keedwell, A.D., When is a (k,n)-arc of PG(2,q) embeddable in a unique algebraic plane curve of order n?, Rend. Mat. (Roma) Serie VI, 12 (1979),397-410.

[10] Keedwell, A.D., Comment on "When is a (k,n)-arc of PG(2,q) embeddable in a unique algebraic plane curve of order n?", Rend. Mat. (Roma) Serie VII, 2 (1982), 371-376.

[11] Lunelli, L. and Sce, M., Considerazione arithmetiche e visultati sperimentali sui $\{K;n\}_q$-archi, Ist. Lombardo Accad. Sci. Rend. A 98 (1964), 3-52.

[12] Segre, B., Introduction to Galois Geometries, Atti. Accad. Naz. Lincei Mem. 8 (1967), 133-236.

[13] Tallini Scafati, M., {k,n}-archi di un piano grafico finito con particolare riguardo a quelli con due caratteri (Nota I), Atti. Accad. Naz. Lincei Rend. 40 (1966), 812-818.

[14] Tallini Scafati, M., {k,n}-archi di un piano grafico finito con particolare riguardo a quelli con due caratteri (Nota II), Atti. Accad. Naz. Lincei Rend. 40 (1966), 1020-1025.

[15] Tallini Scafati, M., Catterizzazione grafica delle forme hermitiane di un $S_{r,q}$. Rend. Mat. e Appl. 26 (1967), 273-303.

[16] Tallini Scafati, M., Graphic Curves on a Galois plane, Atti del convegno di Geometria Combinatoria e sue Applicazioni Perugia (1971), 413-419.

[17] Tallini Scafati, M., k-insiemi di tipo (m,n) di uno spazio affine $A_{r,q}$, Rend. Mat. (Roma) Serie VII, 1 (1981), 63-80.

[18] Tallini Scafati, M., d-Dimensional two-character k-sets in an affine space AG(r,q), J. Geometry 22 (1984), 75-82.

[19] Thas, J.A., Some results concerning $\{(q+1)(n-1)-1,n\}$-arcs and $\{(q+1)(n-1)+1,n\}$-arcs in finite projective planes of order q, J. Combinatorial Theory A 19 (1975), 228-232.

Annals of Discrete Mathematics 30 (1986) 363–372

COMBINATORIAL STRUCTURES CORRESPONDING TO REFLECTIVE CIRCULANT (0,1)-MATRICES

N. Zagaglia Salvi
Dipartimento di Matematica
Politecnico di Milano, Milano, Italy

Let C be a circulant (0,1)-matrix and let us arrange the elements of the first row of C regularly on a circle. If there exists a diameter of the circle with respect to which 1's are symmetric, we call C reflective. In this paper we prove some properties of the reflective circulant (0,1)-matrices and of certain corresponding combinatorial structures.

INTRODUCTION

A matrix C of order n is called circulant if C P = P C, where P represents the permutation (1 2 ... n).
Let C be a circulant (0,1)-matrix and let us arrange the elements of the first row regularly on a circle, so that they are on the vertices of a regular polygon. If there exists a diameter of the circle with respect to which 1's are symmetric, we call C reflective. In this paper we prove some properties of the reflective circulant (0,1)-matrices and of certain corresponding combinatorial structures.
In particular, it is proved that a circulant (0,1)-matrix C of order n satisfies the equation $C P^h = C_T$, $0 \leq h \leq n-1$, if and only if it is reflective. Moreover we determine the number of such C for every h.
It is proved in certain cases the conjecture of the non-existence of circulant Hadamard matrices and, therefore, of the non-existence of certain Barker sequences.
We also give a sufficient condition that the automorphism group of a directed graph is C_n, the cyclic group of order n.
Finally we determine a characterization for the tournaments with reflective circulant adjacency matrix.
For the notations, I and J denote, as usual, the unit and all-one matrices; the matrix C_T denote the transpose of C.

1. Let C be a circulant matrix. If $[c_0, c_1, \dots, c_{n-1}]$ is the first row of C, it follows [2] that the eigenvalues of C are

$$\lambda_r = \sum_{j=0}^{n-1} c_j \omega^{jr} \tag{1}$$

where $0 \leq r \leq n-1$ and $\omega = \exp(\frac{2 \pi i}{n})$.
Consider the circulant matrix A = C P. The first row of A is obtained from the

first row of C by shifting it cyclically one position to the right.
Since the eigenvalues of A are $\mu_r = \sum_{j=0}^{n-1} c_{j-1}\omega^{jr}$, where $0 \leq r \leq n-1$ and the subscripts are reduced mod n, it follows $\mu_0=\lambda_0$, $\mu_1=\lambda_1\omega$, ..., $\mu_{n-1}=\lambda_{n-1}\omega^{n-1}$.
So the matrix $B = C P^h$, $1 \leq h \leq n-1$, is circulant with eigenvalues $\nu_r=\lambda_r(\omega^h)^r$.

PROPOSITION 1.1 - The eigenvalues λ_r and ν_r, $0 \leq r \leq n-1$, of the circulant matrices C and CP^h, $0 \leq h \leq n-1$, as in (1), satisfy the relations $\nu_r=\lambda_r(\omega^h)^r$.

PROPOSITION 1.2 - The eigenvalues λ_r and μ_r, $0 \leq r \leq n-1$, of the circulant matrices C and D, as in (1), satisfy the relations $\mu_r=\bar{\lambda}_r$ if and only if $D = \bar{C}_T$.

Proof. If $\Delta = \mathrm{diag}(1, \omega, \omega^2, \ldots, \omega^{n-1})$, where $\omega=\exp(\frac{2\pi i}{n})$, then there exists an unitary matrix U such that $P = U \Delta U^*$.
If $[c_0, c_1, \ldots, c_{n-1}]$ and $[d_0, d_1, \ldots, d_{n-1}]$ are the first rows of C and D, we have $C = \sum_{j=0}^{n-1} c_j P^j = U \Gamma U^*$ where $\Gamma = \sum_{j=0}^{n-1} c_j \Delta^j$ and $D = U(\sum_{j=0}^{n-1} d_j \Delta^j)U^*$.

If μ_r equals $\bar{\lambda}_r$ for all r, it follows $\sum_{j=0}^{n-1} d_j \Delta^j = \bar{\Gamma}$ and hence $D = \bar{C}_T$.
The converse is easily proved. □

DEFINITION 1.3 - Let ω^r, ω^p two nth roots of unity, $0 \leq r,p \leq n-1$, and $\omega = \exp(\frac{2\pi i}{n})$. If $d = |r-p|$, we say that such roots are at distance d and we write $\mathrm{dist}(\omega^r, \omega^p)=d$. Clearly, then, $\mathrm{dist}(\bar{\omega}^r, \bar{\omega}^p)=d$.

THEOREM 1.4 - A circulant (0,1)-matrix C satisfies the equation $CP^h=C_T$, $0 \leq h \leq n-1$, if and only if it is reflective.

Proof. Let C be a reflective circulant (0,1)-matrix of order n; let λ_r, $0 \leq r \leq n-1$, be the eigenvalues of C as in (1).
Suppose that $\lambda_1 = \omega^{p_1} + \omega^{p_2} + \ldots + \omega^{p_s}$, $0 \leq p_1 < p_2 < \ldots < p_s \leq n-1$.
Then the elements 1 on the first row of C are in the positions p_1+1, $p_2+1, \ldots, p_s+1$.
Since C is reflective, it follows that also for the s nth roots of unity ω^{p_i}, $1 \leq i \leq s$, there exists a diameter $\underline{a}$ of the unit circle with respect to which such roots are symmetric.
If $\underline{a}$ contains at most one of the above roots, consider an ordering $H = \{\omega^{\alpha_1}, \omega^{\alpha_2}, \ldots, \omega^{\alpha_s}\}$ of such roots obtained by traversing the unit circle counterclockwise, so that the diameter $\underline{a}$ is encountered only once. In case $\underline{a}$ does not contain any roots, s is even and $\mathrm{dist}(\omega^{\alpha_i}, \omega^{\alpha_{i+1}})=\mathrm{dist}(\omega^{\alpha_{s-(i-1)}}, \omega^{\alpha_{s-i}})$, $1 \leq i \leq \frac{s}{2} - 1$.

In case $\underline{a}$ contains one root, s is odd and that root is $\omega^{\alpha_{s+1/2}}$.
Then $\mathrm{dist}(\omega^{\alpha_i}, \omega^{\alpha_{i+1}})=\mathrm{dist}(\omega^{\alpha_{s-(i-1)}}, \omega^{\alpha_{s-i}})$, $1 \leq i \leq \frac{s-1}{2}$.

If $\underline{a}$ contains two roots, we ignore one of these, then we proceed as before.

So the two roots on $\underline{a}$ are $\omega^{\alpha_{s/2}}$ and ω^{α_s}.

It is easy to see that the roots $\bar{\omega}^{\alpha_1}, \bar{\omega}^{\alpha_2}, \ldots, \bar{\omega}^{\alpha_s}$ are symmetric for the diameter $\underline{b}$, where $\underline{b}$ is the reflection of $\underline{a}$ in the real axis.

There are two cases:

1) the diameter $\underline{a}$ contains at most one of the roots ω^{α_i}, $1 \leq i \leq s$.

Let h be the minimum positive integer such that $\omega^{\alpha_s+h} = \bar{\omega}^{\alpha_1}$.

By the symmetry of the roots with respect to $\underline{a}$, we obtain: $\mathrm{dist}(\omega^{\alpha_s}, \omega^{\alpha_{s-1}}) = \mathrm{dist}(\omega^{\alpha_1}, \omega^{\alpha_2}) = \mathrm{dist}(\bar{\omega}^{\alpha_1}, \bar{\omega}^{\alpha_2})$. It follows that $\omega^{\alpha_{s-1}+h} = \bar{\omega}^{\alpha_2}$, $\omega^{\alpha_{s-2}+h} = \bar{\omega}^{\alpha_3}, \ldots, \omega^{\alpha_1+h} = \bar{\omega}^{\alpha_s}$ and hence $\lambda_1 \omega^h = \bar{\lambda}_1$. Thus $\lambda_r(\omega^h)^r = \bar{\lambda}_r$, $1 \leq r \leq n-1$.

Consider the matrix $D = CP^h$. Let μ_r be the eigenvalues of D, as in (1). By Prop. 1.1, we have $\mu_r = \lambda_r(\omega^h)^r$ so that, by the above statements, $\mu_r = \bar{\lambda}_r$.

By Prop. 1.2 $D = C_T$.

2) The diameter $\underline{a}$ contains two of the roots ω^{α_i}, $1 \leq i \leq s$.

Let h be the minimum positive integer such that $\omega^{\alpha_{s-1}+h} = \bar{\omega}^{\alpha_1}$. By the above statements we have that $\omega^{\alpha_i+h} = \bar{\omega}^{\alpha_{s-i}}$, $1 \leq i \leq s-1$, and, in particular, $\omega^{\alpha_{s/2}+h} = \bar{\omega}^{\alpha_{s/2}}$.

Hence $\omega^{\alpha_s+h} = \bar{\omega}^{\alpha_s}$ and $CP^h = C_T$.

Suppose, now, that the circulant (0,1)-matrix C satisfies the equation $CP^h = C_T$, $0 \leq h \leq n-1$; we prove that C is reflective.

If $[c_0, c_1, \ldots, c_{n-1}]$ is the first row of C, then the first rows of CP^h and C_T are respectively $[c_{-h}, c_{1-h}, \ldots, c_{n-1-h}]$ and $[c_0, c_{n-1}, \ldots, c_1]$.

From the relation $CP^h = C_T$, we obtain the equalities $c_{i-h} = c_{n-i}$, where $0 \leq i \leq n-1$ and the indices are mod n, so that C is reflective.□

Recall that, if $A = [a_{ij}]$ and B are two matrices of order m and n respectively, then the direct product of A with B is a matrix of order mn defined by

$$A \times B = \begin{bmatrix} a_{11}B & a_{12}B & \ldots & a_{1m}B \\ a_{21}B & a_{22}B & \ldots & a_{2m}B \\ \ldots & & & \\ a_{m1}B & a_{m2}B & \ldots & a_{mm}B \end{bmatrix}.$$

Let C be a reflective circulant (0,1)-matrix. If there are two diameters of the circle with respect to wich the 1's are symmetric, then there are two integers k_1 and k_2, $k_1 \neq k_2$, such that $CP^{k_i} = C_T$, $i = 1,2$.

If C satisfies such equations and $1 \leq k_1 < k_2 \leq n$, then $CP^h = C$, where $h = k_2 - k_1 \in [1, n-1]$.

In this regard we have the following theorem.

THEOREM 1.5 - A circulant (0,1)-matrix C of order n satisfies $CP^h = C$, $1 \leq h \leq n-1$, if and only if it is $C = 0$ or $C = J$ for $(h,n)=1$ and $C = J_q \times D$ for $(h,n)=t > 1$, where $q = \frac{n}{t}$ and D is circulant (0,1)-matrix of order t.

Proof. The first row of CP^h is obtained from the first row of C by shifting it cyclically h positions to the right.
If $[c_1, c_2, \dots, c_n]$ is the first row of C, from $CP^h=C$ we obtain

$$c_1 = c_{1+h} = \dots = c_{1+(n-1)h} \tag{2}$$

where the indices are mod n.
We have two cases to consider:

a) $(h,n)=1$

b) $(h,n)>1$.

Case a). If $(h,n)=1$, then the n entries of $H_1= \{1, 1+h, \dots, 1+(n-1)h\}$ are all distinct and H_1 is equal to the set $\{1,2,\dots,n\}$.
Hence in the sequence (2) there are all the entries of the first row of C and it is or $C = 0$ or $C = J$.

Case b). Let $(h,n)=t>1$ with $n=tq$. In H_1 only q elements of the set $K_1 =\{ 1,1+h, \dots, 1+(q-1)h\}$ are distinct.
It is easy to prove that also the elements of the set $K_i =\{ i,i+h,\dots, i+(q-1)h\}$, $i \in [1,t]$ are distinct and, moreover, K_i is disjoint from K_j, $i \neq j$.

It follows that the sequence $c_1\ c_2 \dots c_t$ repeated q times determines the first row of C.
Then, if D is the circulant matrix whose first row is $[c_1\ c_2 \dots c_t]$, we have $C = J_q \times D$.
Conversely, if it is $C = J_q \times D$, then it is easy to prove that $CP^h = C$. □

Remark that, if a circulant (0,1)-matrix C of order $n=hq$ satisfies the relation $C P^h = C$, then there are also satisfied the relations $C P^{hs} = C$, $s=1,2,\dots,q$.

Let C be a circulant (0,1)-matrix such that $C P^k = C_T$, $k\in[0,n-1]$.

If k is even, say $k=2r$, then we obtain $CP^r = (CP^r)_T$; denoted $CP^r=E$, it follows $E = E_T$.

If k is odd, $k=2q+1$, $q\geq 0$, then we obtain $(CP^q)P=(CP^q)_T$, where, denoted $CP^q=E$, it is $EP=E_T$.

THEOREM 1.6- A n-square circulant (0,1)-matrix $C = J_q \times D$, where D has order t and $n=tq$, is reflective if and only if D is reflective.

Proof. Let $C = J_q \times D$, of order $n=tq$, $t,q>1$, be a circulant (0,1)-matrix; then D is (0,1)-circulant, of order t, and by Theorem 1.5 it is satisfied $CP^t=C$.

If C is reflective, there exists an integer r such that, denoted $CP^r=E$, we have $E = E_T$ or $E P = E_T$.

Since it is $EP^t = E$, by Theorem 1.5, there exists a circulant matrix F of order t such that $E = J_q \times F$.

It is easy to prove that $E = E_T$ iff $F = F_T$ and $E\,P = E_T$ iff $F\,\hat{P} = F_T$, where $\hat{P}$ represents the permutation ($1\ 2\ \dots\ t$).

Since $C\,P^r = E$, then it is $D\,\hat{P}^r = F$ and C is reflective iff D is reflective.□

DEFINITION 1.7 - A set $D = \{d_1, d_2, \dots, d_k\}$ of k integers mod n is reflective if the circulant (0,1)-matrix, with respect to which the 1's are in the positions $d_1, d_2, \dots, d_k$, is reflective.

THEOREM 1.8 - If the set of integers mod n $D = \{d_1, d_2, \dots, d_k\}$ is reflective, then it is reflective the set $tD = \{td_1, td_2, \dots, td_k\}$, for every t prime to n.

Proof. Let $D = \{d_1, d_2, \dots, d_k\}$ be a reflective set and let C be the circulant (0,1)-matrix with respect to which the 1's are in the positions $d_1, d_2, \dots, d_k$. If we arrange the elements of the first row of C regularly on a circle, there exists a diameter $\underline{a}$ of the circle with respect to which the 1's are symmetric. Let $H = \{\alpha_1, \alpha_2, \dots, \alpha_k\}$ an ordering of the elements of D obtained by traversing the unit circle counterclockwise, so that the diameter $\underline{a}$ is encountered only once.
Let α_r be the last element of H before traversing $\underline{a}$ (α_r can belong to $\underline{a}$).
Let α_i' the symmetric of α_i as regards to $\underline{a}$, $i \in [1,r]$.
Consider the set $tH = \{t\alpha_1, t\alpha_2, \dots, t\alpha_k\}$. Because t is prime to n, the elements of tH are distinct and $t\alpha_i \not\equiv t\alpha_i'$ (mod n), but perhaps for i=r.
Moreover, as we have $\alpha_r - \alpha_i = \alpha_i' - \alpha_r'$, then $t\alpha_r - t\alpha_i = t(\alpha_r - \alpha_i) = t(\alpha_i' - \alpha_r') = t\alpha_i' - t\alpha_r'$, where the differences are mod n and $i \in [1,r-1]$.
Then $t\alpha_i'$, $1 \leq i \leq r$, is symmetric of $t\alpha_i$ as regards to the diameter that reduces to half the distance between $t\alpha_r$ and $t\alpha_r'$; hence tD is reflective.□

2. Let C be a reflective circulant (0,1)-matrix.
If $[c_0, c_1, \dots, c_{n-1}]$ is the first row of C, from the relation $CP^h = C_T$, $h \in [0,n-1]$, by the above remarks we obtain

$$c_{i-h} = c_{n-i} \tag{3}$$

where $0 \leq i \leq n-1$ and the indices are mod n.
The relation (3) is an equality between two different elements except for $i-h \equiv n-i$ (mod n); in that case there exists an integer $k \geq 0$ such that

$$2i = h + kn. \tag{4}$$

Let n and h odd, n=2t+1, h=2r+1 and $r \leq t$; then we have $2i=2r+1+k(2t+1)$.
For k=1 we obtain i=r+t+1 and clearly this is the only possible value.

If n is odd and h even, n=2t+1 and h=2r, we have $2i=2r+k(2t+1)$ and i=r is the only possible value.

If n is even and h odd, n=2t and h=2r+1, we have $2i=2r+1+k(2t)$; this equation is clearly impossible.

For n and h even, $n=2t$ and $h=2r$, the equation $2i=2r+k(2t)$ has two solutions: $i=r$ and $i=r+t$, corresponding to $k=0$ and $k=1$.

THEOREM 2.1 - Let h,n be two integers, $0 \leq h \leq n-1$. When n is odd, there are $2^{\frac{n+1}{2}}$ circulant (0,1)-matrices C that satisfy the equation $C P^h = C_T$, while, when n is even, there are $2^{\frac{n}{2}}$ such matrices for h odd and $2^{\frac{n+2}{2}}$ for h even.

Proof. Let n be odd. By the above remarks, from (3) we obtain $\frac{n-1}{2}$ equalities and one identity. So in the first row of C exactly $\frac{n-1}{2} + 1 = \frac{n+1}{2}$ elements are arbitrary.

As every element can be 0 or 1, there are $2^{\frac{n+1}{2}}$ circulant (0,1)-matrices that satisfy the equation $C P^h = C_T$, for n odd.
If n is even, the relations (3) correspond to $\frac{n}{2}$ equalities for h odd and to $\frac{n-2}{2}$ equalities and two identities for h even.

Hence, there are $2^{n/2}$ distinct matrices when h is odd and $2^{\frac{n+2}{2}}$ such matrices when h is even.□

3. Recall that a difference set $D = \{d_1, d_2, \ldots, d_k\}$ is a k-set of integers mod v such that every $a \not\equiv 0 \pmod v$ can be exspressed in exactly λ ways in the form $d_i - d_j \equiv a \pmod v$, where d_i and d_j are in D. We suppose $0<\lambda<k<v-1$ to exclude certain degenerate configurations.
It is well known that a difference set D is equivalent to a (v,k,λ)-configuration with incidence matrix a circulant.

PROPOSITION 3.1 - If C is a circulant incidence matrix of a (v,k,λ)-configuration, it never satisfies the equation $CP^h = C_T$.

Proof. If $[c_0, c_1, \ldots, c_{n-1}]$ is the first row of a circulant matrix C, the first row of C_T is $[c_0, c_{n-1}, \ldots, c_1]$, i.e. c_i has been replaced by c_{-i}, where the indices are mod n.
Let $p_1, p_2, \ldots, p_s$ be the positions of the elements equal to 1 on the first row of C. If C is an incidence matrix of a (v,k,λ)-configuration, then $\{p_1, p_2, \ldots, p_s\}$ is a difference set.
Since -1 is not a multiplier for a difference set [3], then there exists no integer $h \in [0, n-1]$ such that $\{p_1+h, p_2+h, \ldots, p_s+h\} = \{-p_1, -p_2, \ldots, -p_s\}$, i.e. C does not satisfy $C P^h = C_T$.□

Recall that a (v,k,λ)-matrix is a (0,1)-matrix A of order v which satisfies

$$A A_T = (k - \lambda)I + \lambda J$$

where $0 < \lambda < k < v-1$. It follows that $\det A = k(k-\lambda)^{\frac{v-1}{2}}$.

A Hadamard matrix H of order v is a matrix of +1's and -1's such that $H H_T = vI$. It has been conjectured [6] that a Hadamard matrix cannot be circulant for $v > 4$. Remark that we can define the notion of reflective circulant Hadamard matrices, analogous to the corresponding notion for the (0,1)-matrices.

PROPOSITION 3.2- A Hadamard matrix of order v cannot be reflective circulant or coincident block circulant, for $v > 4$.

Proof. If H is a reflective circulant or coincident block circulant Hadamard matrix, then $K = \frac{1}{2}(H + J)$ is a reflective or coincident block circulant (v,k,λ)-matrix, with $k = \frac{1}{2}(v \pm \sqrt{v})$ and $\lambda = \frac{1}{4}(v \pm \sqrt{v})$.
By Prop. 3.1 a reflective circulant matrix cannot be the incidence matrix of a (v,k,λ)-configuration, then cannot be a (v,k,λ)-matrix.
Moreover, since it is clear that the determinant of a coincident block circulant matrix is zero, such a matrix cannot be a (v,k,λ)-matrix. ∎

The Prop. 3.2 is a partial answer to the problem, in digital communications, of the existence of finite sequences of 1 and -1 $\{b_i\}_1^v$ with the property that their aperiodic auto-correlation coefficients

$$c_j = \sum_{i=1}^{v} b_i b_{i+j} \quad ,$$

$1 \leq j \leq v-1$, should be 1,0,-1.
It is well known that there is a one-to-one correspondence between such sequences and circulant Hadamard matrices [1,pg. 98].
We note that the results of the Prop. 3.1 and the Prop. 3.2 were implicitely proved in [3].

4. Let G be a directed graph with n vertices;we can suppose that its adjacency matrix is not symmetric.
We call cyclic a graph with circulant adjacency matrix.
For a cyclic graph the mapping $i \to i+1$, and its powers, are clearly automorphisms by definition. Hence, these graphs always have the cyclic group of order n, C_n, as a subgroup of their automorphism group.
In [5] the problem is posed how to give a procedure for determining the automorphism group of a cyclic graph.
The following Theorem 4.2 gives a subset of the graphs for which the automorphism group is exactly C_n.

LEMMA 4.1- If $C \neq 0,J$ is a circulant matrix of order prime n,then $\det C \neq 0$.

Proof. If $[c_0, c_1, \ldots, c_{n-1}]$ is the first row of C, the eigenvalues of such matrix are as in (1).
If $\det C = 0$, there exists an eigenvalue $\lambda_q = \sum c_j \omega^{jq} = 0$.
Hence $F(\omega)=0$, where $F(x)= \sum c_j x^{jq}$ with iq mod n, is a rational polynomial of degree <n.
Then, F(x) is divisible by the minimum polynomial of ω over the rational numbers. This minimum polynomial is the cyclotomic polynomial of order n, which for n prime is known to be $\psi_n(x)= x^{n-1}+\ldots+x+1$. We obtain $F(x)=k\psi_n(x)$, with k=0 or k=1; then the first row of C is $k[1\ 1\ \ldots 1]$ and C is or 0 or J. ∎

THEOREM 4.2 - Let G be a directed cyclic graph having a prime number n of vertices and reflective circulant adjacency matrix. Then the automorphism group of G is precisely C_n.

Proof. Let G be a directed graph with a prime number n of vertices.
If A is the adjacency matrix of G and Q is the permutation matrix corresponding to an automorphism of G, we have

$$Q_T\, A\, Q = A. \tag{5}$$

If A is reflective circulant, there exists a $k \in [1,n-1]$ for which $A\, P^k = A_T$.
Then we have

$$Q_T\, A\, P^k\, Q\, P^k{}_T = A. \tag{6}$$

By Lemma 4.1, (5) and (6) we obtain

$$P^k\, Q\, P^k{}_T = Q. \tag{7}$$

It follows $P^{rk} Q P^{rk}{}_T = Q,\ r \in [1,n]$.
Since n is prime, the integers rk, $r \in [1,n]$, mod n are distinct; then there exists $j \in [1,n]$ such that $jk \equiv 1 \pmod n$. Hence we have $P\, Q\, P_T = Q$, i. e. Q is circulant. ▯

In [5] it is proved that if a directed graph with a prime number of vertices has C_n as automorphism group, then the adjacency matrix A of G has distinct eigenvalues.

COROLLARY 4.3- If A is the reflective circulant adjacency matrix of a directed graph G with a prime number of vertices, then the eigenvalues of A are distinct.

If R is the matrix corresponding to the permutation $i \to n-i$, we have the following

COROLLARY 4.4 - Let G be a directed graph with a prime number of vertices and reflective circulant adjacency matrix. Then G is self-converse and has exactly n isomorphisms with the converse, corresponding to the matrices RP^h, $h \in [1,n]$.

Proof. Let G be a directed graph with a prime number of vertices and reflective circulant adjacency matrix.
Since $R_T\, A\, R = A_T$ holds, R corresponds to an isomorphism of G with the converse G'. By Theorem 4.2, every automorphism of G corresponds to P^h, $h \in [1,n]$; then every matrix RP^h corresponds to an isomorphism of G with G'.
It is well known that the number of isomorphisms of a self-converse graph G with the converse G' is equal to the number of automorphisms of G. ▯

5. Recall that a tournament T of order n is a directed graph in which every pair of vertices is joined by exactly one arc. A tournament matrix A is the adjacency matrix of a tournament T and satisfies $A + A_T = J - I$.

Let $\mathcal{A}$ the reflective circulant tournament matrix, of odd order n, that satisfies $\mathcal{A}P = \mathcal{A}_T$.
It is easy to prove that the 1's of the first row of $\mathcal{A}$ are in even positions, the 0's in odd positions.

THEOREM 5.1 - The automorphism group of the tournament $\mathcal{T}$ corresponding to $\mathcal{A}$ is C_n.

Proof. Let Q be the permutation matrix corresponding to an automorphism of $\mathcal{T}$; then we have

$$Q_T \mathcal{A} Q = \mathcal{A} \tag{8}$$

and, being $\mathcal{A} P = \mathcal{A}_T$, it follows

$$Q_T \mathcal{A} P Q P_T = \mathcal{A}. \tag{9}$$

Since $\mathcal{T}$ is regular, it follows that $\mathcal{A}$ is nonsingular [9]; by (8) and (9) we obtain $P Q P_T = Q$, i. e. Q is circulant. □

In [8] the following theorems are proved.

THEOREM 5.2 - If a tournament matrix A satisfies $A Q = A_T$, where Q is a permutation matrix, then Q corresponds to a n-cycle.

PROPOSITION 5.3 - Every tournament matrix A for which AQ, where Q corresponds to a n-cycle, is also a tournament matrix, is permutationally similar to $\mathcal{A}$.

Now, we have the following

THEOREM 5.4 - A reflective circulant tournament matrix is reflective circulant if and only it is permutationally similar to $\mathcal{A}$.

Proof. Let A be a reflective circulant tournament matrix of order n. Since the row sums of A are constant, it follows that the tournament corresponding to A is regular; hence n is odd.
Because A is reflective circulant, there exists an integer $k \in [1, n-1]$ for which $A P^k = A_T$.
By Theorem 5.2 P^k corresponds to a n-cycle; then, by Proposition 4.3 A is permutationally similar to $\mathcal{A}$.
The converse is easily proved. □

COROLLARY 5.5 - The automorphism group of a tournament with reflective circulant adjacency matrix is C_n.

Proof. The proof follows immediately from Theorems 5.1 and 5.4.

COROLLARY 5.6 - The determinant of a reflective circulant tournament matrix of order n is $\frac{n-1}{2}$.

Proof. Since the matrix of odd order n $\mathcal{A}$ satisfies the relations $\mathcal{A} + \mathcal{A}_T =$

$J - I$ and $\mathcal{A}P = \mathcal{A}_T$, it follows $\mathcal{A}(I + P) = J - I$; then we have $\det \mathcal{A} \cdot \det(I+P) = \det(J - I)$.
It is immediate to prove that $\det(I + P) = 2$ and $\det(J - I) = n-1$ for n odd.
By Proposition 5.3, the proof is completed.$\square$

REFERENCES

[1] Baumert, L.D., Cyclic Difference Sets (Lecture Notes in Mathematics, n. 182, Springer Verlag, 1971).

[2] Biggs, N., Algebraic Graph Theory (Cambridge University Press,1974).

[3] Brualdi, R.A., A note on multipliers of difference sets, J. of Res. Nat. B. of Standards, vol. 69 B (1965) 87-89.

[4] Davis, P.J., Circulant matrices (A Wiley-Interscience Publication, 1979).

[5] Elspas, B. and Turner, J., Graphs with circulant adjacency matrices, J. Combinatorial Theory 9(1970) 297-307.

[6] Ryser, H.J., Combinatorial Mathematics (Carus Math. Monograph, N. 14, New York, 1963).

[7] Turner, J., Point-Symmetric Graphs with a Prime Number of Points, J. Combinatorial Theory 3 (1967) 136-145.

[8] Zagaglia Salvi, N., Sulle matrici torneo associate a matrici di permutazione, Note di Matematica, vol. II (1982) 177-188.

[9] Zagaglia Salvi, N., Alcune proprietà delle matrici torneo regolari, in: Atti del Convegno "Geometria Combinatoria e di incidenza: fondamenti e applicazioni" La Mendola, 4-11 Luglio 1982 (Editr. Vita e Pensiero) 635-643.

Annals of Discrete Mathematics 30 (1986) 373–382

OVALS IN STEINER TRIPLE SYSTEMS

Herbert Zeitler
Mathematisches Institut, Universität Bayreuth
D-8580 Bayreuth, West Germany

This article is a report on some joint works with H. Lenz, Berlin [2,7,8]. Using the notion of a regular oval, Steiner triple systems of order $v = 9$ or $13 + 12n$ with $n \in \mathbb{N}_0$ are constructed.

INTRODUCTION

Let V with $|V| = v > 3$ be a finite set and B with $|B| = b$ a set of 3-subsets of V. The elements of V are called points, the elements of B lines or blocks. If any 2-subset of V is contained in exactly one line, then the incidence structure $(V,B,\in)$ is called a Steiner triple system of order v, briefly STS(v). Each point of an STS(v) lies on exactly $r = \frac{1}{2}(v-1)$ lines and there are exactly $b = \frac{1}{6}v(v-1)$ lines. The condition $v = 7$ or $9 + 6n$, $n \in \mathbb{N}_0$ is necessary and sufficient for the existence of an STS(v). The set of these "Steiner numbers" v will be denoted by STS.

Any non empty subset $H \subset V$ in a STS(v) is called a *hyperoval* iff each line of STS(v) has exactly either two points *(secant)* or no point *(external line)* in common with H. Thus we obtain $|H| = 1 + r$. The complement $\bar{H} = V \setminus H$ together with all external lines of H forms a subsystem STS(r) and, vice versa, the complement of a subsystem STS(r) is a hyperoval in STS(v). Precisely for any $v = 7$ or $15 + 12n$, $n \in \mathbb{N}_0$ there exists an STS(v) with at least one hyperoval [2], [4], [7]. The set of all these special Steiner numbers will be denoted by HSTS. The remaining Steiner numbers in RSTS = STS $\setminus$ HSTS are $v = 9$ or $13 + 12n$, $n \in \mathbb{N}_0$.

1. OVALS

1.1 DEFINITIONS

Any non empty subset $O \subset V$ in a STS(v) will be called an *oval* iff there exists exactly one *tangent* at each point of O (this tangent meets O in exactly one point, the so-called point of contact) and each line of STS(v) has at most two points in common with O. Sometimes the complement $\overline{O} = V \setminus O$ together with all the external lines to O will be referred to as the "*counter oval*". The points of O will be called the *on-points*, the points on the tangents to O with the points of contact deleted will be called the *ex-points* and all the remaining points the *in-points*.

1.2 ENUMERATION THEOREMS

1.2.1 CLASSES OF POINTS

Each oval has exactly $r = \frac{1}{2}(v-1)$ *points and the corresponding counter oval exactly* $r+1 = \frac{1}{2}(v+1)$ *points.*

Proof: Each point on the oval lies on r lines, and precisely one of them is a tangent.

1.2.2 CLASSES OF LINES

With respect to an oval O *there exist exactly* $r = \frac{1}{2}(v-1)$ *tangents, exactly* $\binom{r}{2} = \frac{1}{8}(v-1)(v-3)$ *secants and exactly* $\frac{1}{6}r(r-1) = \frac{1}{24}(v-1)(v-3)$ *external lines.*

Proof: An easy counting argument.

1.3 THEOREM

The number of tangents through an ex-point of an oval O *is even or odd, according to* r *being even or odd.*

Proof: Let s be the number of secants, t the number of tangents through an ex-point of O. By 1.2.1, $t + 2s = r$, and the statement follows.

1.4 THEOREM

For any $v \in$ HSTS *there exists an* STS(v) *with at least one oval.*

Proof: When $v \in HSTS$ there exists an STS(v) with at least one hyperoval. Any r points of this hyperoval form an oval. These ovals are special ones, because all the tangents meet in exactly one point, the knot of the oval.

1.5 DEFINITION

Ovals all whose tangents are concurrent will be called knot ovals. If through each ex-point of an oval there pass exactly two tangents, then the oval will be called a regular oval.

1.6 THEOREM

In an STS(v) *with* $v \in HSTS$ *there are no regular ovals, in an* STS(v) *with* $v \in RSTS$ *there exist no knot ovals.*

Proof: If $v \in HSTS$, then $r = 3$ or $7 + 6n$, $n \in \mathbb{N}_0$, and r is odd. By 1.3, regular ovals cannot exist in such an STS(v). If in an STS(v) with $v \in RSTS$ a knot oval existed, then adding the knot to the points of the oval we would obtain a hyperoval; therefore, $v \in HSTS$, a contradiction.

1.7 THEOREM

For any regular oval O *in an* STS(v) *there exists exactly one in-point.*

Proof: Let e be the number of ex-points and i the number of in-points of O. The ex-points and in-points together are the points on the counter oval $\overline{O}$. Therefore, $e + i = |\overline{O}| = r + 1$. By 1.5 each ex-point lies on exactly two tangents and on each of the r tangents to O there are exactly two ex-points. Consequently, $e = r$ and $i = 1$.

1.8 THEOREM

For a regular oval the numbers of lines in the different classes through points of different classes are as follows:

	secants	tangents	external lines
ex-point	$\frac{1}{2}(r-2) = \frac{1}{4}(v-5)$	2	$\frac{1}{2}(r-2) = \frac{1}{4}(v-5)$
in-point	$\frac{1}{2}r = \frac{1}{4}(v-1)$	0	$\frac{1}{2}r = \frac{1}{4}(v-1)$
on-point	$r-1 = \frac{1}{2}(v-3)$	1	0

The proof follows immediately from the preceding theorems.

2. THE MAIN THEOREM

For any v ∈ RSTS *there exists an* STS(v) *with at least one regular oval.*

Now the Steiner numbers v ∈ HSTS and v ∈ RSTS will be classified in a geometrical way using the existence of STS(v)'s with knot ovals and with regular ovals, respectively.
Disregarding this distinction between ovals we can summarize:
For any Steiner number v ∈ STS there exists an STS(v) with at least one oval.
In [2] this theorem is proved by contradiction.
We now give a direct proof by construction [8].
Throughout the proof by an oval we always mean a regular oval.
Taking into account theorem 1.6, we start with a Steiner number v ∈ RSTS and construct an STS(v) with at least one regular oval.

2.1 THE POINTS

With v ∈ RSTS, $|O| = r = \frac{1}{2}(v-1) = 4$ or $6+6n$, $n \in \mathbb{N}_o$, and $|\bar{O}| = r+1 = \frac{1}{2}(v+1) = 5$ or $7+6n$, $n \in \mathbb{N}_o$. Therefore, the number of points on the oval O is even and the number of points on the counter oval is odd.

We consider two regular r-gons with the same centre M whose edges have different lengths. Furthermore, the r-gons are rotated with respect to each other by an angle $\frac{\pi}{r}$. The vertices $P_o, P_1, \ldots, P_{r-1}$ of the inner polygon are taken as on-points, the vertices $Q_o, Q_1, \ldots, Q_{r-1}$ of the outer polygon as ex-points and finally the point M is taken as the in-point of the oval O. We count the vertices of each polygon counter clockwise. Figure 1 shows this interpretation for $r = 16$.

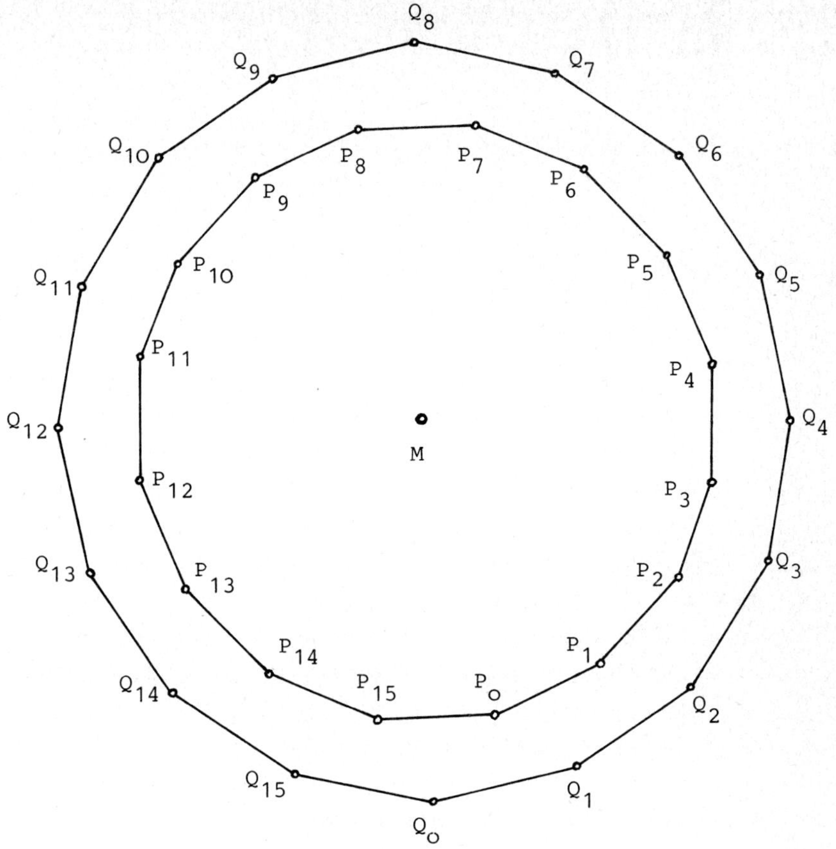

Figure 1

It turns out useful to represent the numbers r modulo 12. Consequently, $r = 12$ or 4 or 10 or $6 + 12n$, $n \in \mathbb{N}_0$.
In this paper we restrict ourself to the case $r = 4 + 12n$. The proofs in the other cases may be done in a similar way.

2.2 CONSTRUCTION OF SPECIAL 3-SUBSETS

First of all we construct special point sets having no connection with an oval.

Let a regular r-gon be given with vertices $0, 1, \ldots, r-1$ and centre M.

(a) <u>First class of subsets</u>

There are $\frac{1}{2}r = \frac{1}{4}(v-1)$ diameters in the given polygon. The two endpoints of each diameter together with the centre M form a 3-subset of the first class.

(b) Second class of subsets

Let i,j with $i < j$ be two vertices of the given polygon. Then the minimum of $j - i$, $r + i - j$ is called the "difference" of the two vertices. By this definition, the following differences are possible: $1,2,\ldots,\frac{1}{2}r - 1 = \frac{1}{4}(v-5)$. (The difference $\frac{1}{2}r$ is already eliminated with the first class!) Now we form ordered "difference triples" (d_1,d_2,d_3) in such a way that each of the mentioned differences occurs at most once and $d_1 + d_2 = d_3$.

Which differences can occur in the triples? How many difference triples of this kind are possible? In [1], [2], [3], [6] these questions are answered. Here we give the result only - without any proof.

$r = 4 + 12n,\ n \in \mathbb{N}$

$$\begin{array}{llllll}
(1, & 5n+1, & 5n+2), & (2, & 3n, & 3n+2), \\
(3, & 5n, & 5n+3), & (4, & 3n-1, & 3n+3), \\
 & \vdots & & & \vdots & \\
(2n-1, & 4n+2, & 6n+1), & (2n, & 2n+1, & 4n+1).
\end{array}$$

The difference $3n + 1 = \frac{1}{4}r$ is missing, any other difference occurs exactly once as required. Altogether we have 2n difference triples.

The least value for r, i.e. $r = 4$, does not provide difference triples. We obtain the affine plane AG(2,3). It is easy to show that in this system there exist regular ovals (for instance the set of points (x,y) with $x,y \in GF(3)$ and $x^2 - y^2 = 1$). In the following we omit this special case.

2.3 THE EXTERNAL LINES

Now the points $0,1,\ldots,r-1$ in 2.2 are the ex-points $Q_o,Q_1,\ldots,Q_{r-1}$ and M is the in-point of an oval O.

Next, the external lines to O are constructed using the difference triples given in 2.2.

For better typing we frequently write (P,x) and (Q,x) instead of P_x and Q_x.

(a) The external lines containing M

The subsets of the first class determined by the polygon diameters are the external lines of O containing M. We can write $\{M, (Q,x), (Q,x+\frac{1}{2}r)\}$ with $x \in \{0,1,...,r-1\}$. We always understand the number x modulo r.

(b) <u>The remaining external lines</u>

$$\{(Q,x),(Q,x+1),\quad (Q,x+5n+2)\},\{(Q,x),(Q,x+2),\quad (Q,x+3n+2)\},$$
$$\{(Q,x),(Q,x+3),\quad (Q,x+5n+3)\},\{(Q,x),(Q,x+4),\quad (Q,x+3n+3)\},$$
$$\vdots \qquad\qquad \vdots$$
$$\{(Q,x),(Q,x+2n-1),(Q,x+6n+1)\},\{(Q,x),(Q,x+2n),(Q,x+4n+1)\}.$$

$x \in \{0,1,...,r-1\}$

In this way we obtain altogether $r \cdot 2n = \frac{1}{24}(v-1)(v-9)$ additional 3-subsets of points. These lines are external lines of O,too. Together with the $\frac{1}{2}r = \frac{1}{4}(v-1)$ external lines containing M we have got $\frac{1}{4}(v-1) + \frac{1}{24}(v-1)(v-9) = \frac{1}{24}(v-1)(v-3)$ external lines. By 1.2.2, this is the set of all external lines.

In the difference triples 2.2 the difference $3n+1$ is missing. The associated pair of ex-points $((Q,x), (Q,x+3n+1))$ with $x \in \{0,1,...,r-1\}$ will be investigated later on in connection with the tangents.

2.4 THE SECANTS

In figure 2 we have the secants of an oval containing the ex-point Q for $r = 16$.

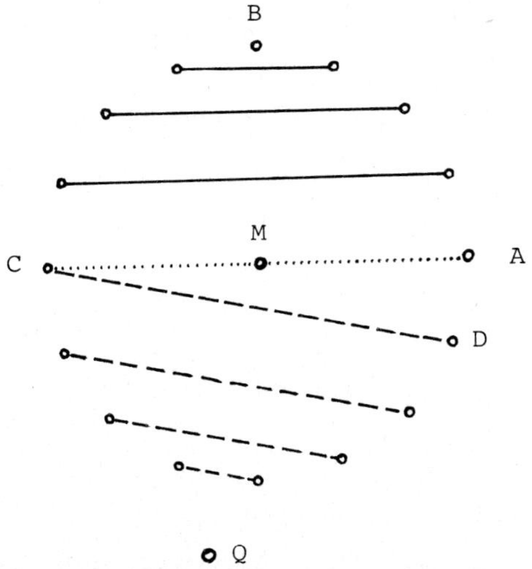

Figure 2

Secants containing an ex-point

(a) First class of polygon chords

We start with the polygon vertices A and B (from B to A clockwise!) having the difference $\frac{1}{4}r$. In the set of polygon chords orthogonal to the diameter through B we choose those $\frac{1}{4}(r-4)$ chords which are nearest to B (full lines). Then the next polygon chord of this kind is a polygon diameter CA. The endpoints of the polygon chords choosen in this way have the differences $2,4,\dots,\frac{1}{2}r-2$, respectively.

(b) Second class of polygon chords

The polygon vertex which follows A going clockwise from B to A is D. In the set of polygon chords parallel to CD we choose those $\frac{1}{4}r$ chords which are farthest from B (dotted lines). The endpoints of these polygon chords have differences $1,3,\dots,\frac{1}{2}r-1$, respectively. One of these chords is an edge of the polygon. The ex-point lying close to this edge is Q.

The endpoints of each polygon chord, both in (a) and in (b), together with Q form an oval secant. In this way we obtain altogether $\frac{1}{4}(r-4)+\frac{1}{4}r = \frac{1}{2}(r-2)$ secants containing Q. By 1.8 we have found all secants of this kind.

There is no secant containing Q and the two on-points A and B as well. Therefore, these two points must be the points of contact on the two tangents through Q. By our construction the difference of A and B is exactly $\frac{1}{4}r$.

In the same way it is possible to construct the secants through any ex-point (Q,x).

Now we can explicitly write down all the secants containing an ex-point (Q,x):

$$\begin{array}{ll}
\{(Q,x),(P,x),\quad (P,x+r-1)\}, & \{(Q,x),(P,x+\frac{1}{4}r+1),(P,x+\frac{3}{4}r-1)\},\\
\{(Q,x),(P,x+1),\quad (P,x+r-2)\}, & \{(Q,x),(P,x+\frac{1}{4}r+2),(P,x+\frac{3}{4}r-2)\},\\
\vdots & \vdots \\
\{(Q,x),(P,x+\frac{1}{4}r-1),(P,x+\frac{3}{4}r)\}, & \{(Q,x),(P,x+\frac{1}{2}r-1),(P,x+\frac{1}{2}r+1)\}.
\end{array}$$

Furthermore, $(A,x) = (P,x+\frac{1}{4}r)$, $(B,x) = (P,x+\frac{1}{2}r)$, $(C,x) = (P,x+\frac{3}{4}r)$.

We further see that the tangent containing (Q,x) and $(Q,x+\frac{1}{4}r)$ has the point of contact (B,x). The difference of these two ex-points is $\frac{1}{4}r$.

Again, $x \in \{0,1,\dots,r-1\}$.

Secants containing the in-point

The endpoints of any diameter of the inner polygon form an oval secant together with M. In this way we obtain exactly $\frac{1}{2}r$ secants containing M. By 1.8 we have found all the secants of this kind; namely, they are $\{M, (P,x), (P,x+\frac{1}{2}r)\}$, $x \in \{0,1,\ldots,r-1\}$.

Adding the numbers of oval secants in the two different classes we obtain $r\cdot\frac{1}{2}(r-2) + \frac{1}{2}r = \frac{1}{2}r(r-1) = \frac{1}{8}(v-1)(v-3)$.

2.5 THE TANGENTS

When constructing the external lines in 2.3 some pairs of ex-points were left without a connecting line. The difference between two points of such a pair was $d = \frac{1}{4}r = 1+3n$.
The construction of the oval secants in 2.4 was done in such a way that the difference of the two ex-points on each tangent gives the number d.
Therefore, the oval tangents are completely determined:
$\{(Q,x), (Q,x+\frac{1}{4}r), (P,x+\frac{1}{2}r)\}$, $x \in \{0,1,\ldots,r-1\}$.
Our construction in case $r = 4+12n$ with $n \in \mathbb{N}$ is completed.

CONCLUDING REMARKS

Long ago P. Erdös asked the following questions:
Which cardinality is the maximal one for a point set M in an STS(v) containing no line?
For which Steiner numbers v may this maximal case occur?
All these questions were already answered in [5]. Here we relate these problems to hyperovals and ovals. The required maximal cardinality occurs if each point of M lies on secants only. This yields $|M| \leq 1+r$. Therefore, in case of maximal cardinality M is a hyperoval. Precisely, for any $v \in$ HSTS there exists an STS(v) with such a set of maximal size $1+r$. If such hyperovals do not exist, then $|M| \leq r$. The upper bound is achieved if M is a regular oval. For all Steiner numbers $v \in$ RSTS there exist STS(v)'s with sets of the required maximal cardinality r.

Another matter is the investigation of finite affine and projective spaces with regard to ovals. We mention some results:
In PG(d,2) with $d \geq 2$ there exist exactly $2^d(2^{d+1}-1)$ knot ovals and no regular ovals. In AG(2,3) there exist exactly 54 regular ovals and no knot ovals. In AG(d,3) with $d \geq 3$ there exist neither knot

ovals nor regular ovals.

We notice that there are still many unsolved problems about ovals in an STS(v). (The total number of ovals in a given STS(v); the number of non-isomorphic STS(v)'s with ovals if the Steiner number is given; investigation of automorphism groups;...)

For Steiner systems S(k,v) with $k \geq 3$ there are only a few results concerning ovals, hyperovals and similar notions [4]. Extensions of the Erdös problems to general systems of this kind are unknown, too.

REFERENCES

[1] Hilton, A.J.W., On Steiner and similar triple systems, *Math. Scand. 24 (1969) 208-216*

[2] Lenz, H. and Zeitler, H., Arcs and Ovals in Steiner Triple Systems, *Combinatorial Theory, Lecture Notes in Mathematics 969 (1982) 229-250* (Springer Verlag, Berlin, Heidelberg, New York)

[3] Pelteson, R., Eine Lösung der beiden Heffterschen Differenzenprobleme, *Compositio Math. 6 (1939) 251-257*

[4] Resmini, M. de, On k-sets of type (m,n) in a Steiner system S(2,l,v), *Finite Geometries and Designs, L.M.S. Lecture Note Series 49 (1981) 104-113*

[5] Sauer, N. and Schönheim, J., Maximal subsets of a given set having no triple in common with a Steiner triple system on the set, *Canad. Math. Bull. 12 (1969) 777-778*

[6] Skolem, T., Some remarks on the triple systems of Steiner, *Math. Scand. 6 (1958) 273-280*

[7] Zeitler, H. and Lenz, H., Hyperovale in Steiner-Tripel-Systemen, *Math. Sem. Ber. 32 (1985) 19-49*

[8] Zeitler, H. and Lenz, H., Regular Ovals in Steiner Triple Systems, *Journal of Combinatorics, Information and System Sciences, Delhi* (to appear)

PARTICIPANTS

L.M. Abatangelo	Bari
V. Abatangelo	Bari
S. Antonucci	Napoli
E.M. Aragno Marauta	Milano
R. Artzy	Haifa
L. Bader	Napoli
G. Balconi	Pavia
A. Barlotti	Firenze
U. Bartocci	Perugia
L. Bénéteau	Toulouse
W. Benz	Hamburg
L. Berardi	L'Aquila
C. Bernasconi	Perugia
L. Bertani	Reggio Emilia
A. Beutelspacher	Mainz
A. Bichara	Roma
M. Biliotti	Lecce
P. Biondi	Napoli
P. Biscarini	Perugia
F. Bonetti	Ferrara
A. Bonisoli	München
A.A. Bruen	Western Ontario
A. Caggegi	Napoli
I. Candela	Bari
G. Cantalupi Tazzi	Milano

R. Capodaglio di Cocco	Bologna
M. Capursi	Bari
B. Casciaro	Bari
P.V. Ceccherini	Roma
N. Cera	Pescara
N. Civolani	Milano
A. Cossu	Bari
C. Cotti Ferrero	Parma
M. Crismale	Bari
F. De Clerck	Gent
E. Dedò	Milano
M. De Finis	Roma
M.L. De Resmini	Roma
M. De Soete	Gent
M.M. Deza	Paris
C. Di Comite	Bari
V. Dicuonzo	Roma
L. Di Terlizzi	Bari
B. D'Orgeyal	Dijon
J. Doyen	Bruxelles
F. Eugeni	L'Aquila
G. Faina	Perugia
L. Faggiano	Bari
M. Falcitelli	Bari
A. Farinola	Bari
G. Ferrero	Parma
O. Ferri	L'Aquila
P. Filip	München
S. Fiorini	Malta
M. Funk	Potenza
M. Gionfriddo	Catania
M. Grieco	Pavia
W. Heise	München
A. Herzer	Mainz

R. Hill	Salford
J. Hirshfeld	Sussex
D. Hughes	London
V. Jha	Glasgow
H. Kellerer	München
H. Karzel	München
A.D. Keedwell	Guildford
G. Korchmaros	Potenza
H.J. Kroll	München
F. Kramer	Cluj-Napoca
H. Kramer	Cluj-Napoca
P. Lancellotti	Modena
B. Larato	Bari
D. Lenzi	Lecce
A. Lizzio	Catania
G. Lo Faro	Messina
P.M. Lo Re	Napoli
G. Luisi	Bari
G. Lunardon	Napoli
H. Lünenburg	Kaiserslautern
N.A. Malara	Modena
M. Marchi	Brescia
A. Maturo	Pescara
F. Mazzocca	Caserta
N. Melone	Napoli
M. Menghini	Roma
G. Menichetti	Bologna
G. Micelli	Lecce
G. Migliori	Roma
F. Milazzo	Catania
S. Milici	Catania
A. Maniglia	Lecce
D. Olanda	Napoli
A. Palombella	Bari

A.M. Pastore	Bari
S. Pellegrini	Brescia
C. Pellegrino	Modena
G. Pellegrino	Perugia
C. Perelli Cippo	Brescia
M. Pertichino	Bari
G. Pica	Napoli
F. Piras	Cagliari
A.I. Pomilio	Roma
L. Porcu	Milano
L. Puccio	Messina
G. Quattrocchi	Catania
P. Quattrocchi	Modena
G. Raguso	Bari
L. Rella	Bari
T. Roman	Bucaresti
L.A. Rosati	Firenze
H.G. Samaga	Hamburg
M. Scafati	Roma
R. Scapellato	Parma
E. Schröder	Hamburg
R. Schulz	Berlin
D. Senato	Napoli
H. Siemon	Ludwigsburg
C. Somma	Roma
R. Spanicciati	Roma
A.G. Spera	Palermo
R. Stangarone	Bari
K. Strambach	Erlangen
G. Tallini	Roma
A. Terrusi	Bari
J. Thas	Gent
M. Ughi	Perugia
V. Vacirca	Catania

K. Vedder	Gissen
A. Venezia	Roma
F. Verroca	Bari
R. Vincenti	Perugia
A. Ventre	Napoli
H. Wefelscheid	Essen
B. Wilson	London
N. Zagaglia Salvi	Milano
H. Zeitler	Bayreuth
E. Zizioli	Brescia
B. Zucchetti	Pavia

ANNALS OF DISCRETE MATHEMATICS

Vol. 11: Studies on Graphs and Discrete Programming
edited by P. HANSEN
1981 viii + 396 pages

Vol. 12: Theory and Practice of Combinatorics
edited by A. ROSA, G. SABIDUSI and J. TURGEON
1982 x + 266 pages

Vol. 13: Graph Theory
edited by B. BOLLOBÁS
1982 viii + 204 pages

Vol. 14: Combinatorial and Geometric Structures and their Applications
edited by A. BARLOTTI
1982 viii + 292 pages

Vol. 15: Algebraic and Geometric Combinatorics
edited by E. MENDELSOHN
1982 xiv + 378 pages

Vol. 16: Bonn Workshop on Combinatorial Optimization
edited by A. BACHEM, M. GRÖTSCHELL and B. KORTE
1982 x + 312 pages

Vol. 17: Combinatorial Mathematics
edited by C. BERGE, D. BRESSON, P. CAMION and F. STERBOUL
1983 x + 660 pages

Vol. 18: Combinatorics '81: In honour of Beniamino Segre
edited by A. BARLOTTI, P.V. CECCHERINI and G. TALLINI
1983 xii + 824 pages

Vol. 19: Algebraic and Combinatorial Methods in Operations Research
edited by R.E. BURKARD, R.A. CUNINGHAME-GREEN and U. ZIMMERMANN
1984 viii + 382 pages

Vol. 20: Convexity and Graph Theory
edited by M. ROSENFELD and J. ZAKS
1984 xii + 340 pages

Vol. 21: Topics on Perfect Graphs
edited by C. BERGE and V. CHVÁTAL
1984 xiv + 370 pages

Vol. 22: Trees and Hills: Methodology for Maximizing Functions of Systems of Linear Relations
R. GREER
1984 xiv + 352 pages

Vol. 23: Orders: Description and Roles
edited by M. POUZET and D. RICHARD
1984 xxviii + 548 pages

Vol. 24: Topics in the Theory of Computation
edited by M. KARPINSKI and J. VAN LEEUWEN
1984 x + 188 pages

Vol. 25: Analysis and Design of Algorithms for Combinatorial Problems
edited by G. AUSIELLO and M. LUCERTINI
1985 x + 320 pages

Vol. 26: Algorithms in Combinatorial Design Theory
edited by C.J. COLBOURN and M.J. COLBOURN
1985 viii + 334 pages

Vol. 27: Cycles in Graphs
edited by B.R. ALSPACH and C.D. GODSIL
1985 x + 468 pages

Vol. 28: Random Graphs '83
edited by M. KAROŃSKI and A. RUCIŃSKI
1986 approx. 510 pages

Vol. 29: Matching Theory
L. LOVÁSZ and M.D. PLUMMER
1986 in preparation

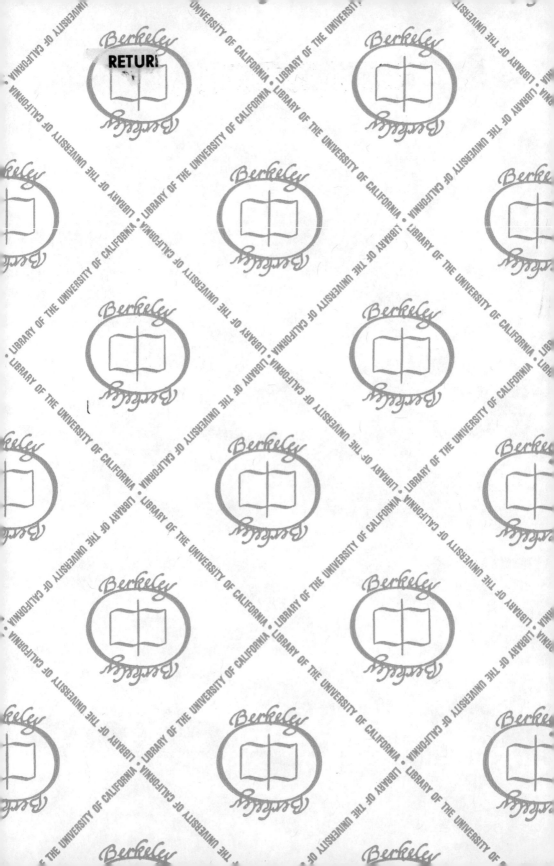